Extractive Metallurgy of Nickel, Cobalt and Platinum-Group Metals

Extractive Metallurgy of Nickel, Cobalt and Platinum-Group Metals

Frank K. Crundwell
CM Solutions (Pty) Ltd

Michael S. Moats
University of Utah

Venkoba Ramachandran
Consultant

Timothy G. Robinson
Freeport McMoRan Mining Company

William G. Davenport
University of Arizona

ELSEVIER

AMSTERDAM • BOSTON • HEIDELBERG • LONDON
NEW YORK • OXFORD • PARIS • SAN DIEGO
SAN FRANCISCO • SINGAPORE • SYDNEY • TOKYO

Elsevier
The Boulevard, Langford Lane, Kidlington, Oxford OX5 1GB, UK
Radarweg 29, PO Box 211, 1000 AE Amsterdam, The Netherlands

British Library Cataloguing in Publication Data
A catalogue record for this book is available from the British Library

Library of Congress Cataloging-in-Publication Data
A catalog record for this book is available from the Library of Congress

ISBN: 978-0-08-097478-1

Contents

Part IV
Recycling Nickel, Cobalt and Platinum-Group Metals

Preface

This book describes extraction of nickel, cobalt and platinum-group metals. The starting point is ore-in-place and the finishing point is high-purity metals and chemicals.

We have combined the description of these metals in one book because they very often occur together, are extracted together and have similar properties.

The objectives of the book are to:

(a) describe how these metals occur and are extracted;
(b) explain why these extraction processes have been chosen;
(c) indicate how the processes can be operated most efficiently, with minimal impact on the environment; and,
(d) suggest future improvements.

Much of the information in the book was obtained by visiting many of the world's nickel-, cobalt- and platinum-group metal extraction plants. We thank our hosts profusely for so graciously and expertly guiding us around their facilities.

Our book could not have been written without enormous help from our colleagues. We thank them all most deeply.

The platinum industry is traditionally secretive about publishing details of the refining processes. Both Impala Platinum and Lonmin Platinum openly shared information, and for that we are grateful. Paul Lessing, in particular, provided freely and openly of his knowledge and deep experience in the platinum refining industry.

We have used SI-based units throughout. However, we use °C for temperature, bar for pressure, and g/L for concentration because they are so common. The unit for gas volume is Nm^3, defined as 1 m^3 of gas at a temperature of 273 K and a pressure of 1 atmosphere (1.01325 bar). Lastly, unless otherwise stated, dollar always means US dollar, and % always means % by mass.

Two abbreviations are used throughout the book: PGM and PGE for platinum-group metals and platinum-group elements, respectively.

Margaret Davenport and Kathy Sole read every word of our manuscript at different stages and made many useful suggestions. We thank them for their devotion to the clarity of this book. We also thank Ayesha Osman, Nicholas du Preez, Justin Lloyd, Trevor Chagonda and Nardo Mennen for assisting with proof-reading and drawing diagrams. We thank Emilie Allemand who so kindly helped WD while he was writing at the University of Cambridge.

Frank K. Crundwell
Johannesburg, South Africa

Michael S. Moats
Salt Lake City, Utah

Venkoba Ramachandran
Phoenix, Arizona

Tim G. Robinson
Phoenix, Arizona

William G. Davenport
Tucson, Arizona

Acknowledgments

The following people hosted us on site visits, replied to requests for information and read parts of the manuscript. We are grateful to them for giving freely of their knowledge to enhance our understanding of the world of extractive metallurgy.

Name	Company
Paul Lessing	Impala Platinum, South Africa
Ian Bratt	Impala Platinum, South Africa
Ricardo Diedericks	Lonmin Platinum, South Africa
Michael Turner-Jones	Lonmin Platinum, South Africa
Alan Keeley	Lonmin Platinum, South Africa
Jacques Eksteen	Lonmin Platinum, South Africa
Nico Steenekamp	Lonmin Platinum, South Africa
Les Bryson	Anglo Ferrous, Brazil
Mark Gilmore	Anglo American Platinum, South Africa
Kathy Sole	Anglo American Research, South Africa
Maurice Solar	Hatch, Canada
Ole Morten Dotterud	Xstrata Nikkelverk, Norway
David White	Consultant
Mike Collins	Sherritt International, Canada
David Muir	CSIRO, Australia
Robbie MacDonald	CSIRO, Australia
Indje Mihaylov	Vale, Canada
Jim Finch	McGill University
Takashi Nakamura	Tohoku University, Japan
Marinda Jacobs	Anglo American Platinum, South Africa
Rodney Jones	Mintek, South Africa
Jay Robie	Phoenix Autocores, U.S.A.
Joe Westerhausen	Multimetco, U.S.A.
Robert Jacobsen	Sabin Metal, U.S.A.
Jim Hicks	Sabin West, U.S.A.
Dan Turk	Stillwater Mining, U.S.A.
Fiona Buttrey	Vale, Wales
Hira Singh	Chambishi Metals, Zambia
Ian Skepper	BHP Billiton, Australia
Christophe Zyde	Umicore, Belgium
Christian Hageluken	Umicore, Germany
Tony Storey	Vale, Canada

(*Continued*)

Name	Company
Bryan Salt	Xstrata, Canada
Pierre Louis	PEL Consult, Democratic Republic of Congo
Lene Hansen	Haldor Topsoe, Denmark
Ernie Mast	Falcondo, Dominican Republic
Richard Lea	Vale, England
Paivi Suikkanen	Boliden, Finland
Esa Lindell	Norilsk Nickel, Finland
Jukika Rimmisto	Norilsk Nickel, Finland
Sinichi Heguri	Sumitomo Nickel, Japan
Jean-Charles Didier	Societe Le Nickel, New Caledonia
Chris Rule	Anglo American Platinum, South Africa
David Jollie	Johnson Matthey, England
Martin Wells	CSIRO, Australia
Martyn Fox	Impala Platinum, South Africa
Debbie Erasmus	Anglo American Platinum

Chapter 1

Overview

Our world is becoming increasingly dependent on nickel, cobalt and platinum-group metals. Ever since Harry Brearley discovered that alloying steel with nickel and chromium produced stainless steel, the demand for nickel has outpaced the global rate of economic growth. Cobalt is a key ingredient in rechargeable batteries, a field that has grown rapidly as a result of mobile phones and other consumer electronic goods, and is bound to grow more with the increased use of electric vehicles. The catalytic properties of the platinum-group elements ensure that the demand for these metals will continue as the demand for automobiles and vehicles grows.

Nickel is mostly used in alloys, for example, in stainless steels, aerospace alloys and specialty steels. Cobalt is also used in specialty alloys, but its largest application is currently in battery chemicals. Platinum-group metals (Pt, Pd, Rh, Ru, Ir and Os) are mostly used as catalysts that minimize emissions from cars and trucks. All of these metals also have many other uses.

Primary production of nickel, cobalt and platinum-group metals in the world in 2009 is shown in Table 1.1. The combined economic value of mining and refining these metals is about US$ 50 billion per annum.

The purpose of this book is to describe how these metals are produced from the ores that are mined.

In the next four sections, the most important processes for the production of nickel, cobalt and platinum-group metals and their recycling are presented. An outline of the structure of this book is then given.

1.1. EXTRACTION OF NICKEL AND COBALT

Nickel is mined from two types of ores: laterites and sulfides. Although about 70% of the ore reserves are found in laterite ores, only about 40% of the nickel production is from laterites. As shown in Figure 1.1, laterites are mostly used to produce ferronickel, which is used directly in steelmaking. Some laterite ores are used to make melting-grade nickel and nickel matte. Sulphides are refined to produce high-grade nickel.

Extractive Metallurgy of Nickel, Cobalt and Platinum-Group Metals. DOI: 10.1016/B978-0-08-096809-4.10001-2

TABLE 1.1 Worldwide Primary Production and Value of Nickel, Cobalt and Platinum-Group Metals in 2009

Commodity	Production, tonnes/year	Value, million $
Nickel, including nickel in ferronickel	1 500 000	30 000
Cobalt	60 000	3000
Platinum-group metals	450	15 000

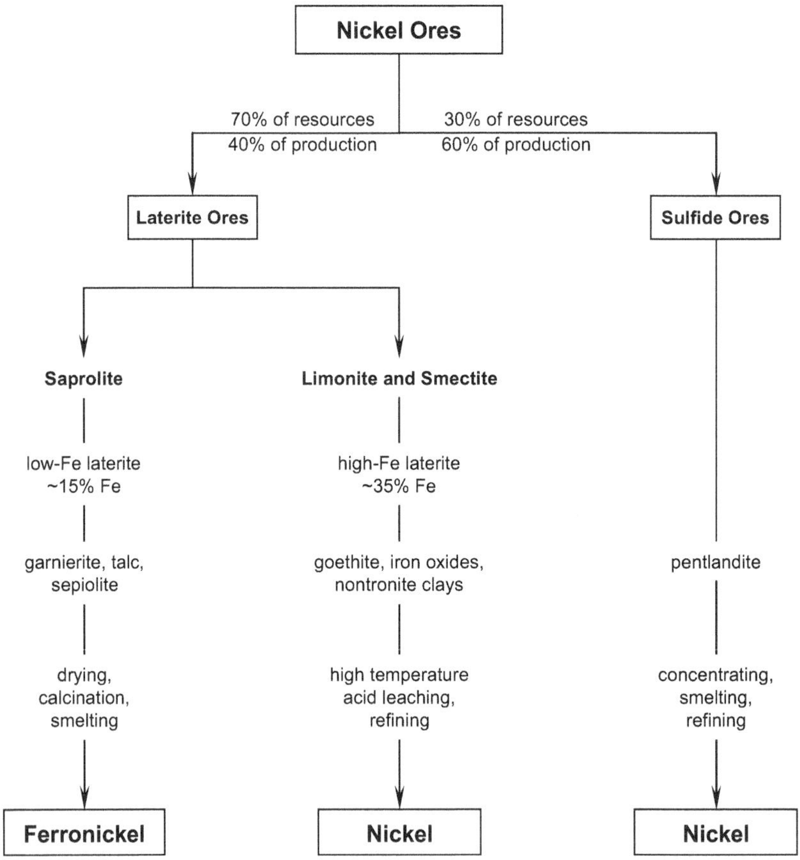

FIGURE 1.1 The extraction of nickel and ferronickel from laterite and suphide ores. Laterite ores occur as saprolite, smectite and limonite layers. Because of their different compositions and mineralogy, they require different methods of extraction. Saprolite, which has a relatively low iron content, is smelted. Limonite and smectite ores, which have a high iron content, are leached and refined. The minerals that represent each type of ore are also shown.

FIGURE 1.2 Hill-top laterite mine in New Caledonia. The overburden has been removed to expose the saprolite ore, which is mined as shown, upgraded and smelted to ferronickel. The ore delivered to the process plant typically contains about 2% Ni. *Photograph courtesy W.G. Davenport.*

1.1.1. Extraction of Nickel and Cobalt from Laterite Ores

Laterites are found mostly in tropical regions and are mined for their nickel and cobalt in countries like Indonesia, the Philippines and Cuba. Sulphides are found mostly in Canada and northern Siberia. Both laterite and sulphide ores are mined for nickel and cobalt when the concentrations exceed about 1.3% Ni and 0.1% Co.

Laterites occur near the surface, and as shown in Figure 1.2, they are mined by surface-mining methods.

Laterites are complex ores formed by the weathering of ocean floor that has been pushed up by tectonic forces. This weathering, which has occurred over millions of years, has resulted in a profile of different minerals from the surface to the bedrock. Three nickel-bearing layers are commonly identified within this profile[1]:

---

1. The classification of the laterites is often unclear and inconsistent. *Limonite* is a mineralogical term, and *saprolite* is a textual rock term. Many of the ores encountered in extractive metallurgy are referred to by host, such as sulfide or oxide. However, the variable nature of laterites makes this difficult.

(a) limonite, which occurs near the surface;
(b) smectite layer, which also occurs near the surface; and,
(c) saprolite layer, which occurs below the limonite and smectite layers.

There may be several other identifiable layers, such as ferricrete and a mottled zone, in the laterite profile. These layers are shown in Figure 1.3.

The limonite layer consists of a mixture of minerals that have a high iron and low MgO content. One group of minerals is limonite or hydrated iron oxides, $FeO(OH) \cdot nH_2O$. Nickel substitutes for iron in these minerals (Carvalho-e-Silva *et al.*, 2003). The generic formula of goethite, (Fe,Ni)OOH, is used to represent limonite ores in chemical reactions.

A layer of smectite clay, such as nontronite, is found in some deposits, for example, at Murrin-Murrin in Western Australia.

The saprolite layer is below the limonite layer. The saprolite ores have a low iron and a high MgO content. Because it is further from the surface, it is less

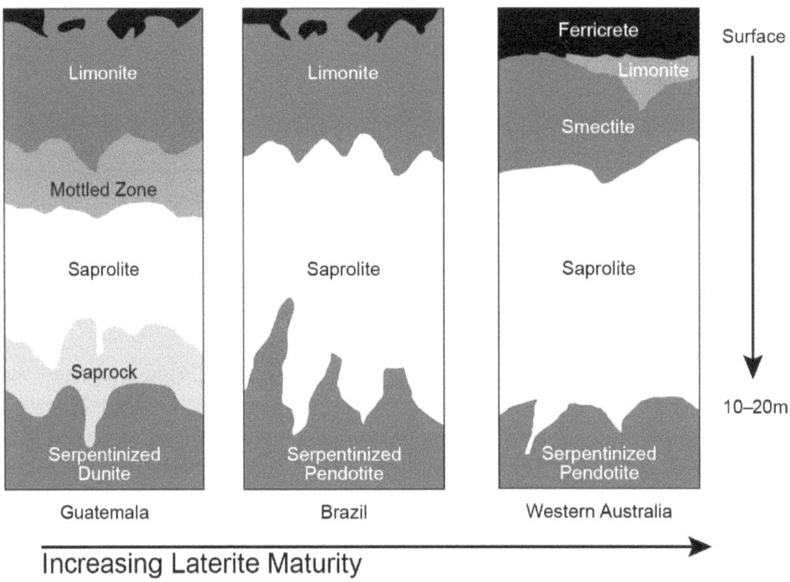

FIGURE 1.3 Laterite profiles from different climatic zones and maturity of weathering (Harris & Magee, 2003, Freyssinet 2005). Each layer is 2–5 m thick. Rain and vegetative acids have leached MgO and SiO_2 from the original igneous peridotite rock – enriching the leached layers in iron, nickel and cobalt. Dissolved nickel has also percolated down into the saprolite layer, where it is most concentrated. The natural leaching process is most advanced at the top and least at the bottom (million and 10 million years). The unleached peridotite rock is ocean floor that has been pushed above sea level by tectonic forces. Most of New Caledonia (for example) is made up of peridotite ocean floor rock – which explains why this region is such an important source of nickel. Diagram after Harris and Magee (2003), courtesy of CM Solutions (Pty) Ltd.

weathered, or chemically altered, than limonite. The minerals found in this layer are magnesium hydroxysilicates,[2] such as chysotile, $Mg_3Si_2O_5(OH)_4$. The generic formula of garnierite, $(Mg[Ni,Co])_3Si_2O_5(OH)_4$, is used to represent saprolite ores in chemical reactions.

The identification of these layers in the laterite ore body is important because different methods of extraction are used. There are two main routes for processing laterites: smelting to produce ferronickel and leaching and refining to produce nickel metal. The iron content of limonite and smectite ore is too high for it to be economically smelted, while the MgO content of the saprolite ore is too high for it to be leached economically. Thus, the method of extraction is matched to the type of ore.

Both saprolite and limonite ores, which typically contain 1.3%–2.5% Ni and 0.05–0.15% Co, are upgraded before metallurgical treatment. This upgrading of the mined ore rejects hard, precursor rock using the techniques of mineral processing, such as crushing, grinding and screening. The precursor rock is generally low in nickel and cobalt. The nickel content of the ore might be doubled during upgrading.

Method of Extraction from Saprolite Ore

Saprolite ores are mostly smelted to ferronickel, which typically contains 30% Ni and 70% Fe, for use in stainless steel and other ferrous alloys. A schematic diagram of this processing route is shown in Figure 1.4. Cobalt is present in the ferronickel in small amounts but it is not of any economic value to end-users.

A small amount of saprolite ore is smelted to sulphide matte by adding sulphur during processing. Most of this matte is made into alloying-grade nickel by oxidation and then by reduction roasting. Some of this matte is refined hydrometallurgically to high-purity nickel and cobalt.

Method of Extraction from Limonite Ore

Limonite ores are mostly leached with hot sulfuric acid at a temperature of about 250°C and a pressure of about 40 bar. A schematic diagram of the process is given in Figure 1.5. Nickel and cobalt are recovered from the pregnant solutions by the following steps:

(a) purification by precipitation and concentration of the solution by redissolution;

---

2. Examples of minerals found in the saprolite layer are as follows: (i) minerals of the serpentine group, like chysotile (lizardite), $Mg_3Si_2O_5(OH)_4$, and nepouite, $Ni_3Si_2O_5(OH)_4$; (ii) talc, $Mg_3Si_4O_{10}(OH)_2$, and willemseite, $(Ni,Mg)_3Si_4O_{10}(OH)_2$; (iii) clinochlore, $(Mg,Fe)_5Al(Si_3Al)O_{10}(OH)_8$, and nimite, $(Ni,Mg,Al)_6(Si,Al)_4O_{10}(OH)_8$; and, (iv) sepiolite, $Mg_4Si_6O_{15}(OH)_2 \cdot 6H_2O$, and falcondite, $(Ni,Mg)_4Si_6O_{15}(OH)_2 \cdot 6H_2O$.

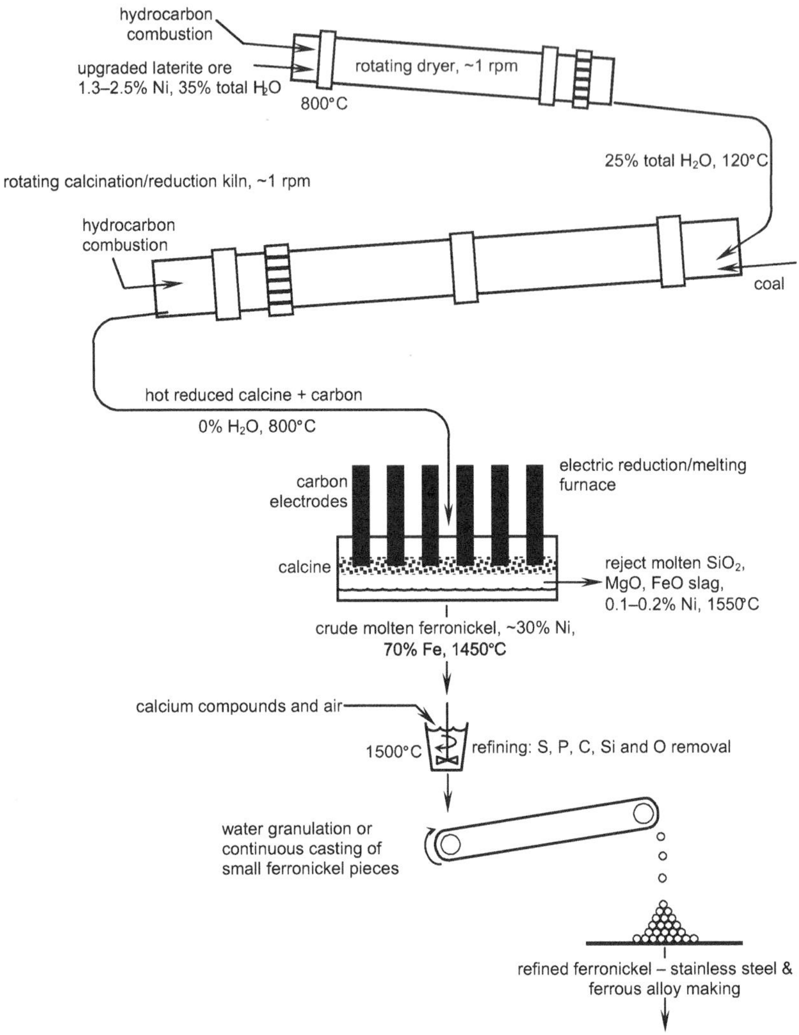

FIGURE 1.4 Representative flowsheet for smelting moist 1.5%–2.5% Ni and low-iron saprolite ore to ferronickel containing 20%–40% Ni and 80%–60% Fe. Reduction is production of metal by removing oxygen from the oxide mineral, for example, $NiO + C \rightarrow Ni + CO$. Ferronickel electric furnaces are round (15–20 m diameter) or rectangular (10 m wide, 25–35 m long). They each produce 100–200 tonnes of ferronickel per day. Drying kilns are typically 4 m diameter × 30 m long. Calcination/reduction kilns are typically 5 m diameter × 100 m long.

(b) separation of nickel and cobalt in solution is often carried out by solvent extraction; and finally,
(c) hydrogen reduction or electrowinning from solution to produce 99.9% pure nickel and cobalt.

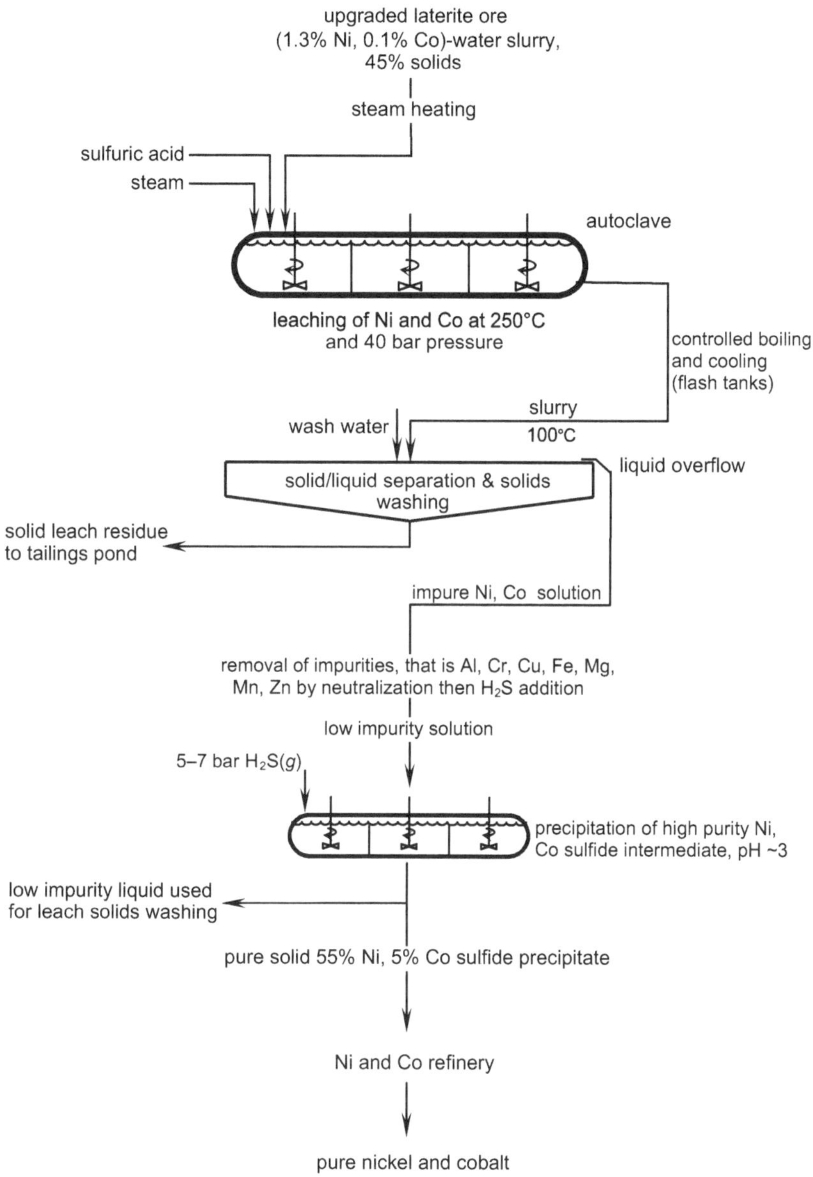

FIGURE 1.5 Representative flowsheet for high-temperature sulphuric acid laterite leaching. The feed is mainly limonite, an ore that is low in magnesium. The product is high-purity nickel–cobalt sulphide precipitate, 55% Ni, 5% Co and 40% S, ready for refining to high-purity nickel and cobalt. A leach autoclave is typically 4.5 m diameter × 30 m long with six stirred compartments. It treats about 2000 tonnes of solid feed per day.

Rationale for the Method of Extraction

Saprolite ore is smelted for the following reasons:

(a) saprolite contains little iron (15%), so it produces nickel-rich ferronickel (20%–30% Ni); and,
(b) saprolite contains considerable MgO (20%), which consumes excessive sulphuric acid if it is leached.

Limonite ore is leached in sulphuric acid for the following reasons:

(a) goethite in the limonite ore dissolves efficiently in hot sulphuric acid and, if the temperature is sufficiently high, iron, which constitutes about 40% of the ore, precipitates as hematite or jarosite; and,
(b) limonite ore has a low content of MgO, usually less than 3% in the ore, which means that the consumption of sulphuric acid is sufficiently low.

A major research thrust is to develop a process that is 'omnivorous', that is, it can treat both limonite and saprolite ores (Duyvensteyn, Wicker, & Doane, 1979; Harris & Magee, 2003; Steyl *et al.*, 2008).

1.1.2. Extraction of Nickel and Cobalt from Sulphide Ores

Sulphide ores were mostly created by (i) the intrusion of molten magma from the mantle into the crust of the earth; and, (ii) the formation of localized metal-rich sulphide ore bodies by sequential solidification and/or hydrothermal leaching and precipitation.

The source of the sulphur in these ore bodies is either from the magma itself or from the pre-existing rock where it was present as sulphate.

The most common nickel mineral in these ores is pentlandite, $(Ni,Fe)_9S_8$. The atomic ratio of nickel to iron in pentlandite ranges between 0.34 and 2.45. Most often, though, it is about 1.15. Common minerals that accompany pentlandite are pyrrhotite, Fe_8S_9, and chalcopyrite, $CuFeS_2$. Cobalt and platinum-group elements are also present, either dissolved in the pentlandite or present as distinct minerals.

Sulphide ores that are mined for nickel and cobalt typically contain 1.5%–3% Ni and 0.05%–0.1% Co.

Concentration, Smelting and Converting to Matte

The sulphide ores are concentrated, smelted and converted to metal-rich matte. A schematic diagram of this process is shown in Figure 1.6. The processing steps are as follows:

(a) the valuable minerals in an ore are concentrated by froth flotation of crushed or ground ore; and,

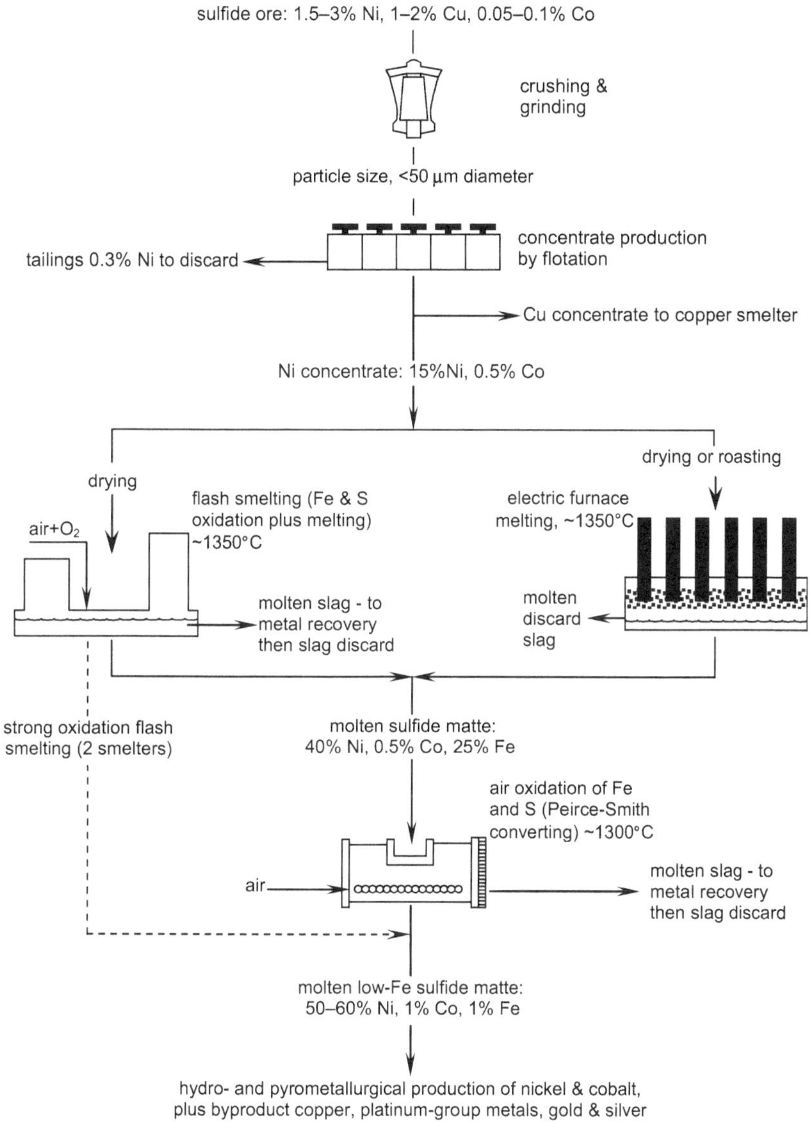

FIGURE 1.6 Main processes for extracting nickel and cobalt from sulphide ores. Parallel lines indicate alternative processes. The dashed line indicates a developing process. The compositions are for ores from Sudbury, Canada.

(b) this concentrate is smelted and converted into an even richer, low-iron sulphide matte.

Representative concentrations before and after each of these steps are given in Table 1.2.

TABLE 1.2 The Change in Concentration with Each Step of the Extraction Process

Element	In mined ore, %	After concentration, %	After smelting and converting, %
Ni	1.5–2.5	10–20	40–70
Co	0.05–0.1	0.3–0.8	0.5–2

Production of Nickel and Cobalt from Matte

Nickel and cobalt are mostly recovered from low-iron matte by the following steps:

(a) the matte is leached using either chlorine gas in a chloride solution, oxygen in an ammonia solution or oxygen in a solution of sulphuric acid;
(b) the pregnant solution is purified;
(c) separate nickel and cobalt solutions are produced, usually by solvent extraction; and,
(d) high-purity nickel and cobalt are produced from the solutions either by electrowinning or by hydrogen reduction (see Figure 1.7).

Several other processes are also used, for example, carbonyl refining and electrolytic refining.

1.2. EXTRACTION OF COBALT FROM COPPER–COBALT ORES

Cobalt (but not nickel) also occurs in the copper–cobalt oxide ores from the Central African Copperbelt. These ores typically contain about 3% Cu and 0.3% Co.

The main cobalt mineral is heterogenite [CoOOH]. Copper occurs as chrysocolla [$CuOSiO_2 \cdot 2H_2O$] and malachite [$CuCO_3 \cdot Cu(OH)_2$]. Some sulphide ores are also mined. The main gangue minerals are siliceous dolomite [$MgCO_3 \cdot CaCO_3$] and quartz (SiO_2).

Cobalt and copper metals or chemicals are produced by the following hydrometallurgical steps:

(a) reductive leaching using sulfur dioxide gas or sodium metabisulphite as a reductant;
(b) solid/liquid separation and clarification of the solution;
(c) separation of copper from cobalt by solvent extraction;
(d) purification of the aqueous solution;

FIGURE 1.7 Nickel briquettes produced from powder obtained from the hydrogen reduction of nickel in solution (see Chapter 27). An important alternative process is electrowinning, which is discussed in Chapter 26. Photograph by M. Fox, courtesy of Impala Platinum.

(e) precipitation of cobalt hydroxide or electrowinning of cobalt; and,
(f) copper electrowinning.

1.3. EXTRACTION OF PLATINUM-GROUP METALS FROM SULFIDE ORES

Virtually all of the production of platinum-group metals is from sulphide ores. A small amount is produced from alluvial deposits of metallic platinum in Russia (Kendall, 2004).

Platinum-group elements occur mainly in or near sulphide minerals in an ore deposit. They are typically dissolved in pentlandite [$(Ni,Fe)_9S_8$], or present as distinct mineral grains, for example, as braggite [(Pt,Pd)S]. The minerals containing platinum-group elements are often attached to or occluded by grains of pentlandite, pyrrhotite [Fe_8S_9] and chalcopyrite [$CuFeS_2$]. The platinum-group elements are extracted by the following process steps:

(a) crushing and grinding of the ore;
(b) production of a bulk sulphide concentrate that is rich in platinum, predominately by froth flotation;

(c) smelting and converting of the concentrate to matte; and,
(d) refining of platinum-group metals from the matte by hydrometallurgical techniques.

A schematic diagram of the process is shown in Figure 1.8.

1.3.1. Matte Leaching

The matte from the smelting and converting of platinum-group element concentrates contains about 0.3% total platinum-group elements (PGE). This matte is leached in sulphuric acid with oxygen to produce an undissolved solid residue containing about 60% total PGE. The solid residue, sometimes called a bullion concentrate, is the feed to the platinum-group metal refinery.

1.3.2. Refining of Platinum-Group Metals

The first step in the refining of platinum-group metals is the dissolution of the solid residue, described in Section 1.3.1, which contains about 60% platinum-group metals, in hydrochloric acid using chlorine as the oxidant. This dissolves all the precious metals except silver. In other words, Pt, Pd, Rh, Ir, Ru and Au dissolve. Osmium forms osmium tetroxide, OsO_4, which is volatile and is captured from the gas phase.

The platinum-group elements are separated and purified into the different metals by the following steps:

(a) sequential removal of each element from the leach solution by processes such as precipitation, solvent extraction, ion exchange and vaporization;
(b) purification of each element in solution by a similar set of processes; and,
(c) forming the metal by ignition or reduction.

The purities of the product metals from a platinum-group metal refinery are typically between 99.9% and 99.99%.

1.4. RECOVERING NICKEL, COBALT AND PLATINUM-GROUP METALS FROM END-OF-USE SCRAP

About 40% of nickel consumption, 20% of cobalt consumption and 30% of platinum-group metal consumption comes from recycled end-of-use scrap.

Production of these metals from scrap has the advantage that it slows the depletion of the earth's resources, uses less energy than metal production from ore, avoids the production of mine waste products and slows the usage of valuable land for wasteland fills.

Most recycled nickel and cobalt are recovered by recycling alloy scrap. It is made into alloy similar in composition to the scrap.

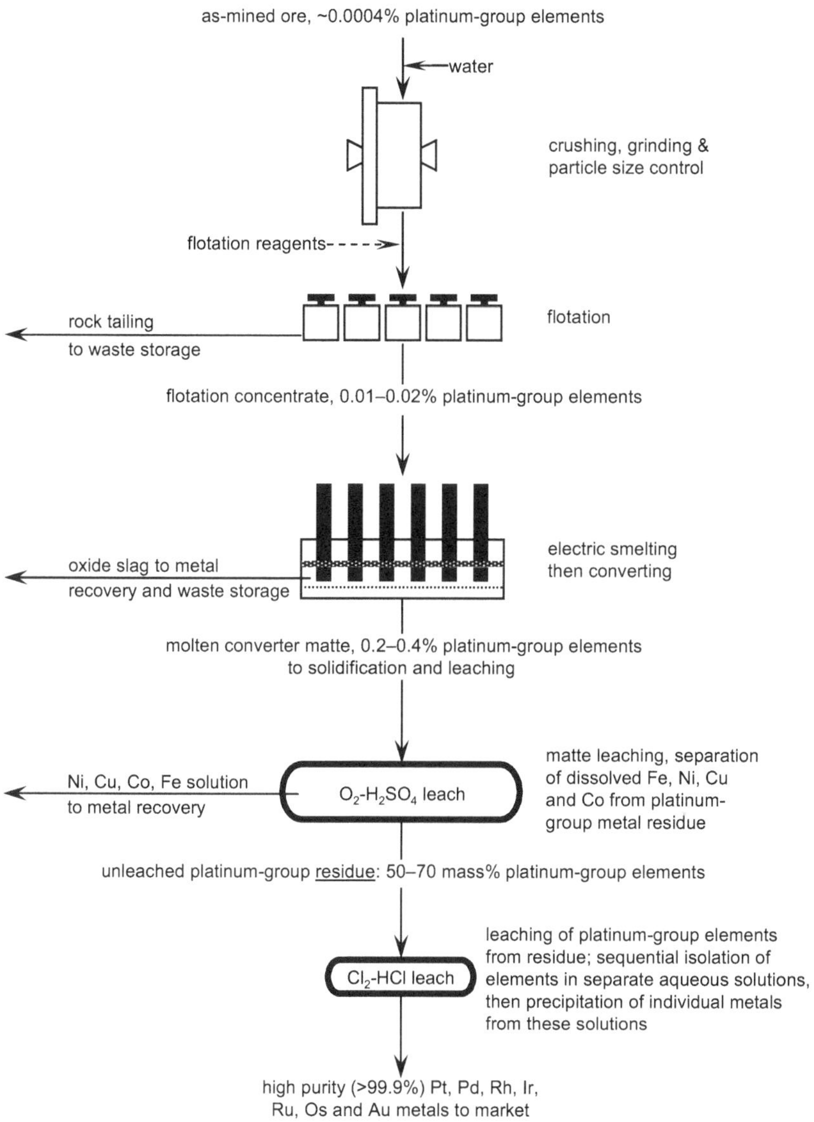

FIGURE 1.8 Generalized flowsheet for producing pure platinum-group metals and gold from ore in South Africa. The major steps are concentrate production, matte production, residue production and high-purity metal production. The numerical values are from Jones (2004) and Jones (2006).

More complex, dilute scrap is often smelted along with sulphide concentrate to make matte, which is subsequently refined to high-purity nickel, cobalt and platinum-group metals. Figure 1.9 shows a recycle flowsheet, using platinum, palladium and rhodium automobile catalyst recycle as an example.

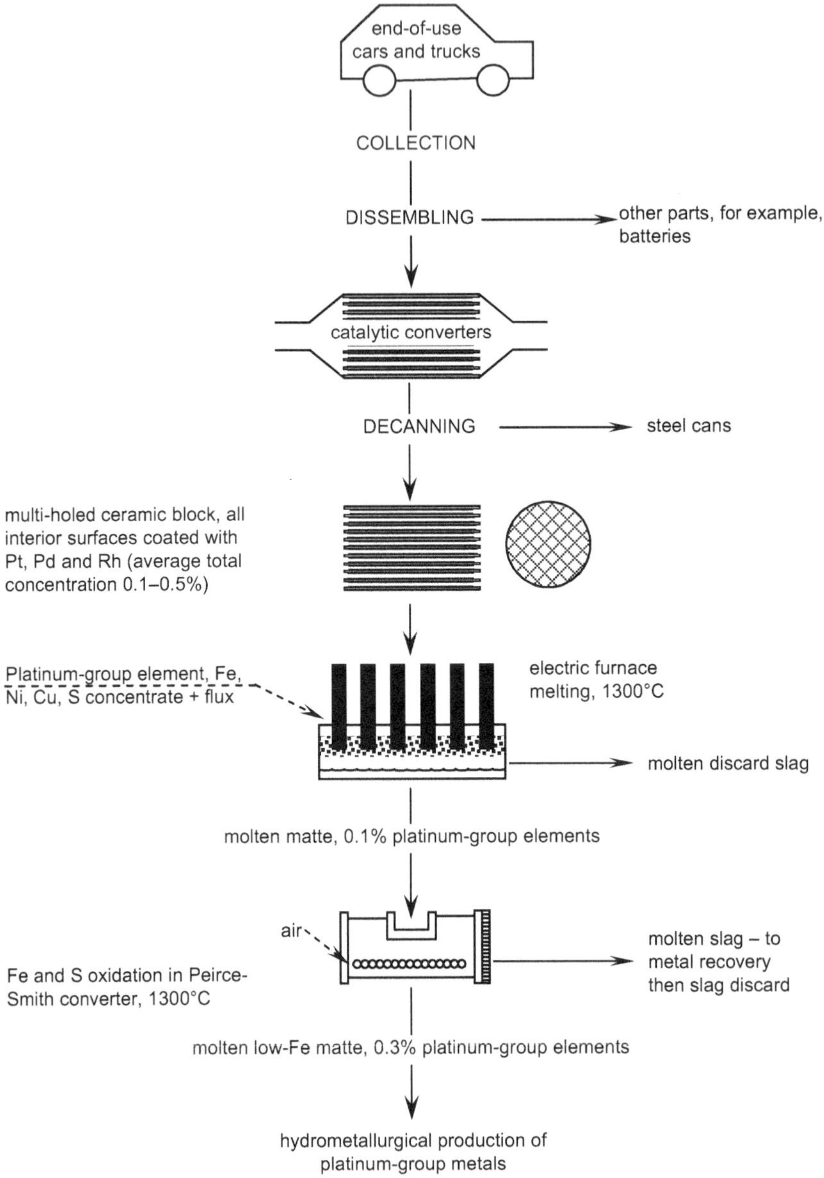

FIGURE 1.9 A schematic flowsheet for the recycle of catalytic converters from cars and trucks. The platinum-, palladium- and rhodium-coated ceramic blocks are treated in a primary smelter along with concentrates containing platinum-group metals. The numerical values are for South African practice. Car and truck catalysts are also recycled through purpose-built secondary smelters, which are discussed in Chapter 38.

Complex scrap is also smelted to metallic alloy in purpose-built secondary (scrap) smelters. This alloy is then refined to high-purity metals.

1.5. ORGANIZATION OF MAJOR THEMES AND TOPICS

The extractive metallurgy of each of the metals is discussed in four parts expressing the major themes of this book:

Part I Nickel;
Part II Cobalt;
Part III Platinum-Group Metals; and,
Part IV Recycling.

Each of these parts begins with a chapter on the major producers, markets, prices and costs of production. The topics within each of these are organized along the lines of the methods of extraction. The arrangement of topics within each theme is shown in Figures 1.10–1.12.

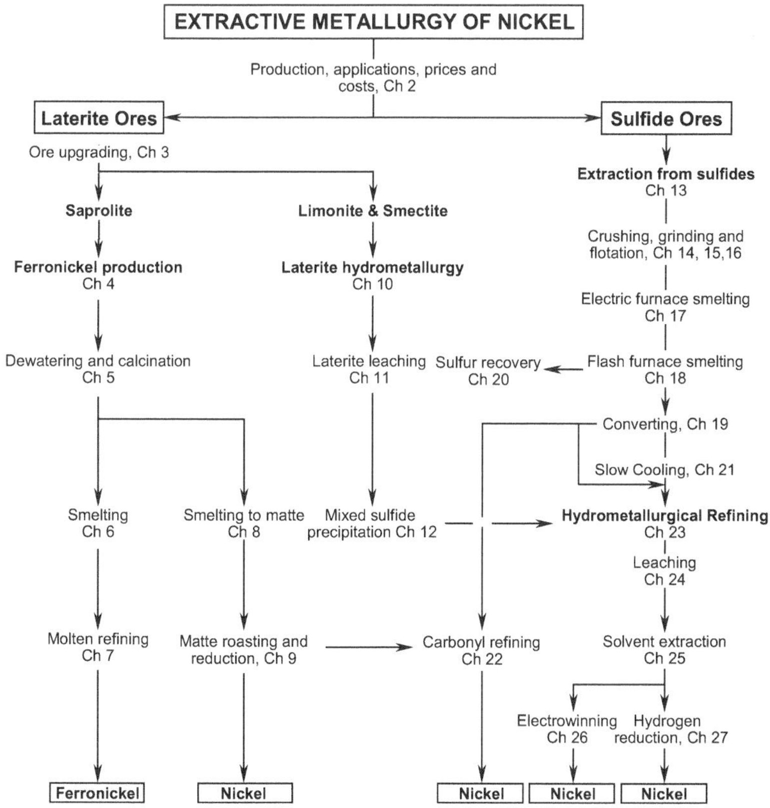

FIGURE 1.10 Organization of the topics describing the extractive metallurgy of nickel.

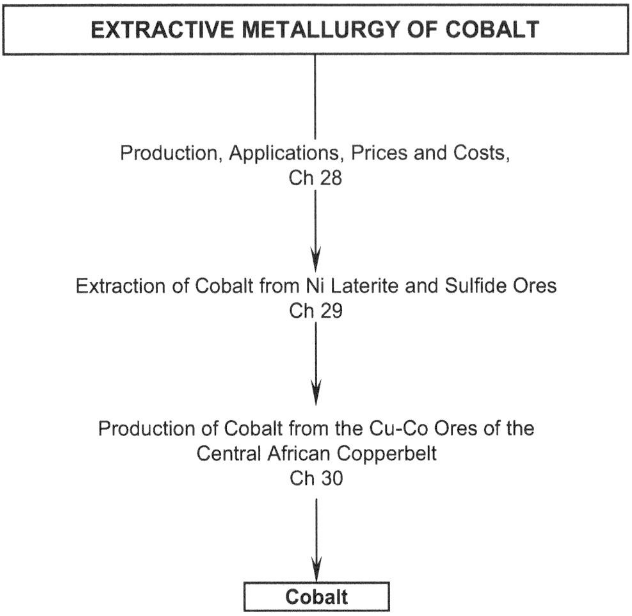

FIGURE 1.11 Organization of the topics describing the extractive metallurgy of cobalt.

The extractive metallurgy of nickel is the most complex of the themes of this book since there are two major ore types, the laterites and the sulfides, different marketable products and a variety of process routes to these marketable products. For the most part, the description is a part-by-part comparison, that is, the details of a single process step are compared across different operations. In addition, there are four short chapters that provide an overview so that sight of the entire process is not lost in the details of the parts. The organization of these topics is shown in Figure 1.10.

Cobalt is produced in conjunction with nickel, and a separate discussion of cobalt in this context is not warranted. However, a significant amount of cobalt originates from the copper ores of the Central African Copperbelt. The organization of these topics in this theme is shown in Figure 1.11.

The extractive metallurgy of the platinum-group metals is the penultimate theme of this book. The organization of the topics on the extraction of these metals is shown in Figure 1.12.

The final theme of the book is concerned with recycling. The recycling of nickel, cobalt and platinum-group metals is described in Chapter 38.

1.6. SUMMARY

Nickel and cobalt are produced from laterite and sulphide ores. Cobalt is also produced from Central African copper–cobalt ores.

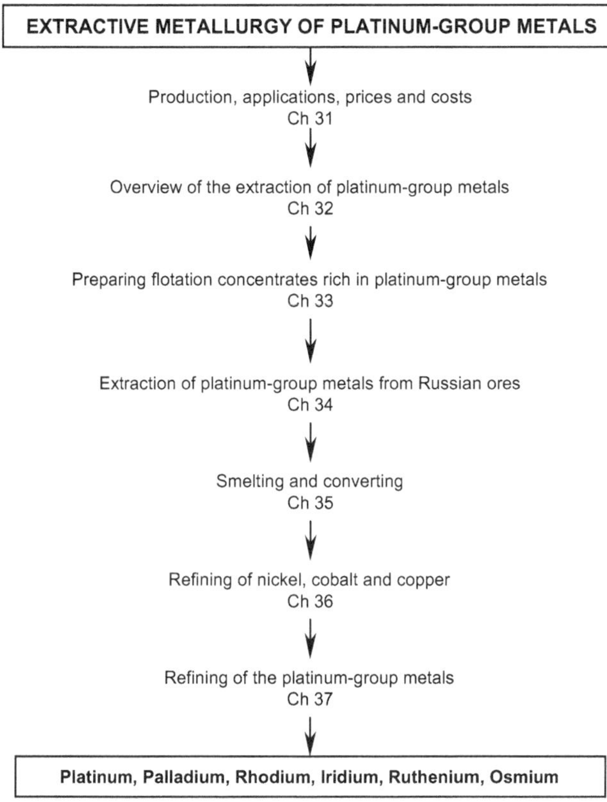

FIGURE 1.12 Organization of the topics describing the extractive metallurgy of platinum-group metals.

Platinum-group metals, that is, Pt, Pd, Rh, Ir, Ru and Os, are predominantly extracted from sulfide ores from South Africa and Siberia. Their extraction always involves smelting followed by hydrometallurgical refining.

Nickel is mostly used in stainless steels; cobalt is used mainly in aerospace alloys and platinum-group elements are used in emission reduction catalysts for cars and trucks. All of these metals also have many other uses.

Significant quantities of nickel, cobalt and platinum-group metals are recovered by treating recycled end-of-use scrap. High-quality alloy scrap is usually melted to form a new alloy of similar composition. Complex and low-purity scrap is usually smelted and then hydrometallurgically refined.

REFERENCES

Carvalho-e-Silva, M. L., Ramo, A. Y., Tolentino, H. C. N, et al. (2003). Incorporation of Ni into natural goethite: An investigation by X-ray absorption spectroscopy. *American Mineralogist, 88*, 876–882.

Duyvensteyn, W. P. C., Wicker, G. R., & Doane, R. E. (1979). An omnivorous process for laterite deposits. In *International laterite symposium* (pp. 553–569). AIME.

Freyssinet, P., Butt, C. R.M., Morris, R. C., & Piantone, P. (2005). Ore-forming processes related to lateritic weathering. In J. W. Hedenquist, J. F. H. Thompson, R. J. Goldfarb & J. P. Richards (Eds.), *Economic geology 100th anniversary volume* (pp. 681–722). Society of Economic Geologists, Inc.

Harris, B., & Magee, J. (2003). Atmospheric chloride leaching: The way forward for nickel laterites. In C. A. Young, A. M. Alfantazi, C. G. Anderson, D. B. Driesinger, B. Harris & A. James (Eds.), *Hydrometallurgy 2003 – Fifth international conference in honor of professor Ian Ritchie* (pp. 502–515). TMS.

Jones, R. T. (2004). JOM world nonferrous smelter survey, part II: Platinum group metals. *Journal of Metals, 56*, 59–63.

Jones, R. T. (2006). Southern African pyrometallurgy 2006 international conference. *The South African Institute of Mining and Metallurgy.*

Kendall, T. (2004). PGM mining in Russia. In *Platinum 2004.* (pp. 16–21). Johnson Matthey.

Steyl, J. D. T., Pelser, M., & Smit, J. T. (2008). Atmospheric leach process for nickel laterite ores. In C. A. Young, P. R. Taylor, C. G. Anderson, & Y. Choi (Eds.), *Hydrometallurgy 2008, Proceedings of the 6th international symposium* (pp. 570–579). SME.

SUGGESTED READING

Bergman, R. A. (2003). Nickel production from low-iron laterite ores: Process descriptions. *CIM Bulletin, 96*, 127–138.

Boldt, J. R., & Queneau, P. (1967). *The winning of nickel.* Longmans.

Budac, J. J., Fraser, R., & Mihaylov, I. et al. (Eds.). (2009). *Hydrometallurgy of nickel and cobalt 2009.* CIM.

Burkin, A.R., Extractive metallurgy of nickel. Society of (British) Chemical Industry.

Cabri, L. J. (2002). The geology, geochemistry, mineralogy and mineral beneficiation of platinum-group elements [Special volume]. *CIM, 54.*

Donald, J., & Schonewille, R. (Eds.). (2005). *Nickel and cobalt: Challenges in extraction and production.* Metallurgical Society of CIM.

Habashi, F. (2009). A history of nickel. In J. Liu, J. Peacey & M. Barati., et al. (Eds.), *Pyrometallurgy of nickel and cobalt 2009 Proceedings of the international symposium* (pp. 77–98). Metallurgical Society of CIM.

Imrie, W. P., & Lane, D. M. (Eds.). (2004). *International laterite nickel symposium.* TMS.

Jones, R. T. (2004). JOM world nonferrous smelter survey, part II: Platinum group metals. *Journal of Metals, 56*, 59–63.

Jones, R. T. (2006). Southern African pyrometallurgy 2006 international conference. *Southern African Institute of Mining and Metallurgy.*

Liu, J., Peacey, J., & Barati, M., et al. (Eds.). (2009). *Pyrometallurgy of nickel and cobalt 2009.* Metallurgical Society of CIM.

Warner, A. E. M., Diaz, C. M., Dalvi, A. D., et al. (2006). JOM world nonferrous smelter survey, part III: Laterite. *Journal of Metals, 58*, 11–20.

Warner, A. E. M., Diaz, C. M., Dalvi, A. D., et al. (2007). JOM world nonferrous smelter survey, part IV: Nickel sulfide. *Journal of Metals, 59*, 58–72.

Part I

Extractive Metallurgy of Nickel and Cobalt

Chapter 2

Nickel Production, Price, and Extraction Costs

Nickel was identified as an element in 1751 by Alex Cronstedt. Before this, the Chinese used a copper–nickel alloy as a substitute for silver. The first major application of nickel in the West was in the 1820s in an alloy of nickel, copper and zinc, called German Silver. Like the Chinese, this was also used as a substitute for silver. In 1857, the United States introduced a one-cent coin made of a copper−nickel alloy, the first significant use of nickel.

Then, the use of nickel grew rapidly in the twentieth century, as shown in Figure 2.1, where its main application was in stainless steel. This growth continues in the twenty-first century, spurred on by rapid industrialization in China and India.

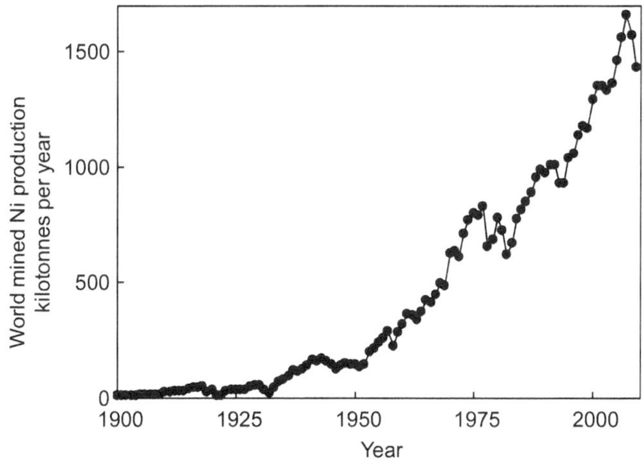

FIGURE 2.1 Global production of mined nickel since 1900 (Kuck, 2010). There has been substantial growth in the production of nickel since 1950.

Extractive Metallurgy of Nickel, Cobalt and Platinum-Group Metals. DOI: 10.1016/B978-0-08-096809-4.10002-4

TABLE 2.1 The Uses of Nickel.
World 2009 nickel consumption (all forms) was approximately two million tonnes. Most of this nickel was provided by mine production, as shown in Figure 2.1, and the remainder was obtained from recycling end-of-use scrap

Category	Proportion of total consumption, %
Stainless steel	60
Nickel-based alloys (for example, superalloys)	14
Alloy steels	09
Electroplating	09
Foundry (castings)	03
Batteries (for example, nickel cadmium)	03
Copper-based alloys (for example, monel)	01
Catalysts	01

This chapter discusses the production, application and price of nickel. The objectives of this chapter are to describe the following aspects of the supply of and demand for nickel:

(a) the major uses of nickel;
(b) the locations of nickel mines and processing plants around the world;
(c) the price of nickel; and,
(d) the costs of nickel extraction.

2.1. APPLICATIONS OF NICKEL

Nickel is mainly used in alloys, as shown in Table 2.1. It is most commonly used in the production of stainless steels. As an alloying ingredient, it imparts the following properties:

(a) corrosion resistance;
(b) workability (that is, ease of manufacture);
(c) strength at high temperatures; and,
(d) attractiveness.

All these qualities make nickel a valuable alloying ingredient.

2.2. LOCATION OF NICKEL MINES AND EXTRACTION PLANTS

The location of major nickel mines and extraction plants are shown in Figure 2.2 and listed in Table 2.2.

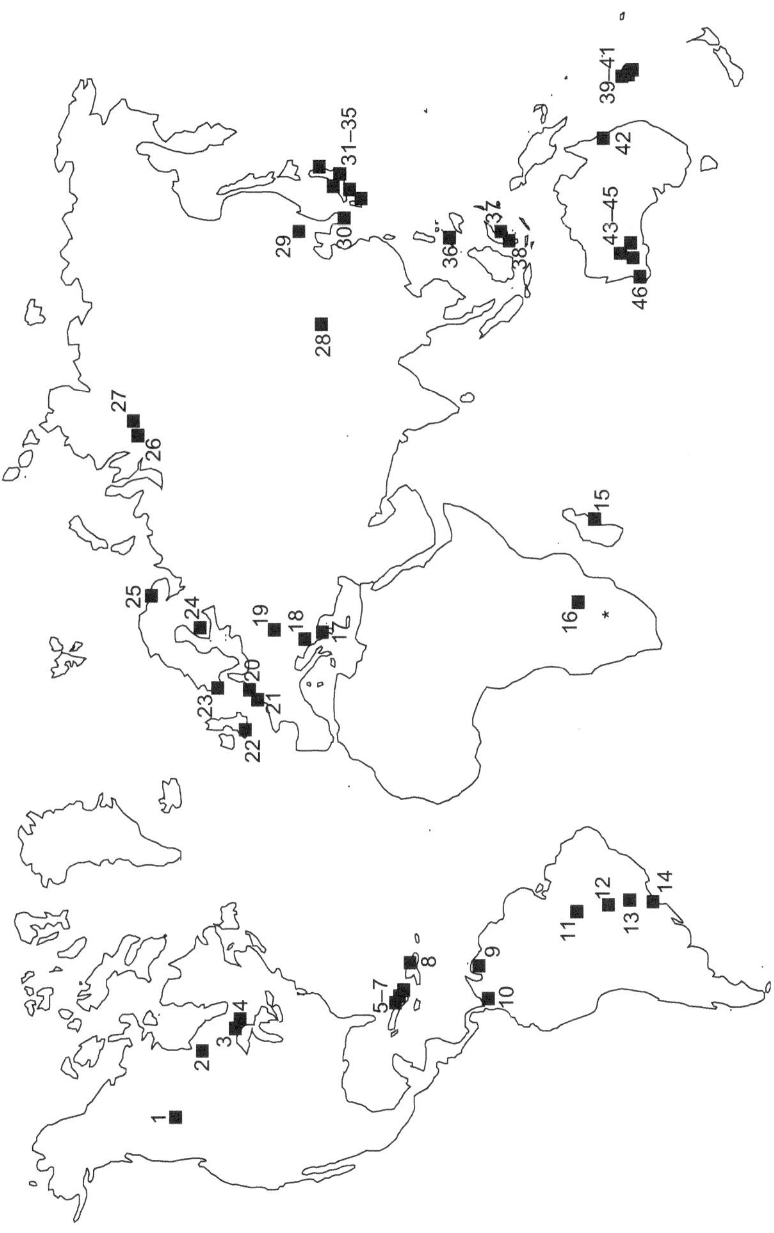

FIGURE 2.2 Major nickel extraction plants of the world. Plant types and locations are listed in Table 2.2. *Platinum group metal plants in South Africa and Zimbabwe produce considerable by-product nickel.

TABLE 2.2 Major Nickel Extraction Plants of the World[a]

	Operation	Location
1	Sherritt, refinery	Ft. Saskatchewan, Canada
2	Vale, matte smelter and refinery	Thompson, Canada
3	Vale, matte smelter and refinery	Sudbury, Canada
4	Xstrata, matte smelter	Falconbridge, Canada
5	Cuban government, Caron roast leach	Punta Gorda, Cuba
6	Cuban government, Caron roast leach	Nicaro, Cuba
7	Sherritt, high-temperature sulphuric acid leach	Moa Bay, Cuba
8	Xstrata, ferronickel smelter (moth-balled)	Bonao, Dominican Republic
9	Anglo American, ferronickel smelter	west of Caracas, Venezuela
10	BHP Billiton, ferronickel smelter	Montelibano, Colombia
11	Vale, ferronickel smelter	Ourilandia, do Norte, Brazil
12	Votorantim Metals, Caron roast leach	Niquelandia, Brazil
13	Votorantim Metals, matte smelter	Fortaleza de Minas, Brazil
14	Votorantim Metals, refinery	Sao Paulo, Brazil
15	Sherritt, high-temperature sulfuric acid leach	Toamasina, Madagascar
16	BCL Ltd., matte smelter	Selebi-Phikwe, Botswana
17	Larco, ferronickel smelter	Larymna, Greece
18	Feni Industries, ferronickel smelter	Kavadarci, Macedonia
19	Pobuzhsky Combine, ferronickel smelter	Pobuzhie, Ukraine
20	Umicore, secondary smelter	Hoboken, Belgium
21	Eramet, refinery	Sandouville, France
22	Vale, refinery	Clydach, Wales
23	Xstrata, refinery	Kristiansand, Norway
24	Boliden and Norilsk, matte smelter and refinery	Harjavalta, Finland
25	Norilsk, matte smelter	Pechenga, Russia
26	Norilsk, matte smelter	Nadezhda, Russia
27	Norilsk, matte smelter and refinery	Norilsk, Russia
28	Jinchuan Group, smelter and refinery	Jinchang (Gansu), China

TABLE 2.2 Major Nickel Extraction Plants of the World[a]—cont'd

	Operation	Location
29	Jinlin Jien Nickel Co., smelter and refinery	Pan Shi (Jilin), China
30	Posco, ferronickel smelter	Gwangyang, Korea
31	Sumitomo, ferronickel smelter	Hyuga (Kyushu), Japan
32	Sumitomo, refinery	Niihama (Shikoku), Japan
33	Vale, refinery	Matsuzaka (Honshu), Japan
34	Nippon Yakin, ferronickel smelter	Miyazu (Honshu), Japan
35	Pacific Metals, ferronickel smelter	Hachinohe (Honshu), Japan
36	Sumitomo, high-temperature leach plant	Coral Bay, Philippines
37	Vale, laterite to matte smelter	Sorowako (Sulawesi), Indonesia
38	Antam, ferronickel smelter	Pomalaa (Sulawesi), Indonesia
39	Vale, laterite leach	Goro, New Caledonia
40	Xstrata, ferronickel smelter	Koniambo, New Caledonia
41	Eramet (Le Nickel), ferronickel smelter	Noumea, New Caledonia
42	Queensland Nickel, Caron roast leach	Townsville, Australia
43	Minara Resources, high-temperature sulfuric acid leach	Murrin-Murrin, W. Australia
44	BHP Billiton, smelter	Kalgoorlie, W. Australia
45	First Quantum, high-temperature sulphuric acid leach (closed)	Ravensthorpe, W. Australia
46	BHP Billiton, refinery	Kwinana, W. Australia
47	Ufaleynickel smelter	Ufaley, Chelyabinsk region, Russia
48	Rezhnickel smelter	Rezh, Sverdlov, Russia
49	Yuzhural nickel smelter	Yekaterinburg, Sverdlov, Russia

[a]Platinum metal plants in South Africa and Zimbabwe also produce considerable by-product nickel. These plants are described in Chapters 31 through 37.

2.2.1. Mine Production of Nickel

The mine production of nickel is shown in Table 2.3. Russia is the major producer of nickel, followed by Indonesia, Australia, Canada and the Philippines.

TABLE 2.3 Mine Production and Ore Reserves of Nickel in 2009

Country	Production, tonnes	Reserves, tonnes
Australia	165 000	24 000 000
Botswana	28 600	490 000
Brazil	54 100	6 700 000
Canada	137 000	3 800 000
China	79 400	3 000 000
Colombia	72 000	1 600 000
Cuba	67 300	5 500 000
Dominican Republic	—	960 000
Indonesia	203 000	3 900 000
Madagascar	—	1 300 000
New Caledonia	92 800	7 100 000
Philippines	137 000	1 100 000
Russia	262 000	6 000 000
South Africa	34 600	3 700 000
Venezuela	13 200	490 000
Other	51 700	4 500 000
Total	1 400 000	76 000 000

2.2.2. Location of Nickel Mines

Nickel operations that mine laterite ores are located mainly on tropical islands (such as New Caledonia, Philippines, Indonesia and Cuba) and in tropical South America.

Nickel operations that mine sulfide ores are located mainly in Northern Canada and Siberia.

Other significant sources of nickel are Southern Africa and Western Australia.

2.2.3. Nickel Extraction Plants

Extraction of nickel from ore usually takes place near where the ore is mined. There are some exceptions to this, such as the following:

(a) shipment of upgraded laterite ore from New Caledonia, Philippines and Indonesia to ferronickel smelters in Japan and Korea;

(b) shipment of sulfide concentrates from Africa and Australia to China and Finland; and,
(c) shipment of sulfide concentrates from Arctic Canada and Labrador to Sudbury, Ontario and Thompson, Manitoba.

In addition, smelting and leaching operations produce nickel sulfide intermediates, that is, mattes and sulfide precipitates. These intermediates are often shipped long distances for refining to metal. Examples of the shipment of intermediates are the following:

(a) the shipment of matte from Indonesia to Japan and from Canada to Norway; and,
(b) the shipment of sulfide precipitates from Cuba to Canada and from Philippines to Japan.

These shipments either take the nickel closer to the markets and/or make use of already existing smelting and refining facilities.

2.3. PRICE OF NICKEL

The spot price of nickel on the London Metals Exchange is shown in Figure 2.3. Overall, the price of nickel has increased since 2000. A spike in the price was experienced in 2007 when demand outstripped supply. However, prices have since returned to values similar to those seen in 2005.

Ferronickel prices are negotiated between producer and consumer. However, the price typically reflects the value of the nickel contained in the ferronickel.

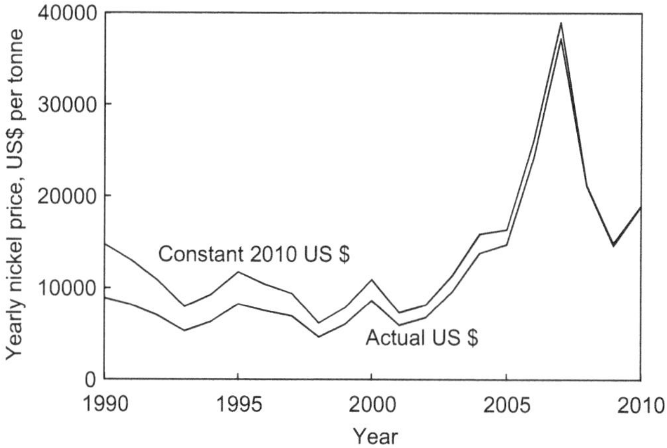

FIGURE 2.3 London Metals Exchange cash price for nickel since 1990 (Kuck, 2010). The peak in 2007 was due to a rapid increase in demand (especially in China) and a slow response in supply. The prices are from Goonan and Kuck (2009). The constant dollar conversions are from Sahr (2009).

High-grade nickel, usually produced by electrowinning, attracts a premium of about $3300 per tonne above the London Metals Exchange price. Melting-grade nickel is discounted about $200 per tonne to the London Metals Exchange price.

2.4. COSTS OF NICKEL EXTRACTION

2.4.1. Accuracy of the Cost Data

In this chapter, the investment and operating costs are at the 'study estimate' level, which is equivalent to an accuracy of ± 30%. Data with these accuracies can be used to examine the economic feasibility of a project before spending significant funds for piloting, market studies, land surveys and property acquisition (Christensen & Dysert, 2005).

2.4.2. Investment and Operating Costs for Laterite Smelting Projects

About half of the global production of primary nickel is produced by smelting of laterite concentrate, mostly to ferronickel. This section gives estimated investment[1] and operating costs for ferronickel smelting.

Investment Costs

Three recent greenfield ferronickel projects are as follows: Onca Puma, Brazil (1.7% Ni in ore) (Onca Puma, 2009); Koniambo, New Caledonia (2.1% Ni in ore) (Pearce, Usmar, Lachance, & Romaniuk, 2009) and Pearl, East Indonesia (2.4% Ni in ore) (Freed, 2007).

The estimated investment costs for these three projects are shown in Table 2.4. The major components of these costs are shown in Table 2.5.

Investment Cost per Annual Tonne of Nickel

The investment cost is expressed in Table 2.4 in US$ per annual tonne of nickel in product ferronickel. This is defined by the equation as follows:

$$\begin{matrix}\text{investment}\\ \text{cost}\end{matrix} = \begin{matrix}\text{investment cost per}\\ \text{annual tonne of}\\ \text{nickel in ferronickel}\end{matrix} \times \begin{matrix}\text{plant capacity, tonnes of}\\ \text{nickel in ferronickel}\\ \text{product per year}\end{matrix} \quad (2.1)$$

This equation shows, for example, that a ferronickel mine-to-market project which (i) costs US$ 60 000 per annual tonne of nickel in ferronickel and (ii) produces 50 000 tonnes of nickel in ferronickel per year will cost:

1. Also referred to as the capital cost.

TABLE 2.4 Projected Investment Costs for Three Laterite-to-Ferronickel Smelting Projects. The Estimates Are at the *Study Estimate* Level*

Project	Projected nickel production, tonnes of nickel in ferronickel per year	Projected total investment cost, US$	Investment cost, US$ per annual tonne of nickel in ferronickel
Onca Puma, Brazil	60 000	3 000 000 0000	50 000
Pearl, Indonesia	50 000	2 300 000 0000	50 000
Koniambo, New Caledonia	60 000	4 000 000 0000	70 000

* *Christensen & Dysert, 2005.*

TABLE 2.5 Breakdown of Laterite-to-Ferronickel Investment Costs into Major Components. The Largest Components Are for Calcination + Smelting + Refining

Item	Contribution to total investment cost, %
Infrastructure (including power lines and roads)	15
Mine and mine facilities	10
Site preparation	05
Yard and plant facilities	15
Ore preparation plant	05
Coal preparation plant	05
Ore drier	05
Calcination kilns including dust agglomeration	10
Smelting and refining	30

$$\text{investment cost} = \begin{matrix}\text{US\$ 60 000 per annual tonne} \\ \text{of nickel in ferronickel}\end{matrix} \times \begin{matrix}\text{50 000 tonnes of nickel in} \\ \text{ferronickel per year}\end{matrix}$$

$$= \text{US\$ 3 000 000 000}$$

Therefore, the investment cost for this operation is expected to be three billion dollars.

Variations in Investment Costs

Investment costs of mines vary considerably between mining operations. This is due to differences in ore grades, mine sizes, mining methods, topography and ground conditions.

The costs for constructing an underground mine are considerably higher than those for constructing an open-pit mine, expressed per annual tonne of mined nickel. This and the high cost of working underground explain why underground orebodies must contain higher grades of nickel ore than open-pit orebodies.

Ore grade has a direct effect on mine investment costs expressed in US$ per annual tonne of nickel. Consider two identical ore bodies, one containing 1.25% Ni ore and the other containing 2.5% Ni ore. Achievement of an identical annual production of nickel in ferronickel requires that the 1.25% Ni ore be mined at twice the rate of the 2.5% Ni ore. This, in turn, requires about twice as much plant and equipment (such as trucks) and, consequently, about double the investment.

The same is true for a minerals processing plant – it will have to treat 1.25% Ni ore twice as fast as 2.5% Ni ore to achieve the same annual production of nickel-in-concentrate. This will require about twice the amount of concentrator equipment and about double the investment.

Smelter investment costs, per annual tonne of nickel in ferronickel production, are influenced by *smelter feed* grade rather than by ore grade. The higher the grade of nickel in the smelter feed, the smaller the smelter (and smelter investment) for a given annual tonne of nickel in ferronickel production. This explains why operating laterite mines strive to maximize grade of the smelter feed.

Economic Sizes of Plants

Ferronickel smelters are often based on one smelting furnace, which can typically produce about 50 000 tonnes of nickel in ferronickel per year. This is the minimum economic size of a ferronickel operation. Additional production requires a second furnace, together with ancillary equipment, requiring considerably more capital investment.

Operating Costs

Projected operating costs for producing ferronickel in the above-mentioned projects are given in Table 2.6. These figures are for cash costs, which are direct operating costs. In other words, they do not include items, such as depreciation or amortization.

These costs can be compared with the May 1, 2010 London Metals Exchange cash price of about US$ 26 000 per tonne of nickel in ferronickel.

TABLE 2.6 Cash Costs for Producing Nickel in Ferronickel

Project	Estimated cash operating cost, US$ per tonne of nickel in ferronickel
Onca Puma, Brazil	4000
Pearl, Indonesia	6000
Koniambo, New Caledonia	5000

These operating costs are divided between plant areas. This itemization is given in Table 2.7. The costs of calcining, smelting and refining are significant.

The operating costs may also be divided into consumables, as shown in Table 2.8. The expenditures on power and coal are the largest.

TABLE 2.7 Breakdown of Operating Costs/Area

Area	Contribution to cost, %
Mining	10
Concentration	5
Drying	25
Calcining (including coal handling)	20
Smelting	30
Refining	15
General and administrative (including yard services)	15

TABLE 2.8 Breakdown of Costs by Category

Category	Contribution to cost, %
Electric power	30
Mine consumables, including fuel and lubricants	10
Coal and other fossil fuels	25
Smelting/refining consumables (for example, electrodes, chemicals)	15
Maintenance	5
Labor	15

As with investment costs, cash operating costs (per tonne of nickel) are lowered by high grades of nickel in ore and concentrate.

2.4.3. Investment and Operating Costs for Laterite Leaching Projects

This section describes the costs of mining/leaching/refining high-purity nickel (and cobalt) from laterite ore. Three recent projects are considered:

(a) Ravensthorpe, Australia (1% Ni and 0.04% Co);
(b) Ambatovy, Madagascar (1.2% Ni and 0.12% Co); and,
(c) Goro, New Caledonia (1.5% Ni and 0.1% Co).

Investment Costs

The estimated investment costs for the laterite leach projects are shown in Table 2.9. The project costs are consistent at about US$ 60 000 per annual tonne of nickel plus cobalt.

These investment costs may be allocated by area, as shown in Table 2.10.

Operating Costs

Operating cash costs for an existing sulfuric acid goethite leach plant are about US$ 10 000 per tonne of nickel and cobalt product (Minara Resources, 2009). About 15% of these costs are for mining and 85% are for metal production. The operating costs by consumable are given in Table 2.11.

TABLE 2.9 Investment Costs of Three Recent High-Temperature Sulfuric Acid Laterite Leach Projects

Project	Projected nickel production, tonnes per year	Projected cobalt production, tonnes per year	Projected total investment cost, US$	Investment cost, US$ per annual tonne of nickel plus cobalt
Ravensthorpe, Australia	60 000 tonnes nickel in mixed hydroxide	3000 tonnes cobalt in mixed hydroxide	3 000 000 000	50 000
Ambatovy, Madagascar	60 000 tonnes nickel as metal	6000 tonnes cobalt as metals	4 000 000 000	60 000
Goro, New Caledonia	60 000 tonnes nickel in oxide	5000 tonnes cobalt in carbonate	4 000 000 000	60 000

TABLE 2.10 Breakdown of Investment Cost by Area*

Area	Contribution to capital cost, %
Mine	15
Feed preparation and concentrator	5
Leach (autoclaves) and solution purification plants	20
Mixed sulfide precipitation and H_2S plants	5
Metal winning plant (H_2 reduction)	15
Spare parts and first fill	5
Infrastructure and other construction costs	15
Sulfuric acid plant	10
Oxygen and hydrogen plants	5
Steam plants	5

** Dolan & Nendick, 2004.*

TABLE 2.11 Breakdown of the Operating Costs by Area*

Item	Contribution to operating costs, %
Labor	15
Sulfur	15
Acid plant operation	5
Limestone	5
Power	5
Maintenance supplies	20
Other reagents	10
Other operating supplies	5
Marketing and sales	10
Miscellaneous	10

** Neudorf & Huggins, 2004.*

2.4.4. Comparison of Costs of Laterite Smelting with Costs of Laterite Leaching

The data on the investment and operating costs indicate that the investment costs for ferronickel smelting and high-temperature sulfuric acid laterite leaching are similar, per annual tonne of nickel.

However, the operating costs of ferronickel smelting are significantly lower than the operating costs of leaching/metal production. The operating costs of a ferronickel smelter are about US$ 6000 per tonne of nickel in ferronickel, whereas those for a leaching plant are about US$ 10 000 per tonne of nickel and cobalt.

Smelting thus has a significant advantage over leaching.

Much of this difference is due to the higher nickel grades of the saprolite ores that are smelted (~2% Ni) compared with the grades of limonite ores that are leached (1%–1.5% nickel plus cobalt). Also, ferronickel production does not require a separate metal production plant.

The Coral Bay operation seems to have the best of both worlds. It sells saprolite ore to ferronickel smelters and leaches stockpiled limonite ore that was previously removed to reach the saprolite.

2.4.5. Effect of Nickel Grade on the Viability of Laterite Projects

Dalvi, Bacon and Osborne (2004) indicated the following cut-off grades for laterite projects:

(a) smelting projects with ores containing less than 1.7% Ni are not likely to be economic; and,
(b) leaching projects with ores containing less than 1.3% Ni are not likely to be economic.

These conclusions seem to have been confirmed over the years.

2.4.6. Costs of Producing Nickel from Sulfide Ore

Mine and Concentrator Investment Costs

Two recent greenfield nickel sulfide mines are Raglan, Quebec (typical ore: 2.5% Ni, 0.7% Cu and 0.05% Co) and Voisey's Bay, Labrador (typical ore: 3% Ni, 1.7% Cu and 0.1% Co).

The investment costs for these operations are given in Table 2.12. The values are for a concentrator and one or more mines.

Investment Costs for Smelter and Refinery

No new data are available for the investment cost for a nickel smelter because no new nickel sulfide smelters have been built for many years.

TABLE 2.12 Investment Costs of Two Greenfield Nickel Sulphide Mines/Concentrators. Both Also Produce Large Amounts of Copper and Cobalt in Concentrate*

Project	Nickel production, tonnes of nickel in concentrate per year	Estimated total investment cost, US$, mine plus concentrator	Investment cost, US$ per annual tonne of nickel in concentrate
Raglan, Quebec (underground mines)	30 000	600 000 000	20 000
Voisey's Bay, Labrador (open-pit mine)	50 000	1 000 000 000	20 000

* *Fiewelling, 2006.*

Operating Costs

The operating cost, after by-product credits for copper and cobalt, for making nickel from sulfide ore is about US$ 10 000 per tonne of refined nickel. This cost may be compared with the May 1, 2010 London Metal Exchange price of US$ 26 000 per tonne of nickel metal.

These operating costs may be allocated by area, as shown in Table 2.13.

These operating costs are in stark contrast with those of laterite processing. Mining is the major cost in sulfide operations, whereas processing is the major operating cost of laterite operations. There are two reasons for this. Sulfide mines are hard rock mines, which are much more expensive to operate; however, the minerals in these ores can be concentrated by factors of approximate 20 times. Laterite ores are upgraded by a factor of between 1.3 and 2. Although this is important to the economics of laterite processing, this upgrading of the ore is very slight compared with sulfides. Consequently, the operating costs of the extraction plant are proportionally higher than those for sulfides.

TABLE 2.13 Breakdown of Operating Costs by Area

Area	Contribution to operating costs, %
Mining	70
Concentration	10
Smelting	10
Refining	10

Cut-Off Grades for Sulfide Operations

Nickel sulfide ores that are mainly mined for nickel, cobalt and copper need a grade of greater than 3% (nickel + copper + cobalt) for underground mining and greater than 1% (nickel + copper + cobalt) for surface mining to be profitable. Most sulfide deposits are mined by underground methods.

2.5. SUMMARY

Nickel imparts corrosion resistance, workability, high-temperature strength and attractiveness to most of its applications. It is mainly used in alloys, especially stainless steel.

Nickel is produced from laterite and sulfide ores. Laterite ores are mainly mined on tropical islands and tropical South America, while sulfide ores are mainly mined in northern Canada and northern Siberia.

Nickel production plants are mostly near the mines. However, significant quantities of laterite ores, sulfide concentrates, sulfide mattes and hydrometallurgical precipitates are shipped around the world.

Worldwide, approximately two million tonnes of nickel are used each year. About two-thirds of this comes from ore while the remaining one-third is from the recycling of end-of-use scrap.

The investment costs for a laterite mine and smelter are similar to those of a laterite mine and leaching plant, that is, of the order of US$ 60 000 per annual tonne of nickel in product.

The operating costs of a laterite smelter are about US$ 5000 per tonne of nickel in product, whereas those of the laterite leaching plant are about US$ 10 000 per tonne of product nickel plus cobalt.

Laterite mining/smelting is generally profitable for ores containing greater than 1.7% nickel, whereas laterite leaching operations are generally profitable with ores containing greater than 1.3% Ni and 0.1% Co.

Sulfide operations with underground mines and concentrators containing greater than 2% (nickel + copper + cobalt) are generally profitable. Surface mines containing greater than 1% (nickel + copper + cobalt) also tend to be profitable.

REFERENCES

Christensen, P., & Dysert, L. R. (2005). *Cost estimate classification system – as applied in engineering, procurement and construction for the process industries*. The Association for the Advancement of Cost Engineering.

Dalvi, A. D., Bacon, W. G., & Osborne, R. C. (2004). The Past and Future of Nickel Laterites. In Prospectors and Developers Association of Canada 2004 International Convention, Trade Show & Investors Exchange. March 7, 2004.

Dolan, D. S., & Nendick, R. M. (2004). Beating US$10 per pound of installed capacity for a laterite nickel plant. In W. P. Imrie, D. M. Lane & S. C. C. Barnett (Eds.), *International laterite nickel symposium 2004, economics and project assessments* (pp. 55–62). Warrendale, Pa: TMS.

Goonan, T. G., & Kuck, P. H. (2010). Nickel mineral commodity report. *United States Geological Survey*. Washington, DC.

Flewelling, S. (2006). Analyst site visit Koniambo, 2006.

Freed, J. (2007). BHP plans nickel project in Indonesia. *Sydney Morning Herald*. December 30, 2007.

Kuck, P. H. (2010). Nickel mineral commodity summaries. *United States Geological Survey*. Washington, DC.

Minara Resources (2009). Annual report 2009.

Neudorf, D. A., & Huggins, D. A. (2004). An alternative nickel laterite project development model. In W. P. Imrie, D. M. Lane & S. C. C. Barnett (Eds.), *International laterite nickel symposium 2004, economics and project assessments* (pp. 63–76). Warrendale, Pa: TMS Publication.

Onca Puma. (2009). Vale to postpone startup of Onca Puma on slow demand (update1). *Bloomberg News*. April 16, 2009.

Pearce, I., Usmar, S., Lachance, D., & Romaniuk, M. (2009). Accelerated transformation into a robust nickel business. *Presentation to analysts visit. Koniambo: New Caledonia. May 2009.*

Sahr, R. C. (2009). *Consumer price index (CPI) conversion factors 1774 to estimated 2019 to convert to dollars of 2009*. Oregon State University. http://oregonstate.edu/cla/polisci/faculty-research/sahr/cv2007.pdf.

SUGGESTED READING

Anglo Platinum (2009). In *Annual report* 2008. pp. 102 & 183.

Crundwell, F. K. (2008). *Finance for engineers. The evaluation and funding of capital projects.* Springer.

Guo, X. J. (2009). Chinese nickel industry – projects, production and technology. In J. Liu, J. Peacey & M. Barati (Eds.), *Pyrometallurgy of nickel and cobalt 2009* (pp. 3–21). Canada, Montreal: Metallurgical Society of CIM.

Habashi, F. (2009). A history of nickel. In J. Liu, J. Peacey & M. Barati., et al. (Eds.), *Pyrometallurgy of nickel and cobalt 2009, proceedings of the international symposium* (pp. 77–98). Canada, Montreal: Metallurgical Society of CIM.

Solar, M. Y., Candy, I., & Wasmund, B. (2008). Selection of optimum ferronickel grade for smelting nickel laterites. *CIM Bulletin, 101*, 1–8, Canada, Montreal.

The Nickel Institute. <http://www.nickelinstitute.org> Accessed 27.05.09.

Tonks, A. (2009). Nickel the wildcard. In: Presented at the 12th World Stainless Steel Conference, organized by Commodities Research Unit and Indian Stainless Steel Development Association. Mumbai: India. November 22–24, 2009.

United States Geological Survey Mineral Information – Nickel <http://minerals.usgs.gov/minerals/pubs/commodity/nickel>.

Chapter 3

Upgrading of Laterite Ores

The first step in the processing of laterites deposits is the upgrading of the ore. The upgraded ore is then smelted to ferronickel or processed to nickel in a hydrometallurgical refinery.

The objectives of this chapter are the following:

(a) to describe laterite ores and their mineralogy;
(b) to describe the upgrading of laterite ore; and,
(c) to indicate the benefits of upgrading to subsequent smelting or leaching operations.

3.1. LATERITE ORES

3.1.1. Laterite Profile

Laterite ores are a heterogeneous mixture of hydrated iron oxides and hydrous magnesium silicates. These deposits were formed by weathering of peridotite rocks. Peridotite consists mainly of olivine [$(Mg,Fe)_2SiO_4$] with a small amount of pyroxene [$(Mg,Fe)_2Si_2O_6$].

Water, containing organic acids and carbon dioxide, percolates down through the weathered material. Iron, nickel, magnesium and silica dissolve in this water. Toward the top of the deposit, iron is oxidized by air and precipitates as hydrated iron oxides, such as goethite. Nickel and cobalt co-precipitate with the iron, substituting for iron in the structure of goethite. The term 'limonite' is usually used to describe this part of the laterite deposit.

Closer to the bedrock, magnesium and silica precipitate, forming magnesium silicates of the serpentine group, such as $Mg_3Si_2O_5(OH)_4$. Nickel precipitates as nepouite [$Ni_3Si_2O_5(OH)_4$]. Mixtures of these two minerals are called garnierite. Other minerals found in this layer are as follows: (i) talc [$Mg_3Si_4O_{10}(OH)_2$] and willemseite [$(Ni,Mg)_3Si_4O_{10}(OH)_2$]; (ii) clinochlore [$(Mg,Fe)_5Al(Si_3Al)O_{10}(OH)_8$] and nimite [$(Ni,Mg,Al)_6(Si,Al)_4O_{10}(OH)_8$] and (iii) sepiolite [$Mg_4Si_6O_{15}(OH)_2\ 6H_2O$] and falcondite [$(Ni,Mg)_4Si_6O_{15}(OH)_2\ 6H_2O$]. This layer of the laterite is frequently referred to as 'saprolite'.

There may be a third layer of clay material. These clays generally belong to the group of minerals called smectites. An example of a clay mineral found in

Extractive Metallurgy of Nickel, Cobalt and Platinum-Group Metals. DOI: 10.1016/B978-0-08-096809-4.10003-6

Laterite Profile

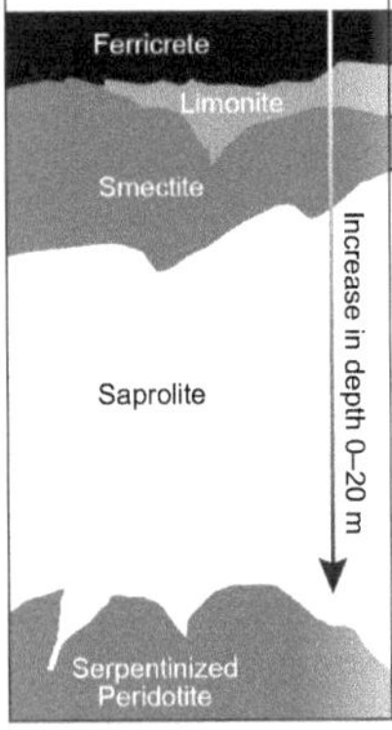

Common Name	Minerals	Approximate analysis (%)			
		Fe	Mg0	Ni	Co
Ferricrete	Goethite	>50	<0.5	<0.8	<0.1
Limonite	Hydrated Fe0(0H)	40–50	0.5–5	0.8–1.5	0.1–0.2
Smectite	Nontronite	10–30	5–15	0.6–2	0.02–0.1
Saprolite	Serpertine Talc Sepiolite Nontronite	10–25	15–35	1.5–4	0.02–0.1
Bedrock	Peridotite	5	35–40	0.3	0.01

FIGURE 3.1 An idealized profile of a laterite deposit. At the surface, the iron content is high and the MgO content is low. With increasing depth, this position reverses so that at the bottom of the deposit, the MgO content is high and the iron content is low. Ferricrete is a hard layer of soil cemented by iron oxides.

nickel laterites is nontronite. Clays can be present in either the limonite or saprolite layers or may be present as a separate and distinct layer.

An idealized profile of a laterite deposit is shown in Figure 3.1. It is emphasized that these layers are not necessarily distinct. Rather, there is a continuous variation with depth.

3.1.2. Mineralogy of Nickeliferous Laterites

The limonite layer is generally uniform, composed mainly of goethite. In contrast, the saprolite layer is very heterogeneous. It is composed of a variety of silicates, such as serpentines, talcs, chlorites and sepiolites.

3.2. UPGRADING OF LATERITE ORES[1]

The ore type and metal contained in the ore are summarized in Table 3.1 for several operating mines. Laterite ores that are mined typically contain 1.3%–2.5% Ni. Saprolite ores are notably richer in nickel than limonite ores.

The weathered material in the laterite deposit is generally soft and enriched in nickel, whereas the unweathered precursor rock is hard and lean in nickel.

1. The term *upgrading* is used for this treatment. *Concentrating* is reserved for a separation process that specifically targets a particular physical or chemical property. As a result, magnetic separation produces a magnetic concentrate because the magnetic property of the minerals is targeted. Similarly, with flotation, which targets the hydrophobicity of the sulfide minerals. This is not to say that upgrading is insignificant.

TABLE 3.1 The Concentrations of Nickel and Cobalt in the Laterite Ores from Operating Mines.
The saprolite ores are mostly smelted to ferronickel. The limonite ores are mostly leached

Mine	Location	Predominant ore type[a]	Ni, %	Co, %
Murrin-Murrin	Australia	L, SM	1.2	0.08
Onca Puma	Brazil	S	1.7	
Cerro Matoso	Colombia	S	2.3	
Moa Bay	Cuba	L	1.3	0.14
Nicaro	Cuba	L	1.3	0.1
Punta Gorda	Cuba	L	1.3	0.10
Larco	Greece	L	1.3	0.06
PT Antam	Indonesia	S	2.4	
PT Inco	Indonesia	S	2.0	
Ambatovy	Madagascar	M	1.3	0.11
Koniambo	New Caledonia	S	2.5	
Nepoui	New Caledonia	S	1.8	
Coral Bay	Philippines	L	1.3	0.09
Loma de Niquel	Venezuela	S	1.5	

[a]*S = saprolite; L = limonite; SM = smectite; M = mixed.*

The ores also contain hard by-products of laterization that are lean in nickel, such as quartz.

The nickel content of the mined ores is upgraded by

(a) crushing and/or grinding the ore;
(b) separating small, soft, nickel-rich particles from larger, hard, nickel-lean particles by using screens, hydrocyclones and spiral classifiers; and,
(c) rejecting the nickel-lean material to waste.

The advantage of these upgrading steps is that it cheaply rejects nickel-lean minerals by physical means rather than expensively removing them chemically by smelting or leaching.

Upgrading also has the advantage that it often rejects minerals, such as magnesium-rich olivine $(Mg,Fe)_2SiO_4$, that adversely affect subsequent smelting (Doyle, 2004) or leaching (Tuffrey, Chalkley, Collins, & Iglesias, 2009).

FIGURE 3.2 Upgraded saprolite ore from New Caledonian, containing about 3% Ni. This material has been upgraded from ore containing 1.8% Ni. It is destined to be smelted and refined to produce ferronickel, containing 30% Ni and 70% Fe, which is used mainly for making stainless steel. *Source: Photograph courtesy of Société Le Nickel-SLN/Eramet.*

3.3. EXTENT OF UPGRADING

Recently installed upgrading plants produce upgraded ores with up to twice the nickel content of the original ore. The feed material for Nepoui mine in New Caledonia, for example, is an ore that contains 1.8% Ni (Le Nickel, 2005). The upgraded product contains 3% Ni. The upgraded ore is shown in Figure 3.2.

The upgrading methods and the extents of upgrading at four laterite mines are given in Table 3.2.

3.4. ECONOMIC JUSTIFICATION FOR UPGRADING LATERITES

The economic justification for upgrading is that, per tonne of nickel, less material has to be transported and smelted or leached. This lowers the energy and reagents requirements, the equipment needed for shipping, smelting and leaching and the effluents produced while increasing the production rate of a given plant expressed as tonnes of nickel per year. All of these factors result in lower costs for the production of nickel.

The downside of the upgrading of the laterite ore is the loss of nickel in the rejected material from the upgrading plant. This loss is the consequence of (i) the intrinsic nickel content of the rejected minerals and (ii) the accidental loss of some nickel-rich minerals with the rejects.

Each mine must balance the benefits of upgrading against these losses.

TABLE 3.2 Methods for Upgrading Laterite Ores.
The ravensthorpe and new caledonian plants work with very fine particles (~75 μm) and achieve considerable upgrading. Falcondo and coral bay work with larger particles and achieve limited upgrading

Mine	Nickel in ore, %	Nickel in upgraded ore, %	Method of upgrading	Reference
Ravensthorpe, Australia	1	2	Makes ore–water slurry by scrubbing in rotating rotary scrubbers and attritioners, Figure 3.3. Separates small Ni-rich particles from large, Ni-lean particles in the slurry by means of screens, hydrocyclones and spiral classifiers. Reject particles are typically >75 μm in size.	Adams *et al.*, 2004
Falcondo, Dominican Republic	1.2	1.4	Ore is passed through wobbler feeders and a ball mill. Product is screened. Pieces larger than 0.01 m are rejected to tailings.	Mast, Fanas, Frias, & Ortiz, 2005
Nepoui and Tiebaghi mines, New Caledonia	1.8	3	Ore slurried with water. Small, Ni-rich particles separated from large Ni-lean particles by screening, hydrocyclones and filters. Principle: small, low-density particles are richer in Ni than large, dense particles. Particle sizes estimated to be ~50 μm.	Le Nickel, 2008
Coral Bay, Philippines	1.26	~1.5	Ore is screened, vigorously washed then screened several more times. Only <2-mm particles go forward to leaching.	Nakai, Kawata, Kyoda, & Tsuchida, 2006

3.5. PRINCIPLES AND METHODS OF UPGRADING LATERITES

All upgrading of laterites is based on the principle that laterized nickel minerals, such as goethite and garnierite, are softer than the unlaterized precursors, such as olivine, and the hard products of laterization that are lean in nickel, such as quartz.

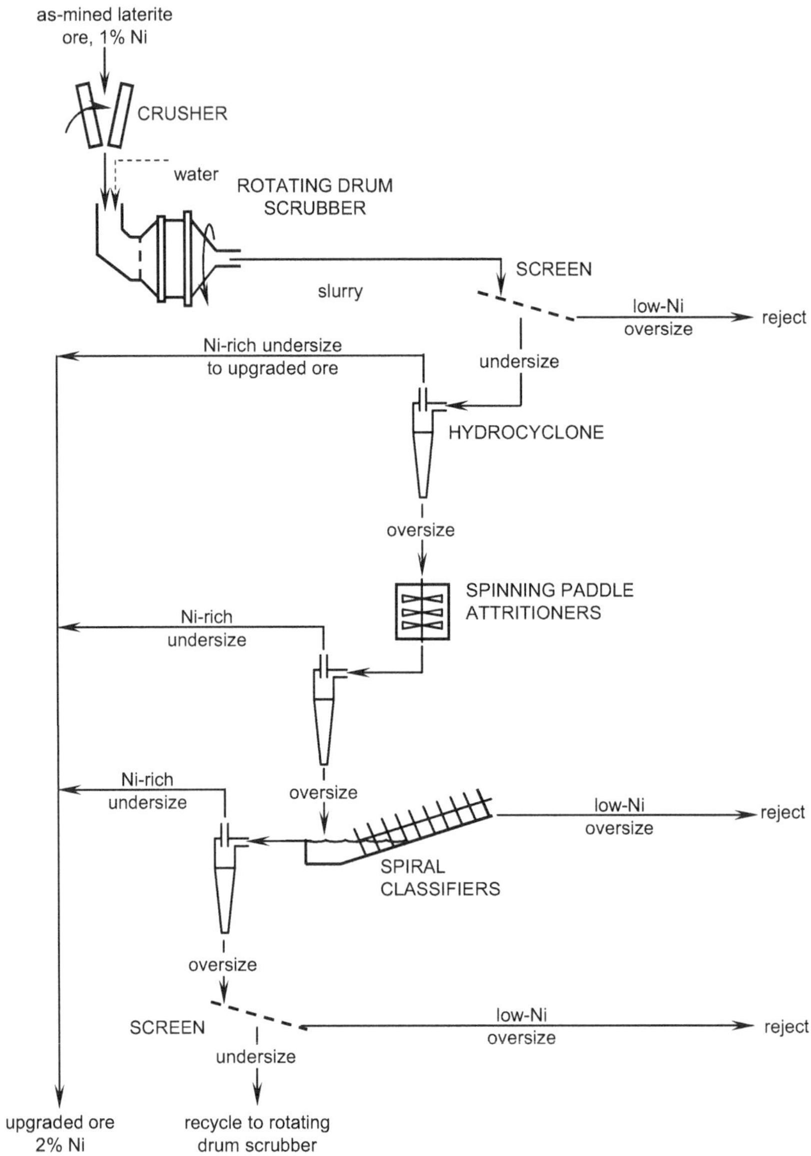

FIGURE 3.3 Flowsheet used at Ravensthorpe for upgrading laterite ore from 1% Ni to 2% Ni. Small, soft, nickel-rich particles are scrubbed from nickel-lean large, hard particles in rotating scrubbers and attritioners. The particles are then separated from each other with screens, hydrocyclones and spiral classifiers. The small nickel-rich particles are typically ~75 μm diameter. A general rule is that small, low-density particles are richer in nickel than large, dense particles. Ravensthorpe is temporarily closed. It reportedly reached full production just before closure.

This means that crushing and grinding at the correct intensity can produce relatively small particles that are rich in nickel and larger particles that are lean in nickel.

These different sized particles can then be separated using screens, filters, hydrocyclones and other size classifiers.

3.5.1. Industrial Flowsheet

A schematic diagram of the flowsheet for the upgrading of the ore at Ravensthorpe in Australia is shown in Figure 3.3. This plant produced upgraded ore containing 2% Ni from an ore containing 1% Ni. The process entails two operations: (i) scrubbing and (ii) classification. At this mine, no grinding was required.

3.5.2. Industrial Equipment

Some particles in saprolite ore deposits consist of an unlaterized, hard core and a laterized, soft weathered skin. A piece of dunite rock that is partially laterized is shown in Figure 3.4. The core is lean in nickel, while the skin is rich in nickel.

FIGURE 3.4 Partially laterized dunite rock, consisting mainly of olivine [$(Mg,Ni)_2SiO_4$]. The outer skin is typically garnierite [$(Mg,Ni)_3Si_4O_5(OH)_4$]. Garnierite can be removed from the dunite by gentle crushing and grinding. *Source: Photograph by W.G. Davenport.*

Upgrading this material requires that the weathered skin be broken into small pieces without breaking up the nickel-lean core. This requires controlled grinding such as that provided by an autogenous mill with a small diameter.

Some deposits are already made up of fine particles (Adams *et al.*, 2004). In this case, the fine particles need only be scrubbed off the larger particles and then screened and cycloned to separate them from the nickel-lean oversize particles. Rotating-drum scrubbers and attritioners are used for this scrubbing.

3.6. EVALUATION

All laterite orebodies are different. For this reason, every ore must be thoroughly tested to determine the extent to which it can be upgraded.

The strongest advocate of upgrading is Le Nickel in New Caledonia – and they are somewhat secretive about their upgrading processes. It is believed that their final 3% Ni upgraded ore is obtained by filtering very fine waste rock from even finer garnierite particles.

3.7. SUMMARY

Laterite ores are always upgraded before smelting or leaching. Upgrading of these ores entails the following steps:

(a) gently crushing and/or grinding the ore; and,
(b) separating the resulting small, soft, low-density nickel-rich laterized mineral particles from the large, hard, dense, nickel-lean, unlaterized precursor rock and by-product laterization products, such as quartz.

Upgrading minimizes the amount of material that has to be transported, smelted and/or leached per tonne of product nickel. It thereby minimizes the energy, reagent and equipment requirements for shipping, smelting and leaching, and the gas, liquid and solid effluents that are produced.

REFERENCES

Adams, M., van der Meulen, D., Czerny, C., et al. (2004). Piloting of the beneficiation and EPAL circuits for Ravensthorpe nickel operations. In W. P. Imrie, D. M. Lane & S. C. C. Barnett, et al. (Eds.), *International Laterite Nickel Symposium 2004, Process Development for Prospective Projects* (pp. 193–202). Warrendale, Pa: TMS.

Doyle, C. (2004). The steps required to meet production targets at PT Inco, Indonesia: A new innovative business strategy. In W. P. Imrie, D. M. Lane & S. C. C. Barnett (Eds.), *International Laterite Nickel Symposium 2004* (pp. 667–684). Warrendale, Pa: TMS.

Nickel, Le (2005). Les mineurs, ces chercheurs d'or, remuent des montagnes pur trouver le precieux nickel. *Jour et Nuit, Le Nickel, New Caledonia.* August 1, 2005.

Le Nickel (2008). Les unites de traitement du minerai. January 17, 2008. Accessed at http://www.sln.nc/content/view/76/44/lang,french/ on May 19, 2011.

Mast, E. D., Fanas, J. J., Frias, J. R., & Ortiz, D. (2005). Process improvements at Falconbridge Dominicana. In J. Donald & R. Schonewille (Eds.), *Nickel and Cobalt 2005, Challenges in Extraction and Production* (pp. 427–439). Canada, Montreal: CIM.

Nakai, O., Kawata, M., Kyoda, Y., & Tsuchida, N. (2006). Commissioning of Coral Bay nickel project. In *ALTA 2006 Nickel/Cobalt Conference Proceedings*. A. Taylor (Ed.) Melbourne, Australia: ALTA Metallurgical Services.

Tuffrey, N. E., Chalkley, M. E., Collins, M. J., & Iglesias, C. (2009). The effect of magnesium on HPAL – comparison of Sherritt laboratory studies and Moa plant operating data. In J. J. Budac, R. Fraser & I. Mihaylov (Eds.), *Hydrometallurgy of Nickel and Cobalt 2009* (pp. 421–432). Canada, Montreal: The Metallurgical Society of CIM.

SUGGESTED READING

Adams, M., van der Meulen, D., Czerny, C., et al. (2004). Piloting of the beneficiation and EPAL circuits for Ravensthorpe nickel operations. In W. P. Imrie, D. M. Lane & S. C. C Barnett, et al. (Eds.), *International Laterite Nickel Symposium 2004, Process Development for Prospective Projects* (pp. 193–202). Warrendale, Pa: TMS.

Chapter 4

Overview of the Smelting of Nickel Laterite to Ferronickel

Half of all primary nickel, approximately 700 000 tonnes per year, is produced by smelting laterite ores. About 90% of this smelting produces ferronickel and about 10% produces sulfide matte.

The objectives of this chapter are to present an overview of this process and the principles of the ferronickel smelting.

4.1. FEED TO FERRONICKEL SMELTING

The feed material for ferronickel production is moist saprolite, a type of laterite. This ore typically contains 1.5%–3% Ni. The nickel is present as nickel silicate minerals, such as garnierite, $(Mg,Ni)_3Si_2O_5(OH)_4$. Nickel may also be present in other forms, such as nickel–talc and nickel–sepiolite.

The product is refined ferronickel, which contains 20%–40% Ni and 80%–60% Fe. The ferronickel is suitable for making stainless steel and other ferrous alloys.

The feed is predominantly saprolite, rather than limonite. This is because the iron content of the saprolite ore, about 15% Fe, is lower than that of limonite ore. The choice of saprolite ore ensures that the ferronickel product is not excessively diluted by iron. As discussed in Chapter 3, these ores are upgraded wherever possible.

The typical composition of the feed to a ferronickel smelter is shown in Table 4.1. It also contains 25%–30% mechanically entrained water (Warner *et al.*, 2006).

4.2. FERRONICKEL PRODUCT

The typical composition of the ferronickel product from the smelter is given in Table 4.2. These compositions show that the SiO_2, MgO and Al_2O_3 content of the laterite feed report to the slag, while the nickel and iron oxides are reduced to metal.

The ferronickel is produced in the form of small ingots or water-granulated 'beans' that are convenient for adding to alloy-making furnaces.

Extractive Metallurgy of Nickel, Cobalt and Platinum-Group Metals. DOI: 10.1016/B978-0-08-096809-4.10004-8

TABLE 4.1 Typical Composition of the Feed to a Ferronickel Smelter

Component	Upgraded laterite, %
Ni	1.5–3
Co	0.04–0.08
Fe	15%
O	5 (bonded to Ni, Co and Fe)
SiO_2	40
MgO	25
Al_2O_3	1
Chemically bonded H_2O	11

TABLE 4.2 Typical Composition of the Product from a Ferronickel Smelter

Component	Ferronickel product, %
Ni	20–40
Co	0.3–1
C	0.01–5
Si	0.03–5
Cr	0.01–2
S	0.02–0.05
P	0.01–0.04
Fe	Remainder

4.3. PRINCIPLES OF FERRONICKEL SMELTING

The smelting of laterite ores to produce ferronickel is based on two principles, which are discussed in the following sections.

4.3.1. First Principle of Ferronickel Smelting

The first principle of ferronickel smelting is that nickel and iron oxides are easily reduced to metallic nickel and iron, while other oxides, for example,

Al_2O_3 and MgO, are not reduced. Examples of the reactions that occur are the following:

$$\underset{}{C(s)} + \underset{\text{in calcined ore}}{NiO(s)} \xrightarrow{800°C} CO(g) + \underset{\text{metal}}{Ni(s)} \quad (4.1)$$

$$\underset{}{C(s)} + \underset{\text{in calcined ore}}{FeO(s)} \xrightarrow{800°C} CO(g) + \underset{\text{metal}}{Fe(s)} \quad (4.2)$$

Al_2O_3 and MgO are not easily reduced because the bonds between oxygen and aluminium or magnesium are much stronger than those between oxygen and nickel or iron.

The behavior of SiO_2 and Cr_2O_3 is between that of these two extremes. This means that ferronickel often contains several percent of silicon and chromium.

4.3.2. Second Principle of Ferronickel Smelting

The second principle of ferronickel smelting is that laterite ores cannot be smelted to pure nickel metal, that is, without iron. The reasons for this are as follows:

(a) laterite ores contain approximately 15% Fe but only contain about 1.5%–3% Ni; and,
(b) the bond between oxygen and iron is only slightly stronger than that between oxygen and nickel, which means that FeO in calcine is reduced almost as readily as NiO.

This means that the product of laterite smelting is always ferronickel, which contains between 20% and 40% Ni and between 80% and 60% Fe.

Of course, the iron that is contained in the ferronickel is not a major disadvantage because it is a key ingredient in stainless steels and other ferrous alloys. However, the costs of production increase with higher content of iron because proportionately more energy is used for reduction, heating and melting. For this reason, ore selection, blending and upgrading are vitally important.

4.4. BRIEF PROCESS DESCRIPTION

A schematic diagram of the flowsheet of the process is shown in Figure 4.1. The four main steps are the following:

(a) dewatering: the removal of mechanically entrained water from the concentrate;
(b) calcination: the removal of chemically bonded water from the dried ore;
(c) reduction: the removal of oxygen from the nickel and iron oxide in the calcine; and,
(d) refining: the removal of impurities, such as sulfur and phosphorus, from the molten ferronickel.

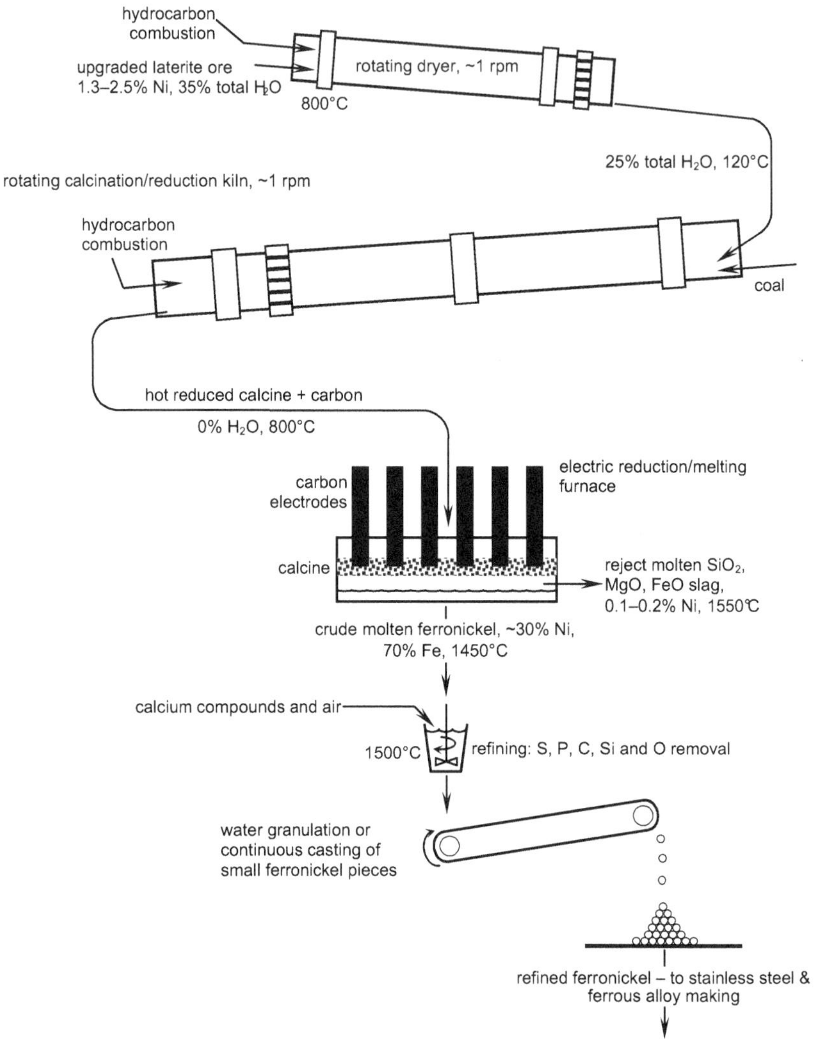

FIGURE 4.1 Schematic diagram of the flowsheet for smelting moist 1.5%–2.5% Ni, low-iron laterite (saprolite) to ferronickel.

Dewatering, or drying, is typically performed in a rotating kiln that is 4 m diameter × 30 m long. The kiln is heated with pulverized coal, fuel oil or natural gas. The moisture of the dried ore is typically about 22%, although in some operations this is significantly lower. The dried material is either sent directly to the calcination kiln or first to a crushing and screening plant and then to a calcination kiln.

The calcination kilns are much larger than the dewatering kilns, typically 5 m diameter by 100 m long. The purpose of the kilns is to produce a bone dry, partially reduced ore. The kiln is heated with pulverized coal, fuel oil or natural gas.

The purpose of ferronickel smelting is to reduce the nickel and iron to metal and to reject the MgO, SiO_2 and Al_2O_3 to slag. Ferronickel electric furnaces are round or rectangular. Round furnaces are 15–20 m in diameter, while rectangular furnaces are 10 m wide and 25–35 m long. These furnaces produce 100–200 tonnes of ferronickel per day.

Refining is performed in electrically heated ladles in batch mode in four steps: charging, phosphorus removal, sulfur removal and casting.

Dewatering, calcination and reduction are described in Chapter 5, ferronickel smelting is described in Chapter 6 and refining of molten ferronickel is described in Chapter 7.

As mentioned earlier, about 10% of laterite smelting produces sulfide matte. The process is similar to the calcination and smelting of laterite, except that molten sulfur is added to the calcination kiln. This matte is roasted in oxidizing and reducing conditions to produce nickel.

The production of the sulfide matte is described subsequently in Chapter 8, and the roasting of the matte to metal is described in Chapter 9.

REFERENCE

Warner, A. E. M., Diaz, C. M., Dalvi, A. D., et al. (2006). JOM world nonferrous smelter survey, part III: Laterite. *Journal of Metals, 58*, 11–20.

SUGGESTED READING

Bergman, R. A. (2003). Nickel production from low-iron laterite ores: Process descriptions. *CIM Bulletin, 96*, 127–138.

Warner, A. E. M., Diaz, C. M., Dalvi, A. D., et al. (2006). JOM world nonferrous smelter survey, part III: Laterite. *Journal of Metals, 58*, 11–20.

Chapter 5

Dewatering and Calcination of Laterite Ores

Smelting of laterites ores accounts for about half of all primary nickel. The first step in the smelting process is dewatering, which is then followed by calcination of the upgraded laterite ore.

The objectives of this chapter are to describe and discuss the dewatering and calcination of the laterite ore.

5.1. DEWATERING OF THE UPGRADED LATERITE ORE

Nickel laterite ores are always mined from the surface and usually in the tropics.[1] They are almost always wet and sticky and their moisture content varies from the wet season to the dry season. The upgraded ores are also wet and sticky.

All the water in the upgraded ore must be removed before it is fed to the electric smelting furnace to avoid explosions in the furnace. The water is removed in two processing steps: dewatering followed by calcination.

Hot rotating kilns are used for both of these processing steps, as shown in Figure 4.1. The internal structure of a rotary calcination kiln is shown in Figure 5.1.

The goals of dewatering are to make a product that has the following properties:

(a) it has a consistent moisture content of approximately 20% entrained water;
(b) it does not stick or adhere to conveyor belts, machinery, *etc.*; and,
(c) it is not too dusty.

Wet ore is continuously fed down a sloped rotating kiln. At the same time, hot air and combustion gas from oil, gas or coal burners are blown down the kiln. The temperature of the gas is about 800°C. Hot gas from the smelting furnace, the calcination kiln and from slag cooling are also used. The ore and gas travel co-currently through the kiln because counter-current flow tends to make 'mud rings' in the kiln (Boldt & Queneau, 1967).

The operating details of the dewatering operations at seven sites are given in Table 5.1.

1. Some ferronickel is produced from non-tropical sources. See Appendix A.

Extractive Metallurgy of Nickel, Cobalt and Platinum-Group Metals. DOI: 10.1016/B978-0-08-096809-4.10005-X

FIGURE 5.1 Inside a new rotary calcination kiln. The internal lifters are notable. This kiln is 6 m diameter and 130 m long. It removes water from the partially dried feed and partially reduces nickel oxide to nickel metal. *Photograph courtesy of Hatch.*

The main reaction is the evaporation of water:

$$\underset{30^{\circ}\mathrm{C}}{H_2O(\ell)} \rightarrow \underset{120^{\circ}\mathrm{C}}{H_2O(g)} \quad \Delta H = 2260 \text{ kJ per kg of } H_2O \tag{5.1}$$

The evaporation of water is endothermic, that is, it requires heat. This heat is provided by the hot combustion gas.

The kiln contains 'lifters' that (i) lift the wet ore as the kiln rotates; and, (ii) cascade it through the passing hot gas. The cascading of the wet ore exposes all the ore to hot gas, so that the water in the ore is evaporated evenly and efficiently.

The products are dusty offgas at a temperature of 120°C and partially dewatered ore. The gas is collected, de-dusted in electrostatic precipitators and then released to the atmosphere. The dust is recycled.

The dewatered ore continuously exits the lower end of the kiln where it is screened and crushed before being transferred to calcination (see Figure 4.1).

5.2. CONTROL OF THE DEWATERING KILN

Dewatering kilns are designed to produce dried ore with a specified water content (approximately 20% H_2O) at a specified rate from upgraded ore.

TABLE 5.1 Details of Seven Laterite Ore Dewatering Kiln Operations
The objective of the dewatering kilns is to produce a partially dried, non-sticky product for easy materials handling. The kilns are 3–5 m diameter and up to 40 m long. They rotate at about 1 rpm. Each kiln treats 100–200 tonnes of ore/h (dry basis)

Smelter	Codemin, Brazil	Cerro Matoso, Columbia	PT Antam , Indonesia
Type and number of dryers	One rotary dryer	Two rotary dryers	Two rotary dryers
Outside diameter × length, m	3.4 × 22	5.1 × 45	3.2 × 30
Capacity, tonnes/h dry ore	104	Up to 200 each	50 each
Ore moisture in, %[a]	25–27	22–30 (seasonal)	30
Ore moisture out, %[a]	23–24	10–12 (seasonal)	22
Fossil fuel, type	Fuel oil	Natural gas	Pulverized coal
Average fossil fuel consumption per tonne of dry ore	9 kg	12–18 Nm^3 seasonal	35 kg
Dust production, % of ore feed	0.5	4	3
Dust destination	Recycle to dryer	To calcination kilns	Recycle to dryer
Dryer product destination	To calcination kilns	To calcination kilns	Screening, hammer milling then to calcination kilns

(Continued)

TABLE 5.1 Details of Seven Laterite Ore Dewatering Kiln Operations—cont'd

The objective of the dewatering kilns is to produce a partially dried, non-sticky product for easy materials handling. The kilns are 3–5 m diameter and up to 40 m long. They rotate at about 1 rpm. Each kiln treats 100–200 tonnes of ore/h (dry basis)

Smelter	Hyuga Smelting, Japan	Pacific Metals Co., Japan	Le Nickel, New Caledonia	Loma de Niquel, Venezuela
Type and number of dryers	1 rotary dryer	1 rotary dryer, 2 impact dryers	2 rotary dryers	1 rotary dryer
Outside diameter × length, m	5 × 40	Rotary 5 × 35	4 × 32	4.8 × 34
Capacity, tonnes/h dry ore	160	Rotary 105, impact 110	200 (each)	230
Ore moisture in, %[a]	22–30	30	26	25–30
Ore moisture out, %[a]	22–23	24	18	Minimum of 15
Fossil fuel, type	Coal, bunker C oil, waste electric furnace offgas	Waste calcination kiln offgas, slag heated gas	Heavy fuel oil	Natural gas
Average fossil fuel consumption per tonne of dry ore	12–13 L of oil including coal equivalent	None		10–11 Nm^3
Dust production, % of ore feed	2–5	1		
Dust destination	Recycled to dryer	Recycled to dryer	To dryer discharge	Pelletized and to calcination/reduction
Dryer production destination	Screening and crushing then to calcination kilns	Screening, crushing and blending then to calcination kilns	Screening and crushing then to calcination kilns	Crushing then to calcination kilns

[a]*plus 11% chemically bonded H_2O.*
Source: Data are from Bergman (2003) and Warner *et al.* (2006).

These design criteria are achieved by adjusting the feed rate of the fuel and the feed rate of ore. More combustion of fuel and a lower feed rate of ore give a drier product (and *vice versa*).

5.3. CALCINATION AND REDUCTION OF DEWATERED LATERITE

The partial removal of entrained water from moist saprolite ore was described in the previous section. This section describes (i) the removal of the remaining water in the ore; and (ii) the partial removal of oxygen from the nickel and iron minerals in the ore, that is, partial reduction of the ore.

The objectives of calcination are the following:

(a) to remove the remainder of the water in the ore (to avoid explosions during subsequent electric furnace smelting);
(b) to reduce about a quarter of the nickel in the ore to nickel metal;
(c) to reduce most of the Fe^{3+} minerals to Fe^{2+} minerals and about 5% of the iron to metallic iron;
(d) to add enough coal in the kiln so that some remains in the ore when it is transferred to the electric furnace for final reduction of nickel and iron; and,
(e) to provide the calcine at a temperature of about 900°C to the ferronickel smelting furnace, so that the energy used is minimized.

The operating details of seven calcination kilns are given in Table 5.2.

5.3.1. Rotating Kilns for Calcination and Reduction

The calcination of laterite ores is performed in rotating kilns. The kilns are long, up to 185 m (Walker *et al.*, 2009) and are fired by hydrocarbon fuels. The kilns are typically sloped about 4° from horizontal and rotate at approximately 1 rotation/minute.

Upgraded and dewatered ore is fed continuously to the upper end of the rotating kiln. Coal is also added continuously along with recycled dust from the calcination kiln that has been pelletized. The feed rate of coal is about 5% of the feed rate of upgraded ore. These materials flow slowly down the rotating kiln and out of the hot discharge end of the kiln.

5.3.2. Heat and Reducing Gas Supply

Energy and reducing gas are continuously supplied to the rotating kiln by partially combusting the hydrocarbon fuel at the discharge end of the kiln. This combustion produces hot reducing gas which (i) heats and dries the feed; and, (ii) partially reduces the nickel and iron minerals in the feed as it travels in a counter-current direction up the kiln.

TABLE 5.2 Details of seven calcination kiln operations

The main objectives of *calcination* are to remove all water from the feed, partially reduce nickel and iron minerals and heat the feed in preparation for electric furnace melting/reduction. The kilns are 4–6 m diameter and 70–185 m long. They rotate at about 1 revolution/min. They treat 50–150 tonnes of partially dried ore/h and produce partially reduced product at ~900 °C. Offgas leaves the feed end of the kiln at 250–450 °C

Smelter	Codemin, Brazil	Cerro Matoso, Columbia		PT Antam, Indonesia	
Equipment type and number	Two rotary kilns	Two rotary kilns		Two rotary kilns	
Outside diameter × length, m	3.6 × 70	6.1 × 185	6.1 × 135	4 × 90	4.2 × 90
Capacity, tonnes/h dry ore	40 each kiln	165 each kiln		32	35
Calcine discharge temp., °C	900	800–850		800–1000	
Fuel type	Heavy oil	Natural gas		Pulverized coal	
Fuel consumption/tonne of ore	52 kg	50–55 Nm^3		115 kg	
Reductant	Woodchips	Anthracite		Anthracite and coal	
Average reductant consumption per tonne of dry ore	180 kg	50–60 kg		67 kg	
Dust production, % of ore feed	20	12	22	8	
Dust destination	Recycled to calcination kilns	Extruded and fed to calcination kilns		Pelletized and fed to calcination kilns	

Smelter	Hyuga, Japan	Pacific Metals, Japan	Le Nickel, New Caledonia	Loma de Niquel, Venezuela
Equipment type and number	2 rotary kilns	3 rotary kilns	5 rotary kilns	2 rotary kilns
Outside diameter × length, m	4.8 × 105	5.25 × 100, 5.5 × 115, 4.6 × 131	4 × 95	5.4 × 120
Capacity, tonnes/h dry ore	60–65 each kiln	90, 110, 90, respectively		65 each kiln
Calcine discharge temp., °C	800–900	1050	900	850
Fuel type	55–65% pulverized coal & bunker C oil	Pulverized coal	Pulverized coal	Natural gas
Fuel consumption/tonne of ore	60–62 L (including equivalent amount of coal)	30–50 kg		80–85 Nm^3
Reductant	Coal	Coal	Anthracite	Coal
Average reductant consumption per tonne of dry ore	70–80 kg	110 kg	50 kg	55 kg
Dust production, % of ore feed	15–20	25	10	15
Dust destination	Pelletized and recycled to calcination kilns	Pelletized and recycled to calcination kilns	Recycled to calcination kilns	Pelletized and recycled to calcination kilns

5.4. CHEMISTRY

The reactions that occur in the calcination kiln are as follows:

(a) evaporation of the remaining mechanically entrained water in the ore, given as follows:

$$H_2O(\ell) \rightarrow H_2O(g) \tag{5.2}$$

(b) thermal dissociation of laterite minerals to oxides and $H_2O(g)$, given by reactions such as the following:

$$\underset{\text{garnierite}}{Ni_3Mg_3Si_4O_{10}(OH)_8(s)} \xrightarrow{700^\circ C} 3NiO(s) + 3MgO(s) + 4SiO_2(s) + 4H_2O(g) \tag{5.3}$$

$$\underset{\text{goethite}}{2FeOOH(s)} \xrightarrow{700^\circ C} Fe_2O_3(s) + H_2O(g) \tag{5.4}$$

(c) reduction of the resulting oxides by coal and reducing gases by the following reactions:

$$\underset{\text{in coal}}{C(s)} + NiO(s) \xrightarrow{800^\circ C} CO(g) + \underset{\text{metal}}{Ni(s)} \tag{5.5}$$

$$CO(g) + NiO(s) \xrightarrow{800^\circ C} CO_2(g) + \underset{\text{metal}}{Ni(s)} \tag{5.6}$$

$$CO_2(g) + C(s) \xrightarrow{800^\circ C} 2CO(g) \tag{5.7}$$

$$CO(g) + \underset{\text{ferric oxide}}{Fe_2O_3(s)} \xrightarrow{800^\circ C} CO_2(s) + \underset{\text{ferrous oxide}}{2FeO(s)} \tag{5.8}$$

$$\underset{\text{from hydrocarbons}}{H_2(g)} + \underset{\text{ferric oxide}}{Fe_2O_3(s)} \xrightarrow{800^\circ C} H_2O(g) + \underset{\text{ferrous oxide}}{2FeO(s)} \tag{5.9}$$

At the same time that these reactions occur, the solid products are heated to approximately 900°C for hot charging to an electric furnace, which is discussed in Chapter 6.

5.4.1. Reaction Mechanisms

Water vapor is driven off the hydrated minerals in the feed to the kiln at about 700°C. This produces highly reactive nickel and iron oxides (Warner *et al.*, 2006). The rate of reduction of these oxides is rapid in the rotating kiln.

At temperatures above 900°C, nickel and iron oxides fuse with SiO_2, forming unreactive silicates. As a result, the temperature is kept below this value.

5.4.2. Kiln Internals

Calcination kilns almost always contain internal lifters. They lift the ore and cascade it through the hot reducing gas (Kashani-Nejad, Candy, & Kozlowski, 2005).

5.4.3. Kiln Externals

The kilns are also equipped with (i) onboard air fans, which blow air into the kiln; and, (ii) onboard coal scoops, which add coal, along the length of the kiln. Power for the fans is provided through copper 'slip rings' that rotate with the kiln (outside). On-board measurements are radioed to a central control room.

These fans and scoops are used to adjust gas temperature and the concentrations of carbon monoxide, CO, and hydrogen, H_2, in the gas along the length of the kiln. This is done to ensure that there is complete removal of water from the ore, that the extents of reduction of nickel and iron in the calcine product are controlled and that the levels of coal in the final calcine product are controlled.

Controlled levels of nickel, iron and coal in the calcine ensure efficient production of ferronickel with a constant composition during smelting.

5.5. PRODUCTS

The products of calcination kilns are the following:

(a) dusty offgas composed mainly of carbon dioxide, CO_2, water, H_2O, and nitrogen, N_2, at a temperature of 250°C; and,
(b) bone dry, partially reduced, coal-bearing calcine at a temperature of 900°C.

The offgas is de-dusted in electrostatic precipitators. This dust, which amounts to about 15% of the kiln feed, is pelletized and recycled to the calcination kiln. The de-dusted offgas is usually released to the atmosphere. In some operations, the offgas is used to heat the dewatering kiln in some operations.

The hot calcine continuously discharges from the low end of the kiln. It drops into 20 tonne, brick-lined transfer containers from which it is fed (hot) into a ferronickel smelting furnace.

The typical composition of the calcine is given in Table 5.3.

5.6. APPRAISAL

Rotary kilns for the dewatering and calcination of laterite ores are essential components for efficient production of ferronickel (Walker *et al*., 2009). Other static methods have been attempted (Mast *et al*., 2005) but with limited success.

Most new ferronickel smelters are installing rotary kilns. One new smelter is installing static shaft calciners.

TABLE 5.3 Typical Composition of the Calcine from the Calcination and Reduction Kiln for Laterites

Component	Calcine, %
Coal	2
Ni	1.5–3 (25% as metal, 75% as oxide)
Co	0.4–0.08 (20% as metal, 80% as oxide)
Fe	15 (about 5% as metallic iron, 95% as FeO)
SiO_2	40
MgO	25
Al_2O_3	1

Design emphasis is currently being placed on minimizing dust production (Souza, Aquino, & Goncalves, 2009; Vahed *et al.*, 2009).

5.7. SUMMARY

The smelting of laterites that are low in iron and rich in nickel (1.5–3% Ni) accounts for about half of the global production of primary nickel.

About 90% is smelted to ferronickel (20–40% Ni, 80–60% Fe) for use in making stainless steel and other ferrous alloys. The remainder is smelted to sulphide matte for refining to nickel metal and other nickel products.

Dewatering and calcination are key steps in ferronickel production. Together these processes produce the hot, bone dry, partially reduced calcine that is essential for successful ferronickel smelting.

REFERENCES

Bergman, R. A. (2003). Nickel production from low-iron laterite ores: process descriptions. *CIM Bulletin, 96*, 127–138.

Boldt, J. R., & Queneau, P. (1967). *The winning of nickel*. Longmans. p. 426.

Kashani-Nejad, S., Candy, I., & Kozlowski, M. (2005). Modelling of nickel laterite kiln processing – a conceptual review. In J. Donald & R. Schonewille (Eds.), *Nickel and cobalt 2005 challenges in extraction and production* (pp. 231–246). CIM.

Mast, E. D., Fanas, J. J., Frias, J. R., & Ortiz, D. (2005). Process improvements at Falconbridge Dominicana. In J. Donald & R. Schonewille (Eds.), *Nickel and cobalt 2005, challenges in extraction and production* (pp. 427–439). CIM.

Souza, A. G., Aquino, R. M., & Goncalves, D. J. (2009). Evaluation of Onca-Puma laterite dusting behavior and dust recovery techniques. In J. Liu, J. Peacey & M. Barati (Eds.), *Pyrometallurgy of nickel and cobalt 2009* (pp. 381–390). Metallurgical Society of CIM.

Vahed, A., Liu, J., Prokesch, M., et al. (2009). Testing of nickel laterite smelter dust insufflation – part I. In J. Liu, J. Peacey & M. Barati (Eds.), *Pyrometallurgy of nickel and cobalt 2009* (pp. 221–232). Metallurgical Society of CIM.

Walker, C., Kashani-Nejad, S., Dalvi, A. D., et al. (2009). Nickel laterite rotary kiln-electric furnace plant of the future. In J. Liu, J. Peacey & M. Barati (Eds.), *Pyrometallurgy of nickel and cobalt 2009* (pp. 381–390). Metallurgical Society of CIM.

Warner, A. E. M., Diaz, C. M., Dalvi, A. D., et al. (2006). JOM world nonferrous smelter survey, part III: Laterite. *Journal of the Minerals, Metals and Materials Society, 58*, 11–20.

SUGGESTED READING

Bergman, R. A. (2003). Nickel production from low-iron laterite ores: process descriptions. *CIM Bulletin, 96*, 127–138.

Warner, A. E. M., Diaz, C. M., Dalvi, A. D., et al. (2006). JOM world nonferrous smelter survey, part III: Laterite. *Journal of the Minerals, Metals and Materials Society, 58*, 11–20.

Walker, C., Kashani-Nejad, S., Dalvi, A. D., et al. (2009). Nickel laterite rotary kiln-electric furnace plant of the future. In J. Liu, J. Peacey & M. Barati, et al. (Eds.), *Pyrometallurgy of nickel and cobalt 2009* (pp. 381–390). CIM.

Chapter 6

Smelting of Laterite Ores to Ferronickel

Objectives of the chapter

The objectives of this chapter are the following:

(a) to describe industrial electric furnace laterite-to-ferronickel smelting; and,
(b) to indicate how the smelting is controlled and optimized.

The product of the calcination operation, which was described in Chapter 5, is bone dry, partially reduced calcine at 900°C. This calcine contains 1.5–3% Ni, a quarter of which is present as metallic nickel and three quarters as nickel oxide, 15% Fe, which occurs mostly as ferrous oxide with a small amount as iron metal and 1–2% C. Magnesium and silicon oxides are also present.

Ferronickel smelting makes this calcine into hot, molten ferronickel that is suitable for refining and use in making stainless steel and other ferrous alloys.

The calcine is fed continuously into an electric furnace where it is reduced and melted to give the following products:

(a) molten crude ferronickel at 1450°C with a composition of 20–40% Ni, 80–60% Fe;
(b) molten slag, composed of SiO_2, MgO and FeO, at 1550°C, which contains between 0.1 and 0.2% Ni; and,
(c) hot offgas at 900°C, composed of carbon monoxide, CO and nitrogen, N_2.

The internals of the electrical furnace are shown in Figure 6.1 and a sketch of the furnace is given in Figure 6.2.

Since the slag and ferronickel are immiscible,[1] the dense ferronickel settles at the bottom of the furnace bath while the slag floats on top. (The densities of the ferronickel and the slag are about 7 tonnes/m^3 and 3 tonnes/m^3, respectively.)

1. Their immiscibility is due to their different molecular structures. Ferronickel is metallically bonded; slag is ionically bonded. Ferronickel is denser than slag because its elements (Fe, Ni) are heavier than the constituents of the slag (O, Mg, Si).

Extractive Metallurgy of Nickel, Cobalt and Platinum-Group Metals. DOI: 10.1016/B978-0-08-096809-4.10006-1

FIGURE 6.1 Inside of refurbished laterite-to-ferronickel smelting furnace. Six suspended carbon electrodes, inside their steel 'cans', are shown behind the group. Each electrode has a diameter of 1.4 m. Calcine feed pipes can be seen between the electrodes. Note the scrap ferronickel on the floor. It is used to start up the furnace. *Photograph courtesy of Société Le Nickel-SLN/Eramet.*

The molten ferronickel and slag are tapped separately from the furnace through low and high tapholes. The molten ferronickel is tapped intermittently into a ladle. It is transported by ladle to refining, casting and use. The slag is tapped continuously or intermittently and sent to a waste dump, except in Japan where it is granulated in huge water sprays and sold as building material, metallurgical flux and grit (Hyuga, 2007; Pamco, 2007).

The offgas is transferred to an afterburner for combustion of carbon monoxide, de-dusted in cyclones and a baghouse and released to the atmosphere. Occasionally, the offgas is used to evaporate water in the dewatering kilns that were discussed in Chapter 5. The dust is recycled to the calcination/reduction kilns.

Operating data for several ferronickel furnaces are given in Table 6.1.

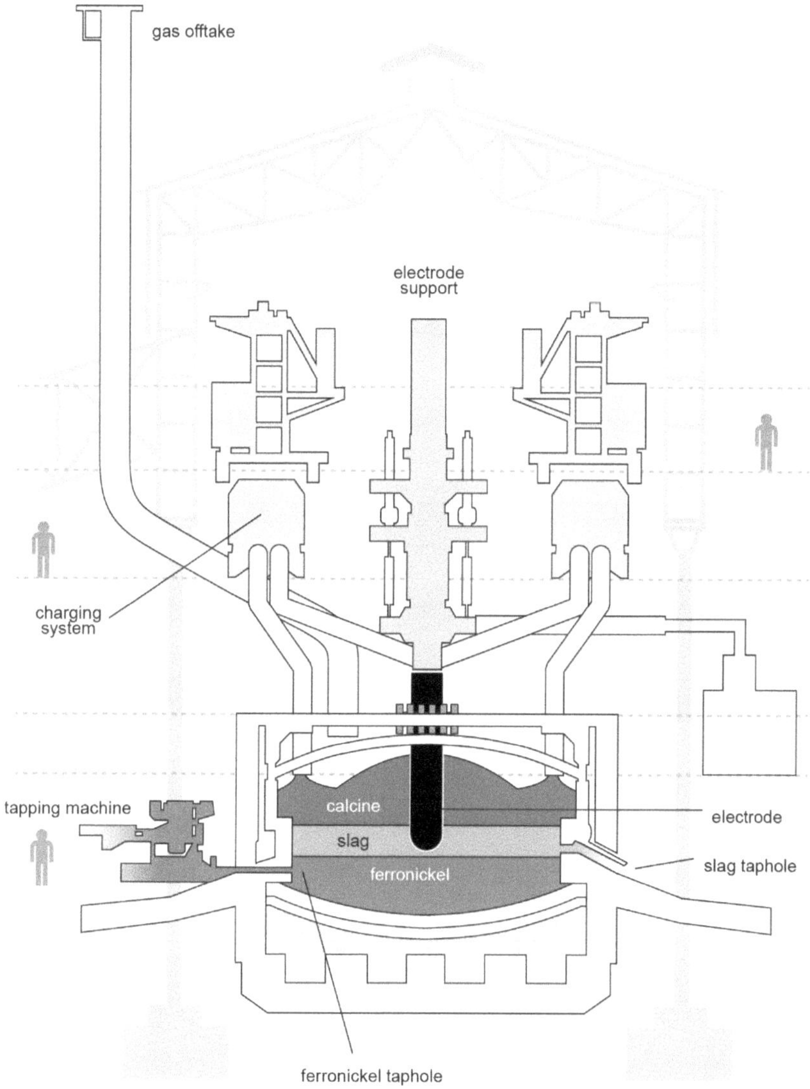

FIGURE 6.2 End-on sketch of laterite-to-ferronickel smelting furnace. This furnace is approximately 33 m long (into sketch), 13 m wide and 6 m high. Note the electrode (one of six), solid calcine, molten slag and molten ferronickel. Note also the electrode support system, charge pipes, gas offtake and slag and ferronickel tapholes. *Drawing courtesy of CM Solutions (Pty) Ltd.*

Table 6.1 Physical Details and Operating Data of Ferronickel Smelting Furnaces

Smelter	Falcondo Bonao, Dominican Republic	Hyuga Smelting Co. Miyazaki, Japan	Pacific Metals Co. Hachinohe, Japan	Le Nickel Doniambo, New Caledonia
Smelting furnaces	Temporarily shut			
Number	2	2	3	3
Annual production rate of Ni as FeNi, all furnaces, tonnes	28 500	22 000	41 000, average 2002–2004	48 000
Ni recovery, %	91	98	98	
Shape	Rectangular	Round	Round	Rectangular
Hearth dimensions, outside				
w × l × h, m	24 × 9 × 7			33 × 13 × 5.5
Diameter × h, m		#1 18.5 × 5.5 #2 17.5 × 5.4	#1 18 × 5.6 #2 19 × 6.2 #3 20 × 6.6	
Electrodes per furnace				
Number	6 in line	3	3	6 in line
Diameter, m	1.0	1.7	#1 1.7 m #2 and #3 2 m	1.4
Material	Prebake		Söderberg	Söderberg
Wall cooling system	Copper cooling fingers	Water sprays on outside shell	Water sprays on outside shell	
Electrical (per furnace)				
Maximum power, kW	80 000	#1 60 000 #2 40 000	#1 43 000, #2 and #3 54 000	50 000
Average power, kW	56 000	60 000 day, 80 000 night, 2 furnaces	45 000	36 000
Hearth power density, kW/m^2	329	#1, 170; #2, 140	160	94 (average)
Average voltage, V	1500	400–900	#1 664 #2 and #3 760	300
Average current, kA	12	28–32	#1: 35 #2 and #3: 42	20
Operating details				
Slag tapholes	1			

TABLE 6.1 Physical Details and Operating Data of Ferronickel Smelting Furnaces—cont'd

Smelter	Falcondo Bonao, Dominican Republic	Hyuga Smelting Co. Miyazaki, Japan	Pacific Metals Co. Hachinohe, Japan	Le Nickel Doniambo, New Caledonia
Ferronickel tapholes	1			
Calcine depth, m	2			
Slag depth, m	1			
Ferronickel depth, m	1			
Feed, nominal tonnes/h				
Calcine (per furnace)	140	60–65	#1 80 #1 and #2 100	76 @ 36 000 kW
Temperatures, °C				
Calcine feed	900	850	1000	900
Slag	1550	1575	1550	1600
Ferronickel	1450	1425	1450	1500
Consumptions, per tonne of Calcine feed				
Electrical energy, kWh	379	475	450	475
Electrode material, kg	1.1	~1	1.5	
Production details				
Ni in ferronickel production, tonnes/d per furnace	45	30	40	45
Composition, mass%				
Ni	38	22	19	22–28
Fe	60	73	75	
Co	0.9	0.6	0.5	
C	Trace	1.55	2.14	1.2–1.9
Si	Trace	0.65	1.63	1–3
P	0.033	0.012	0.023	
S	0.26	0.38	0.3	0.23
Cr	Trace	0.49	1.44	
Slag composition, mass%				
Ni	0.14	0.07	0.07	0.14
Fe	14	11	12	6
Co	0.01	0.002	0.02	0.02
MgO	29	31	33	32
SiO_2	43	51	50	56
CaO		0.5		
Al_2O_3	2.6		1.6	0.9

Source: Bergman (2003); Walker *et al.* (2009); Warner *et al.* (2006)

6.1. REACTIONS IN THE ELECTRIC FURNACE

The four main reactions that occur during ferronickel smelting are the following:

(a) the reduction of the nickel oxide in the calcine to metallic nickel, which is given as:

$$\underset{\text{in calcine feed}}{\overset{900^\circ C}{NiO(s)}} + \underset{\text{carbon in calcine feed}}{\overset{900^\circ C}{C(s)}} \rightarrow \underset{\text{metal}}{\overset{1450^\circ C}{Ni(\ell)}} + \underset{\text{in offgas}}{\overset{900^\circ C}{CO(g)}} \tag{6.1}$$

(b) the reduction of the iron oxide in the calcine to metallic iron, which is given as:

$$\underset{\text{in calcine feed}}{\overset{900^\circ C}{FeO(s)}} + \underset{\text{carbon in calcine feed}}{\overset{900^\circ C}{C(s)}} \rightarrow \underset{\text{metal}}{\overset{1450^\circ C}{Fe(s)}} + \underset{\text{in offgas}}{\overset{900^\circ C}{CO(g)}} \tag{6.2}$$

(c) the melting and alloying of nickel and iron to form molten ferronickel by the following reaction:

$$\overset{1450^\circ C}{Ni(\ell)} + \overset{1450^\circ C}{Fe(s)} \rightarrow \overset{1450^\circ C}{(Ni, Fe)(\ell)} \tag{6.3}$$

and

(d) the reaction and melting of the unreduced oxides, mainly SiO_2, MgO, FeO, in the calcine to form molten slag by reactions such as:

$$\underbrace{\overset{900^\circ C}{2MgO(s)} + \overset{900^\circ C}{SiO_2(s)}}_{\text{solid in calcine (from ore)}} \rightarrow \underbrace{\overset{1550^\circ C}{2Mg^{2+}} + \overset{1550^\circ C}{SiO_4^{4-}}}_{\text{in molten ionic slag}} \tag{6.4}$$

Overall, these reactions are highly endothermic, that is, they require energy. The energy is supplied by applying electrical power to the smelting furnace. Ferronickel smelters typically have their own captive power stations. For example, PT Inco has a hydroelectrical power plant, which is the cheapest way to generate electricity. Several ferronickel smelters are converting their power stations from fuel oil to coal.

6.2. NICKEL RECOVERY

The recovery of nickel to ferronickel is between 90% and 98%. The portion not recovered to ferronickel is lost in the slag. About half the nickel lost to the slag

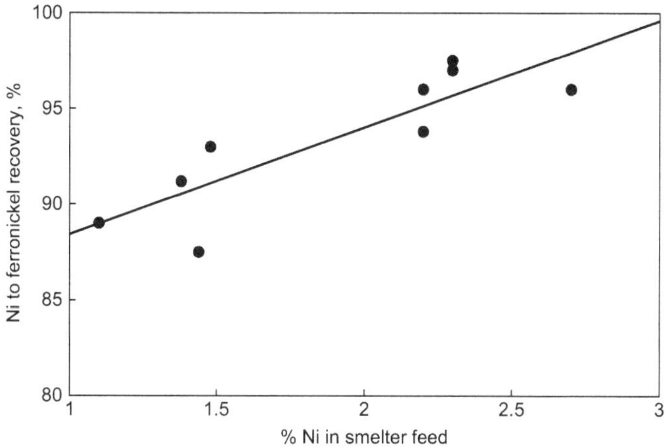

FIGURE 6.3 Recovery of nickel from smelter feed to ferronickel against the nickel content (% Ni) in the feed. Recovery is shown to increase significantly with increasing % Ni in feed. *Data from Bergman (2003) and Warner et al. (2006).*

is dissolved in the slag and the other half entrained as ferronickel droplets (Solar, 2009b). Ferronickel slags typically contain 0.1–0.2% Ni.

The loss of nickel is minimized by minimizing slag mass, for example, by maximizing the nickel content of the feed to the smelter. The effect of the nickel content of the laterite feed on the recovery of nickel is shown in Figure 6.3. As the nickel content in the feed increases, the recovery of nickel to the ferronickel product increases.

Other measures that are taken to minimize the loss of nickel are the following:

(a) ensuring that the slag is fully molten and fluid by smelting at a high temperature (~1550°C);
(b) ensuring that the calcine feed to the furnace contains enough carbon for complete reduction of nickel oxide (without reducing excessive iron); and,
(c) ensuring that ferronickel is not carelessly tapped with slag.

6.3. MELTING TEMPERATURES

The melting (liquidus) temperature of crude industrial ferronickel is in the range of 1400–1450°C (Solar, Candy, & Wasmund, 2008). The ferronickel is tapped at temperatures that are between 25°C and 50°C hotter than this to ensure that it taps easily and remains fluid during transport to refining. Ferronickel fluidity increases with increasing temperature (Sato, Sugisawa, Aoki, & Yamamura, 2005).

Ferronickel smelting slags are ionic. They melt at about 1500°C (Solar *et al.*, 2008). They are made up of cations, mainly Mg^{2+} and Fe^{2+}, and anions, mainly SiO_4^{4-}. They typically contain 40–55% SiO_2, 20–35% MgO, 5–20% FeO, 1–7% CaO and 1–2% Al_2O_3, depending on the composition of the feed to the smelter.

Some feed blending is done to control slag composition, particularly the ratio of MgO:SiO_2 in the slag. The Hyuga smelter (Sumitomo) blends the feed to obtain a mass ratio of MgO:SiO_2 in the slag of 0.63:1 (Hyuga, 2007). More MgO than this in the slag tends to increase the corrosion of the refractory, while more SiO_2 raises the melting point of the slag.

The slag from a ferronickel furnace is tapped at about 50°C above the melting point of the slag. This ensures that the slag is fluid, that good separation of the metal and the slag is obtained and that the slag taps easily.

6.4. INDUSTRIAL SMELTING FURNACES

Laterite smelting is done in electrically heated furnaces with suspended carbon electrodes. The furnace is shown in Figures 6.1 and 6.2. Energy for melting the furnace charge is provided by passing high electrical currents (about 30 000 A) between suspended carbon electrodes through electrically resistant air and slag.

Two types of furnaces are used, either a rectangular furnace with six suspended electrodes or a circular furnace with three suspended electrodes.

6.4.1 Furnace Construction

The interior of an electric ferronickel furnace consists of high-purity burnt magnesite bricks. The bricks are backed on the bottom and sides by a 0.03 m thick welded steel shell.

The furnace sits on a concrete foundation, often cooled by blowing air. The furnace is kept in compression at all times with adjustable springs and tie rods to keep the refractories tight and coherent (Koehler, 2009; Stober *et al.*, 2009).

The carbon electrodes hang through holes in the furnace roof, supported by moveable clamp-and-hoist systems. The electrodes are discussed in more detail in Section 6.6.

6.4.2 Calcine Feed System

Hot calcine is fed to the furnace through refractory-lined pipes positioned strategically throughout the roof. The calcine is released from insulated overhead bins as needed to keep the molten slag covered with calcine.

6.4.3. Liquids Tapping System

Ferronickel and slag are tapped separately through low and high tapholes at opposite sides of a furnace (Koehler, 2009). The holes are lined with graphite and typically have an inside diameter of 0.1 m. They are water cooled with copper chill blocks.

The tapholes are plugged with moist fireclay, which solidifies when it is pushed into the hot taphole.

The holes are opened by either chipping or drilling out the clay or by 'burning' the clay out with ignited oxygen-in-steel lances (hot, molten iron oxide from the oxidizing lance dissolves the clay). Mechanical tapping and plugging machines are used to assist with these operations.

The molten ferronickel flows down a refractory-lined steel trough into a ladle, in which it is transported to refining (see Figure 4.1).

The molten slag flows down a water-cooled steel trough and into ladles on a slag-removal train or into a huge torrent of water for granulation.

6.4.4. Furnace Wall Cooling

The slag layer in a ferronickel furnace is hot and corrosive. It is also in constant motion due to the passage of large electrical currents through it. These conditions promote refractory erosion with the potential to cause liquid breakouts. The erosion by the slag and liquid breakouts are prevented by cooling the furnace walls with water.

This cooling water extracts heat from the slag and causes a solid layer of slag to freeze on the sidewall. The solid layer of slag on the sidewall is referred to as the *freeze layer.*

The cooling is done in three different ways:

(a) cascading cool water down the outside of the furnace shell (mildest cooling);
(b) imbedding water-cooled copper slabs (fingers) through the steel shell into the refractory; and,
(c) replacing parts of the furnace wall with water-cooled 'waffle'-shaped copper plates (strongest cooling) (Voermann *et al.,* 2004).

Great care is taken to avoid possible explosions by ensuring that water does not come into contact with molten ferronickel or slag.

6.5. METHOD OF HEATING THE FURNACE

A rectangular electric furnace used for ferronickel smelting is shown in Figure 6.1. It has three pairs of 1.8 m diameter carbon electrodes.

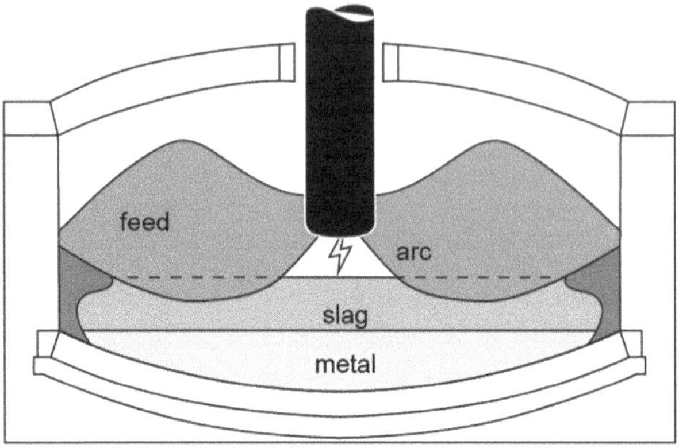

FIGURE 6.4 Sketch showing the smelting system of shielded arc electric furnace (Voermann *et al.*, 2004). The tip of the electrode is ~0.3 m above the molten slag. The arc is surrounded by bone-dry calcine. This maximizes melting rate and minimizes energy requirement. It also prevents overheating of the roof and sidewall refractories. The slag and metal layers are typically 1 m and 0.75 m thick, respectively. *Drawing courtesy of CM Solutions (Pty) Ltd.*

The energy for smelting is provided by I^2R heating, where I represents the current and R represents the resistance (Koehler, 2009). I^2R heating is also called Joule heating.

Shielded arc heating is used to heat the charge. The energy is obtained by passing electric current from one electrode of a pair to the other electrode down through an air gap between the first electrode and the slag layer, through the molten slag and up through the air gap from the slag to the second electrode. The air gap between the electrode and the slag is shown in Figure 6.4. About 80% of the energy is provided by I^2R heating in the air gaps and the other 20% by I^2R heating in the slag.

The main advantage of shielded arc heating is that most of the heat is generated within the solid calcine charge where it is needed for heating and melting.

6.6. ELECTRODES

6.6.1. Electrode Construction

Three or six massive carbon electrodes are used to conduct electricity into a ferronickel smelting furnace. These electrodes are 1–2 m in diameter and 20–25 m high. Almost all ferronickel smelting electrodes consist of a 0.005 m thick steel casing around a solid carbon core. There are also support fins on the inside of the casing to hold the carbon in place.

Current density in the carbon core is 10 000–15 000 A/m^2 of cross-sectional area.

6.6.2. Electrode Consumption

The tip of each electrode slowly oxidizes as smelting proceeds. The oxidation of the carbon is given by the following reaction:

$$\underset{\text{electrode}}{C(s)} + \underset{\text{in calcine}}{NiO(s)} \rightarrow \underset{\substack{\text{in molten}\\\text{ferronickel}}}{Ni(\ell)} + \underset{\text{in furnace gas}}{CO(g)} \tag{6.5}$$

The steel casing of the electrode dissolves into the slag and the ferronickel. As a result, the electrode is consumed during furnace operation.

The electrodes are lowered individually to compensate for this electrode consumption. Each electrode is rebuilt at the top by (i) welding a new 2-m tall section of casing to the top of the electrode, about 15 m above the furnace; and, (ii) filling the new casing with blocks of electrode paste.

As the electrode continues to be lowered, heat from the furnace liquefies the paste and drives out the volatile organic components, eventually forming a self-baked carbon electrode that is solid, coherent and strong. These types of electrodes are called *Söderberg electrodes.*

Great care is taken with all electrode maneuvers to avoid worker electrocution.

6.6.3. Electrode Support

Each electrode is supported by two circular mechanical clamps and a circular electrical clamp, all within a moveable hoist (Southall *et al.*, 2005).

This arrangement allows (i) automatically controlled up and down electrode movement by means of the hoist; and, (ii) controlled slipping of the electrode down through the hoist as the electrode tip oxidizes away.

6.7. FURNACE OPERATION

Ferronickel furnaces smelt up to 4000 tonnes/day of hot calcine, produce up to 250 tonnes/day of molten ferronickel and 3700 tonnes/day of molten slag. They consume about 500 kWh of electrical energy/tonne of calcine feed. This section describes how the furnaces operate.

6.7.1. Start-up

Operation of a ferronickel smelting furnace is begun by laying down a protective castable refractory layer on the furnace hearth, then covering it with a meter-thick layer of steel and ferronickel scrap pieces (Stober *et al.*, 2009).

The furnace is then heated slowly, initially with auxiliary oil or gas burners then later by arcing the electrodes to the steel and ferronickel scrap at low power. Electrical power is then gradually increased and calcine feeding is begun. Steady-state smelting is achieved in 4–5 weeks.

The furnace is kept under appropriate tie-rod compression throughout this start-up.

6.7.2. Shutdown

Long-term shutdown of a furnace for major repairs is accomplished by (i) stopping calcine feeding, (ii) tapping slag and then ferronickel through their tapholes, (iii) turning off furnace power and then raising the electrodes about 0.5 m above the residual slag layer; and, (iv) allowing the furnace to cool at its natural rate.

The furnace is kept under appropriate tie-rod compression throughout this shutdown.

The typical life of a furnace before it needs to be re-bricked or rebuilt is 10–15 years.

6.7.3. Restart

The restart of the furnace is similar to the start-up of a new furnace except that coke pieces are placed between the electrodes instead of scrap. The coke pieces are about 0.25 m thick.

6.7.4. Steady-State Operation

Operation of the ferronickel furnace at steady state entails the following:

(a) feeding calcine at a constant rate;
(b) providing electrical energy at just the right rate to (i) reduce and melt the calcine feed; and, (ii) keep the ferronickel and slag products at their prescribed temperatures;
(c) tapping ferronickel from the furnace on a scheduled basis; and,
(d) tapping slag from the furnace on a scheduled basis or when it reaches a prescribed level in the furnace.

The next section describes how steady-state smelting is attained and controlled.

6.8. CONTROL

The ferronickel furnace operator must smelt calcine at a steady specified rate while (i) producing ferronickel of prescribed composition of 20–40% Ni, 80–60% Fe, (ii) producing ferronickel and slag at their specified temperatures, that is, at 1450°C and 1550°C, respectively; and, (iii) maintaining a protective coating of solid slag on the furnace walls (referred to as the freeze lining).

6.8.1. Control of the Composition of the Ferronickel

The nickel content in the ferronickel product is adjusted by altering the amount of carbon in the feed to smelting furnace. If the content of carbon in the calcine charge is high, there will be considerable reduction of FeO and hence the content of iron will be high in the ferronickel product. A higher iron content means lower content of nickel in the ferronickel.

If the nickel content in the ferronickel decreases (due, say, to a change in %Ni in ore), the amount of carbon in the calcine feed can be decreased. The amount of carbon in the calcine feed is altered by adding more or less coal to calcination kilns.

6.8.2. Choice of Ferronickel Composition

Ferronickel compositions vary from 20% Ni, 80% Fe (high iron reduction) to 40% Ni, 60% Fe (low iron reduction). They are adjusted by changing the amount of carbon in the calcine feed, as described above.

An evaluation of operations with high and low amounts of iron reduction is shown Table 6.2. Generally, high reduction of iron (lower content of nickel in ferronickel) is favored when (i) high recovery of nickel is critical, (ii) a reasonable price is realized for the Fe in the ferronickel; and, (iii) the costs of transportation of the ferronickel to market are low. The opposite conditions favor higher nickel content in the ferronickel.

6.8.3. Control of the Rate of Smelting

The rate at which calcine is smelted in a ferronickel furnace is set by adjusting the feed rate of calcine to the prescribed value while simultaneously adjusting the power to match.

For example, smelting of calcine needs 500 kWh of electrical energy/tonne of calcine for heating, reduction and melting. At a feed rate of 100 tonnes/h calcine, the required energy is calculated as follows: 100 tonnes/h calcine × 500 kWh/t calcine = 50 000 kWh/h. This means that the power to the furnace is 50 000 kW.

6.8.4. Adjusting Energy Input Rate

The rate at which electrical energy is supplied to a furnace by a pair of electrodes is given by the following equation:[2]

$$\text{Electrical energy supply rate, kW} = \frac{V_t^2}{R_t} \tag{6.6}$$

2. More precisely, this equation should be $P = (V^2/R) \times \Phi$, where Φ is the electric furnace power factor ~0.9.

Table 6.2 Evaluation of Low and High Levels of Reduction of Iron in Ferronickel Smelting Furnaces

Category	Low iron reduction (generally high % Ni in ferronickel)	High iron reduction (generally low % Ni in ferronickel)
Typical % Ni in ferronickel	up to 40	20–25
Ni recovery to ferronickel, %	~90	Up to 98%, Table 6.1
Fe recovery to ferronickel, % of Fe in smelting furnace feed	15–30	45–65
Requirements per tonne of Ni in ferronickel		
Carbon	Low	High (for Fe reduction)
Electric power	Low	High (for Fe reduction)
Refining	Low	High (more ferronickel, higher impurity levels)
Ferronickel transportation	Low	High (more ferronickel per tonne of Ni)
Furnace productivity, tonnes of Ni per day	High	Low (extensive Fe reduction)

Negotiated payments for iron (as in Japan) offset the higher costs of high iron reduction. High iron reduction is also favoured by custom smelters that pay mines per tonne of nickel in concentrate, so need high recovery of nickel into ferronickel.
Source: Solar *et al.* (2008)

where V_t represents the total voltage (potential) in kV across the electrodes and R_t represents the total electrical resistance in ohms between the electrodes.

It can be seen from Equation (6.6) that the total rate of energy input is increased by (i) increasing the electrical potential between the electrodes; and/or by, (ii) decreasing the electrical resistance between the electrodes and *vice versa*.

The potential is adjusted by changing transformer voltage tap position.

The resistance is adjusted by changing electrode tip position, that is, by raising or lowering the electrodes. This changes the length of the air gap shown in Figure 6.4 and hence changes the electrical resistance.

The two are adjusted together to give the energy input rate required for smelting calcine at its designated rate and the appropriate ferronickel and slag temperatures.

6.8.5. Adjusting Slag Temperature at a Constant Smelting Rate

Suppose, for example, that a new calcine feed with a high content of SiO_2 is expected. The slag from this calcine is expected to have a high melting point, so temperature of the slag must be raised.

Industrially, the temperature of the slag is increased by (i) lowering the carbon electrodes toward the slag (which decreases air gap resistance and thereby increases current through the slag) while simultaneously; (ii) adjusting the electrode-to-electrode voltage (which restores furnace power to its setpoint).

6.8.6. Adjusting Ferronickel Temperature

The temperature of the ferronickel increases with increasing temperatures of the slag and *vice versa*. The temperature of the ferronickel can be controlled somewhat independently by varying slag thickness.

Thinning the slag layer brings the electric arc nearer to the ferronickel and thereby raises ferronickel temperature – by about 20–30°C for each 0.25 m of thinning (Voermann *et al.,* 2004).

6.8.7. Control of Wall Accretion

Ferronickel furnaces are operated with a protective layer of frozen slag on the inside of their walls, referred to as the freeze lining. The slag is frozen by removing heat from the slag through the furnace wall into the cooling water.

The thickness of the protective layer around the furnace can be inferred by measuring cooling water exit temperatures around the furnace or by acoustic wall thickness measurement. Long-term operation of a furnace requires that a protective layer always be present on the furnace walls, especially at the slag–metal interface.

The presence of the freeze lining is ensured by closely monitoring cooling water flowrates and temperatures and by controlling the temperature of the slag.

6.9. APPRAISAL AND FUTURE TRENDS

Smelting of laterites to ferronickel in electric furnaces recovers nickel efficiently with little adverse impact on the local environment.

The only disadvantage of smelting is the large amount of energy required by the electric furnaces. This energy requirement is minimized by maximizing the nickel content of the feed and by charging hot calcine.

Nevertheless, the high temperature of smelting will always require considerable energy. This is offset somewhat by the small energy requirement of mining laterite ores – which are always near the surface – and by locating the furnace close to low-cost sources of power.

6.10. SUMMARY

The last step in ferronickel production is reduction smelting of bone dry, partially reduced nickel-rich calcine at 900°C from the calcination/reduction kilns.

The reduction smelting is always done in electric furnaces at about 1550°C. Power is supplied to the furnaces through three to six huge (1.5 m diameter, 25 m tall) carbon electrodes.

The products of the process are the following:

(a) molten slag at a temperature of 1550°C, which is discarded (except in Japan where it is granulated and sold as building material, metallurgical flux and grit); and,
(b) molten ferronickel containing 20–40% Ni at a temperature of 1450°C, which is transferred to pyrometallurgical refining.

REFERENCES

Bergman, R. A. (2003). Nickel production from low-iron laterite ores: process descriptions. *CIM Bulletin, 96*, 127–138.

Hyuga. (2007). *Company brochure*. Hyuga Smelting Co., Ltd.

Koehler, T. (2009). Electric furnaces – design aspects. In *Nickel Pyrometallurgy short course notes*. CIM.

Pamco. (2007). *Corporate profile*. Pacific Metals Company, Ltd.

Sato, Y., Sugisawa, K., Aoki, D., & Yamamura, T. (2005). Viscosity of molten Fe–Ni binary alloy. *Measurement Science and Technology, 16*, 363–371.

Solar, M. Y. (2009b). Mechanical slag losses in laterite smelting – nickel. In J. Liu, J. Peacey & M. Barati et al. (Eds.), *Pyrometallurgy of nickel and cobalt* (pp. 277–292). CIM, 2009.

Solar, M. Y., Candy, I., & Wasmund, B. (2008). Selection of optimum ferronickel grade for smelting nickel laterites. *CIM Bulletin, 101*, 1–8.

Southall, S., Darini, M., Voermann, N., et al. (2005). Hatch electrode column – latest developments. In J. Donald & R. Schonewille (Eds.), *Nickel and cobalt 2005 challenges in extraction and production* (pp. 323–332). CIM.

Stober, F., Jastrzebski, M., Walker, C., et al. (2009). Start-up and ramp-up of metallurgical furnaces to design production rate. In J. Liu, J. Peacey & M. Barati et al. (Eds.), *Pyrometallurgy of nickel and cobalt 2009* (pp. 487–507). CIM.

Voermann, N., Gerritsen, T., Candy, I., et al. (2004). Furnace technology for ferronickel production – an update. In W. P. Imrie, D. M. Lane & S. C. C. Barnett et al. (Eds.), *International laterite nickel symposium – 2004* (pp. 563–577). TMS.

Walker, C., Kashani-Nejad, S., Dalvi, A. D., et al. (2009). Nickel laterite rotary kiln-electric furnace plant of the future. In J. Liu, J. Peacey & M. Barati et al. (Eds.), *Pyrometallurgy of nickel and cobalt 2009* (pp. 33–50). CIM.

Warner, A. E. M., Diaz, C. M., Dalvi, A. D., et al. (2006). JOM world nonferrous smelter survey, part III: Laterite. *Journal of the Minerals, Metals and Materials Society, 58*, 11–20.

SUGGESTED READING

Nelson, L. R., Geldenhuis, M. M. A., Miraza, T., et al. (2007). Role of operational support in ramp-up of the FeNi-II furnace at PT Antam in Pomalaa. In Das & Sunderesan et al. (Eds.), *Proceedings of innovations in Ferro alloy industry conference (INFACON XI)* (pp. 798–813). India: Macmillan.

Solar, M. Y. (2009a). Some operating considerations in laterite smelters. In *Nickel Pyrometallurgy short course notes*. CIM.

Chapter 7

Refining Molten Ferronickel

The ferronickel product from electric furnace smelting contains up to 0.06% P and 0.4% S from the ore and coal feeds to the smelter.[1] These elements lower the strength, toughness and corrosion resistance of ferrous alloys, which means that they must be removed before the ferronickel is used for alloying.

Typical market requirements for ferronickel are that the phosphorus content must be less than 0.02% and the sulfur content must be less than 0.03%.

It is optimal to remove phosphorus and sulfur from ferronickel just after it has been tapped from the smelting furnace – when it is molten and hot. All ferronickel smelters do this. Typical flowsheets for the refining operations are shown in Figures 4.1 and 7.1 (Marin and Vahed, 2009).

The objective of this chapter is to describe the removal of phosphorus and sulfur from molten ferronickel. The removal of carbon, cobalt, chromium and silicon and the casting of ferronickel are also described.

7.1. PHOSPHORUS REMOVAL

Phosphorus is removed from molten ferronickel into molten slag by reacting the ferronickel with lime, CaO and oxygen.

The reaction can be represented as (Simeonov *et al.*, 1997):

$$\underset{\text{in molten ferronickel}}{2[\mathrm{P}]} + \underset{\text{added to molten ferronickel by injecting oxygen}}{5[\mathrm{O}]} + \underset{\text{mixed into molten ferronickel}}{4\mathrm{CaO}(s)} \xrightarrow{1550^\circ\mathrm{C}} \underset{\text{molten slag that floats and is decanted from the molten ferronickel}}{(\mathrm{CaO})_4\mathrm{P_2O_5}(\ell)} \quad (7.1)$$

The reaction indicates that the removal of phosphorus from ferronickel is favored by (i) efficient mixing of lime into the molten ferronickel and (ii) deep injection of oxygen into the molten ferronickel.

1. Some smelters use low-phosphorus, low-sulfur fossil fuels to minimize these concentrations (Hyuga, 2007; Kohga *et al.*, 1997).

Extractive Metallurgy of Nickel, Cobalt and Platinum-Group Metals. DOI: 10.1016/B978-0-08-096809-4.10007-3

Efficient mixing of lime into the molten ferronickel is obtained by adding it to the ferronickel ladle before and during tapping from the smelting furnace.

Deep injection of oxygen into the molten ferronickel is obtained by blowing the oxygen through refractory-coated stainless steel lances.

Finally, the phosphorus is removed from the ladle by carefully decanting off the floating slag.

Lime removes phosphorus down to less than 0.03% P and occasionally down to 0.01% P (Warner *et al.*, 2006). This range is acceptable to virtually all stainless steel and ferrous alloy makers.

7.2. SULFUR REMOVAL

Sulfur is removed from molten ferronickel most commonly by mixing calcium carbide, CaC_2, into the molten ferronickel. A representative reaction is given as follows:

$$\underset{\text{in molten ferronickel}}{[S]} + \underset{\text{mixed into molten ferronickel}}{CaC_2(s)} \xrightarrow{1550^\circ C} \underset{\text{in molten ferronickel}}{2[C]} + \underset{\text{molten slag, decanted from molten ferronickel}}{CaS(\ell)} \tag{7.2}$$

This reaction shows that the removal of sulfur into slag as CaS is favored by (i) efficient mixing of calcium carbide into the ferronickel; and, (ii) a low initial carbon content (%C) in the ferronickel.

The calcium carbide is mixed into the molten ferronickel by induction or mechanical stirring. It forms calcium sulfide slag that floats to the top, from where it is decanted and discarded.

Sulfur can also be removed with CaO, $CaCO_3$, CaSi, Na_2O and Na_2CO_3 (Bergman, 2003; Warner *et al.*, 2006; Young, Brosig, & Candy, 2009).

7.3. INDUSTRIAL REFINING

The removal of phosphorus and sulfur from molten ferronickel is regarded as '*ladle metallurgy*'. All removal of phosphorus and sulfur is done in a ladle of freshly tapped ferronickel, which is physically moved through dephosphorization, desulfurization and casting stations as shown in Figure 7.1.

The methods used at a variety of industrial operations are summarized in Table 7.1.

7.4. REMOVING OTHER IMPURITIES

Ferronickel that arises from the smelting of laterites also contains the following impurities:

cobalt 0.3%–1%
carbon trace to 2.4%

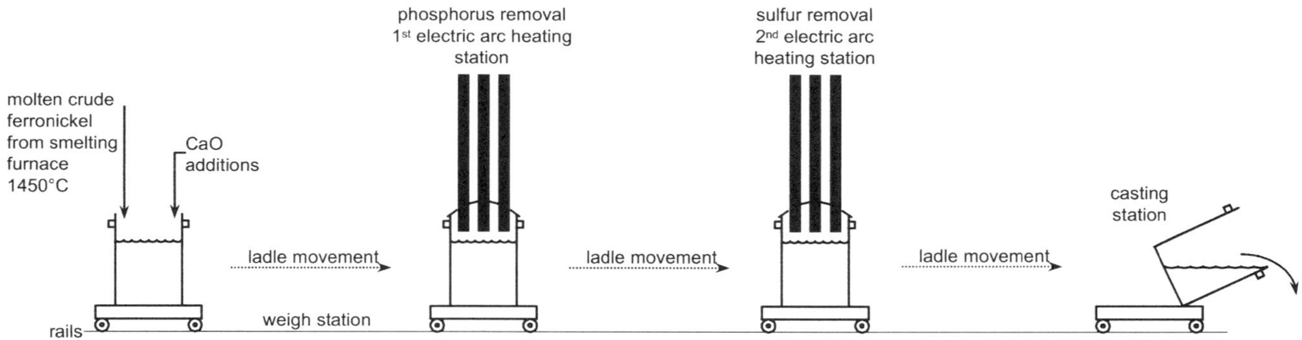

1. CaO is added to ladle.
2. Crude molten ferronickel is tapped into ladle while additional CaO is added.
3. Ladle + ferronickel are weighed.
4. Roof + electrodes are lowered and ferronickel is heated to 1550°C.
5. Roof and electrodes are raised, ferronickel is stirred, and oxygen is injected.
6. Calcium phosphate slag is removed by tipping ladle.
7. CaC_2 is added.
8. Roof + electrodes are lowered and ferronickel is heated to 1550°C then stirred.
9. Calcium sulfide slag is removed by tipping ladle.
10. More CaC_2 is added and ferronickel is re-heated.
11. Slag is removed by tipping ladle.
12. Refined ferronickel is cast into huge streams of water (granulation) or continuously into small molds.

FIGURE 7.1 Representative ferronickel refining flowsheet. It depicts a ladle (2 m diameter, 3 m high, 40 tonnes of ferronickel) moving through four refining stations: Crude ferronickel tapping from smelting furnace, phosphorus removal, sulfur removal and refined ferronickel casting. The phosphorus content of the ferronickel is lowered from 0.03% to less than 0.02%. The sulfur content of the ferronickel is lowered from 0.3% to less than 0.03%. Other heating and stirring methods include induction furnaces, induction stirring, shaking ladles, and refractory-lined stirrers. Most smelters use the same station for dephosphorization and for desulfurization.

TABLE 7.1 Details of the Refining of Ferronickel.* The First Step Always Begins with Crude Molten Electric Furnace Ferronickel, 1450°C

Smelter	Falcondo, Dominican Republic	Hyuga Smelting, Japan	Le Nickel, New Caledonia	Loma de Niquel, Venezuela
Refining				
First step	**Dephosphorization**	**Desulfurization**	**Desulfurization**	**Dephosphorization**
Equipment	Two 4 MW ASEA–SKF ladles	Low frequency induction furnace	Shaking ladle	Electrically heated ladles
Reagents	Basic oxidizing slag	CaC_2	CaC_2	CaO + oxygen
Process temperature, °C	1500–1550	1400–1450	–	1650–1700
Start and finish compositions	0.03%–0.01% P	0.3%–0.03% S	0.2%–0.03% S	0.06%–0.03% P
Second step	**Deoxidation**	**Decarburization and Desiliconization**	**Decarburization**	**Desulfurization**
Equipment	Same as above	LD converter	Shaking ladle	Same as above
Reagents	Ferrosilicon	Oxygen	Oxygen	CaO, CaSi, FeSi, $CaCO_3$
Process temperature, °C	1500–1550	1600–1650		1600

Third step	Desulfurization, if required					
Equipment	Same as above					
Reagents	Basic reducing slag					
Process temperature	1500–1550					
Casting method	Belt casting	Mostly water granulation		Casting and granulation		
Product	0.1 kg ferrocones	Water-granulated shot		Shot, 15–40 kg ingots		Water-granulated shot
Ferronickel composition, %		High C & S	Low C & S	FN 1	FN 4	
Ni	39	16	17–28	24–30	22–28	20–25
Co	0.93	< %Ni × 0.05				0.5
C	0.06	3	<0.02	0.03	1.2–1.9	<0.04
S	0.04	>0.03	<0.03	0.03	0.23	<0.06
Si	0.35	<5	<0.3	0.03	1.0–3.0	<0.2
P	0.01	<0.05	<0.02			<0.03
Cr	0.02	<2.5	<0.3			

* *Bergman, 2003; Warner et al., 2006; Young et al., 2009*

silicon trace to 3%
chromium trace to 1.7%

This section discusses the impact of each of these impurities.

7.4.1. Cobalt

Cobalt is never removed from ferronickel. The chemical properties of molten cobalt and nickel are so similar that cobalt cannot be removed from ferronickel without also removing considerable amounts of nickel.

In small quantities, cobalt is not deleterious to stainless or alloy steels.

7.4.2. Carbon

The ferronickel from some laterite smelting furnaces contains as little as 0.02% C (Bergman, 2003; Solar, Candy, & Wasmund, 2008). The ferronickel from others contains up to 2.4% C. These concentrations are purposefully obtained by controlling the amount of carbon in the calcine feed to the smelting furnace.

A high-carbon product tends to be produced by smelters that smelt calcine that is rich in nickel and low in iron (Warner *et al.*, 2006). It is the product of strongly reducing conditions in the smelting furnace. Strongly reducing conditions give recoveries of nickel as high as 98% into the ferronickel.

A low-carbon product tends to be produced by smelters that smelt calcine that is high in iron and low in nickel. Mildly reducing conditions are used in the smelting furnace. Such conditions prevent excessive iron reduction and hence ferronickel that is excessively dilute in nickel (Solar *et al.*, 2008).

Carbon can be oxidized from ferronickel, which lowers the content of carbon down to 0.02% without significant oxidation of nickel. Oxygen is injected into ladles or into small oxygen-steelmaking furnaces to burn the carbon (Bergman, 2003; Warner *et al.*, 2006). A representative reaction is given as follows:

$$\underset{\substack{\text{in molten}\\\text{high-carbon}\\\text{ferronickel}}}{2[\mathrm{C}]} + \underset{\text{injected oxygen}}{\mathrm{O_2}(g)} \xrightarrow{1550^\circ\mathrm{C}} \underset{\text{offgas}}{2\mathrm{CO}(s)} \tag{7.3}$$

Industrially, the process is done either during dephosphorization or after dephosphorization and desulfurization.

7.4.3. Chromium

Chromium is present in ferronickel arising from smelting furnaces that operate with high contents of carbon in the feed. The chromium content in this type of ferronickel can be up to 1.7% (Bergman, 2003). It is removed during carbon oxidation with little loss of nickel.

7.4.4. Silicon

The concentration of silicon in low-carbon ferronickel is low enough (<0.1% Si) to be used directly in stainless steel manufacture. However, the concentrations of silicon are relatively high in ferronickel that is high in carbon. In these types of ferronickel, the silicon content can be as high as 3%.

Silicon is readily oxidized. As a result, it is removed during the oxidation of carbon.

7.5. CASTING OF FERRONICKEL

All ferronickel is used for making stainless steel and other ferrous alloys. It is therefore cast into shapes that are convenient for adding it to molten metal in furnaces and ladles.

It is usually cast in small molds on a moving conveyor belt, as shown in Figure 4.1, or granulated in water, as shown in Figure 7.2.

FIGURE 7.2 Production of ferronickel 'beans' by controlled pouring of molten ferronickel into a spinning refractory cup and on into water. The product is flat, ~0.03 m diameter × 0.005 m thick 'beans' suitable for controlled addition into alloy-making ladles or furnaces. *Source: Photograph courtesy of Pacific Metals Co., Ltd., Hachinohe, Japan.*

The products from casting vary from 0.1 kg 'ferrocones' through 20 kg ingots (and occasionally 90 kg ingots). They are used for all types of melting and alloying. The product from granulation is about 0.03 m diameter, 0.005 m thick flat ferronickel 'beans'. This product is particularly useful for controlled-rate feeding to argon–oxygen and vacuum-oxygen stainless steelmaking furnaces (Kohga, Yamagiwa, Kubo, & Akada, 1997).

7.6. APPRAISAL

Most impurities are readily removed from molten ferronickel without losing nickel. The exception is cobalt, which behaves so much like nickel that it cannot be preferentially removed.

7.7. SUMMARY

The ferronickel product from laterite smelting always contains phosphorus and sulfur from ore and hydrocarbon fuels.

These elements are detrimental to the properties of ferrous alloys. As a result, they must be removed from ferronickel before it is used to make stainless steel and other alloys. The impurity elements are typically removed down to less than 0.02% P and less than 0.03% S.

Phosphorus is removed by oxidation from molten ferronickel in the presence of lime, CaO. Sulfur is then removed by reaction with calcium carbide, CaC_2, and/or other calcium and sodium compounds.

Molten refined ferronickel is cast in small ingots or water granulated to 0.03 m diameter flat beans. The latter are especially useful for controlled-rate additions to argon- and vacuum-oxygen stainless steelmaking furnaces.

REFERENCES

Bergman, R. A. (2003). Nickel production from low-iron laterite ores: Process descriptions. *CIM Bulletin, 96*, 127–138.

Hyuga. (2007). Hyuga Smelting Co., Ltd. www.smm.co.jp\E\business\metal\hyuga.html. Accessed 19.05.2011.

Kohga, T., Yamagiwa, M., Kubo, N., & Akada, A. (1997). Recent ferronickel production at Hyuga smelter. In C. Diaz, I. Holubec & C. G. Tan (Eds.), *Pyrometallurgical operations, the environment and vessel integrity in nonferrous smelting and converting. Proceedings of the Nickel/Cobalt 97 International Symposium, Vol. III* (pp. 217–228). CIM.

Marin, T., & Vahed, A. (2009). Thermodynamic model of FeNi refining, Mineracao Onca Puma. In J. Liu, J. Peacey, M. Barati, S. Kashani-Nejad & B. Davis (Eds.), *Pyrometallurgy of Nickel and Cobalt 2009* (pp. 431–447). CIM.

Simeonov, S., Bergman, R. A., Uceda, D., et al. (1997). Study of the behavior of sulfur, phosphorus and oxygen during refining of ferronickel melts. In C. Diaz, I. Holubec & C. G. Tan (Eds.), *Pyrometallurgical operations, the environment and vessel integrity in nonferrous smelting and converting processess. Proceedings of the Nickel/Cobalt 97 International Symposium, Vol. III* (pp. 229–236). CIM.

Solar, M. Y., Candy, I., & Wasmund, B. (2008). Selection of optimum ferronickel grade for smelting nickel laterites. *CIM Bulletin, 101*, 1–8.

Warner, A. E. M., Diaz, C. M., Dalvi, A. D., et al. (2006). JOM world nonferrous smelter survey, part III: Laterite. *JOM, 58*, 11–20.

Young, J., Brosig, D., & Candy, I. (2009). Comparison of ladle refining in the ferronickel and steel industries. In J. Liu, J. Peacey, M. Barati, S. Kashani-Nejad & B. Davis (Eds.), *Pyrometallurgy of nickel and cobalt 2009* (pp. 463–474). CIM.

SUGGESTED READING

Bergman, R. A. (2003). Nickel production from low-iron laterite ores: Process descriptions. *CIM Bulletin, 96*, 127–138.

Warner, A. E. M., Diaz, C. M., Dalvi, A. D., et al. (2006). JOM world nonferrous smelter survey, part III: laterite. *JOM, 58*, 11–20.

Chapter 8

Smelting Laterite Concentrates to Sulfide Matte

About 90% of laterite smelting produces molten ferronickel. This process route was discussed in Chapters 4 through 7. The remaining portion of the laterite is used to produce a Ni–Fe–S matte, which is discussed in this chapter.

The matte is mostly made into melting-grade nickel (95%–97% Ni), which is used for ferrous alloy manufacture. The production of melting-grade nickel is discussed in Chapter 9. Some of the matte is refined hydrometallurgically to high-purity nickel and nickel chemicals (Le Nickel, 2010).

The objective of this chapter is to describe smelting of laterites to produce molten matte.

8.1. MATTE PRODUCTION FLOWSHEETS

Two companies produce Ni–Fe–S matte from laterite concentrate:

PT Inco in Indonesia (Doyle, 2004; Vahed *et al.*, 2009) and Le Nickel in New Caledonia (Le Nickel, 2010).

PT Inco produces about 80 000 tonnes per year of nickel in matte. This matte is produced by spraying molten sulfur into the calcination/reduction kilns and then smelting the sulfided calcine from the kiln into molten matte.

Le Nickel produces about 15 000 tonnes per year of nickel in matte. This matte is produced by injecting liquid sulfur into some of the molten refined ferronickel.

The PT Inco process is described first and then the Le Nickel process.

8.2. PT INCO PROCESS

PT Inco produces high-grade matte that contains about 78% Ni. The process route used is shown in Figure 8.1.

The process employs the following steps:

(a) upgrading laterite ore to 1.8% Ni;
(b) dewatering of the upgraded ore in hot rotating kilns;
(c) calcining, reducing and sulfiding the dewatered material in hot rotating kilns;
(d) smelting the sulfided calcine in an electric furnace to molten matte that contains about 26% Ni; and,

Extractive Metallurgy of Nickel, Cobalt and Platinum-Group Metals. DOI: 10.1016/B978-0-08-096809-4.10008-5

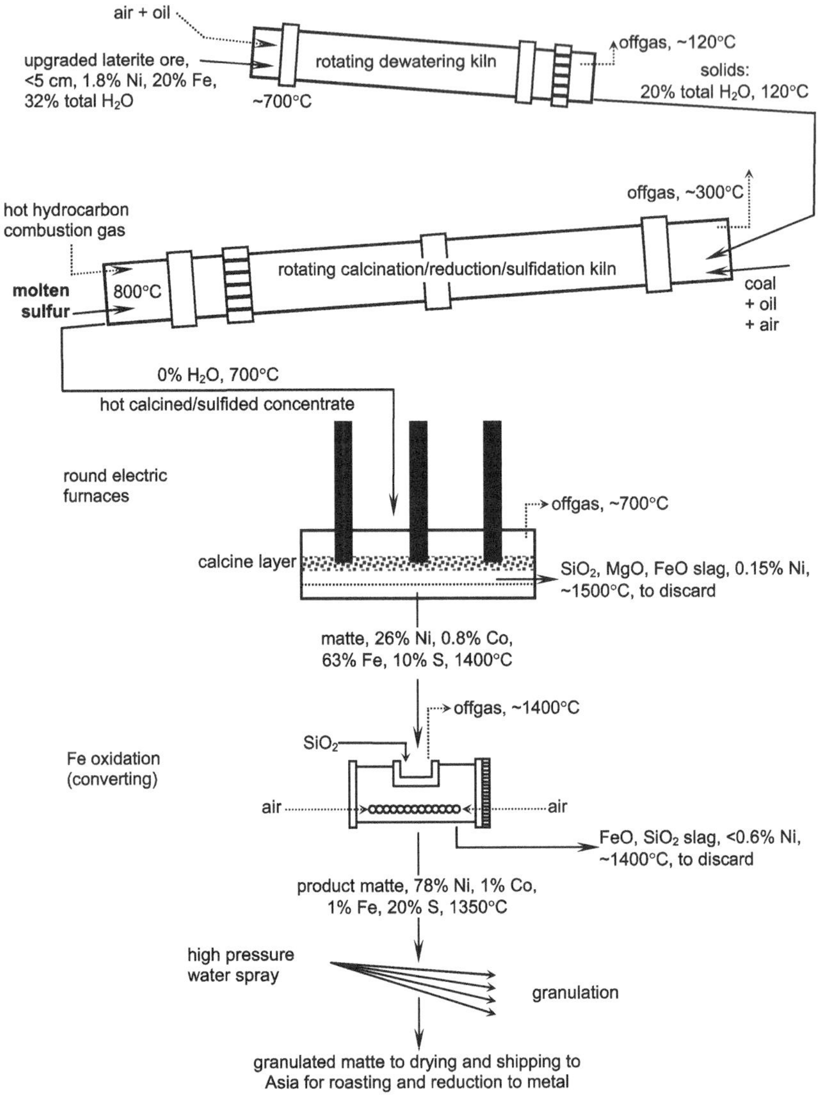

FIGURE 8.1 PT Inco (Indonesia) laterite-to-matte smelting flowsheet (Doyle, 2004). Sulfidation of metal at the discharge end of the calcination/reduction/sulfidation kiln is notable. The smelter has three dewatering kilns, five calcination/sulfidation kilns, four round electric melting furnaces and three Peirce–Smith converters (Bergman, 2003). It produces 80 000 tonnes of nickel in matte per year. Overall recovery of nickel is ~90%. *Further details are given by Bergman (2003), Doyle (2004) and Warner et al. (2006).*

(e) converting (oxidizing) the molten matte to product matte that contains 78% Ni and is ready for refining.

The first two steps are carried out much as described in detail in Chapters 3–6. The last three steps are discussed in this section.

8.2.1. Feed Composition

The feed to the dewatering kiln, shown in Figure 8.1, is a nickel–silicate ore, consisting of magnesium–nickel silicates.[1] It typically contains 1.8% Ni, 0.06% Co, 20% Fe, 36% SiO_2, 17% MgO, and 32% total H_2O.

This feed is dewatered to about 20% total water in a hot rotating kiln and then transferred to the calcination kiln where the ore is sulfidized and partially reduced.

8.2.2. Sulfidation in a Rotary Kiln

The calcination process step in the production of ferronickel produces calcine that is partially metallized. This was discussed in Chapter 5. A similar process is used in the production of matte, except that liquid sulfur is sprayed into the discharge end of the calcination kiln at a rate of about 14 kg of sulfur per tonne of product calcine. The addition of the sulfur results in the sulfidization of the metallic nickel and iron in the calcine.

The sulfur vaporizes in the kiln and reacts with the nickel and iron. The reactions are as follows:

$$\underset{\text{nickel metal reduced from oxide}}{3Ni(s)} + \underset{\text{vaporised liquid sulfur}}{2S(g)} \xrightarrow{700^\circ C} \underset{\text{sulfided Ni}}{Ni_3S_2(s)} \quad (8.1)$$

$$\underset{\text{iron metal reduced from oxide}}{Fe(s)} + \underset{\text{vaporised liquid sulfur}}{S(g)} \xrightarrow{700^\circ C} \underset{\text{sulfided Fe}}{FeS(s)} \quad (8.2)$$

Both reactions are exothermic. The next step in the process is the smelting of the sulfided calcine.

8.2.3. Smelting Sulfided Calcine in an Electric Furnace

Sulfided calcine typically contains 1.3% C, 2% Ni, 0.08% Co, 20% Fe, 1% S, 40% SiO_2, and 20% MgO. The carbon originates from the coal. The sulfur is present as a coating of metal sulfide on calcine particles.

The sulfided calcine is fed hot (about 700°C) to electric furnaces where it reacts to form molten matte, molten slag and offgas. Typical operating conditions for the furnace are given in Table 8.1.

1. Referred to as saprolite ore.

TABLE 8.1 Matte-from-Laterite Smelting Data for PT Inco

Smelter	PT Inco, Sorowako, Indonesia
Smelting Furnaces	
Number	4
Annual Ni in matte production, all furnaces, tonnes	80 000
Shape	Round
Diameter × height, m	18 × 6
Electrodes	
Number, per furnace	3
Diameter, m	1.8
Material	Selfbaking (Söderberg)
Wall-cooling system	Copper cooling fingers in refractory walls
Electrical (per Furnace)	
Maximum power, kW	70 000–80 000
Average power, kW	55 000–60 000
Hearth power density, kW/m^2	240
Average voltage, V	1000–1800
Average current, A	23 000–35 000
Operating Details	
Slag, matte tapholes	Two each at opposite ends
Calcine depth, m	~1 m
Slag depth, m	0.25 m above slag taphole
Matte depth, m	0.25 m above matte taphole
Feed, nominal tonnes of calcine per day per furnace	3000
Calcine composition, %	
C	1.3
Ni	2.0
Co	0.08

TABLE 8.1 Matte-from-Laterite Smelting Data for PT Inco—cont'd

Smelter	PT Inco, Sorowako, Indonesia
Fe	19.2
S	1.0
SiO_2	39.5
MgO	19.8
Consumption, per tonne of calcine feed	
Electrical energy, kWh	465
Electrode material, kg	1.4
Temperature, °C	
Offgas	~700
Slag	1500–1550
Matte	1350–1400
Slag and Matte Details	
Slag, tonnes per day per furnace	2500
Slag composition, mass %	
Ni	0.15
Co	
FeO	23
MgO	23
SiO_2	48
Ni in Matte, tonnes per day per furnace	~60
Matte, tonnes per day per furnace	~200
Matte composition, %	
Ni	26
Co	0.8
Fe	63
S	10

Note the round furnaces and the large slag-to-matte ratio. The matte product goes to Peirce–Smith converters where the nickel content is increased to 78% Ni, shown in Figure 8.1.

**Bangun et al., 1997; Warner et al., 2006.*

The composition of the matte is typically 26% Ni, 63% Fe, 0.8% Co and 10% S. The temperature of the molten matte is 1400°C.

The composition of the slag is typically 0.15% Ni, 23% FeO, 23% MgO, 48% SiO_2, and 6% Al_2O_3 plus CaO. The temperature of the molten slag is 1500°C. The offgas, which is at a temperature of 700°C, consists mainly of air that has leaked into the furnace, carbon dioxide, CO_2 and some sulfur dioxide, SO_2.

Carbon in Calcine

The amount of carbon in calcine feed is chosen so that all the remaining NiO in the calcine is reduced to metallic nickel in the smelting furnace, while at the same time producing just enough metallic iron to give the required matte composition (Bangun, Prenata, & Dalvi, 1997).

The metallic nickel and iron are quickly sulfided by the sulfur in the calcine feed to the smelter. The sulfides then combine to form a layer of molten matte, while the unsulfided oxides form a layer of molten slag.

Tapping Matte and Slag

The molten matte and slag are immiscible. They are tapped separately through low and high tapholes. Matte, which has a density of about 4.5 tonnes/m^3 (Sheng, Irons, & Tisdale, 1997), is tapped through the low taphole. Slag, which has a density of about 3.2 tonnes/m^3, is tapped through the high taphole.

The tapped matte is transferred to the converter for oxidation, Section 8.2.4. The tapped slag is discarded and the offgas is released to the atmosphere.

Minimizing Nickel Loss in the Discard Slag

All the electric furnace slag is discarded. It is crucial, therefore, that it contain as little nickel as possible. This is achieved by minimizing the amount of slag and by minimizing the nickel content in the slag.

Slag mass is minimized by maximizing the nickel content in the feed to the smelter. Typically, the slag mass is greater than 40 tonnes per tonne of nickel in the matte.

The nickel content in the slag is minimized by (i) ensuring that the calcine feed contains enough carbon for complete reduction of NiO; (ii) keeping the content of nickel in the matte at a reasonably low level, at approximately 26% Ni;[2] (iii) producing fluid slag through which matte droplets can settle quickly and efficiently; and, (iv) avoiding accidental tapping of matte with slag.

PT Inco keeps the content of nickel in the slag below about 0.15% by following these procedures. This is equivalent to about 93% recovery of nickel into matte (Warner, Diaz, Dalvi, Mackey, & Tarasov, 2006).

2. %Ni in slag increases with increasing %Ni in matte (Diaz, Landolt, Vahed, Warner, & Taylor, 1988).

Protection of the Furnace Wall

As with ferronickel smelting, the walls of the electric furnace used for smelting matte are protected by the use of cooling water. This causes needles of solid olivine, $(Mg,Fe)_2SiO_4$, to deposit on the sidewall refractories (Doyle, 2004). These deposits, referred to as a freeze lining, prevent refractory erosion and allow the furnaces to operate for between 10 and 20 years without major rebuilds.

The rate of deposition of olivine is controlled by adjusting the MgO/SiO_2 ratio of the furnace feed by ore blending. The higher the MgO/SiO_2 ratio, the higher the rate of deposition of olivine, and *vice versa*.

8.2.4. Converting – removing Iron from Electric Furnace Matte

The first step in making nickel metal from furnace matte is the removal of most of the iron in the matte. Iron is removed by blowing air into molten matte that has been freshly tapped. Peirce–Smith converters are used. Schematic diagrams of a Peirce–Smith converter are shown in Figures 8.2 and 8.3.

Operating details for converting are given in Table 8.2.

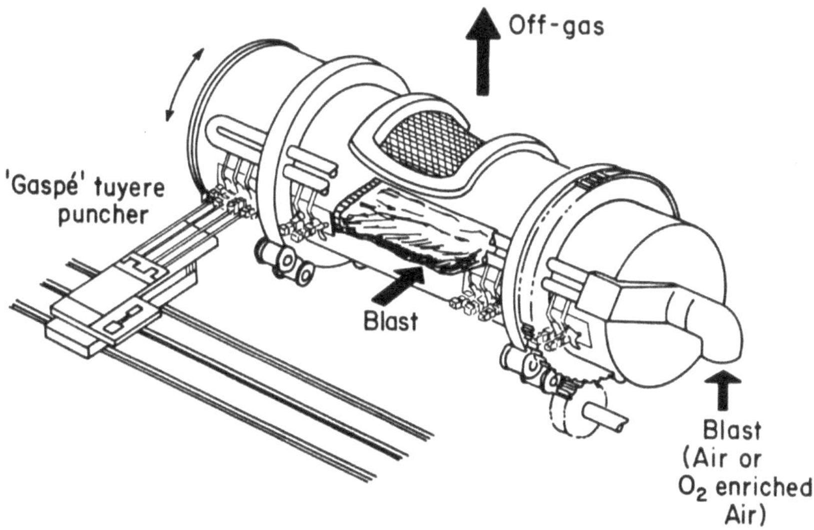

FIGURE 8.2 Peirce–Smith converter for producing molten low-iron matte from molten electric-smelting furnace matte. The main process is iron oxidation, which is accomplished by blowing air through submerged tuyeres at the back of the converter. Silica flux (to make molten iron silicate slag) is added through the converter mouth, added by air gun or added by conveyor through the end-wall. The tuyeres tend to block with frozen matte and slag. They are cleared by 'punching' from the back, as shown. Le Nickel also uses Peirce–Smith converters for making nickel-rich matte from refined molten ferronickel. *Drawing from Boldt and Queneau (1967), courtesy of Vale.*

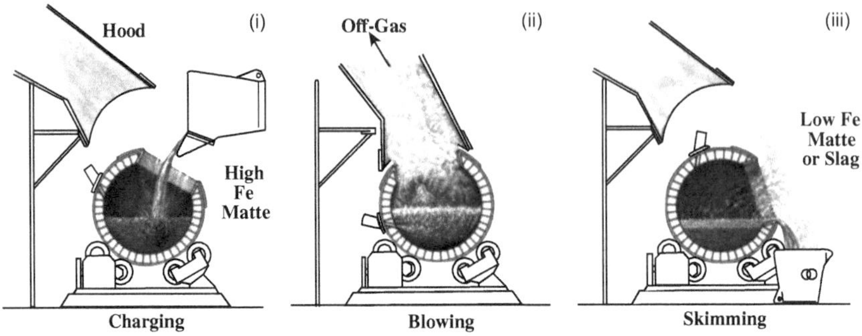

FIGURE 8.3 Positions of Peirce–Smith converter for (i) charging molten electric furnace matte, (ii) blowing air through the molten matte to oxidize Fe and (iii) pouring out slag and then matte. The converting is done in batches rather than continuously. The product high-grade matte is granulated or cast and sent to pyrometallurgical or hydrometallurgical refining. *Drawing from Boldt and Queneau (1967), courtesy of Vale.*

TABLE 8.2 Details of Matte Converting at PT Inco

Smelter	PT Inco, Sorowako, Indonesia
Converters	
Type	Peirce–Smith, Figures 8.2 and 8.3
Number	3
Outside dimensions, diameter × length, m	7.3 × 11.8 (one); 7.3 × 12.7 (two)
Tuyeres	
Number	20–28
Diameter, m	0.051
Blowing details	
Average blowing rate per converter, Nm^3/h	18000
% O_2	Air
Feed details	
Matte composition, %	
Ni	26
Co	0.8
Fe	63
S	10

TABLE 8.2 Details of Matte Converting at PT Inco—cont'd

Smelter	PT Inco, Sorowako, Indonesia
Product details	
Molten high-grade matte, total tonnes per day, all converters	280
Matte temperatures, °C	
Into converter	1400
Leaving converter	1400
Matte composition, %	
Ni	78
Co	1
Fe	1
S	20
High-grade matte disposition	Water granulated, dried, and shipped to Asia
Molten slag, total tonnes per day, all converters	900
Slag temperature leaving converter, °C	1400
Slag composition, %	
Ni	0.6
SiO_2	25
Fe	53
Slag disposition	%Ni < 0.6 is discarded %Ni > 0.6 is recycled to calcination kilns

Note: Iron in the matte is oxidized with air to produce a matte with high-nickel content. The process is autothermal, that is, no additional energy is required.

The main reaction is the oxidation of the iron in the matte:

$$\underset{\substack{\text{in molten electric}\\ \text{furnace matte}}}{\overset{1350^\circ\text{C}}{2\text{Fe}\,(\ell)}} + \underset{\substack{\text{in injected}\\ \text{air}}}{\overset{30^\circ\text{C}}{\text{O}_2\,(g)}} + \underset{\substack{\text{in added crushed}\\ \text{rock flux}}}{\overset{30^\circ\text{C}}{\text{SiO}_2\,(s)}} \rightarrow \underset{\substack{\text{molten fayalite}\\ \text{slag}}}{\overset{1350^\circ\text{C}}{\text{Fe}_2\text{SiO}_4\,(\ell)}} \quad (8.3)$$

Silica, SiO_2, is added as flux to give an immiscible low-melting point slag.

During the oxidation of iron, some sulfur is also oxidized especially when the iron in the matte is nearly depleted (Kellogg, 1987). The reaction is as follows:

$$\underset{\substack{\text{in molten electric}\\\text{furnace matte}}}{\overset{1350^\circ C}{S\ (\ell)}} + \underset{\text{in injected air}}{\overset{30^\circ C}{O_2\ (g)}} \rightarrow \underset{\text{in offgas}}{\overset{1400^\circ C}{SO_2\ (g)}} \tag{8.4}$$

Both reactions are highly exothermic, which means that the process requires no external heat. In fact, the reactions supply so much heat that the converter is used to melt nickel-bearing solids that are inadvertently produced during tapping and ladle transfers.

Final product from the converting stage is typically 78% Ni, 20% S, 1% Fe, and 1% Co.

Behavior of Nickel During Converting

Nickel has less affinity for oxygen than iron and sulfur. As a result, only a small amount of nickel is oxidized during converting (Kellogg, 1987). The resulting nickel oxide, NiO, reports to the slag. Most converter slag contains less than 0.6% Ni.

If the slag contains less than 0.6% Ni, it is discarded. The nickel lost to the slag during converting accounts for a loss of about 3% of the total nickel in the feed to the smelter. Slag with more than 0.6% Ni, which arises from the later stages of converting when the iron content of the matte is low, is granulated using water and recycled to the calcination/reduction/sulfidation kilns for recovery of nickel (Chen, Dutrizac, Krause, & Osborne, 2004).

8.2.5. Solidifying High-grade Matte

PT Inco granulates the molten matte by pouring the matte into a huge stream of water. This process produces solid granules with a diameter of about 0.3 mm. These granules are sent to Asia where they are used to produce nickel oxide and melting-grade nickel (95%–97% Ni). This is discussed in Chapter 9.

8.3. LE NICKEL PROCESS – MAKING MATTE FROM MOLTEN REFINED FERRONICKEL

About 20% of the molten refined ferronickel produced at Le Nickel's smelter in New Caledonia is made into molten matte. The process is illustrated in Figure 8.4. The matte is made by (i) injecting liquid sulfur into molten refined ferronickel through two tuyeres in a Peirce–Smith converter to make matte and

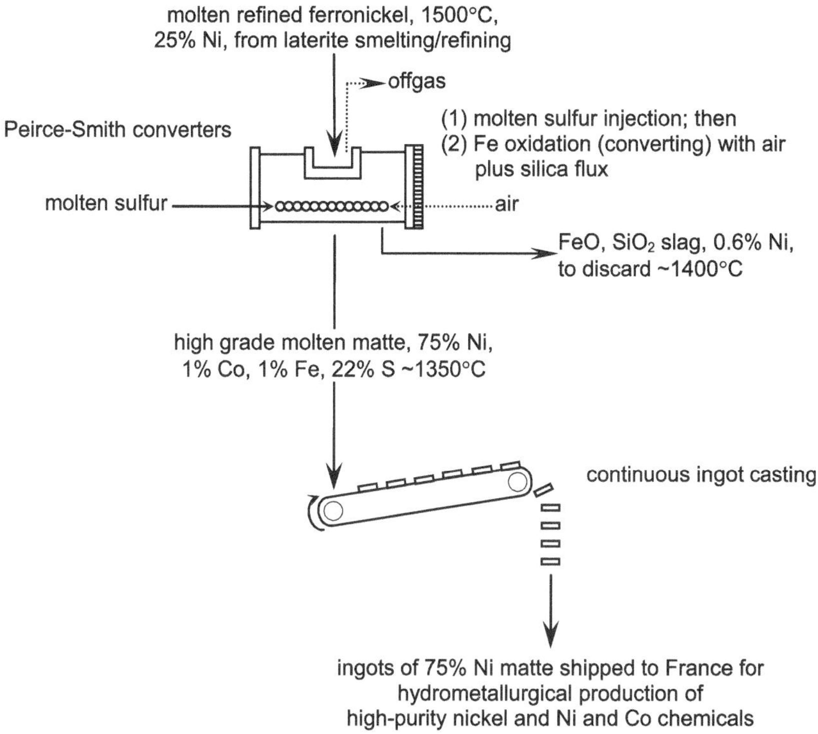

FIGURE 8.4 Le Nickel matte production flowsheet (Le Nickel, 2010). Sulfidation of molten ferronickel then oxidation of iron in Peirce–Smith converters is notable. The plant has two Peirce–Smith converters. It produces ~15 000 tonnes of nickel in matte per year.

then; (ii) converting (oxidizing) the iron in the resulting matte with air, as described in Section 8.2.

The final matte product typically contains 75% Ni, 23% S, 1% Fe and 1% Co. The matte is poured from the Peirce–Smith converter and cast.

Le Nickel casts matte as 0.1 m × 0.3 m × 0.4 m ingots on a continuous-belt mold caster. These ingots are shipped to France where they are crushed, ground and leached using chlorine in preparation for the production of high-purity nickel metal and nickel and cobalt chemicals (Eramet, 2010).

8.4. PROCESS APPRAISAL

PT Inco and Le Nickel produce nickel-rich matte from laterite ore rapidly and efficiently. However, introduction of sulfur into laterite smelting/converting inevitably leads to problems of emission of sulfur dioxide that are not found with conventional ferronickel smelting. For this reason, it is unlikely that laterite-to-matte smelting will be expanded.

8.5. SUMMARY

About 90% of laterite smelting produces ferronickel and about 10% produces matte. The ferronickel is used directly for making stainless steel and other ferrous alloys. The matte is refined (i) pyrometallurgically to melting-grade nickel; and, (ii) hydrometallurgically to high-purity nickel and nickel and cobalt chemicals.

Matte is made from laterite ore by either (i) sulfidizing dewatered laterite concentrate then smelting the resulting sulfided calcine; or, (ii) suphiding molten refined ferronickel, which is done to a lesser extent.

Most of the iron in the matte is then oxidized to produce a final content of 75%–78% Ni, and 1% Fe in the matte product. The matte is refined to metal and chemicals in Asia and France.

REFERENCES

Bangun, C. D., Prenata, W., & Dalvi, A. D. (1997). Sidewall design and refractory wear mechanism in electric furnaces at P.T. Inco. In C. Diaz, I. Holubec & C. G. Tan (Eds.), *Proceedings of the Nickel/Cobalt 97 International Symposium, Vol. III. Pyrometallurgical operations, the environment and vessel integrity in nonferrous smelting and converting* (pp. 115–132). CIM.

Bergman, R. A. (2003). Nickel production from low-iron laterite ores: Process descriptions. *CIM Bulletin, 96*, 127–138.

Chen, T. T., Dutrizac, J. E., Krause, E., & Osborne, R. (2004). Mineralogical characterization of nickel laterites from New Caledonia and Indonesia. In W. P. Imrie, D. M. Lane & S. C. C. Bartlett (Eds.), *International Laterite Nickel Symposium – 2004* (pp. 79–99). TMS.

Diaz, C. M., Landolt, C. A., Vahed, A., Warner, A. E. M., & Taylor, J. C. (1988). A review of nickel pyrometallurgical operations. In G. P. Tyroler & C. A. Landolt (Eds.), *Extractive Metallurgy of Nickel and Cobalt* (pp. 211–239). TMS.

Doyle, C. (2004). The steps required to meet production targets at PT Inco, Indonesia: A new innovative business strategy. In W. P. Imrie, D. M. Lane & S. C. C. Bartlett et al. (Eds.), *International Laterite Nickel Symposium – 2004* (pp. 670). TMS.

Eramet. (2010). Le Havre-Sandouville refinery [Company brochure].

Kellogg, H. H. (1987). Thermochemistry of nickel-matte converting. *Can. Metall. Q., 26*, 285–298.

Le Nickel, (2010). Nickel Production. www.sln.nc/content/view/75/44/lang,french/ Accessed May 19, 2011.

Sheng, Y. Y., Irons, G. A., & Tisdale, D. G. (1997). Power, fluid flow and temperature distributions in electric smelting of nickel matte. In C. Diaz, I. Holubec & C. G. Tan (Eds.), *Proceedings of the Nickel/Cobalt 97 International Symposium, Vol. III. Pyrometallurgical Operations, the Environment and Vessel Integrity in Nonferrous Smelting and Converting* (pp. 45–66). CIM.

Vahed, A., Liu, J., Prokesch, M., Riddle, R., Jafri, M., Barus, R., & Syukirman. (2009). Testing of nickel laterite smelter dust insufflation – part 1. In J. Liu, J. Peacey, M. Barati, S. Kahani-Nejad & B. Davis (Eds.), *Pyrometallurgy of Nickel and Cobalt 2009. Proceedings of the International Symposium* (pp. 221–232). CIM.

Warner, A. E. M., Diaz, C. M., Dalvi, A. D., Mackey, P. J., & Tarasov, A. V. (2006). JOM world nonferrous smelter survey, Part III: Laterite. *JOM, 58*, 11–20.

SUGGESTED READING

Bergman, R. A. (2003). Nickel production from low-iron laterite ores: Process descriptions. *CIM Bulletin, 96*, 127–138.

Daenuwy, A., & Dalvi, A. D. (1997). Development of reduction kiln design and operation at P.T. Inco (Indonesia). In C. Diaz, I. Holubec & C. G. Tan (Eds.), *Proceedings of the Nickel/Cobalt 97 International Symposium, Vol. III. Pyrometallurgical Operations, the Environment and Vessel Integrity in Nonferrous Smelting and Converting* (pp. 93–113). Metallurgical Society of CIM.

Warner, A. E. M., Diaz, C. M., Dalvi, A. D., Mackey, P. J., & Tarasov, A. V. (2006). JOM world nonferrous smelter survey, Part III: Laterite. *JOM, 58*, 11–20.

Chapter 9

Roasting Matte to Nickel Oxide and Metal

The matte produced at PT Inco is sent to Asia for either (a) oxidation roasting to produce nickel oxide, containing 75% Ni, for direct use in stainless and low-alloy steel manufacture or (b) oxidation roasting and reduction roasting to produce melting-grade nickel that contains either 95% or 97% Ni. Melting-grade nickel is used for a variety of purposes in the steelmaking and foundry industries.

The objectives of this chapter are to describe the following topics:

(a) matte roasting objectives; and,
(b) matte roasting techniques.

The description is based mainly on the practice of Tokyo Nickel, situated in Matsusaka, Japan (Ishiyama & Hirai, 1997). A schematic diagram of the flowsheet of the roasting operation is shown in Figure 9.1.

9.1. MATTE ROASTING OBJECTIVES

The typical composition of the matte, in the form of 0.3 mm diameter granules, is 78% Ni, 1% Co, 0.15% Cu, 1% Fe, and 20% S.

The feed to steelmaking must be devoid of sulfur. This is because sulfur in steel decreases ductility and strength and can cause the steel to crack during manufacture, particularly during hot rolling.

The main objective of matte roasting is, therefore, to lower sulfur from 20% in the matte to 0.003% in the roaster product.

9.2. CHEMISTRY

The main roasting reaction is given as follows:

$$\underset{\substack{\text{in solid matte}\\ \text{granules}}}{\overset{30^\circ\text{C}}{NiS\,(s)}} + \underset{\substack{\text{in oxygen-enriched air}\\ \text{blast 28 volume\% } O_2}}{\overset{30^\circ\text{C}}{1.5O_2\,(g)}} \rightarrow \underset{\substack{\text{Ni oxide granules}\\ 0.003\%\ \text{S}}}{\overset{1050^\circ\text{C}}{NiO\,(s)}} + \underset{\text{in } N_2,\, SO_2 \text{ offgas}}{\overset{1050^\circ\text{C}}{SO_2\,(g)}} \tag{9.1}$$

Extractive Metallurgy of Nickel, Cobalt and Platinum-Group Metals. DOI: 10.1016/B978-0-08-096809-4.10009-7

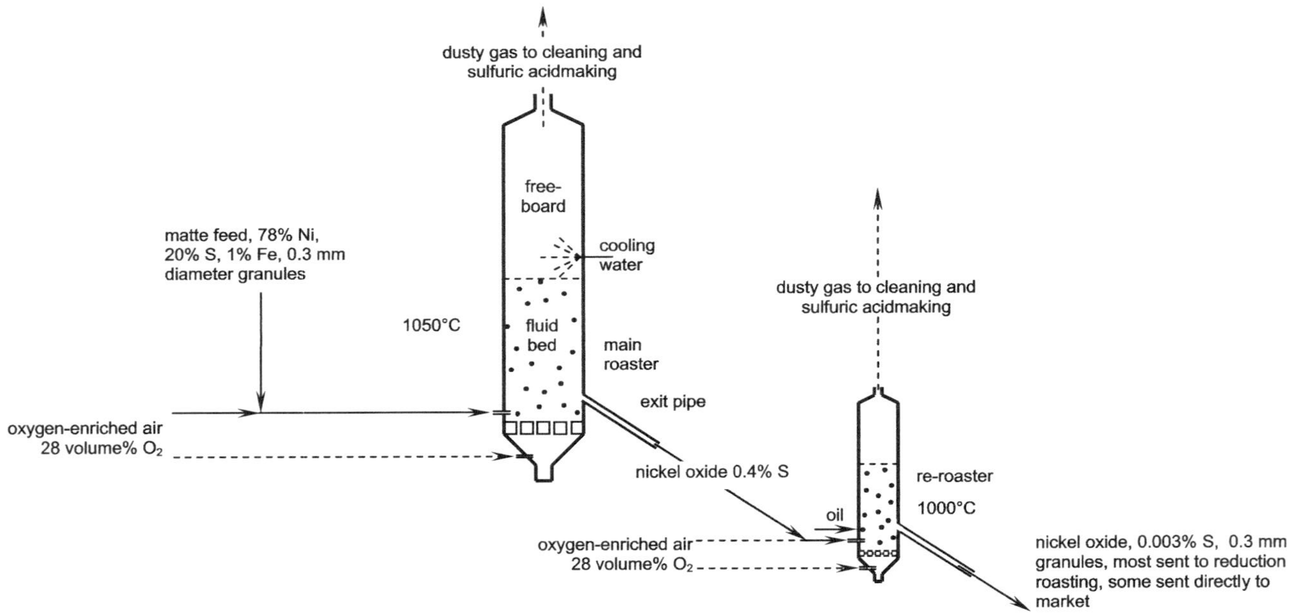

FIGURE 9.1 Flowsheet for producing nickel oxide, 0.003% S from 20% S matte granules at PT Inco. The process is continuous. The main roaster does most of the oxidation and is autothermal. The small re-roaster lowers the sulphur content from 0.4% to 0.003% S. It is not autothermal and requires oil combustion. A 10 m high, 3 m diameter roaster consumes about 150 tonnes of matte per day. Details are given by Ishiyama and Hirai (1997).

It is carried out at 1050°C to ensure rapid, efficient sulfur elimination. The reaction is strongly exothermic and provides all the heating required for roasting. In fact, the roasters must be water-cooled to prevent them from overheating.

9.3. PRODUCTS

The products of oxidation roasting are:

(a) hot, dusty, offgas-containing sulfur dioxide, nitrogen and water vapor; and,
(b) hot nickel oxide granules.

The offgas is sent to gas cooling and de-dusting and on to sulfuric acid manufacture. The dust from the offgas is collected and recycled to roasting.

The nickel oxide granules are air-cooled then either compacted and sold or sent to reduction roasting to produce melting-grade nickel.

9.4. INDUSTRIAL ROASTING

Most industrial roasting is done continuously in fluidized-bed roasters, as illustrated in Figure 9.1. For matte roasting, the process entails the following:

(a) creating a fluidized bed of 0.3 mm diameter matte granules by blowing oxygen-enriched air upwards through evenly spaced tuyeres into the bed;
(b) continuously side-blowing matte granules into the fluidized bed with oxygen-enriched air;
(c) continuously removing dusty offgas from the top of the roaster; and,
(d) continuously withdrawing hot roasted granules into a roaster discharge seal leg,[1] then down a pipe into a re-roaster.

9.5. RE-ROASTING

Matte roasting is done in two sequential fluidized-bed roasters, as shown in Figure 9.1. The first oxidizes most of the sulfur in the matte. The second oxidizes sulfur from granules that have short-circuited through the first roaster.

The product from the first roaster has an average sulfur grade of 0.4%, while the product from the second roaster has an average sulfur grade of 0.003%.

A notable feature of the re-roaster is that it relies on oil combustion for most of its heat. This is because the feed is mostly oxidized and, consequently, the amount of heat that is generated by the reaction given in Equation (9.1) is small.

In other respects, the roasters operate similarly.

---

1. A roaster seal discharge leg (not shown in Figure 9.1) permits steady flow of calcine from a fluidized-bed roaster without accidentally emptying the bed. It is described by Adham and Lee (2004).

9.6. FLUIDIZATION

A bed of solid particles becomes fluidized when gas is blown up into the bed through a set of equidistant tuyeres (Adham & Lee, 2004; Biswas & Davenport, 1980). The bed is referred to as *fluidized* because it behaves much like a liquid.

The Tokyo Nickel roasters are operated with so-called *bubbling* beds with gas velocities that are approximately in the range of 1.5–1.9 m/s (at temperature). Under these conditions, about 95% of the product flows out of a submerged-exit pipe (see Figure 9.1) and only about 5% is removed with the offgas that exits from the top of the roaster. All material that is removed with the offgas is recycled to the roaster.

The blast to the oxidation roasters is air that has been enriched to 28% O_2 (by volume). This enrichment increases the feed rate of the oxygen, and hence results in higher rates of oxidation of the matte, while maintaining a gas velocity that will give the desired bubbling bed.

9.7. ADVANTAGES OF FLUIDIZED BEDS

The particles in a bubbling fluidized bed are completely surrounded by gas, and agitated. These are perfect conditions for rapid oxidation and heat transfer.

9.8. INDUSTRIAL OPERATION

9.8.1. Start-up

The roasters at Tokyo Nickel are started in the following sequence:

(a) the roasters are heated for about 1 day with heavy oil burners;
(b) the air flow through the bottom tuyeres is started and the roaster is filled to operating level with previously roasted (NiO) granules;
(c) the oxidation of the matte is started by side-blowing matte granules and oxygen-enriched air into the bed;
(d) starting withdrawal of roasted products; and,
(e) the matte and air feed rates are gradually increased (while the feed rate of oil is decreased) until the required production rate is achieved and steady.

9.8.2. Shutdown

The procedure for short shutdowns of the roaster is as follows:

(a) matte feeding and product withdrawal are stopped;
(b) the roaster is heated with oil burners; and,
(c) the fluidized bed is collapsed by stopping air flow through the bottom tuyeres.

Restart after a short maintenance shutdown is achieved by reversing these steps while keeping the bed hot with oil burners.

Long-term shutdowns follow a similar procedure, except that the collapsed bed is cooled and dug out through a manhole at tuyere level to prevent agglomeration of the granules. Restarting the roaster after a long-term shutdown follows the same procedure as the initial start-up.

9.8.3. Steady-state Operation

Steady-state operation of the roaster entails the following:

(a) feeding solids and gases at a constant prescribed rate;
(b) monitoring the temperature of the bed and controlling it by automatically adjusting water spray input rate; and,
(c) monitoring sulfur content of the product and adjusting granule residence time accordingly (by adjusting the feed rate of the matte).

9.8.4. Oxidation Roaster Product

The final re-roaster product is the same size (~0.3 mm diameter) as the matte feed. The composition of the final product is 74%–76% Ni, 23% O, 0.8%–1.2% Co, 0.1%–0.2% Cu, 0.2%–0.5% Fe, and 0.003%–0.004% S.

It is air-cooled and either sent to reduction roasting for oxygen removal or, to a lesser extent, bagged and sent to market for direct use in stainless steel and low-alloy steel production.

9.8.5. Sulfur Dioxide By-product

An inevitable by-product of the roasting of sulfide matte is sulfur dioxide gas. This gas is captured and made into a useful product. The sulfur dioxide is almost always converted into sulfuric acid, $H_2SO_4(\ell)$. The manufacture of sulfuric acid is detailed in Chapter 20.

9.9. REDUCTION ROASTING

Most of the product from the oxidation roaster is reduced to metal in reduction roasters. The products of the reduction roasting are metal granules with a diameter of 0.3 mm, known as Tonimet 95 and Tonimet 97. The compositions of these two products are given in Table 9.1.

The process is similar to oxidation roasting but reducing gas is provided by air-deficient combustion of low-sulfur kerosene in the fluidized bed.

The main reactions are:

$$\underset{\text{kerosene}}{H_2C(\ell)} + \underset{\text{in air}}{0.5O_2(g)} \xrightarrow{1000^\circ C} \underset{\substack{\text{reducing gas} \\ \text{plus } N_2}}{CO(g) + H_2(g)} \quad (9.2)$$

$$CO(g) + NiO(s) \longrightarrow Ni(s) + CO_2(g) \quad (9.3)$$

TABLE 9.1 Composition of Two Grades of Melting-Grade Nickel

Component	Tonimet 95 (%)	Tonimet 97 (%)
Ni	93–95	97–97.2
O	~4	~1
Co	1.0–1.5	1.0–1.6
Cu	0.1–0.2	0.1–0.2
Fe	0.3–0.6	0.4–0.7

$$H_2(g) + NiO(s) \rightarrow Ni(s) + H_2O(g) \quad (9.4)$$

The reduced granules are cooled and shipped 'as is', or briquetted, or compacted. All these products are used throughout the steelmaking and foundry industries.

9.9.1. Combustion Control

The presence of kerosene, carbon monoxide and hydrogen in the reduction roasters necessitates careful monitoring and control of combustion conditions. These conditions are monitored optically and controlled by changing the feed rates of kerosene and air.

9.9.2. Control of the Oxygen Content in the Product

About 65% of the product from the reduction roaster at Tokyo Nickel is *Tonimet 97*. This product contains 0.5%–1% oxygen. This low level of oxygen is achieved by increasing the residence time of the solids in the roaster either by decreasing the feed rate of the solids or by recycling some of the product back to the reduction roaster or both.

9.10. NICKEL RECOVERY

Nickel recovery to product at the roasting plant is nearly 100%. This is accomplished by undertaking the following measures:

(a) collecting and cleaning all effluent gases;
(b) recycling all solids from gas cleaning to the oxidation roaster;
(c) using effluent liquids from gas cleaning for cooling the oxidation roasters; and,
(d) thoroughly cleaning the plant and its surroundings (and recycling the 'cleanup' products to the oxidation roasters).

9.11. SULFUR CAPTURE

Sulfur capture at the roasting plant is nearly complete. This is achieved by:

(a) capturing more than 99.9% of the sulfur dioxide from the roasters in a double-contact sulfuric acid plant;
(b) scrubbing the acid plant effluent gas with Na_2O; and,
(c) using all gas-cleaning and scrubbing liquids for cooling the oxidation roasters.

The gypsum, $CaSO_4 \cdot 2H_2O$, and sodium bisulfate by-products of gas cooling and cleaning are sold.

9.12. SUMMARY

Most of matte produced by PT Inco is desulfurized by oxidation roasting in fluidized beds. The sulfur content of the roaster product is about 0.003%, suitable for most steelmaking purposes.

The product of oxidation roasting is nickel oxide granules containing about 75% Ni. These are used directly for stainless steel and low-alloy steel production.

More commonly, however, this nickel oxide product is reduction roasted in fluidized beds to give products with a low oxygen content and containing 95%–97% Ni. These products are suitable for all steelmaking and foundry applications.

REFERENCES

Adham, K., & Lee, C. (2004). Process design considerations for the fluidized bed technology applications in the nickel industry. *CIM Bulletin, 97*, 106–111.

Biswas, A. K., & Davenport, W. G. (1980). *Extractive Metallurgy of Copper.* Chapter 3 (2nd ed.). Pergamon Press.

Ishiyama, H., & Hirai, Y. (1997). Fluid bed roasting at Tokyo Nickel Company. In C. Diaz, I. Holubec & C. G. Tan (Eds.), *Proceedings of the Nickel/Cobalt 97 International Symposium, Vol. III. Pyrometallurgical operations, the environment and vessel integrity in nonferrous smelting and converting* (pp. 133–140). CIM.

SUGGESTED READING

Ishiyama, H., & Hirai, Y. (1997). Fluid bed roasting at Tokyo Nickel company. In C. Diaz, I. Holubec & C. G. Tan (Eds.), *Proceedings of the Nickel/Cobalt 97 International Symposium, Vol. III. Pyrometallurgical operations, the environment and vessel integrity in nonferrous smelting and converting* (pp. 133–140). CIM.

Chapter 10

Overview of the Hydrometallurgical Processing of Laterite Ores

Laterite ore bodies are composed of (at least) two layers: the saprolite layer and the limonite layer. The pyrometallurgical processes for the production of ferronickel and melting-grade nickel, described in Chapters 4–9, cannot process laterite ores that have a high iron content. As a result, these processes are limited to the saprolite ores. The limonite-type laterite ores are processed by hydrometallurgical techniques.

The objective of this chapter is to provide an overview of the hydrometallurgy of laterites.

10.1. INTRODUCTION

Laterite hydrometallurgy produces about 100 000 tonnes per year of high-purity nickel and 10 000 tonnes per year of high-purity cobalt mostly by leaching at high temperature using sulfuric acid.[1]

A similar amount of nickel is produced by the pyrometallurgical reduction of limonitic laterite followed by leaching of the reduced product using NH_3, CO_2 and O_2. This process, called the Caron process after its developer, is described in Appendix B.

Laterite hydrometallurgy is based mainly on leaching of limonite or smectite ores. These ores have a high iron content and a low MgO content. They are leached at high temperature with sulfuric acid. The process is frequently referred to as the high-pressure acid leaching process or *HPAL process*.[2] At these temperatures, the iron precipitates from solution as hematite, leaving a solution containing nickel and cobalt. A generalized flow diagram is shown in Figure 10.1.

Once the ore has been leached, it is cooled through several flash tanks to recover as much of the energy as possible. The cooled slurry is then pre-neutralized and sent to a counter-current decantation circuit, where the

1. Some laterite is also processed by the Caron Process, see Appendix B.

2. *HPAL* is a misnomer because it is the temperature that is of importance in the leaching of laterite ores, not the pressure.

Extractive Metallurgy of Nickel, Cobalt and Platinum-Group Metals. DOI: 10.1016/B978-0-08-096809-4.10010-3

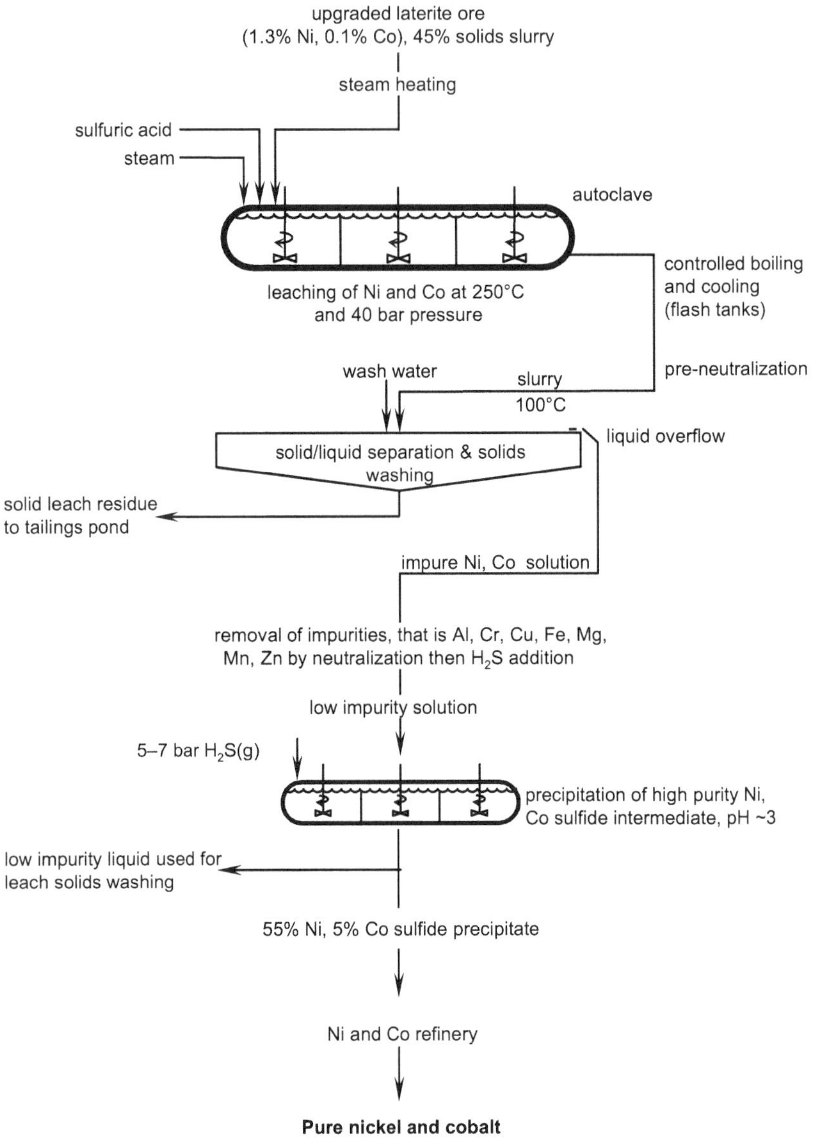

FIGURE 10.1 Flowsheet for high-temperature leaching of laterite ore in sulfuric acid. The feed is mainly laterite that is low in magnesium. The product is high-purity nickel–cobalt sulfide precipitate, 55% Ni, 5% Co and 40% S, ready for refining to high-purity nickel and cobalt. A leach autoclave is typically 4.5 m diameter × 30 m long with six stirred compartments and treats about 2000 tonnes of solid feed per day.

solids and liquids are separated. The impure leach solution is further neutralized to remove more contaminants.

Nickel and cobalt are precipitated from the solution leaving counter-current decantation as a mixed nickel–cobalt sulfide using hydrogen sulfide gas. The advantage of this precipitation is that many of the impurities in the solution do not co-precipitate. For example, the solutions are frequently saturated in calcium after neutralization, which can cause the formation of crud in downstream solvent extraction. This calcium does not co-precipitate with the nickel and cobalt during sulfide precipitation, making downstream operations easier.

Leaching and sulfide precipitation are discussed in more detail in Chapters 11 and 12, respectively. The alternatives to sulfide precipitation are examined in the remainder of this chapter.

10.2. ALTERNATIVES TO MIXED SULFIDE PRECIPITATION

Three alternatives to mixed sulfide precipitation are currently practiced or have recently been practiced: (i) precipitation of nickel carbonate; (ii) direct solvent extraction; and, (iii) the precipitation of mixed nickel–cobalt hydroxides.

10.2.1. Nickel Carbonate Precipitation

Cawse (Norilsk), Western Australia, originally operated a refinery. However, it currently produces nickel carbonate, which is sent to the refinery at Harjavalta, Finland. The Queensland Nickel refinery at Yabulu, near Townsville, Queensland, Australia, also precipitates nickel carbonate from solution, which is then calcined and partially reduced in hydrogen gas (Reid & Price, 1993; Sole & Cole, 2002).

10.2.2. Direct Solvent Extraction

Bulong, also in Western Australia, implemented a process based on direct solvent extraction (Sole & Cole, 2002). There were numerous commissioning problems, not the least of which was the contamination of the solvent extraction circuit with calcium. These problems were, on the whole, solved before the operation closed in 2003.

Goro (Vale), New Caledonia, has also implemented a direct solvent extraction process after extensive testing. Few operating details have been published; however, Goro has suffered from extensive commissioning problems (Finch, 2010). Currently, Goro produces a mixed hydroxide precipitate that is processed at the Yabulu refinery in Queensland, Australia.

10.2.3. Mixed Hydroxide Precipitation

Ravensthorpe, in Western Australia, implemented a process that included the precipitation of a mixed nickel–cobalt hydroxide as part of their enhanced

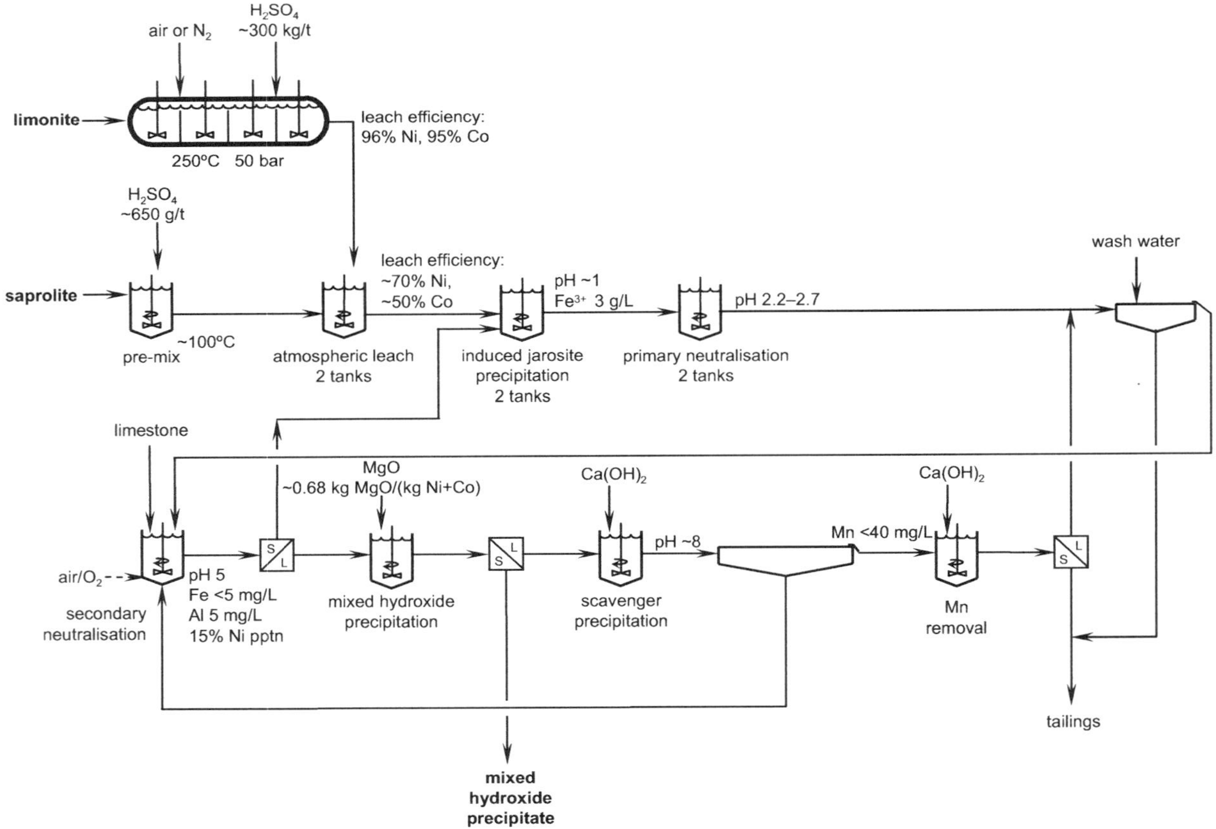

FIGURE 10.2 The Ravensthorpe laterite process (Adams *et al.*, 2004). The process uses both the limonite and saprolite portions of the laterite ore body and precipitates a mixed nickel and cobalt hydroxide.

pressure acid leaching process. The process is interesting because it is an example of an 'omnivorous' process, that is, one that can use both the saprolite and limonite parts of the laterite ore body (Duyvensteyn, Wicker, & Doane, 1979).

The schematic flowsheet for Ravensthorpe Nickel is shown in Figure 10.2. The process is a combination of high-pressure acid leaching of limonite and atmospheric leaching of saprolite. The limonite fraction of the laterite ore body is leached at 250°C and 50 bar in an autoclave. The exit slurry from the autoclave is combined with saprolite to partially neutralize the acid.

The nickel-bearing solution then passes through two neutralization stages prior to precipitation of the product, which is the mixed nickel–cobalt hydroxide, $(Ni,Co)(OH)_2$. The pH of the solution is raised to a value of about 7.5 using MgO.

The solution remaining after the mixed hydroxide precipitation step passes through two further stage of neutralization before being recycled back to the counter-current decantation circuit.

Ravensthorpe operated for about a year before being closed due to the economic downturn of 2008–2009. It has subsequently been sold to First Quantum Minerals, who plan to resume operations in 2011.

10.3. DOWNSTREAM PROCESSING

While there are a number of operations successfully using HPAL and sulphide precipitation, there is no standard process for the refining of nickel and cobalt from the mixed sulfide precipitate. Some of the refining processes that are used are presented in the overview of the hydrometallurgical refining of nickel and cobalt in Chapter 23.

10.4. SUMMARY

About 100 000 tonnes of nickel per annum is produced by the leaching of laterite ores. The laterite ores that are amenable to leaching are low in MgO and high in iron, that is, the limonite layer of the ore body. These ores are leached at high temperature, between 250°C and 270°C using sulfuric acid. Iron dissolves from the goethite, then precipitates as hematite, releasing nickel and cobalt into solution. The reason for the high temperature is to ensure that the iron precipitates to a sufficiently low level.

Nickel and cobalt are precipitated from the solution using hydrogen sulfide. The precipitate, a mixture of nickel and cobalt sulfides, is sent to a nickel–cobalt refinery for further processing.

An alternative to mixed sulfide precipitation is to either treat the solution directly or to precipitate mixed hydroxide or mixed carbonates.

Processing of both the limonite and saprolite layers of the laterite ore body, such as the process implemented at Ravensthorpe, should result in greater utilization of the ore body.

REFERENCES

Adams, M., van der Meulen, D., Czerny, C., et al. (2004). Piloting of the beneficiation and EPAL circuits for Ravensthorpe Nickel Operations. In W. P. Imrie, D. M. Lane, S. C. C. Barnett, R. M. Berezowsky, E. J. M. Jahnsen & P. G. Mason (Eds.), *International Laterite Nickel Symposium – 2004* (pp. 79–99). TMS.

Duyvensteyn, W. P., Wicker, G. R., & Doane, R. E. (1979). An omnivorous process for laterite deposits. In D. J. I. Evans, R. S. Shoemaker & H. Veltman (Eds.), *International Laterite Nickel Symposium* (pp. 553–569). AIME.

Finch, C. (2010). Goro start-up partly stopped after acid spill. *Reuters*. 26 April.

Reid, J. G., & Price, M. J. (1993). *Ammoniacal solvent extraction at Queensland Nickel Process installation and operation. In Solvent extraction in the process industries, Vol. I.* Society of chemical industries. pp. 225–231.

Sole, K. C., & Cole, P. M. (2002). Purification of nickel by solvent extraction. In Y. Marcus, A. K. SenGupta & J. A. Marinsky (Eds.), *Ion Exchange and Solvent Extraction* (pp. 143–195). Marcel Dekker Inc.

SUGGESTED READING

Budac, J. J., Fraser, R. & Mihaylov, I. (Eds.). (2009). *Hydrometallurgy of Nickel and Cobalt 2009.* Metallurgical Society of CIM.

Imrie, W. P., Lane, D. M. & Barnett, S. C. C. (Eds.). (2004). *International Laterite Nickel Symposium – 2004.* TMS.

McDonald, R. D., & Whittington, B. I. (2008). Atmospheric leaching of nickel laterites review Part I. Sulphuric acid technologies. *Hydrometallurgy, 91*, 35–55.

Sole, K. C., & Cole, P. M. (2002). Purification of nickel by solvent extraction. In Y. Marcus, A. K. SenGupta & J. A. Marinsky (Eds.), *Ion exchange and solvent extraction* (pp. 143–195). Marcel Dekker Inc.

Chapter 11

High-Temperature Sulfuric Acid Leaching of Laterite Ores

The first step in the hydrometallurgical processing of laterite ores is leaching at high temperature using solutions of sulfuric acid. (The overall flowsheet was shown in Figure 10.1.) The process is frequently referred to as the high-pressure acid-leaching process or HPAL process.

The leaching plant at Coral Bay is shown in Figure 11.1. Nickel and cobalt are efficiently leached from the laterite feed by solutions of sulfuric acid at temperatures of about 250°C with relatively low consumption of sulfuric acid.

FIGURE 11.1 High-temperature sulfide leach plant at Coral Bay, Philippines. It processes the limonite layer of the laterite deposit and produces a mixed precipitate of nickel and cobalt sulfides, which is shipped to Niihama, Japan for refining to high-purity nickel and cobalt. The autoclave is on the lower right, sulfuric acid storage tank is in the center and the thickeners are on the left. The flash-tank tower is above the autoclave. *Source: Photograph courtesy of Sumitomo Metal Mining Ltd.*

Extractive Metallurgy of Nickel, Cobalt and Platinum-Group Metals. DOI: 10.1016/B978-0-08-096809-4.10011-5

11.1. CHAPTER OBJECTIVES

The objectives of this chapter are as follows:

(a) to describe industrial practice of HPAL process; and,
(b) to indicate how the leaching is controlled and optimized.

11.2. SULFURIC ACID LEACHING

A schematic diagram of the pressure leaching autoclave used in the leaching of laterite ore is shown in Figure 11.2. The autoclave operates at about 250°C and 40 bar pressure. The internal and external features of an industrial autoclave are shown in Figures 11.3 and 11.4.

The feed to the autoclave is the upgraded laterite ore that has been slurried with water and then heated with steam to about 200°C. The product of leaching is a slurry at 250°C containing the following:

(a) a pregnant leach solution containing about 6 g/L Ni, 0.5 g/L Co and 30–50 g/L H_2SO_4; and,
(b) a leach residue that consists of precipitated impurities with a composition of about 0.06% Ni, 0.008% Co, and 51% Fe.

11.2.1. Slurry Flow

The hot feed slurry is pumped into the autoclave with positive displacement slurry pumps (Geho, 2011). The feed pressure to the autoclave is above the operating pressure of about 40 bar.

The slurry then flows left to right as shown in Figure 11.2 through six stirred chambers (over the partitions) toward the leached slurry exit pipe. Sulfuric acid and steam are added to each chamber to leach the concentrate and heat the slurry.

Finally, the leached slurry departs the autoclave up the exit pipe, forced out by the autoclave pressure. The slurry level in the autoclave is controlled by a choke valve that adjusts the exit flow rate.

Slurry residence time is typically between one and two hours, which gives a leaching efficiency for nickel and cobalt of about 95%.

11.2.2. Slurry Destination

After leaving the autoclave, the product slurry is sent to flash cooling (which is discussed in Appendix C), neutralization, liquid/solid separation (which is discussed in Appendix D), solution purification and sulfide precipitation of nickel and cobalt (which is discussed in Chapter 12).

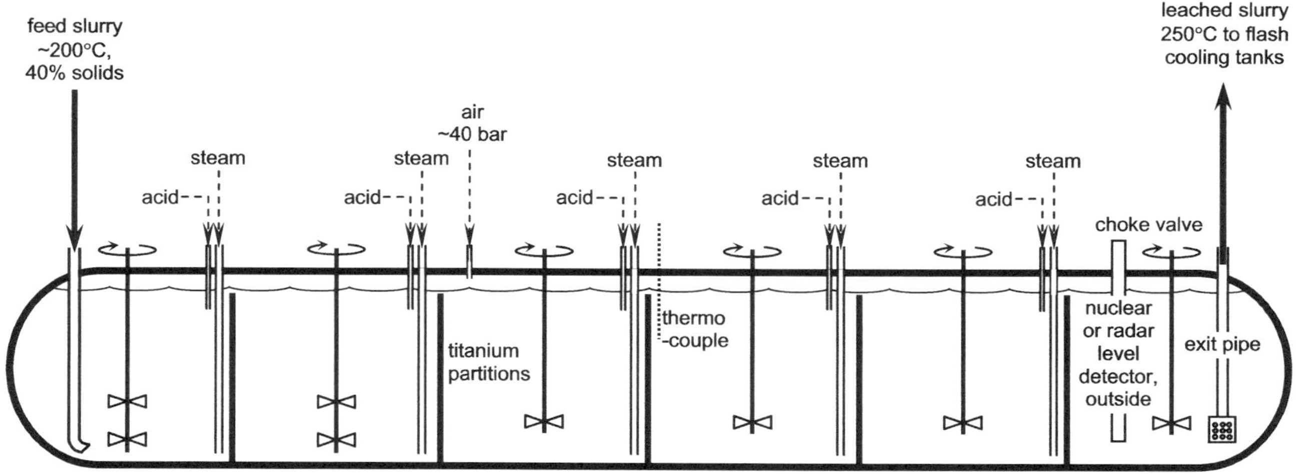

FIGURE 11.2 High-temperature sulfuric acid laterite leaching autoclave. The leaching takes place in a 0.1-m-thick steel, titanium clad (0.008 m) autoclave. Hot laterite/water slurry enters on the left, is leached in six stirred compartments and departs on the right. About 95% of the nickel and cobalt in the feed departs in pregnant solution, while most of the impurities depart in unleached and precipitated solids. Nickel laterite leach autoclaves typically operate at about 250 °C and 40 bar pressure. This autoclave is 29 m long and 4.6 m diameter. It was designed to treat about 2500 tonnes of concentrate per day. Autoclave fabrication and materials of construction are discussed by Banker (2009) and Laermans and Van Roy (2006).

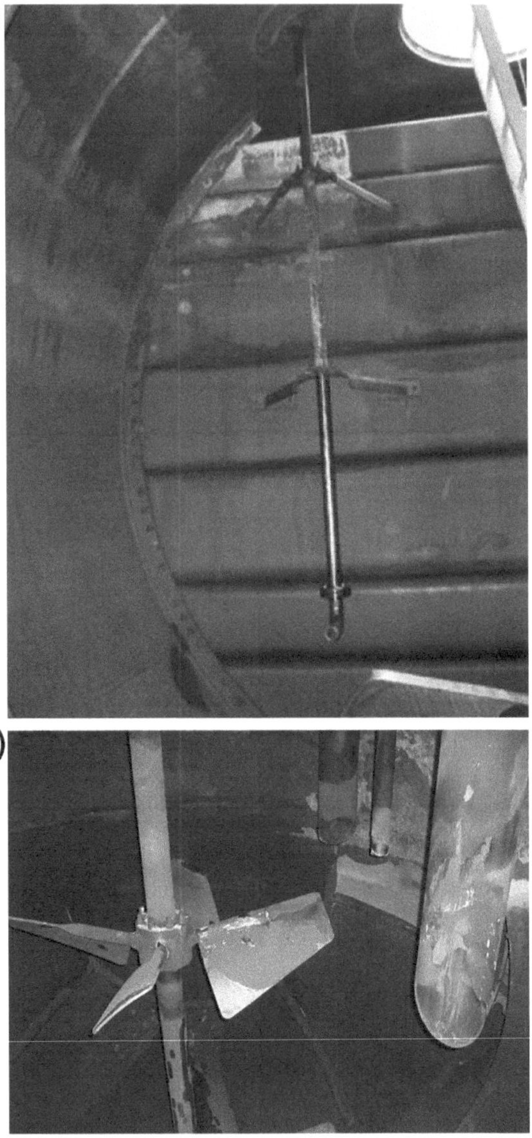

FIGURE 11.3 (a) Interior of high-temperature leaching autoclave. The circular wall is shown on the left. The steam inlet pipe is in the middle and is attached to the compartment partition. Access through the agitator nozzle is shown at the top. All slurry flow is over partitions. The inside diameter is ~4.6 m. (b) Feed end of a high-temperature leaching autoclave. The impeller, slurry input pipes (large) and acid input pipe (small) are shown. Note the hematite scale on the equipment surfaces. The exposed parts are all made of titanium except the inside of the acid input pipe, which is tantalum-lined. *Source: Photograph courtesy of David White.*

FIGURE 11.4 High-temperature sulfuric acid leaching autoclave moving toward laterite leach plant in Ambatovy, Madagascar. Note that almost all access into the autoclave is through the top.

11.3. CHEMISTRY

11.3.1. Leach Feed

The composition of the feed is given in Table 11.1. The feed typically contains small amounts of nickel and cobalt and large amounts of other elements, especially iron. The nickel and cobalt contents of four industrial operations are given in Table 11.2.

11.3.2. Particle Size

Laterite particles are very small, mostly less than 1000 μm, some as small as 5 μm (Carlson & Simons, 1961; Collins *et al.*, 2009). Small particles provide large solid/liquid interfacial areas for leaching, giving rapid, near-complete dissolution of nickel and cobalt in the autoclave.

11.3.3. Chemical Reactions for the Leaching of Nickel and Cobalt

Nickel and cobalt substitute for iron in the goethite (FeOOH) mineral lattice up to a maximum of about 6% (McDonald & Whittington, 2008). The charge balance in the goethite is maintained by the substitution of OH^- ions for O^{2-} so that the nickel cluster in the goethite can be represented by $Ni(OH)_2$.

TABLE 11.1 Composition of Leach Feeds*

Element	Concentrate feed, %	Main minerals (Chen, Dutrizac, Krause, & Osborne, 2004)
Ni	1.2–1.4	Goethite (Fe[Ni,Co]OOH)
Co	0.08–0.14	Goethite (Fe[Ni,Co]OOH)
Fe	30–50	Goethite (Fe[Ni,Co]OOH)
Si	4–20	Lizardite ($Mg_3Si_2O_5(OH)_4$), quartz (SiO_2), talc ($Mg_3Si_4O_{10}(OH)_2$)
Mg	0.5–4	Lizardite, talc
Al	2–5	Gibbsite ($Al(OH)_3$)
Cr	1–2	Spinel ($(Cr,Al,Mg,Fe)_3O_4$)
Mn	0.5–1	Oxide and oxyhydrate
Zn	Up to 0.05	
Cu	Up to 0.05	

**Collins et al., 2009; Kofluk & Freeman, 2006; Nakai et al., 2006; Tsuchida et al., 2004; Tuffrey et al., 2009*

Nickel and cobalt leaching may, therefore, be represented by the following reactions (Kofluk & Freeman, 2006):

$$\underset{\text{in goethite}}{Ni(OH)_2(s)} + \underset{\text{98\% in sulfuric acid}}{H_2SO_4(\ell)} \xrightarrow{250^\circ C} \underset{\text{pregnant solution}}{NiSO_4(aq) + 2H_2O(\ell)} \quad (11.1)$$

$$\underset{\text{in goethite}}{Co(OH)_2(s)} + \underset{\text{98\% in sulfuric acid}}{H_2SO_4(\ell)} \xrightarrow{250^\circ C} \underset{\text{pregnant solution}}{CoSO_4(aq) + 2H_2O(\ell)} \quad (11.2)$$

TABLE 11.2 Nickel and Cobalt Contents of the Feeds to Four Industrial HPAL Plants

	Feed solids composition	
Leach plant	% Ni	% Co
Murrin-Murrin, Australia	1.25	0.08
Moa Bay, Cuba	1.3	0.14
Ambatovy, Madagascar	1.2	0.12
Coral Bay, Philippines	1.3	0.09

Because nickel and cobalt are contained within the goethite structure, complete dissolution of the goethite particles is required to achieve high efficiencies. Complete dissolution of the other minerals found in laterites is not necessary to extract the nickel. For example, nickel is extracted from nontronite (clay) and nepouite (serpentine) by 'congruent' dissolution, that is, by extracting the iron and nickel while leaving the silica structure in place (McDonald & Whittington, 2008).

11.3.4. Behavior of Iron during Leaching

Most of the iron in the goethite also dissolves rapidly by the following reaction:

$$\underset{\text{goethite}}{2FeOOH(s)} + \underset{98\%\ \text{sulfuric acid}}{3H_2SO_4(\ell)} \xrightarrow{250^\circ C} \underset{\text{in pregnant solution}}{Fe_2(SO_4)_3(aq) + 4H_2O(\ell)} \quad (11.3)$$

The products of this reaction hydrolyze rapidly to precipitate hematite or jarosite at high temperatures. These precipitation reactions are given as follows:

$$Fe_2(SO_4)_3(aq) + 3H_2O(\ell) \xrightarrow{250^\circ C} \underset{\text{hematite}}{Fe_2O_3(s)} + 3H_2SO_4(aq) \quad (11.4)$$

$$3Fe_2(SO_4)_3(aq) + 14H_2O(\ell) \xrightarrow{250^\circ C} 2\underset{\text{hydronium jarosite}}{\left[H_3OFe_3(SO_4)_2(OH)_6\right](s)} + 5H_2SO_4(aq) \quad (11.5)$$

$$3Fe_2(SO_4)_3(aq) + \underset{\text{from sea water or brine in the leach slurry}}{Na_2SO_4(aq)} + 12H_2O(\ell) \xrightarrow{250^\circ C} 2\underset{\text{sodium jarosite}}{\left[NaFe_3(SO_4)_2(OH)_6\right](s)} + 6H_2SO_4(aq) \quad (11.6)$$

The net result is that less than 3% of the iron in the feed is dissolved in the leach solution (Carlson & Simons, 1961; Nakai, Kawata, Kyoda, & Tsuchida, 2006; White, Miller, & Napier, 2006). This greatly simplifies downstream purification of the leach solution.

Likewise, only small fractions of aluminum, chromium and silicon report to the leach solution. The extractions of various metals from the feed solids are given in Table 11.3.

11.3.5. Selectivity

The above discussion indicates that high-temperature sulfuric acid leaching is selective for nickel and cobalt. The selectivity is due to the high temperature. At high temperature, most of the iron precipitates, leaving nickel and cobalt in solution.

TABLE 11.3 Extraction of Nickel, Cobalt and Other Elements during High-Temperature (245–260°C) Sulfuric Acid Leaching of Laterite.
Nickel and cobalt extractions are 96%. Iron, aluminum, silicon and chromium extractions are much lower*

	% Extraction			
Elements	Ravensthorpe, Australia	Moa Bay, Cuba	Ambatovy, Madagascar	Coral Bay, Philippines
Ni	96	96	97	95
Co	96	95	97	95
Fe	2	0.4		<3
Al	1.5	11		
Mn	97	57		
Mg	88	60		
Cr	1.5	3		
Cu		100		
Zn		100		
Si		12		

Carlson & Simons, 1961; Collins et al., 2009; Nakai et al., 2006; White et al., 2006; Tuffrey et al., 2009

11.4. AUTOCLAVE OPERATION

11.4.1. Pressure

The equilibrium vapor pressure of the leach solution at 245°C is about 35 bar. High-temperature leaching must, therefore, be done in pressure vessels, referred to as autoclaves. All material is pumped into the autoclave at high pressure and the agitators rotate through high-pressure gland seals. The product slurry from the autoclave is let down to atmospheric pressure by allowing it to boil in 'flash tanks'. (Flashing of the autoclave is discussed in Appendix C.)

Industrially, laterite leach autoclaves are pressurized to 6 bar above the vapor pressure of the slurry. The extra pressure is obtained by pumping air into the headspace above the level of the slurry, as shown in Figure 11.2. As a result of the extra pressure, the autoclave operates below the boiling point of the slurry. In addition, the exit slurry can lose some pressure without boiling in piping and across valves.

The autoclaves that are used are made of 0.1 m thick steel that is internally clad with titanium that is 0.008 m thick. A laterite autoclave was shown in Figure 11.4.

The exit slurry flows up to an elevated location in the leach plant to facilitate subsequent downward flow under gravity. The elevated flash tanks are shown in Figure 11.1.

11.4.2. Steady-State Operation

Steady-state autoclave operation consists of the following actions:

(a) pumping the feed slurry at a temperature of about 200°C into the autoclave at a prescribed rate;
(b) injecting concentrated sulfuric acid into the autoclave so that the exit slurry contains about 30 g/L H_2SO_4. (The acid consumption is about 0.3 tonnes H_2SO_4 per tonne of concentrate feed or about 25 tonnes H_2SO_4 per tonne of nickel product.);
(c) injecting steam at 45 bar to maintain the temperature at 250°C in the autoclave;
(d) maintaining the pressure at 6 bar above the vapor pressure of the slurry; and,
(e) allowing product slurry to exit at a rate that maintains a constant slurry level in the autoclave, which is controlled by opening and closing the choke valve on the exit.

11.4.3. Operating Measurements

The principal measurements used for operating the autoclave are the slurry temperature in each compartment, the slurry level and the concentration of sulfuric acid in the exit slurry.

Slurry temperature is measured with thermocouples in each compartment (enclosed in titanium tubes). The temperature measurements are used to automatically control the rate of steam injection into each compartment.

Slurry level in the last compartment is measured with a nuclear or radar level detector from outside the autoclave. The level measurement is used to manipulate the choke valve on the exit line so as to maintain the slurry level in the autoclave.

The concentration of sulfuric acid is measured with a pH sensor in a cooled sample line after flash cooling. The pH measurement is used to control the rate of injection of acid into the autoclave.

11.4.4. Formation of Scale by Iron Precipitates

Scale tends to form on the autoclave walls. This scale consists mainly of iron precipitates that are formed by the reactions given in Equations (11.4) through (11.6).

At Moa Bay, up to 0.1 m of scale builds up in 3–4 months, necessitating autoclave shutdowns for removing scale from the autoclave walls using jackhammers.

However, the Coral Bay plant reports little or no scale formation as long as the aluminum, magnesium and sodium concentrations in the feed slurry are kept at low levels. This is achieved by blending ores and using 'soft' water, that is, it is low in magnesium and calcium (Tsuchida, Ozaki, Nakai, & Kobayashi, 2004). The use of soft water also minimizes scaling in the slurry pre-heaters.

11.5. PROCESS APPRAISAL

Autoclave leaching takes place under hot, highly corrosive conditions.

Two laterite leach plants have excellent production records – Moa Bay and Coral Bay (Tuffrey, Chalkley, Collins, & Iglesias, 2009; Nakai *et al.*, 2006). One other plant, Murrin-Murrin in Australia, has had difficulties, but it is still operating (Canterford, 2009).

Cawse and Bulong both had significant technical difficulties that were, on the whole, overcome (Donegan, 2006). Ravensthorpe was closed after less than a year of operation. It is due to restart operations in 2011 under the new ownership of First Quantum Minerals.

New plants have started up in Goro, New Caledonia and Ambatovy, Madagascar (Collins *et al.*, 2009; Finch, 2010).

A major problem with a significant number of recent laterite leaching projects has been financial: they exceeded their construction budgets. For example, Ravensthorpe cost twice as much to construct as was originally planned.

Leaching in chloride solutions has been mentioned as an alternative process to leaching in sulfuric acid (Steyl, Pelser, & Smit, 2008). However, this seems unlikely to be implemented commercially in the near future (Canterford, 2009).

11.6. SUMMARY

About 100 000 tonnes per year of nickel and 10 000 tonnes per year of cobalt are produced via hot sulfuric acid leaching of limonitic laterite. Nickel and cobalt extractions into solution are typically 95% at a final sulfuric acid concentration of about 30 g/L.

Steel autoclaves, 0.1 m thick, internally clad with titanium, 0.008 m thick, are used.

The advantages of the high leach temperature are as follows:

(a) most of the nickel and cobalt in the feed dissolves and leaves the autoclave in nickel- and cobalt-rich pregnant solution; and,
(b) most of the impurities in the feed leave the autoclave in nickel- and cobalt-lean precipitated solids.

REFERENCES

Banker, J. G. (2009). Detaclad explosion clad for autoclaves and vessels. In J. J. Budac, R. Fraser & I. Mihaylov, et al. (Eds.), *Hydrometallurgy of nickel and cobalt 2009* (pp. 181–193). CIM.

Canterford, J. H. (2009). Acid leaching of laterites in Australia: Where have we been and where are we going? In J. J. Budac, R. Fraser & I. Mihaylov, et al. (Eds.), *Hydrometallurgy of nickel and cobalt 2009* (pp. 511–522) CIM.

Carlson, E. T., & Simons, C. S. (1961). Pressure leaching of nickeliferous laterites with sulfuric acid. In P. E. Queneau (Ed.), *Extractive metallurgy of copper, nickel and cobalt* (pp. 363–397). AIME.

Chen, T. T., Dutrizac, J. E., Krause, E., & Osborne, R. (2004). Mineralogical characterization of nickel laterites from New Caledonia and Indonesia. In W. P. Imrie, D. M. Lane & S. C. C. Barnett, et al. (Eds.), *International laterite nickel symposium – 2004* (pp. 79–99). TMS.

Collins, M. J., Buban, K. R., Holloway, P. C., Masters, et al. (2009). Ambatovy laterite ore preparation plant and high pressure acid leach pilot plant operation. In J. J. Budac, R. Fraser & I. Mihaylov, et al. (Eds.), *Hydrometallurgy of nickel and cobalt 2009* (pp. 499–510). CIM.

Donegan, S. (2006). Direct solvent extraction of nickel and Bulong operations. *Minerals Engineering, 19*, 1234–1245.

Finch, C. (2010). Goro start-up partly stopped after acid spill. *Reuters*. 26 April 2010.

Geho. (2011). Piston diaphragm slurry pumps. www.weirminerals.com/ Accessed 26.05.11.

Kofluk, R. P., & Freeman, G. K. W. (2006). Iron control in the Moa Bay laterite operation. In J. E. Dutrizac & P. A. Riveros (Eds.), *Iron control technologies* (pp. 573–589). CIM.

Laermans, J., & Van Roy, P. (2006). Large titanium clad pressure vessels: Design, manufacture, and fabrication issues. In *Alta 2006 nickel/cobalt conference proceedings*. ALTA Metallurgical Services.

McDonald, R. D., & Whittington, B. I. (2008). Atmospheric leaching of nickel laterites review Part I. Sulphuric acid technologies. *Hydrometallurgy, 91*, 35–55.

Nakai, O., Kawata, M., Kyoda, Y., & Tsuchida, N. (2006). Commissioning of Coral Bay nickel project. In *Alta 2006 nickel/cobalt conference proceedings*. ALTA Metallurgical Services.

Steyl, J. D. T., Pelser, M., & Smit, J. T. (2008). Atmospheric leach process for nickel laterite ores. In C. A. Young, P. R. Taylor, C. G. Anderson & Y. Choi (Eds.), *Hydrometallurgy 2008, proceedings of the sixth international symposium* (pp. 570–579). SME.

Tsuchida, N., Ozaki, Y., Nakai, O., & Kobayashi, H. (2004). Development of process design for Coral Bay nickel project. In W. P. Imrie, D. M. Lane & S. C. C. Barnett et al. (Eds.), *International laterite nickel symposium 2004, process development for prospective projects* (pp. 152–160). TMS.

Tuffrey, N. E., Chalkley, M. E., Collins, M. J., & Iglesias, C. (2009). The effect of magnesium on HPAL – comparison of Sherritt laboratory studies and Moa plant operating data. In J. J. Budac, R. Fraser & I. Mihaylov (Eds.), *Hydrometallurgy of nickel and cobalt 2009* (pp. 421–432). CIM.

White, D. T., Miller, M. J., & Napier, A. C. (2006). Impurity disposition and control in the Ravensthorpe acid leaching process. In J. E. Dutrizac & P. A. Riveros (Eds.), *Iron control technologies* (pp. 591–609). CIM.

SUGGESTED READING

Budac, J. J., Fraser, R. & Mihaylov, I., et al. (Eds.). (2009). *Hydrometallurgy of nickel and cobalt 2009*. CIM.

Carlson, E. T., & Simons, C. S. (1961). Pressure leaching of nickeliferous laterites with sulfuric acid. In P. E. Queneau (Ed.), *Extractive metallurgy of copper, nickel and cobalt* (pp. 363–397). AIME.

Dutrizac, J. E. & Riveros, P. A. (Eds.). (2006). *Iron control technologies*. Metallurgical Society of CIM.

Imrie, W. P., Lane, D. M. & Barnett, S. C. C., et al. (Eds.). (2004). *International laterite nickel symposium – 2004*. TMS.

McDonald, R. (2009). The importance of mineralogy to the high pressure acid leaching of nickel laterites. In J. J. Budac, R. Fraser & I. Mihaylov, et al. (Eds.), *Hydrometallurgy of nickel and cobalt 2009* (pp. 485–498). CIM.

Chapter 12

Precipitation Of Nickel–Cobalt Sulfide

The leaching of laterite in an autoclave at about 250°C was described in Chapter 11. The pregnant solution in the slurry leaving these autoclaves contains approximately 6 g/L Ni and 0.5 g/L Co.

Currently, most industrial operations recover the nickel and cobalt from the autoclave leach solution by the precipitation of mixed nickel–cobalt sulfides. This precipitate is then refined to high-purity metal.

This chapter describes the reasons why this is done and the methods used to achieve results on an industrial scale.

12.1. REASONS FOR MAKING A MIXED-SULFIDE PRECIPITATE

An alternative processing route (to that of sulfide precipitation) for producing nickel and cobalt from the laterite pregnant solution is to (i) isolate nickel and cobalt into two separate solutions by solvent extraction; and then to, (ii) produce high-purity nickel and cobalt from the two solutions by electrowinning or by hydrogen reduction.

A process route using solvent extraction was implemented on an industrial scale at Bulong in Western Australia. This process was plagued by high levels of impurities, especially calcium, in the autoclave exit solution that impacted negatively on the solvent extraction (Donegan, 2006).

The nickel and cobalt in the leach solution can also be recovered by the precipitation of a mixed hydroxide product, which is then refined to high-purity metal. The processing of cobalt, in particular, uses this processing option to make a transportable intermediate.

The precipitation of a mixed nickel–cobalt sulfide has been used from the earliest days of laterite leaching and is still being installed today (Collins *et al.*, 2009). The essence of this process is that the resulting precipitates contain few impurities, particularly calcium, and are readily leached locally or at a distant refinery to make pure, concentrated solutions of nickel and cobalt suitable for production of high-purity metal (Molina, 2009).

The main advantage of sulfide precipitation over hydroxide precipitation is that the sulfide precipitates have lower levels of impurities, such as

Extractive Metallurgy of Nickel, Cobalt and Platinum-Group Metals. DOI: 10.1016/B978-0-08-096809-4.10012-7

manganese. In addition, the contained water is lower than hydroxide precipitates, which makes it more economical to transport sulfides to distant refineries.

12.2. FLOWSHEET

The flowsheet for the precipitation of the mixed nickel–cobalt sulfide from the hot slurry leaving the autoclave was shown in Figure 10.1. The process is continuous. It entails the following steps:

(a) cooling the slurry;
(b) neutralization of the slurry;
(c) solid/liquid separation and residue washing;
(d) further neutralization of the solution to remove impurities like aluminum, chromium, iron and silicon by precipitation together with their associated solid/liquid separation;
(e) removal of zinc and copper by sulfide precipitation; and finally,
(f) precipitation of the mixed nickel–cobalt sulfide.

The product of this process, the mixed nickel–cobalt sulfide precipitate, is then sent to a local or distant refinery for releaching and metal production.

The details of various industrial operations are given in Tables 12.1–12.3.

Steps (a) and (c) in this process are discussed in Appendices D and E. Steps (b), (d), (e) and (f) are discussed here.

12.3. AUTOCLAVE EXIT SLURRY NEUTRALIZATION

The autoclave slurry, which has been cooled from 250°C to the ambient boiling point, contains between 30 and 50 g/L H_2SO_4. It is continuously neutralized with limestone to a pH of about 1 to prevent downstream corrosion. The neutralization reaction is given as follows:

$$\underset{\text{in autoclave exit slurry}}{H_2SO_4(aq)} + \underset{\text{in limestone slurry}}{CaCO_3(s)} + H_2O(\ell) \xrightarrow{100^\circ C} \underset{\text{precipitated gypsum}}{CaSO_4 \cdot 2H_2O(s)} + CO_2(g) \quad (12.1)$$

The neutralization is performed in a stirred vertical tank with a residence time of about 48 minutes, and then the flocculants are added. A portion of the leach residue is recycled to the tank to ensure sufficient growth of gypsum particles so that they can settle rapidly in the subsequent counter-current decantation circuit. The rate of the addition of limestone is controlled to prevent excessive production of carbon dioxide gas.

The neutralized slurry is transferred to solid/liquid separation and residue washing in the counter-current decantation circuit. The solution from the counter-current decantation circuit is pumped to solution reneutralization.

TABLE 12.1 Compositions of the Leach Solution Feed to Industrial Nickel–Cobalt Sulfide Precipitation Circuits.
The nickel concentration is controlled by dilution

	Concentration, g/L		
Component	Moa Bay, Cuba	Coral Bay, Philippines	Ambatovy, Madagascar (pilot-plant results)
Ni	4.4[a]	5	5
Co	0.4	0.4	0.3
Fe	0.6		0.5
SO_4^{2-}	27		1.6
Cu	0.008[a]		
Zn	0.1	0.001	
Cr	0.2		0.07
Ca	0.1		
Al	1.6		1.6
Si			0.03
Mg	1.93		0.9
Mn	1.4		2.9

[a]Ni + Co concentration controlled to < 4.8 g/L to avoid sulfide pellet formation. Cu concentration controlled to less than 0.008 g/L to avoid excessively fine precipitate (Chalkley & Toirac, 1997).

12.4. SOLUTION RENEUTRALIZATION

The pH of the solution from the counter-current decantation circuit is increased to about 3.3 in order to precipitate impurities, such as aluminum, chromium, iron and silicon. This is achieved by adding limestone. The precipitation of iron occurs by the following reaction:

$$\underset{\text{solution from solid/liquid separation}}{Fe_2(SO_4)_3(aq)} + \underset{\text{in limestone slurry}}{3CaCO_3(s)} + 9H_2O(\ell) \xrightarrow{60^\circ C} \underset{\text{precipitates}}{2Fe(OH)_3(s) + 3(CaSO_4\cdot 2H_2O)(s)} + 3CO_2(g) \tag{12.2}$$

The neutralization is performed in a stirred vertical tank. Flocculants and recycle leach residue are added so that the precipitates that are formed will

TABLE 12.2 Compositions of Nickel–Cobalt Sulfide Precipitates from Industrial Laterite Leach Plants.[a]

The similarity of the nickel and sulfur contents are noteworthy. Particle sizes are 10–50 μm

	Concentration, %		
Element	Moa Bay, Cuba	Coral Bay, Philippines	Ambatovy, Madagascar (pilot-plant results)
Ni	55	57	55
Co	5	4.5	4.2
Fe	1.3	0.4	0.3
S	35	35	34
Cu	<0.1		0.5
Zn	0.9	0.01	1.6
Cr	0.4	0.01	0.1
Ca		0.004	
Al		<0.1	0.2
Si			<0.01

[a]*Precipitate from the Murrin-Murrin (Australia) plant contains ~57% Ni, 5% Co and 36% S.*

settle rapidly in the subsequent thickening step. Air is often bubbled into the tank to ensure that iron is present mostly as Fe^{3+}.

The slurry is transferred to solid/liquid separation. The solution is pumped to zinc and copper removal.

12.5. REMOVAL OF ZINC AND COPPER FROM SOLUTION BY SULFIDE PRECIPITATION

An important advantage of sulfide precipitation is that solution impurities, such as magnesium, calcium, aluminum and manganese, are not precipitated with the nickel and cobalt (Kofluk & Freeman, 2006; Molina, 2009).

The impurities that pose the greatest challenges in the sulfide precipitation of nickel and cobalt are copper and zinc. These impurities have the following characteristics:

(a) they are always present in laterite ores;
(b) they dissolve during leaching; and,

TABLE 12.3 Nickel–Cobalt Sulfide Precipitate Production Details. Production Is Continuous.

Note the different temperatures and pressures[a]. Tsuchida *et al.*, (2004) report that a low operating temperature minimizes scale formation on the reactor walls. Extensive recycle of product precipitate (sometimes ground very fine) also minimizes scale formation. The new ambatovy plant plans to recycle about 3.5 tonnes of precipitate per tonne of final precipitate product

Location	Murrin-Murrin, Australia, Hayward (2008)	Moa Bay, Cuba	Coral Bay, Philippines
Startup date	1998	1959	2005
Precipitation vessels	Four stirred vertical tanks	Four stirred horizontal autoclaves	Two sulfurisation tanks
Precipitation reagent	$H_2S(g)$	$H_2S(g)$	$H_2S(g)$
Temperature, °C	95	120	80
pH		2–2.5	3–3.5
H_2S pressure, bar	1	10	2
Solution residence time, h		~1	~1
Feed solution concentrations, g/L			
Ni		4.8 (maximum)	5
Co		0.5	0.4
Precipitation efficiency, %			
Ni		98.5–99.5	>99
Co		98–99	
Precipitation seed	Recycle precipitation product	Recycle precipitation product	Recycle ground precipitation product, 5 μm median diameter
Recycle extent, tonnes per tonne of final product	Some	2	Up to 4

(*Continued*)

TABLE 12.3 Nickel–Cobalt Sulfide Precipitate Production Details. Production Is Continuous.
Note the different temperatures and pressures[a]. Tsuchida *et al.*, (2004) report that a low operating temperature minimizes scale formation on the reactor walls. Extensive recycle of product precipitate (sometimes ground very fine) also minimizes scale formation. The new ambatovy plant plans to recycle about 3.5 tonnes of precipitate per tonne of final precipitate product—**cont'd**

Location	Murrin-Murrin, Australia, Hayward (2008)	Moa Bay, Cuba	Coral Bay, Philippines
Reaction vessel exit method		Flash tank	Flash tank
Precipitate composition, %			
Ni	57	55	57
Co	5	5	4.5
Fe		~1	0.4
S		35	
Precipitate diameter, μm		~50	~10
Precipitation rate, total, dry t/d	~80	~150	60

[a] *The new Toamasina plant plans to precipitate in a stirred vertical tank at 105°C and 2-bar H_2S pressure.*

(c) they are harmful to nickel–cobalt sulfide precipitation or to subsequent nickel and cobalt refining (Kofluk & Freeman, 2006; Molina, 2009; Nakai, Kawata, Kyoda, & Tsuchida, 2006).

Fortunately, zinc and copper can be removed from the leach solution (with negligible nickel and cobalt precipitation) by adding hydrogen sulfide to the solution at low pressure and temperature (Molina, 2009). This is because the solubility products for zinc and copper sulfides are lower than those for nickel or cobalt sulfide.

Operating conditions for zinc and copper removal (Nakai *et al.*, 2006; Tsuchida, Ozaki, Nakai, & Kobayashi, 2004) are given in Table 12.4.

The concentration of zinc in solution after this treatment is less than 0.001 g/L. The concentration of copper is even lower.

TABLE 12.4 Typical Conditions for the Precipitation of Zinc and Copper from Solution Prior to the Precipitation of Nickel and Cobalt

Variable	Value
Temperature	<60°C
pH	3.0–3.4
$H_2S(g)$ pressure	<0.1 bar
$H_2S(g)$ addition method	From gas space above agitated solution in a baffle tank
Reaction time	~60 min
Product	Mixed Zn–Cu sulfide precipitate

The nickel and cobalt losses during the precipitation of zinc and copper from solution as sulfides are less than 0.1% and less than 0.3%, respectively.

The resulting slurry is filtered. The zinc and copper sulfide precipitate is discarded. The filtrate is sent to nickel and cobalt precipitation.

12.6. PRECIPITATION OF NICKEL–COBALT SULFIDE

Industrially, nickel–cobalt sulfide precipitates are made by contacting purified pregnant leach solution with hydrogen sulfide gas. A sketch of a precipitation vessel is given in Figure 12.1.

12.6.1. Chemistry of Nickel–Cobalt Sulfide Precipitation

The precipitation reactions are as follows:

$$\underset{\substack{\text{in purified}\\ \text{pregnant solution}}}{NiSO_4(aq)} + H_2S(g) \xrightarrow{80-120^\circ C} \underset{\substack{\text{in sulfide}\\ \text{precipitate}}}{NiS(s)} + \underset{\substack{\text{sulfuric acid to}\\ \text{neutralization}}}{H_2SO_4(aq)} \quad (12.3)$$

$$\underset{\substack{\text{in purified}\\ \text{pregnant solution}}}{CoSO_4(aq)} + H_2S(g) \xrightarrow{80-120^\circ C} \underset{\substack{\text{in sulfide}\\ \text{precipitate}}}{CoS(s)} + \underset{\substack{\text{sulfuric acid}\\ \text{neutralization}}}{H_2SO_4(aq)} \quad (12.4)$$

As shown in Table 12.1, the purified feed solutions to the sulfide precipitation typically contain 5 g/L Ni and 0.4 g/L Co. These solutions are less concentrated in nickel and cobalt than the solution leaving the autoclave due to the addition of water during slurry washing.

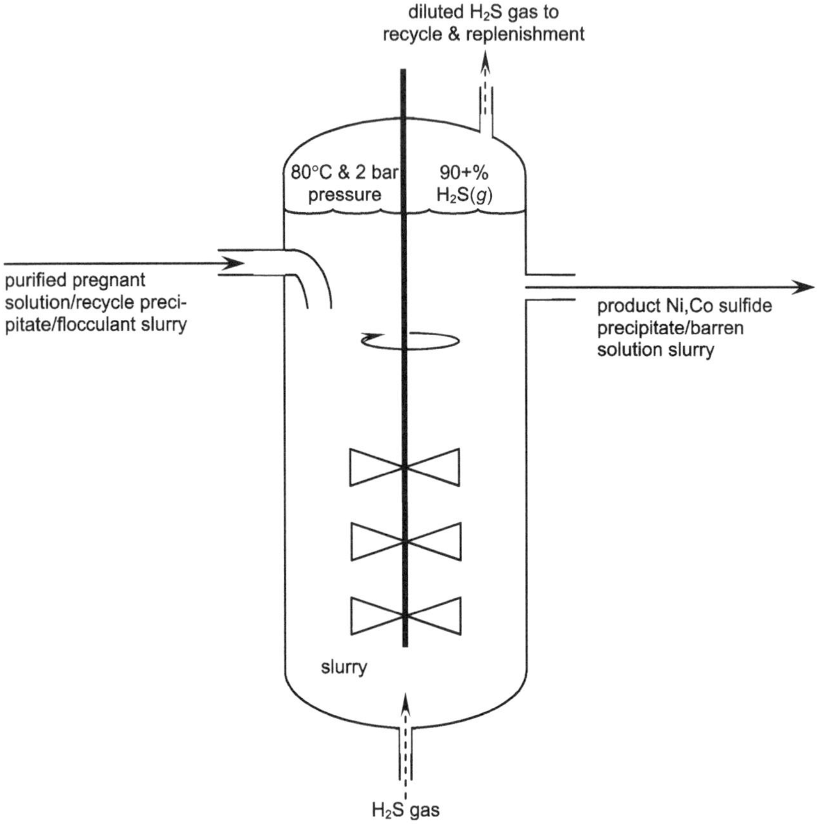

FIGURE 12.1 Sketch of nickel–cobalt sulfide precipitation vessel. The process is continuous. The product slurry is settled and filtered. The barren solution is used for washing of leaching residue or is discarded. The nickel–cobalt sulfide precipitate is washed and sent to an adjacent refinery or washed, dried and bagged then sent to a distant refinery for metal production. The input solution typically contains 5 g/L Ni and 0.4 g/L Co. The product precipitate typically contains 55% Ni and 5% Co. Precipitation is also done in horizontal autoclaves.

The product composition is given in Table 12.2. It typically contains 56% Ni, 5% Co, and 1% Fe. The remainder of the material is sulfur.

About 99% of the nickel and the cobalt in the leach solution are precipitated.

12.6.2. Industrial Practice

A survey of the industrial practice of nickel and cobalt sulfide precipitation is given in Table 12.3. This survey indicates that precipitation temperatures vary between 80 and 120°C and hydrogen sulfide pressures between 1 and 10 bar. The residence time in the precipitation vessels is typically between one and two hours.

12.6.3. Gas–Liquid Contact

The original sulfide precipitation plant is situated at Moa Bay, Cuba. This plant operates horizontal autoclaves at 120°C and 10-bar H_2S pressure. More recent plants operate at between 80 and 105°C and H_2S partial pressures between 2 and 3 bar in vertical pressure vessels, as shown in Figure 12.1. This is a cheaper, safer option (Molina, 2009) that also minimizes scale formation on the reactor walls (Nakai *et al.*, 2006; Tsuchida *et al.*, 2004). The new plant at Ambatovy, Madagascar, plans to operate this way (Collins *et al.*, 2005).

In all cases, the solution is contacted with hydrogen sulfide gas by introducing the gas below the impeller of the agitator, as shown in Figure 12.1, so that the gas is dispersed as small bubbles into the solution. This ensures that the gas–liquid contact and associated mass transfer is optimal.

12.6.4. Precipitate Recycle

A critical aspect of nickel–cobalt precipitation is the recycle of the precipitated product as 'seed'. This seed provides deposition sites for the newly forming precipitate. The result of seeding is that it (i) accelerates the precipitation process; and, (ii) minimizes scaling (precipitation) on the reactor walls. Up to 4 tonnes of precipitate are recycled per tonne of plant product.

Not only is precipitate seed recycled at Coral Bay, but the recycle particles are also ground to a size of about 5 μm in order to prevent the formation of scale (Nakai *et al.*, 2006; Tsuchida *et al.*, 2004). The grinding maintains the thickness of the scale at about 1 mm.

12.7. PRODUCT DESTINATION

Two producers of mixed sulfide precipitate send the precipitate to distant refineries for refining and metal production. They are as: (i) Moa Bay, Cuba – which sends its precipitate to Fort Saskatchewan, Canada; and, (ii) Coral Bay, Philippines – which sends its precipitate to Niihama, Japan.

These plants settle, filter, wash and bag their precipitate for ocean and rail shipment to these distant refineries.

The Murrin-Murrin plant in Australia refines the precipitate on-site. The new Ambatovy plant in Madagascar will also refine the precipitate on-site (Collins *et al.*, 2009). These plants settle, filter and wash the precipitate and then transfer it directly to the adjacent refinery.

12.8. APPRAISAL

Production of concentrated solutions of pure nickel and cobalt by precipitation of mixed nickel–cobalt sulfide and later releaching it in a refinery seems awkward and costly. However, there is no denying its efficacy.

Solvent extraction would seem to be a better process, but high impurity levels, particularly calcium, in autoclave exit solutions have delayed its adoption. However, the operation at Goro, New Caledonia, planned to use solvent extraction to (i) extract nickel and cobalt from its purified pregnant solution; and then, (ii) produce separate nickel and cobalt solutions from which it will make NiO and $CoCO_3$ (Okita, Singhal, & Perraud, 2006). Due to problems in commissioning, Goro currently produces a mixed hydroxide precipitate that is processed at Yabulu, Queensland.

12.9. SUMMARY

Most currently operating laterite leaching plants precipitate nickel and cobalt from solution as a mixed nickel–cobalt sulfide. This sulfide typically contains 56% Ni, 5% Co, and 0.5%–1% Fe, with the remaining mass made up of sulfur.

The mixed nickel–cobalt precipitate is produced by contacting purified pregnant leach solution with hydrogen sulfide gas at temperatures of 80–120°C and partial pressures of hydrogen sulfide of between 2 and 10 bar.

The recovery of the nickel and the cobalt to precipitate is approximately 99%.

Rapid precipitation is promoted by recycling large quantities of product precipitate (sometimes finely ground) to the precipitation tank. This also prevents scale formation on the reactor walls, extending campaign life and minimizing maintenance costs.

The mixed-sulfide precipitate is an intermediate product that is redissolved to make concentrated solutions of nickel and cobalt suitable for high-purity metal production, which is discussed in Chapter 23.

REFERENCES

Chalkley, M. E., & Toirac, I. L. (1997). The acid pressure leach process for nickel and cobalt laterite. In W. C. Cooper & I. Mihaylov (Eds.), *Nickel/cobalt 97, Vol. I, Hydrometallurgy and refining of nickel and cobalt* (pp. 341–353). CIM.

Collins, M. J., Barta, L. A., Buban, K. R., et al. (2005). Process development by Dynatec for the Ambatovy nickel project. *CIM Bulletin, 98*, 90.

Collins, M. J., Buban, K. R., Holloway, P. C., et al. (2009). Ambatovy laterite ore preparation plant and high pressure acid leach pilot plant operation. In J. J. Budac, R. Fraser & I. Mihaylov, et al. (Eds.), *Hydrometallurgy of nickel and cobalt 2009* (pp. 499–510). CIM.

Donegan, S. (2006). Direct solvent extraction of nickel and Bulong operations. *Minerals Engineering, 19*, 1234–1245.

Hayward, K. (2008). Murrin Murrin leads the way. *Sulfuric Acid Today, 14*, 7–10.

Kofluk, R. P., & Freeman, G. K. W. (2006). Iron control in the Moa Bay laterite operation. In J. E. Dutrizac & P. A. Riveros (Eds.), *Iron control technologies* (pp. 573–589). CIM.

Molina, N. (2009). Nickel and cobalt sulphide precipitation, a proven method of selective metal precipitation in laterite process flowsheets. In J. J. Budac, R. Fraser & I. Mihaylov, et al. (Eds.), *Hydrometallurgy of nickel and cobalt 2009* (pp. 271–281). CIM.

Nakai, O., Kawata, M., Kyoda, Y., & Tsuchida, N. (2006). Commissioning of Coral Bay nickel project. In *ALTA 2006 nickel/cobalt conference proceedings*. ALTA Metallurgical Services.

Okita, Y., Singhal, A., & Perraud, J.-J. (2006). Iron control in the Goro nickel process. In J. E. Dutrizac & P. A. Riveros (Eds.), *Iron control technologies* (pp. 573–589). CIM.

Tsuchida, N., Ozaki, Y., Nakai, O., & Kobayashi, H. (2004). Development of process design for Coral Bay nickel project. In W. P. Imrie, D. M. Lane & S. C. C Barnett et al. (Eds.), *International laterite nickel symposium 2004, process development for prospective projects* (pp. 152–160). TMS.

SUGGESTED READING

Lewis, A. E. (2010). Review of metal sulfide precipitation. *Hydrometallurgy, 104*, 222–234.

Molina, N. (2009). Nickel and cobalt sulphide precipitation, a proven method of selective metal precipitation in laterite process flowsheets. In J. J. Budac, R. Fraser, I. Mihaylov, V. G. Papangelakis & D. J. Robinson (Eds.), *Hydrometallurgy of nickel and cobalt 2009* (pp. 271–281). CIM.

Matos, R. R. (1997). Industrial experience with the Ni/Co sulphide precipitation process. In W. C. Cooper & I. Mihaylov (Eds.), *Nickel/cobalt 97, Vol. I, Hydrometallurgy and refining of nickel and cobalt* (pp. 371–378). CIM.

Chapter 13

Extraction of Nickel and Cobalt from Sulfide Ores

About half the global production of primary nickel is extracted from the sulfide ores.

This chapter describes nickel sulfide ores and provides an overview of the methods used to extract nickel and cobalt from these ores. These extraction processes are described in detail in subsequent chapters in the following sequence:

(a) the production of nickel-rich sulfide concentrates is discussed in Chapters 14 through 16;
(b) the smelting and converting of the nickel concentrate to molten sulfide matte is discussed in Chapters 17 through 19; and,
(c) the refining of the sulfide matte to high-purity nickel metal and other products is discussed in Chapters 21 through 27.

13.1. NICKEL SULFIDE ORES

Nickel sulfide deposits occur as massive sulfides or as disseminated ores. Massive sulfides have a mineral sulfide content of 90%–95%. Disseminated sulfides are a mixture of sulfides and siliceous gangue rock.

The sulfide minerals found in nickel sulfide ores are given in Table 13.1.

Virtually, all the nickel in nickel sulfide ores occurs in the mineral pentlandite [$(Ni,Fe)_9S_8$]. Only small amounts of nickel in the ore occur as millerite [NiS], violerite [Ni_2FeS_4] and nickeliferrous pyrrhotite [$(Fe,Ni)_8S_9$].

The proportion of nickel and iron in pentlandite is variable. Pentlandite typically contains approximately 36% Ni, 30% Fe and 33% S (plus about 1% Co). Pentlandite always occurs with other sulfide minerals, mostly pyrrhotite [Fe_8S_9] and chalcopyrite [$CuFeS_2$]. These minerals are hosted by silicate and alumino-silicate minerals and their hydrated derivatives (Kerr, 2002). These silicate rock minerals are referred to as *gangue rock*.

Extractive Metallurgy of Nickel, Cobalt and Platinum-Group Metals. DOI: 10.1016/B978-0-08-096809-4.10013-9

TABLE 13.1 Minerals Found in Nickel Sulfide Ores*

Mineral name	Chemical formula	Nickel content, %
Pentlandite	$Ni_{4.5}Fe_{4.5}S_8$	34.2
Millerite	NiS	64.7
Heazlewoodite	Ni_3S_2	73.3
Polydymite	Ni_3S_4	57.9
Violarite	Ni_2FeS_4	38.9
Siegenite	$(Co,Ni)_3S_4$	28.9
Fletcherite	Ni_2CuS_4	75.9
Niccolite	$NiAs$	43.9
Maucherite	$Ni_{11}As_8$	51.9
Rammelsbergite	$NiAs_2$	35.4
Breithauptite	$NiSb$	32.5
Annabergite	$Ni_3As_2O_8 \cdot 8H_2O$	29.4
Pyrrhotite	$(Ni,Fe)_7S_8$	1-5
Chalcopyrite	$CuFeS_2$	
Magnetite	Fe_3O_4	
Cubanite	$CuFe_2S_3$	
Chromite	$(Mg,Fe)Cr_2O_4$	
Galena	PbS	
Sphalerite	ZnS	
Bornite	Cu_5FeS_4	
Mackinawite	$(Fe,Ni,Co)S$	
Valleriite	$Cu_3Fe_4S_7$	

* *Boldt & Queneau, 1967*

The mineralogical compositions of five ores are given in Table 13.2, and their chemical compositions are given in Table 13.3.

Besides nickel, pentlandite ores contain copper, cobalt, silver, gold and platinum-group elements (Pt, Pd, Rh, Ru, Ir) to a greater or lesser extent.

TABLE 13.2 Major Constituents of Five Nickel Sulfide Ores

	Ore composition, %			
Location	Pentlandite	Chalcopyrite	Pyrrhotite	Gangue rock
Kambalda, Australia	6	0.6	12 + (2% pyrite FeS_2)	79
Raglan, Canada	8	2.4	11	79
Sudbury, Canada (A)	3.6	4.3	23	70
Sudbury, Canada (B)	3–6	2–5	20–30	Remainder
Thompson, Canada	7	0.4	22	70

TABLE 13.3 Chemical Composition of the Ores Given in Table 13.2, with Two Additional Ores from Russia.
Note the large variation in copper content

	Ore composition, %		
Location	Ni	Co	Cu
Kambalda, Australia	2.2	0.06	0.2
Raglan, Canada	2.8	0.08	0.8
Sudbury, Canada (A)	1.3	0.04	1.5
Sudbury, Canada (B)	1.6	0.04	1.2
Thompson, Canada	2.5	0.07	0.1
Norilsk, Russia (average)[a]	1.7	0.06	3.2
Pechenga, Russia[a]	0.7	NA	0.3

[a]*Plus platinum-group metals and Au.*

13.2. EXTRACTION OF NICKEL AND COBALT FROM SULFIDE ORES

The main extraction process for treating nickel sulfide ores is shown in Figure 13.1. The process is broadly divided into three steps:

(a) the production of a nickel–cobalt concentrate;
(b) the production of nickel–cobalt converter matte; and,
(c) the refining of this converter matte into pure metals or chemicals of nickel and cobalt.

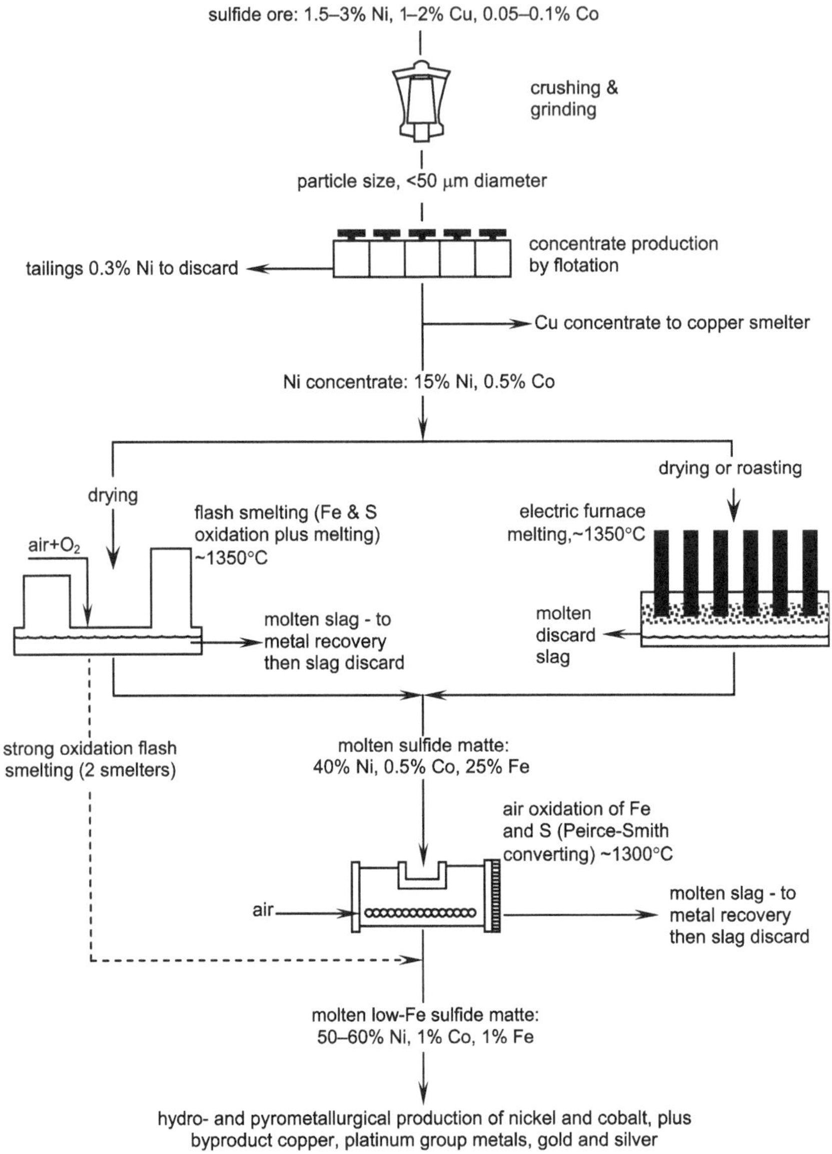

FIGURE 13.1 Main process steps for extracting nickel and cobalt from sulfide ores. Parallel lines indicate alternative processes. The dashed line indicates a developing process. The compositions are for ores from Sudbury, Canada.

13.2.1. Concentrate Production

Mined sulfide ores contain pentlandite, pyrrhotite, chalcopyrite and gangue rock. A high-grade nickel concentrate is produced from these ores by the following minerals-processing techniques:

(a) crushing and grinding the ores to liberate the nickel- and cobalt-bearing minerals from the other minerals and gangue;
(b) froth flotation to separate the nickel-, cobalt- and copper-bearing minerals from the gangue rock and pyrrhotite; and,
(c) froth flotation to separate the nickel- and cobalt-bearing minerals from the copper-bearing minerals.

These process steps result in a high-grade concentrate that is a suitable feed to a nickel–cobalt smelter.

13.2.2. Smelting and Converting

The nickel- and cobalt-rich concentrates are smelted and converted to form a converter matte that has a high nickel–cobalt content and a low iron content. As shown in Figure 13.1, there are two options for smelting and converting: (i) flash smelting; and, (ii) electric furnace smelting.

The smelting of nickel–cobalt sulfides involves the following three processes:

(a) the partial oxidation of the nickel–cobalt sulfide to nickel–cobalt oxide;
(b) the oxidation of iron sulfide to iron oxide and the dissolution of the iron oxide in the silicate slag; and,
(c) the melting of silicate gangue rock to form a silicate slag.

In the electric furnace process, there is a separate roaster for the oxidation of nickel and iron sulfides. The melting of these roasted products then occurs in an electric furnace. Two products are separated in this electric furnace: the furnace matte, which is rich in nickel–cobalt and lean in iron, and the silicate slag, which is rich in iron and lean in nickel and cobalt.

In the flash furnace process, the oxidation and melting reactions both occur in the furnace. The advantage of this process is that the heat of reaction from roasting is used to power the furnace. The disadvantage of this type of operation is that the losses of nickel to the slag are higher than in the electric furnace.

Both the flash furnace and the electric furnace produce a furnace matte that contains 40% Ni, 0.5% Co, and 25% Fe, with the remainder being sulfur. The furnace matte is converted, which is the oxidation of iron and sulfur to produce a converter matte that contains 50%–60% Ni, 1% Co, 1% Fe, and 20%–23% S.

There are two flash furnace installations where strong oxidizing conditions are used. This results in a furnace matte that is similar to the converter matte, thus by-passing the need for a converter.

The converter matte is refined, either using vapometallurgical or hydrometallurgical techniques.

13.2.3. Refining of Converter Mattes to Nickel and Cobalt Metal

Converter mattes are refined to pure nickel and cobalt either in a vapometallurgical refinery (carbonyl process) or in a hydrometallurgical refinery.

The basis for the carbonyl process is that carbon monoxide, CO, reacts preferentially with nickel to form nickel carbonyl, $Ni(CO)_4$, which is a gas, at about 50°C. This gas, which contains only nickel, can be decomposed to nickel metal at much higher temperatures, such as 240°C. The carbonyl process has been used for over a hundred years, and the extractive metallurgy and conditions are well established.

In contrast to the established nature of carbonyl refining, there is no standard flowsheet for the hydrometallurgical processing of converter mattes and nickel sulfides.

Broadly, the hydrometallurgical refining of nickel sulfides occurs in the following steps: (i) leaching; (ii) solvent extraction and one of either; (iii) electrowinning; or, (iv) hydrogen reduction. These topics are examined in separate chapters, starting with Chapter 24 on leaching. However, since this part-by-part comparison of the refining steps oversimplifies the integrated nature of hydrometallurgical operations, a whole-by-whole comparison of these refining processes is given in Chapter 23.

The product of the refining processes is high-purity nickel metal, suitable for applications.

13.3. HYDROMETALLURGICAL ALTERNATIVES TO MATTE SMELTING

An alternative to the smelting of nickel concentrates to matte, which is then refined in a hydrometallurgical refinery, is to treat all of the concentrate hydrometallurgically.

There are three different process chemistries that such a hydrometallurgical treatment facility can be based on: (i) ammoniacal; (ii) sulfate; and, (iii) chloride routes.

13.3.1. Ammoniacal Processes for Nickel Concentrates

Nickel concentrates were pressure leached in ammonia solutions using air in Fort Saskatchewan, Alberta from 1954–1987 and at Kwinana, Western Australia from 1970–1987. Both operations used the Sherritt process.

The mines supplying concentrate to the Fort Saskatchewan operation became depleted. This led to the processing of various feeds, such as nickel–copper mattes, and in 1987 exclusively to leaching of mixed nickel–cobalt sulfide precipitates from Moa Bay, Cuba (Kofluk and Freeman, 2006).

The Kwinana nickel refinery began processing concentrate in 1970. In 1974, low-iron matte from the Kalgoorlie nickel smelter was blended with concentrate feed. The proportion of matte was gradually increased until 1987 when the entire input became matte (Smith, 2004; Woodward & Bahri, 2007).

Both refineries could still process concentrate; however, this would probably lead to lower production rates and increased amounts of residue. Both of these refineries are now in built-up communities making residue disposal more challenging.

13.3.2. Sulfate Processes for Nickel Concentrates

Pentlandite concentrates can be processed in sulfate solutions, in a manner similar to the processing of mixed nickel–cobalt precipitates and converter mattes. However, the concentration of nickel in the concentrate is about 15%, whereas it is about 50% in converter mattes. This means that the rate of processing must be about three or four times more efficient in the hydrometallurgical process that treats concentrates than in a similar process that treats converter matte if the same size of process plant were to be used. Because of the need to minimize capital investment, attention has focused on increasing the rate of leaching.

The rate of leaching can be increased in two ways: (i) increase the intrinsic rate by adding a catalyst, such as chloride ions or increasing the temperature; and, (ii) decrease the particle size (Crundwell, 2005). Both of these strategies have been proposed for the leaching of nickel concentrates.

Vale plans to add hydrochloric acid to the leaching liquor in order to accelerate the rate of leaching for their Voisey's Bay operation (Stevens, Bishop, Singhal, Love, & Mihaylov, 2009). The pentlandite is leached from the gangue in sulfuric acid using oxygen. The construction of the Voisey's Bay operation at Long Harbour, Newfoundland, is well advanced. This process is described in more detail in the next section.

CESL have proposed the same strategy for their process (Jones, Mayhew, & O'Connor, 2009; Jones, Mayhew, Mean, & Neef, 2010).

The addition of hydrochloric acid complicates the processes, particularly the choice of the materials of construction. This may explain why industrial adoption has been slow.

Another process using sulfate solutions is Activox (Norilsk). This process adopts the second strategy mentioned above for increasing the rate of leaching, which is, decreasing the particle size. The process was demonstrated at Tati Nickel, in Botswana, but the full-scale operation was not built. The concentrate was ground very fine, to about 10 μm (Palmer & Johnson, 2005).

13.3.3. Chloride Processes for Nickel Concentrates

Outotec has developed an ambient pressure Cl_2-HCl-O_2 leaching step, called the HydroNic process, for pentlandite concentrate (Karonen, Tiihonen & Haavanlammi, 2009). It, too, has not yet been adopted industrially.

13.3.4. Disadvantages of Hydrometallurgical Processes for Concentrates

The most important disadvantage of hydrometallurgical options for processing nickel concentrates is the difficulty of recovering both the precious metals and the base metals. Smelting and converting are able to remove gangue, sulfur and iron from the concentrate while leaving all the base and precious metals in the matte. This simplifies the hydrometallurgical recovery of the nickel, copper, cobalt and precious metals from the matte.

13.4. VOISEY'S BAY PROCESS FOR LEACHING NICKEL CONCENTRATES

A schematic flowsheet for the Voisey's Bay process is shown in Figure 13.2 (Stevens *et al.*, 2009). The feed to this process is ground concentrate, which is contacted with a mixture of oxygen and chlorine gases from the nickel tankhouse in a series of stirred tank reactors. The step is called the chlorine preleach. The slurry is pumped into the pressure-leaching autoclave where the concentrate is dissolved using nickel spent electrolyte (sometimes called nickel anolyte). The anolyte is acidic and provides the acid required for leaching.

The discharge slurry from the autoclave goes through the counter-current decantation section, where undissolved and precipitated solids are separated from the solution. The solids are neutralized and transferred to disposal. The solution goes to impurity removal where iron and other contaminants are removed. After impurity removal, the solution is transferred to copper solvent extraction.

Copper is removed from the solution by solvent extraction using LIX 84. The stripped solution from solvent extraction is the feed to copper electrowinning, where high-purity copper is produced.

Cobalt is extracted from the solution by solvent extraction using CYANEX 272. Cobalt is electrowon from the strip solution from solvent extraction as cobalt rounds. The raffinate from cobalt extraction is the feed to nickel electrowinning.

The nickel electrowinning occurs in a divided cell, with anode bags for the collection of chlorine and oxygen gases. These gases are used in the leaching section of the process. The nickel spent, or anolyte, is recycled to leaching.

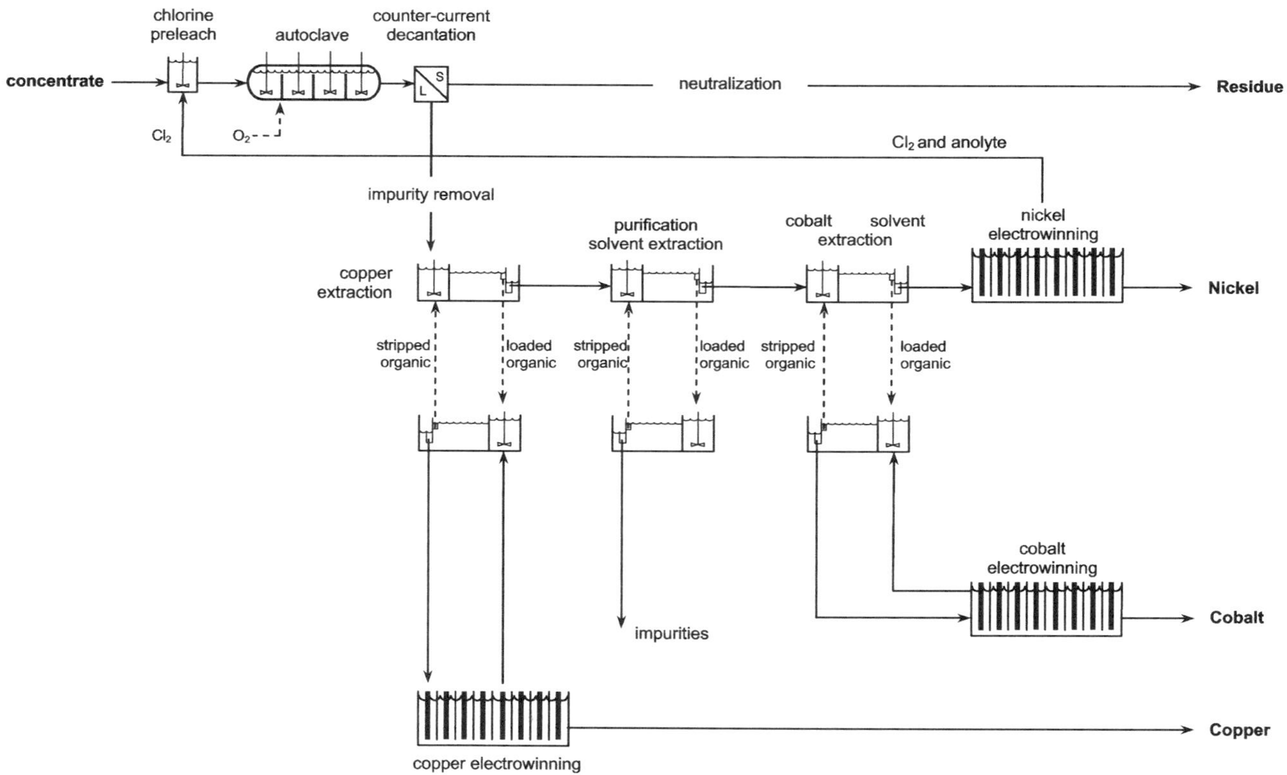

FIGURE 13.2 Schematic diagram of the Voisey's Bay process. The raffinate from the copper solvent extraction is passed through further purification steps, such as cadmium removal by sulfide precipitation, and a second iron-removal step. Then, the solution goes on to impurity solvent extraction were D2EPHA is used to remove calcium, copper, lead, iron, manganese and zinc. The purified solution is transferred to cobalt solvent extraction.

The Voisey's Bay process has been extensively piloted. The demonstration plant showed that both magnesium and silica built up in the circuit. Effective controls measures were implemented.

It is anticipated that this process will be commissioned during 2013. The advantage of this process is that the capital-intensive steps of smelting and converting are avoided. The process cannot, however, recover precious metals.

13.5. HEAP LEACHING OF NICKEL SULFIDE ORE

An alternative to intensifying the rate of leaching of concentrates is to heap leach the whole ore. This has the significant advantage of eliminating the concentration of the sulfides, which is the step that has the lowest recovery on route from ore to metal.

Nickel sulfide ores are processed by heap leaching in Sotkama, eastern Finland (Talvivaara, 2011). The process entails the following:

(a) open-pit mining of 0.2% Ni, 0.5% Zn, 0.1% Cu, and 0.02% Co sulfide ore;
(b) four-stage crushing;
(c) rotary kiln agglomeration with sulfuric acid;
(d) stacking the agglomerate in 8 m high heaps;
(e) leaching nickel and other metals from the agglomerate by sprinkling metal-depleted, purified, recycle solution on the heaps; and,
(f) maintaining adequate temperature and conditions to promote bacterial activity.

Copper, zinc and nickel are recovered from the pregnant leach solution by sequential hydrogen sulfide precipitation of separate metal sulfide precipitates. The heaps are typically leached for 1.5 years.

The planned production of nickel in the sulfide precipitate is approximately 50 000 tonnes per year. Recent production has been considerably lower (Talvivaara, 2010).

13.6. SUMMARY

Nickel sulfide ores occur as disseminated or massive orebodies. The nickel sulfide is separated from the host rock by flotation. The flotation concentrate is smelted to produce a converter matte, which is then refined either by carbonyl refining or in a hydrometallurgical refinery. This route is by far the most dominant.

Alternatives to smelting and converting include various hydrometallurgical processes that treat concentrates, which have been demonstrated in the past, but are not currently practiced.

Heap leaching of the sulfide ore is practiced at only one location.

REFERENCES

Boldt, J. R., & Queneau, P. (1967). *The winning of nickel*. Longmans.

Crundwell, F. K. (2005). The leaching number: Its definition and use in determining the performance of leaching reactors and autoclaves. *Minerals Engineering, 18*, 1315–1324.

Jones, D. L., Mayhew, K., Mean, R., & Neef, M. (2010). Unlocking disseminated nickel sulphides using the CESL nickel process. In A. Taylor (Ed.), *ALTA 2010 Nickel-Cobalt Conference*. ALTA Metallurgical Services.

Jones, D. L., Mayhew, K., & O'Connor, L. (2009). Nickel and cobalt recovery from a bulk copper-nickel concentrate using the CESL process. In J. J. Budac, R. Fraser & I. Mihaylov, et al. (Eds.), *Hydrometallurgy of nickel and cobalt 2009* (pp. 45–57). CIM.

Karonen, J., Tiihonen, M., & Haavanlammi, L. (2009). Hydronic – a novel nickel refining method for nickel concentrates. In J. J. Budac, R. Fraser & I. Mihaylov, et al. (Eds.), *Hydrometallurgy of nickel and cobalt 2009* (pp. 17–26). CIM.

Kofluk, R. P., & Freeman, G. K. W. (2006). Iron control in the Moa Bay operation. In J. E. Dutrizac & P. A. Riveros (Eds.), *Iron control technologies* (pp. 573–589). CIM.

Palmer, C. M., & Johnson, G. D. (2005). The Activox process: Growing significance in the nickel industry. *Journal of Metals, 57*, 40–47.

Smith, R. (2004). *Kwinana nickel refinery*. October 28, 2004.

Stevens, D., Bishop, G., Singhal, A., Love, B., & Mihaylov, I. (2009). Operation of the pressure oxidative leach process for Voisey's Bay nickel concentrate at Vale Inco's hydromet demonstration plant. In J. J. Budac, R. Fraser & I. Mihaylov, et al. (Eds.), *Hydrometallurgy of nickel and cobalt 2009* (pp. 3–16). CIM.

Talvivarra. (2010). Talvivaara falls after hydrogen sulphide leak. *The Guardian (UK)*. May 6, 2010.

Talvivaara. (2011). Talvivaara operations. <www.talvivaara.com>. Accessed 25.05.2011.

Woodward, T. M., & Bahri, P. A. (2007). Steady-state optimisation of the leaching process at Kwinana nickel refinery. In V. Plesu & P. S. Agachi (Eds.), *17th European symposium on computer aided process engineering – ESCAPE17* (pp. 1–6). Elsevier.

SUGGESTED READING

Budac, J. J., Fraser, R., Mihaylov, I., et al. (2009). *Hydrometallurgy of nickel and cobalt 2009*. Metallurgical Society of CIM.

Burkin, A. R. (1987). *Extractive metallurgy of nickel*. Society of (British) Chemical Industry.

Donald, J. & Schonewille, R. (Eds.). (2005). *Nickel and cobalt 2005, challenges in extraction and production*. Metallurgical Society of CIM.

Habashi, F. (2009). A history of nickel. In J. Liu, J. Peacey & M. Barati, et al. (Eds.), *Pyrometallurgy of nickel and cobalt 2009, proceedings of the international symposium(Taylor, A., ed.)* (pp. 77–98). Metallurgical Society of CIM.

Jones, R. T. (2004). JOM world nonferrous smelter survey, part II: Platinum group metals. *Journal of Metals, 56*, 59–63.

Jones, R. T. (2006). *Southern African pyrometallurgy 2006 international conference*. The South African Institute of Mining and Metallurgy.

Liu, J., Peacey, J., Barati, M., et al. (2009). *Pyrometallurgy of nickel and cobalt 2009*. Metallurgical Society of CIM.

Kerr, A. (2002). An overview of recent developments in flotation technology and plant practice for nickel ores. In A. L. Mular, D. N. Halbe & D. J. Barratt (Eds.), *Mineral processing, plant design, practice and control proceedings, Volume 1* (pp. 1142–1158). SME.

Warner, A. E. M., Diaz, C. M., Dalvi, A. D., et al. (2006). JOM world nonferrous smelter survey, part III: Laterite. *Journal of Metals, 58*, 11–20.

Warner, A. E. M., Diaz, C. M., Dalvi, A. D., et al. (2007). JOM world nonferrous smelter survey, part IV: Nickel sulfide. *Journal of Metals, 59*, 58–72.

Chapter 14

Production of Nickel Concentrates from Sulfide Ores

The first step in extracting nickel and cobalt from sulfide ores is crushing and grinding. As shown in Figure 13.1, the crushed and ground ore is pumped to flotation, where sulfide minerals are selectively extracted.

The objective of this chapter is to describe the crushing and grinding of sulfide ores.

14.1. THE ADVANTAGES OF GRINDING AND CONCENTRATION

Nickel sulfide ores are always ground and concentrated, as shown in Figure 13.1. The reasons for this are as follows:

(a) smelting 1%–3% Ni ore is expensive, because it requires excessive amounts of energy and furnace capacity per tonne of product nickel; and,
(b) technology for the production of sulfide concentrates, particularly flotation technology, is inexpensive and relatively efficient.

14.2. CRUSHING AND GRINDING

Crushing and grinding (or comminution) prepares ores for concentrate production by liberating the pentlandite grains from the surrounding rock and pyrrhotite grains. A schematic diagram of the process flowsheet is shown Figure 14.1. Other flowsheets may be used, depending on the ore feed and the history of the operation.

Industrial crushing and grinding data are given in Tables 14.1–14.3.

The size of pieces of blasted ore is mostly smaller than 0.5 m. After crushing and grinding, most of the ore particles are smaller than 100 μm and many are less than 50 μm.

There is an optimum particle size for flotation. The effect of grind size on loss of pentlandite to the flotation tailings for a particular ore is shown in Figure 14.2. These results indicate that there is a particle size that minimizes the amount of pentlandite that is lost in tailings.

Extractive Metallurgy of Nickel, Cobalt and Platinum-Group Metals. DOI: 10.1016/B978-0-08-096809-4.10014-0

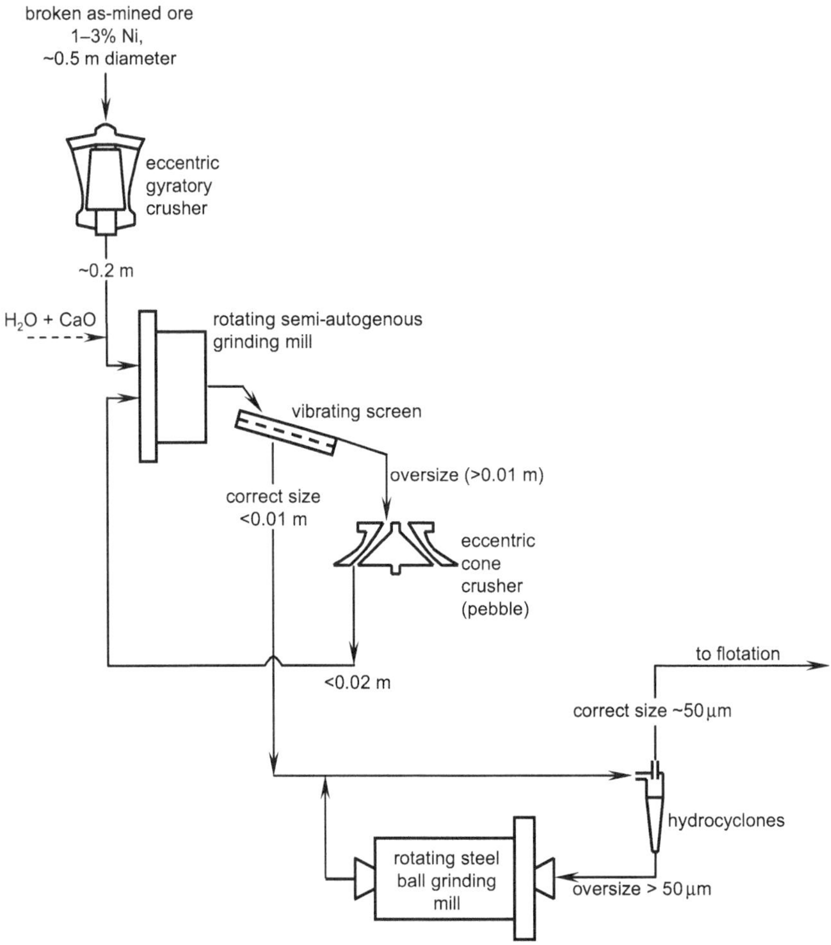

FIGURE 14.1 Flowsheet for preparing small particles for feed to flotation from as-mined ore pieces. Two stages of crushing and two stages of grinding are shown. The crushing is open circuit, that is, there is no recycle loop. The two grinding circuits are closed circuit, that is, oversize material is recycled for re-crushing or re-grinding to specified particle sizes.

There are two reasons for this optimum:

(a) too large a grind size leaves some pentlandite grains combined with or hidden in pyrrhotite and rock grains, preventing them from floating; and,
(b) too fine a grind size causes rising bubbles to push very tiny particles aside, avoiding contact.

In addition, very fine pyrrhotite and rock particles may agglomerate with pentlandite particles, preventing them from floating.

TABLE 14.1 Nickel Concentrator Crushing Details

Concentrator	Thompson, Canada	Sudbury, Canada	Kambalda, Australia
New ore treated, t/h	400 each crusher	Up to 220 each crusher	Up to 500
Crusher type	Standard gyratory	Standard gyratory	Svedala 48/36 single toggle jaw
Number	3	4	1
Size	2 m diameter	2 m diameter	0.9 m jaw, 4 m long
Power rating, kW	220	220	
Feed size, nominal diameter, m	0.15	0.2	0.3
Product size	70% passing through 0.013 m screen	90% passing through 0.02 m screen	<0.09 m
Product destination	To rod mill[a] grinding then ball mill grinding	To rod mill[a] grinding then ball mill grinding	To semi-autogenous grinding mill

Notably, the Australian concentrator uses a jaw crusher while the Canadian concentrators use gyratory crushers.
[a]*Rod mills tumble long steel rods rather than balls (Boldt & Queneau, 1967).*

14.3. COMMINUTION STEPS

Comminution is performed in the following three sequential steps:

(a) breaking of *in situ* ore by explosions in the mine;
(b) crushing of the broken ore by compression in eccentric crushers (refer to Figure 14.3 and Table 14.1); and,
(c) wet grinding of the crushed ore in rotating mills where abrasion, impact and compression all contribute to breaking the ore (refer to Figure 14.4 and Tables 14.2 and 14.3).

Separate crushing and grinding steps are necessary because it is not possible to break ore that is as large as 0.5 m in diameter while at the same time making particles that are approximately 50 μm in diameter that are required for optimal flotation.

TABLE 14.2 Operating Details of Two Semi-Autogenous Grinding Mills and One Autogenous Grinding Mill

Location	Sudbury, Canada	Kambalda, Australia	Raglan, Canada
Material treated, Mt/a	0.3	1.5	1.2
Feed and diameter	Crushed ore, <0.2 m	Contractor crushed ore, <0.09 m	As-mined ore ~0.1 m
Mill type	Semi-autogenous	Autogenous	Semi-autogenous
Number of mills	1	1	1
Diameter × length, m	10 × 4	8 × 5	7 × 4
Power rating, kW	8200		2200
Ore plus ball charge	28% of mill volume		
Ball charge, % total charge	8–10	0	5–6
Initial ball diameter, m	0.13		
Ball consumption, kg/tonne of ore	0.5		
Speed, rpm	11 (maximum)		
Mill liners	Chrome-moly steel		
Product diameter, m	<0.003	<0.004	
Destination	Screen then ball mills	Cyclones then flash flotation	Vibrating screen then ball mill
Oversize treatment	Recycle to semi-autogenous mill	Simons 1.7 m, 300 kW cone crusher	250 kWh Sandvik cone crusher

14.3.1. Blasting

Blasting entails drilling holes in mine walls or benches, filling the holes with chemical explosive and exploding out fragments of rock. The explosions cause cracks to propagate through the rock, releasing multiple fragments of rock. Burger *et al.* (2006) reported that closer drill holes and larger explosive charges make smaller rock fragments. This decreases subsequent crushing and grinding requirements, potentially decreasing cost and increasing concentrator throughput rate.

TABLE 14.3 Details of Four Ball Mill Operations That Produce Ground Ore for Flotation of Pentlandite Concentrate

Location	Thompson, Canada	Sudbury, Canada	Strathcona, Canada	Raglan, Canada
Feed	Rod mill product	Semi-autogenous and rod mill products	Rod mill products	Semi-autogenous mill product
Rate, t/d per ball mill	2500	6000	5000 (maximum)	3500
Diameter, m		0.003		
Slurry density, % solids	76		78	
Number of ball mills	5	6	2	1
Size, diameter × length, m	4 × 5	4 × 5.5	4 × 5.5	4 × 6.4
Power rating, kW	1100	1500	1300	2200
Ball diameter, m	0.038 and 0.051	0.051 and 0.063	0.025	
Ball consumption, kg/tonne of ore	0.4	0.2	0.3	
Mill lining material	Rubber	Rubber	Rubber	
Speed, rpm	16 (70% of critical)	17 (80% of critical)	16 (77% of critical)	
Product size	70% passing 150 μm	88% passing 250 μm	55% passing 75 μm	80% passing 65 μm
Destination	Cyclones (oversize recycle to ball mills; correct size to flotation)	Cyclones (oversize recycle to ball mills; correct size to flotation)	Cyclones (oversize recycle to ball mills; correct size to flotation)	Cyclones (oversize recycle to ball mills; correct size to flotation)

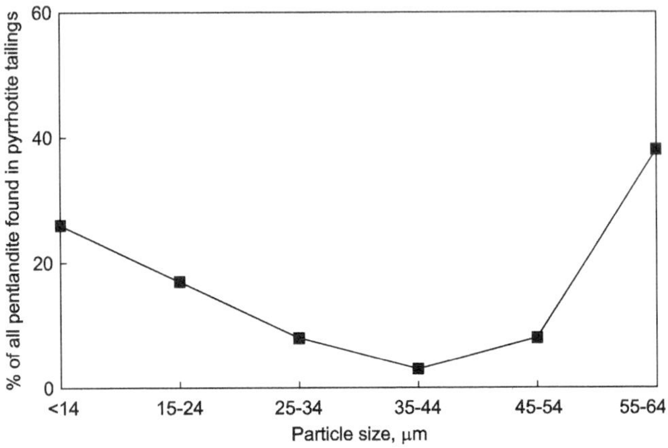

FIGURE 14.2 Loss of pentlandite in tailings as a function of flotation tailing particle size (Kerr *et al.*, 2003). Most of the pentlandite loss is in very large and very small particles. The graph suggests that this ore might beneficially be ground finer – to eliminate loss of nickel in the large particle sizes. However, this might produce more very small particles and greater overall pentlandite loss. Optimum grind size is best determined by well-controlled in-plant experiments.

FIGURE 14.3 Gyratory crusher for crushing as-mined ore to ~0.2-m-diameter pieces. The crushing is done by compression of ore pieces between the eccentrically rotating spindle and the fixed crusher wall. The crushing surface on the spindle can be up to 3 m in height. Crushing rates are 5000–15 000 tonnes of ore per day. *Source: Drawing from Boldt and Queneau (1967), courtesy of Vale.*

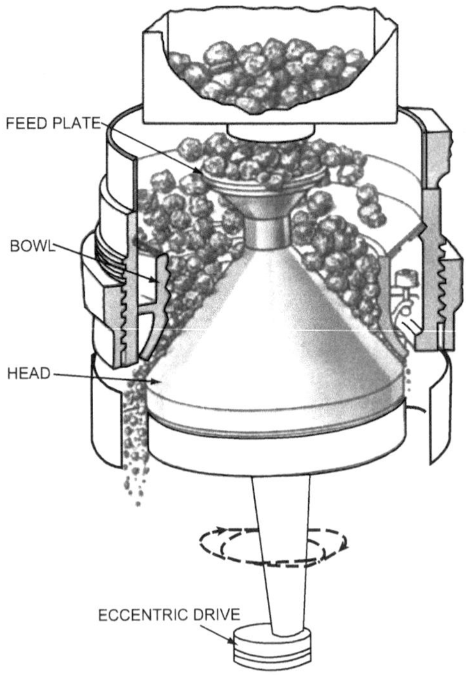

14.3.2. Crushing

Primary crushing can be done in the mine. This permits easy hoisting of ore out of underground mines. It also permits belt conveying of ore out of open pit mines and shallow underground mines.

The crushed ore is stored in a small coarse-ore stockpile. It is transferred from the stockpile by conveyor to the grinding mills. The stockpile permits steady, controlled flow of ore into the concentrator.

14.3.3. Grinding

The ore from the crushing circuit is fed to the grinding circuit. The purpose of grinding is to produce particles that are 'liberated', that is, small enough to consist mostly of one mineral, for example, pentlandite, pyrrhotite or gangue rock. The liberated particles can then be separated by flotation into two streams:

(a) a nickel-rich concentrate consisting of pentlandite; and,
(b) a nickel-lean waste stream, consisting of pyrrhotite and gangue rock.

Grinding is done wet. The ore slurry is composed of about 70% solids and 30% water.

The most common grinding mills are either (a) semi-autogenous and autogenous mills; or, (b) ball mills. A semi-autogenous mill is shown in Figure 14.4.

Semi-autogenous and autogenous mills have the advantage that they accept a wide range of feed sizes. Ball mills have the advantage that they deliver finely ground product to flotation.

14.4. CONTROL OF PARTICLE SIZE

The objective of grinding is to produce ore particles that are small enough for efficient pentlandite flotation, but not excessively small. Size control is universally done with hydrocyclones (Krebs, 2010). A hydrocyclone is shown in Figure 14.5.

The hydrocyclone makes use of the principle that, under a force field, large particles in an ore–water slurry tend to move faster than small particles. This principle is put into practice by pumping product slurry from the grinding mill into hydrocyclones at high speed, for example, at 5–10 m/s. The slurry is pumped in tangentially, giving it a rotational motion inside the cyclone, as shown in Figure 14.5. This creates a centrifugal force that accelerates larger ore particles towards the cyclone wall.

The centrifugal force in the hydrocyclone is such that (i) large particles move to the wall, where they are dragged out by water flow along the wall and through the apex; and, (ii) the path of small particles is hindered by larger particles so that they are forced out through the vortex finder.

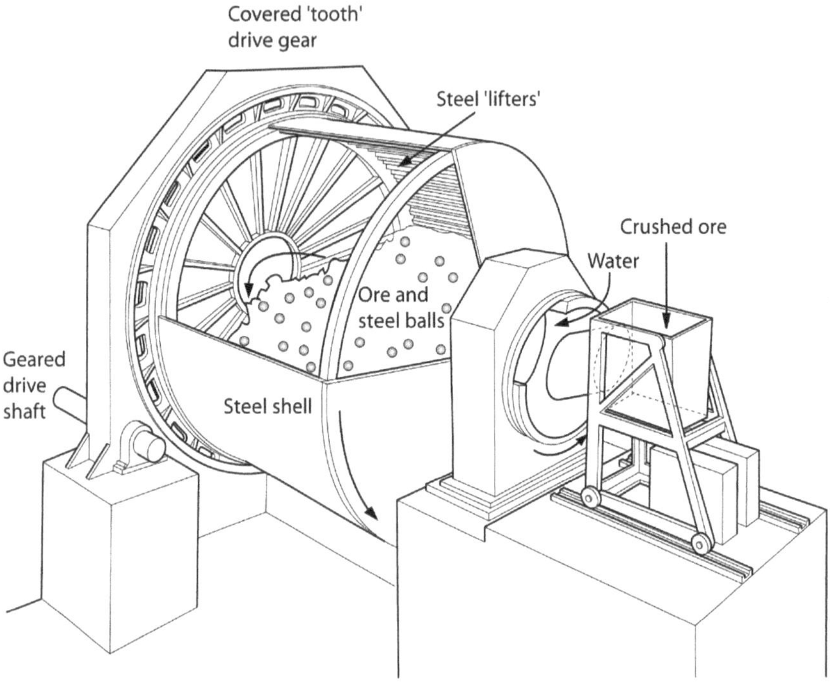

FIGURE 14.4 Semi-autogenous grinding mill. It is a rotating barrel in which ore is broken by ore pieces (0.2 m diameter) and steel balls (0.1–0.14 m diameter) falling on ore as they are lifted and fall off the moving circumference of the barrel. Autogenous grinding mills are similar but without the steel balls. The mills are fed and grind continuously. Their product is typically less than 0.01 m diameter. Autogenous and semi-autogenous mills are up to 10 m diameter with diameter-to-length ratios of 1.6:1–2.5:1. They rotate at about 10 rpm and grind 2000–15 000 tonnes of ore per day. *Source: Drawing courtesy W. G. Davenport.*

The principal parameter for the control of particle size is the water content of the incoming slurry, which is approximately 50% H_2O. An increase in water content gives less-hindered movement of particles and allows a greater fraction of input particles to reach the wall and pass through the apex. As a result, the fraction of particles that are being recycled to re-grinding increases with increasing water content and ultimately results in a more finely ground product.

14.4.1. Instrumentation and Control

Grinding circuits are extensively instrumented and closely controlled. The objectives of the controls are to (i) produce ore particles that are correctly sized for efficient pentlandite flotation; and, (ii) produce these particles with minimum energy consumption.

The most common strategy is to (i) measure the cyclone overflow using an on-stream particle size analyzer to ensure that the product is always correctly

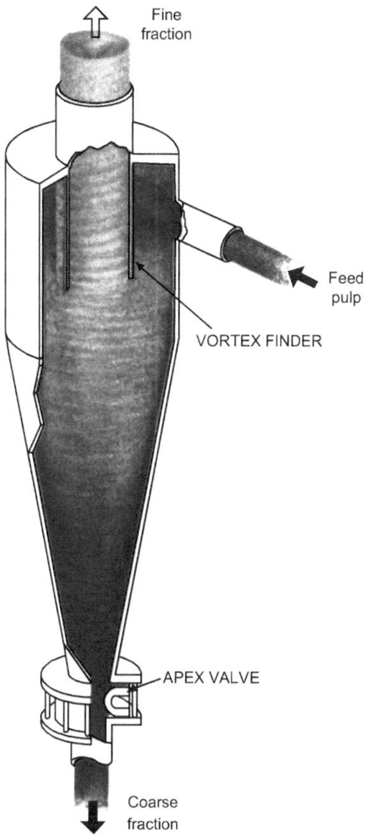

FIGURE 14.5 Cut-away view of hydrocyclone showing (i) tangential input of water-ore particle feed; and, (ii) separation of the feed into fine particle and coarse particle fractions. The cut between fine particles and coarse particles is controlled by adjusting the water content of the feed mixture. *Source: Drawing from Boldt and Queneau (1967), courtesy of Vale.*

sized (Outotec, 2010); and, (ii) optimize grinding rate and energy consumption while maintaining this size.

A typical control system is shown in Figure 14.6. There are two main control loops. The first is the control of the particle size and the second is the control of the feed rate to the ball mill.

14.4.2. Control of Particle Size

The particle size loop, shown in Figure 14.6, controls the size of the ground particles by automatically adjusting the rate at which water is added to the hydrocyclone feed sump.

14.4.3. Control of Feed Rate to the Ball Mill

The second loop shown in Figure 14.6 adjusts the rate at which the material is fed to the ball mill by controlling the level in the hydrocyclone sump level.

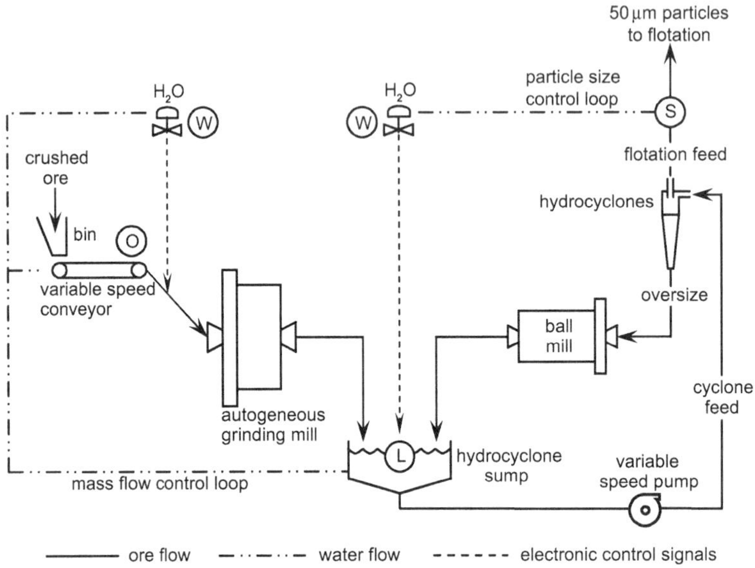

FIGURE 14.6 Control system for grinding mill circuit. The objective is to produce correct-size ore particles for optimal pentlandite flotation. The circled symbols refer to the sensing devices in Table 14.4. A circuit usually consists of an autogenous or semi-autogenous grinding mill, a hydrocyclone feed sump, a hydrocyclone 'cluster' (about six cyclones) and two ball mills. A similar flowsheet is described by Outotec (2010).

If, for example, slurry level sensor (L) detects that the slurry level is rising (due to tougher[1] ore and hence more hydrocyclone recycle), it automatically slows the ore feed conveyor. This decreases flow rates throughout the plant and gives the ball mill enough time to adequately grind its input ore particles.

If, the level in the sump falls, the control system automatically increases feed rate of ore until the sump level is maintained at its set point.

14.5. RECENT DEVELOPMENTS

The primary development in recent years in the comminution of pentlandite ores has been almost universal adoption of semi-autogenous grinding. This technique has proven to be an economic, efficient method of primary grinding.

Other developments have been adoption of high-pressure grinding rolls, stirred grinding mills and ceramic media grinding, which are used in platinum-group metal concentrators. High-pressure grinding rolls increase crushing

1. The resistance of an ore to grinding is usually represented by the *Bond Work Index*, measured in kilojoules per tonne of ore. High-index ore is difficult to grind: lower index ore is easier to grind. The definition, measurement and use of *Bond Work Index* are given by Wills and Napier-Munn (2008).

TABLE 14.4 Sensing and Control Devices for the Grinding Control Circuit Shown in Figure 14.6

Sensing instruments	Symbol in Figure 14.6	Purpose	Type of device	Use in automatic control system
Ore input rate weightometer	O	Senses feed rate of ore into grinding circuit	Load cells, conveyor speed	Controls ore feed rate
Water flow gages	W	Sense water addition rates	Rotameters	Control water-to-ore ratio in grinding mill feed
On-stream particle size analyzer	S	Senses a critical particle size parameter (for example, percent minus 70 μm) on the basis of calibration curves for the specific ore	Measures diffraction of laser beam by particles in an automatically taken slurry sample (Outotec, 2010)	Controls water addition rate to hydrocyclone feed (which controls recycle and the size of the final grinding circuit product)
Hydro-cyclone feed sump level indicator	L	Senses changes of slurry level in sump	Bubble pressure tubes; electric contact probes; ultrasonic echoes; nuclear beam	Controls rate of ore input into grinding circuit (prevents over-loading of ball mills)

throughput with lower energy requirements. Stirred grinding mills with ceramic pellets give very fine particles of pentlandite that are not contaminated with iron.

An important new tool for maximizing pentlandite recovery is quantitative scanning electron microscopy. This technique indicates the makeup of every particle in a sample. It shows liberated, partially liberated and un-liberated pentlandite grains for any assemblage of ground ore. The metallurgist can then determine whether more or less grinding will improve pentlandite recovery. Quantitative scanning electron microscopy is used extensively in the platinum extraction industry and is finding increasing use in nickel extraction (Charland *et al.*, 2006).

14.6. SUMMARY

Pentlandite [$(Ni,Fe)_9S_8$] ore is the largest source of nickel as sulfide in the world.

As-mined nickel sulfide ores typically contain 3%–8% pentlandite (1%–3% Ni), remainder being chalcopyrite, pyrrhotite Fe_8S_9 and gangue rock. These ores are too dilute in nickel for direct smelting. They would require too much energy and furnace capacity. They are, therefore, always made into nickel-enriched (12%–20% Ni) concentrate, mainly by froth flotation, which is described in Chapter 15.

Efficient isolation of the pentlandite in the ore into nickel-enriched concentrate requires that the pentlandite grains be liberated, that is, separated from pyrrhotite and rock grains. This is accomplished by crushing and grinding the ore to a particle size of approximately 50–100 μm. The pentlandite in this ground ore is then concentrated using froth flotation. Gyratory crushers and rotating grinding mills are used, usually with an automatic particle size control system.

REFERENCES

Boldt, J. R., & Queneau, P. (1967). *The winning of nickel.* Longmans.

Burger, B., McCaffery, K., McGaffin, I., et al. (2006). Batu Hijau model for throughput forecast, mining and milling optimization and expansion studies. In S. K. Kawatra (Ed.), *Advances in comminution* (pp. 461–479). SME.

Charland, A., Kormos, L., Whittaker, P., et al. (2006). A case study for integrated use of automated mineralogy in plant optimization: the Falconbridge Montcalm concentrator. In *Paper presented to the Automated Mineralogy Conference*; Brisbane: July 17–18, 2006.

Kerr, A., Bouchard, A., Truskoski, J., et al. (2003). The "Mill Redesign Project" at Inco's Clarabelle Mill. *CIM Bulletin, 96*, 58–66.

Krebs. (2010). Hydrocyclones [Company brochure].

Outotec. (2010). Grinding control solutions [Company brochure].

Wills, B. A., & Napier-Munn, T. J. (2008). *Wills' mineral processing technology* (7th ed.). Elsevier.

SUGGESTED READING

Damjanovic, B., & Goode, J. R. (Eds.), (2000), *Canadian milling pratice, Special Vol. 10.* CIM.

Fuerstenau, M. C., Jameson, G., & Yoon, R.-H. (Eds.), (2006). *Froth flotation, A century of innovation.* SME.

Kawatra, S. K. (Ed.), (2006). *Advances in comminution.* SME.

Kerr, A. (2002). An overview of recent developments in flotation technology and plant practice for nickel ores. In A. L. Mular, D. N. Halbe & D. J. Barratt (Eds.), *Mineral processing, plant design, practice and control proceedings, Vol. 1* (pp. 1142–1158). SME.

Malhotra, D., Taylor, P., Spiller, E., & LeVier, M. (Eds.), (2009). *Recent advances in mineral processing plant design.* SME.

Chapter 15

Production of Nickel Concentrate from Ground Sulfide Ore

The starting material for the production of concentrate is ground sulfide ore, which contains 1%–3% Ni with a particle size of 50–100 μm in a 40% solids slurry. The composition of several concentrates is given in Table 15.1.

TABLE 15.1 Nickel, Copper and Cobalt Contents in Concentrate Feeds at Smelters Around the World*

		Concentrate, %		
Company	Smelter location	Ni	Cu	Co
BHP Billiton	Kalgoorlie, Australia	15	0.3	0.4
BCL	Selebi Phikwe, Botswana	5	3	0.2
Votorantim	Fortaleza, Brazil	7	1	0.1
Xstrata	Falconbridge, Canada	12	4.5	0.5
Vale	Sudbury, Canada	10	12	0.3
Vale	Thompson, Canada	14	0.3	0.3
Jinchuan Group	Jinchang, Gansu, China	9	4	0.2
Boliden	Harjavalta, Finland	14	0.8	0.4
Norilsk	Nadezda, Russia	12	5	0.4
Norilsk	Norilsk Russia	5	2.5	0.2
Norilsk	Pechenga, Russia	9	4	0.3

The feeds are sometimes combinations of several concentrates but they give an indication of the compositions of nickel concentrate around the world.
Warner et al., 2007

Extractive Metallurgy of Nickel, Cobalt and Platinum-Group Metals. DOI: 10.1016/B978-0-08-096809-4.10015-2

The products of the process are flotation concentrate, which contains 5%–15% Ni, and tailings, which contain approximately 0.25% Ni. The tailings are composed of gangue rock and pyrrhotite, Fe_8S_9. The tailings are discarded.

The objectives of this chapter are

(a) to describe the principles of froth flotation concentrate production;
(b) to indicate how these principles are applied to produce concentrate that is rich in Ni from the ground ore; and,
(c) to discuss recent developments in industrial flotation.

A flow circuit for a flotation plant is shown in Figure 15.1.

15.1. NEED FOR CONCENTRATION

As-mined nickel sulfide ores contain between 1% and 3% Ni. These concentrations are too low for direct smelting to be economical. Heating and melting of the large quantity of pyrrhotite and gangue rock in the ore would require excessive energy and furnace capacity. As a result, a concentrate is produced by partially separating the pentlandite from the host rock and pyrrhotite. This concentrate is then smelted and refined to high-purity nickel.

15.2. PRINCIPLES OF FROTH FLOTATION

The indispensable tool for making nickel sulfide concentrate from nickel ore is froth flotation (Fuerstenau, Jameson, & Yoon, 2007; Gomez, Nesset, & Rao, 2009).

The principles of froth flotation can be summarized as follows:

(a) Sulfide minerals in a slurry are normally hydrophilic, that is, they are easy to wet. They can be conditioned with reagents (collectors) that make them hydrophobic, that is, they are more difficult to wet. Hence, these particles prefer an environment that is not wetted;
(b) Chemical conditioning can selectively make particles either hydrophilic or hydrophobic. For example, pentlandite can be conditioned to be hydrophobic, leaving gangue and pyrrhotite hydrophilic;
(c) Collisions between rising air bubbles and the hydrophobic pentlandite particles cause the pentlandite particles to attach to the rising bubbles, as shown in Figure 15.2, because their surfaces are hydrophobic; and,
(d) The particles of gangue rock and pyrrhotite do not attach to the rising bubbles because their surfaces are hydrophilic, hence they prefer to be in the water phase.

Industrially, the slurry is conditioned so that the pentlandite particles are hydrophobic while leaving rock and pyrrhotite hydrophilic. A 'frother' is added to create a more stable foam at the top of the slurry. A cloud of small bubbles, about 1 mm in diameter (Hernandez-Aguilar, 2009), is passed through the slurry.

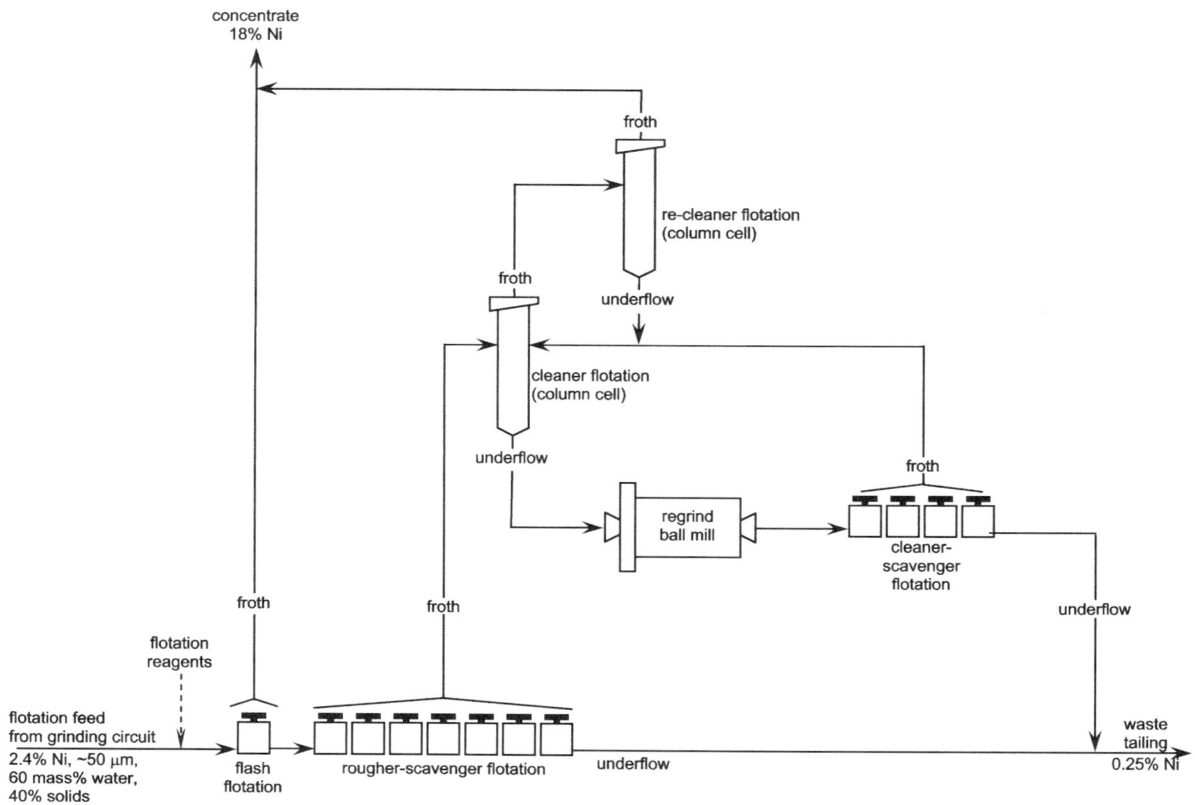

FIGURE 15.1 Schematic diagram of a flotation circuit for floating pentlandite particles from gangue rock and pyrrhotite particles. The process is continuous and treats about 3000 tonnes of ground 2.4% Ni ore and produces about 360 tonnes of 15% Ni concentrate on a dry basis. Note particularly 'flash flotation'. It recovers about half the pentlandite in the ore at above the plant's overall concentrate grade (%Ni). The plant's mechanical cells are ~30 m^3 each. The column (cleaner) cells are ~150 m^3 each. The particle size of the product from the re-grind ball mill product is controlled to ~50 μm by cyclones (data not shown) in closed circuit with the re-grind ball mill.

FIGURE 15.2 Photograph of hydrophobic mineral particles attached to rising bubbles. The input bubbles in industrial flotation are ~1 mm in diameter, which is about the size of the largest bubble in this photograph. *Source: Photograph from Boldt and Queneau (1967), courtesy of Vale.*

The pentlandite particles attach to the rising bubbles, which carry them to the top of the flotation cell. The frother ensures that the bubbles do not burst and prematurely release their attached particles. The froth at the top of the flotation cell overflows into a collection trough and then into a concentrate collection tank. The gangue rock and pyrrhotite do not attach to the rising bubbles, and, as a result, they remain behind in the slurry. They leave the cell through an underflow system.

The mechanism of flotation based on the adhesion of particles to air bubbles described above is referred to as '*true flotation*'. Another mechanism of recovery of minerals to the froth is *entrainment* (Knights and Bryson, 2009). There are three factors that contribute to entrainment. First, mineral particles are carried up in the slurry by the swarm of bubbles in the flotation cell. The transportation of the particles is not selective, and it is envisaged that the concentration of minerals that are entrained is the same as that in the bulk of the flotation cell. Second, the entrained slurry can drain from the froth back into the bulk of the flotation cell. Some selectivity may be introduced in drainage because of the physical characteristics of the particles. Third, the entrained particles are transported by froth transfer from the flotation cell to the concentrate launder.

Flotation circuits are usually designed to maximize true flotation and mimimize entrainment, because entrainment of particles of gangue rock

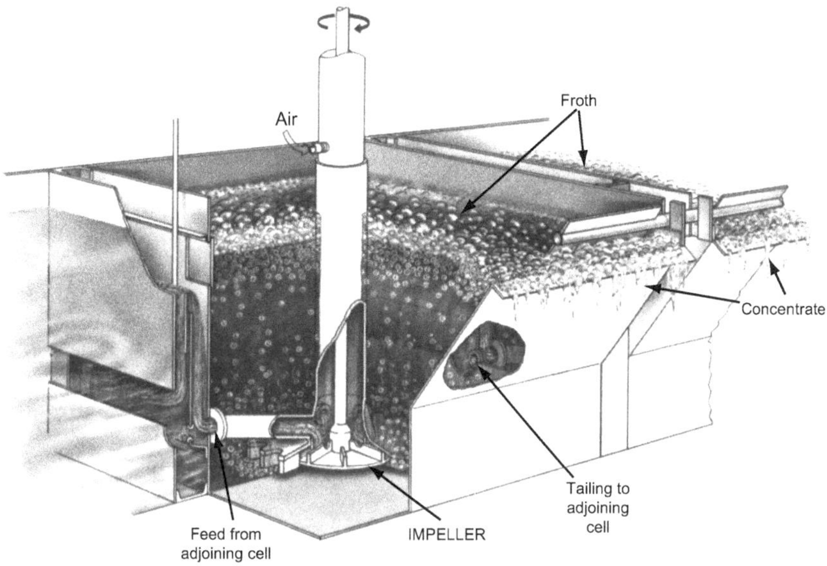

FIGURE 15.3 Cut-away view of cubic mechanical flotation cell. The methods of producing bubbles and gathering froth are shown. *Source: Boldt and Queneau (1967), courtesy of Vale.*

between bubbles results in poor separation efficiencies (Knights and Bryson, 2009). True flotation is maximized by choosing the correct froth residence time, particle size, solids content in the feed slurry and froth stability.

An industrial flotation cell, illustrating these features of flotation, is shown in Figure 15.3. Mechanical flotation cells in recently designed pentlandite concentrators are 20–40 m^3 in size, shaped either as cubic or cylindrical tanks. They operate continuously.

15.3. FLOTATION CELLS

Industrial flotation plants contain a series of flotation cells, chemical addition points and re-grind mills designed to maximize pentlandite recovery and the nickel grade in the concentrate. Two types of cells are used: mechanical cells and column cells.

15.3.1. Mechanical Cells

The flotation cell illustrated in Figure 15.3 is a mechanical cell. In the flowsheet shown in Figure 15.1, mechanical flotation cells are used for flash flotation, rougher flotation and scavenger flotation.

Mechanical cells introduce air into the slurry through a rotating impeller. The impeller agitates the slurry, creates bubbles and disperses the bubbles throughout the cell. These conditions provide frequent bubble–particle collisions, thereby creating the strong flotation conditions needed for flash, rougher and scavenger flotation (Gorain *et al.*, 2007).

15.3.2. Column Cells

The froth cleaning is done with column cells (Finch, Cilliers, & Yianatos, 2007). These cells provide bubble–particle collisions that are gentler and allow loosely attached particles to drain from the froth before it overflows into the collection trough.

The overall result is that particles of gangue and partially liberated pentlandite descend to the underflow of the cell, leaving a froth that is more concentrated in pentlandite than the feed.

15.4. FLOTATION CHEMICALS

15.4.1. Collectors

The collectors that create the hydrophobic surfaces on sulfide minerals are heteropolar molecules (Nagaraj & Ravishankar, 2007). They have a charged polar end and an uncharged non-polar end. The polar end attaches to the mineral surface (which is also charged) leaving the non-polar hydrocarbon end extended outwards into water. This is illustrated in Figure 15.4. It is this non-polar end that imparts the hydrophobic property to the conditioned pentlandite surfaces.

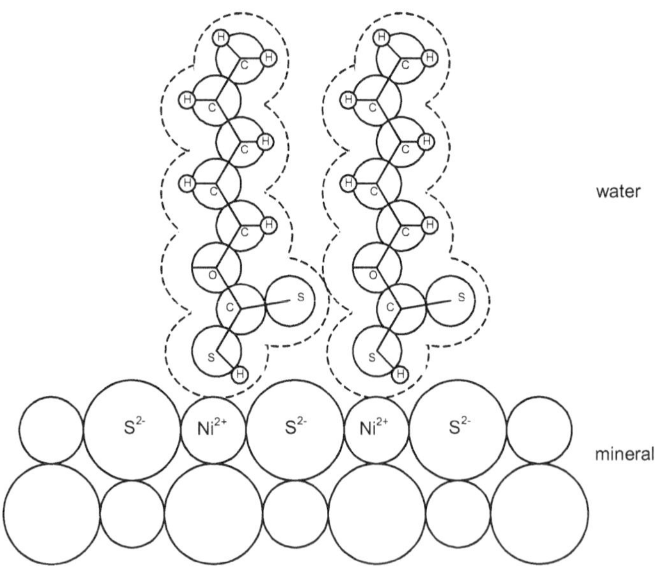

FIGURE 15.4 Sketch of attachment of amyl xanthate ions onto millerite, NiS. There is a hydrogen atom hidden behind each carbon in the hydrocarbon chain (after Hagihara, 1952). Millerite is rare in nickel ores (Kerr, 2002). It is pictured here for simplicity.

The mechanism of attachment of the collector to the sulfide surface is dependent on the charge of the surface. The charge on the surface is dependent on the pH of the solution, since either H^+ ions or OH^- ions may adsorb on the surface, and thereby alter the charge. Consequently, the pH of the solution during flotation is an important parameter.

15.4.2. Selectivity in Flotation

The simplest separation by froth flotation is that of sulfide minerals from waste rock, which is composed of silicates and their hydrated derivatives. It uses collectors that, when dissolved in a water–ore slurry, preferentially attach themselves to sulfides. The collectors have a sulfur group at the polar end, which attaches to the sulfide minerals but not to the oxides.

The collector used for nickel sulfide minerals is xanthate (Table 15.2). Potassium amyl xanthate, used in many of the operations, has the following chemical structure:

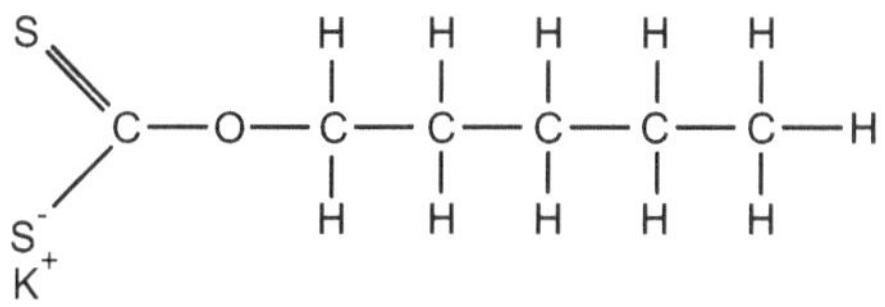

potassium amyl xanthate

It is typically added at a rate of 0.05–0.2 kg/tonne of ore.

15.4.3. Differential Flotation – Modifiers

Separating sulfide minerals from each other, for example, separating pentlandite $[(Ni,Fe)_9S_8]$ from pyrrhotite $[Fe_8S_9]$, is more complex. It relies on modifying the surfaces of the pyrrhotite so that the xanthate collector does not attach to them.

The most common modifier in water–ore slurry is the OH^- ion (hydroxyl ion). The concentration of OH^- ions is varied by adjusting the pH of the slurry using lime, CaO, or soda ash, Na_2CO_3.

The modifying effect of OH^- is due to its competition with xanthate anions for a place on the mineral surface. OH^- ions, for example, absorb preferentially on pyrrhotite. This prevents appreciable xanthate adsorption on pyrrhotite particles, selectively 'depressing' them. However, too high a concentration of OH^- ions will also depress pentlandite – so too much CaO or Na_2CO_3 must be avoided (Kerr, 2002).

Triethylene tetramine, in combination with sodium sulfite, is also used to depress pyrrhotite (Doucet, Price, Barrette, & Lawson, 2009). Guar gum and carboxy methyl cellulose are used to depress talc, which is often present in pentlandite ore and tends to float with pentlandite (Kerr, 2002).

TABLE 15.2 Collectors and Frothers Used in Floating Pentlandite from Gangue and Pyrrhotite

Concentrator	Pentlandite collector	Frother	Pentlandite recovery, %	Gangue rejection to waste, %	Pyrrhotite rejection to waste, %
Kambalda, W. Australia (BHP Billiton)	Sodium ethyl xanthate	Polyfroth H 20	90		
Leinster, W. Australia (BHP Billiton)	Sodium ethyl xanthate				
Mt. Keith, W. Australia (BHP Billiton)	Sodium ethyl xanthate	Polyfroth H 405	65		
Thompson, Manitoba (Vale)	Potassium amyl xanthate	Methyl isobutyl carbinol	96	95	55
Clarabelle, Sudbury, Ontario (Vale)	Potassium amyl xanthate	Unifroth 250 CM	91	99	82
Voisey's Bay, Labrador (Vale)	Flex 31 (sodium isopropyl xanthate plus surfactant)	Dowfroth 250	92		92
Strathcona, Sudbury, Ontario (Xstrata)	Potassium isobutyl xanthate	Dowfroth 250C	95	97	75
Montcalm, Tim-mins, Ontario (Xstrata)	Potassium isobutyl xanthate	Methyl isobutyl carbinol	~85		
Raglan, North Quebec (Xstrata)	Potassium ethyl xanthate	Methyl isobutyl carbinol	87	97	62
Norilsk, Russia	Na butyl xanthate and Na butyl dithiophosphate	T80			

All the collectors are xanthates. Recovery of pentlandite to concentrate and rejection of gangue and pyrrhotite to waste are also shown.

15.4.4. Frothers

Collectors and modifiers give selective flotation of pentlandite particles from rock and pyrrhotite particles. *Frothers* create the strong but short-lived froth, which holds the floated particles at the top of the cell. They give a froth that (i) is strong enough in the flotation cell to support the floated pentlandite particles; and, (ii) breaks down quickly after the pentlandite particles overflow the cell.

Short-chain aliphatic alcohols and polyglycols are the most common pentlandite flotation frothers (Elmahdy & Finch, 2009; Nagaraj & Ravishankar, 2007). They stabilize the froth by absorbing their OH^- polar end in water – while their branch chains form a cross-linked network in air. The use of frother is typically 0.02–0.05 kg/tonne of ore. Detailed discussions of frothers, frothing and froth breakdown are given in the studies by Comeau, Hardie, Pelletier, and Rhabani (2008) and Najaraj and Ravishankar (2007).

15.4.5. Activators

Nickel sulfides tend to float slowly, perhaps because xanthate collector ions tend to attach weakly to their particle surfaces. Many concentrators accelerate, or activate, the flotation of the nickel sulfide by adding copper sulfate solution to the ore slurry (Kerr, 2002).

In simplest terms, the dissolved copper sulfate tends to coat the surfaces of the nickel mineral particles with copper sulfide, for example, by the reaction:

$$\underset{\text{copper sulfate in feed ore slurry}}{CuSO_4(aq)} + \underset{\text{millerite particle surfaces}}{NiS(s)} \rightarrow \underset{\text{copper sulfide coating on millerite particles}}{CuS(s)} + \underset{\text{nickel sulfate in ore slurry}}{NiSO_4(aq)} \quad (15.1)$$

Because of this surface reaction, the nickel minerals float as quickly as copper minerals. Addition of copper sulfate is particularly important when the ore contains little copper, that is, when there is no *natural* activation. Western Australian concentrators are good examples of this (Kerr, 2002).

Copper sulfate is also used to activate pyrrhotite, where desired. The copper sulfate coats the surfaces of the pyrrhotite with CuS, which encourage the pyrrhotite to float with copper and nickel sulfides. This is especially important in the South African platinum industry where the objective is to float all the sulfides.

Of course, excessive addition of copper sulfate must be avoided if the flotation of pyrrhotite is undesired. Excessive addition of copper sulfate may also encourage the flotation of oxide and silicate minerals (Wiese *et al.*, 2007).

The rate of addition of copper sulfate ranges between about 20 (Canada) and 400 g (Western Australia) per tonne of ore.

15.5. SPECIFIC FLOTATION PROCEDURES FOR PENTLANDITE ORES

Flotation of pentlandite particles from particles of rock and pyrrhotite is done at pH values of about 9 with xanthate collectors and short-chain alcohol or glycol frothers. Lime, CaO or soda ash, Na_2CO_3, is used for the control of the pH.

The objectives of the flotation are to

(a) maximize the recovery of pentlandite to concentrate; and,
(b) maximize the grade of nickel (that is, % Ni) in the concentrate.

Achievement of maximum grade, which is objective (b), maximizes rejection of rock and pyrrhotite to tailings[1,2]. It is important to maximize the grade because it minimizes the amount of material that has to be transported and smelted per tonne of nickel.

An industrial flowsheet for achieving these goals is given in Figure 15.1. A slurry of ground ore and water is continuously pumped into the flash flotation cell at the lower left and separated into nickel-rich concentrate and nickel-lean waste tailing. Industrial details of the operation of such flotation circuits are given in Tables 15.3–15.6.

The five types of flotation cells used in the flotation circuit shown in Figure 15.1 are as follows:

(a) a *flash flotation* cell in which the incoming water–ore particle slurry is treated quickly under conditions at which the fresh-surfaced pentlandite particles are selectively floated and sent to the concentrate product (Warder & McQuie, 2005);
(b) *rougher–scavenger* cells where the gangue and pyrrhotite particles are depressed and the remaining pentlandite particles are floated;
(c) a quiescent *cleaner* column cell where the gangue and pyrrhotite particles in the rougher–scavenger froth sink to give a high-grade pentlandite froth;
(d) a *re-cleaner* column cell where the percentage of the pentlandite in concentrate is maximized by giving gangue and pyrrhotite a final depression; and,
(e) *cleaner–scavenger* cells where further grinding and more collector extract the last bit of pentlandite from the cleaner tailing before it is discarded.

Cleaning and re-cleaning is done in column cells where unliberated particles, that is, pentlandite–gangue and pentlandite–pyrrhotite particles, can descend and be re-grounded for further pentlandite liberation. The newly liberated

---

1. Several older concentrators remove magnetic monoclinic pyrrhotite from their flotation circuits by magnetic separation (Doucet *et al.*, 2009). This process has not been installed in recent concentrators.

2. Smelters occasionally need some pyrrhotite in their concentrate feed to offset high melting-point talc. It is obtained by controlling pH in the flotation cells.

TABLE 15.3 Details of Industrial Flash Flotation Plants

Concentrator	Kambalda, Australia	Raglan, Canada	Thompson, Canada
Ore treated, Mt/a	1.5	1.1	3.5
Concentrate produced, Mt/a	0.24	0.11	0.6
Ore grade, % Ni	2.3	2.4	2.4
Concentrate grade, % Ni	13	18	12
Tailings grade, % Ni	0.3	0.4	0.3
Ni recovery to concentrate, %	90	87	90
Flash flotation			
Feed	Semi-autogenous grinding mill oversize	Ball mill cyclone undersize	
Feed size	−53 μm diameter	80% passing 65 μm	
Number of flash flotation cells	1	1	None
Cell volume	50 m^3 cylindrical	30 m^3 cubic	
Cell type	OKSK 1200 skim air	OK 28	
pH	Natural ~9	Natural ~9	
Collector	Sodium ethyl xanthate	Potassium ethyl xanthate	
Frother	Polyfroth H 20	None: natural frothing with recycle H_2O	
Other	$CuSO_4$ to activate pentlandite, guar gum to depress talc	Carboxy methyl cellulose to depress talc	
Concentrate slurry destination	To final concentrate via cycloning and cleaner flotation	To concentrate thickener	
Underflow slurry destination	Semi-autogenous mill re-grind	Rougher flotation	

Kambalda's addition of copper sulfate to the ore slurry leads to the formation of copper sulfide on the surface of the pentlandite, thereby increasing the kinetics of the flotation of pentlandite (Kerr, 2002)

TABLE 15.4 Details of Three Industrial Rougher Flotation Plants

Concentrator	Kambalda, Australia	Raglan, Canada	Thompson, Canada
Feed	Autogenous grinding mill middlings	Flash flotation underflow	Ball mill cyclone undersize
Feed size		80% Passing 65 μm	70% Passing 150 μm
Number of cells	7	7	90
Cell volume (each cell)	16 m^3 cubic	30 m^3 cubic	3 m^3 cubic
Cell type	OK 16	OK 28	Denver DR 30
% Solids in feed			50
pH	Natural ~9	Natural ~9	10
Collector	Sodium ethyl xanthate	Potassium ethyl xanthate	Potassium amyl xanthate
Frother	Polyfroth H 20	None	Methyl isobutyl carbinol
Other	Guar gum to depress talc	Carboxy methyl cellulose to depress talc	Soda ash to adjust pH and $CuSO_4$ to activate pentlandite
Froth destination	Cleaner flotation	Cleaner flotation	Cleaner flotation
Underflow destination	Tailings thickener	Tailings thickener	Ball mill re-grind then scavenger flotation

pentlandite particles are then recovered in cleaner–scavenger froth and sent to cleaning and on to the final concentrate.

The rougher–scavenger and cleaner–scavenger cells maximize pentlandite recovery to concentrate. The cleaner and re-cleaner cells maximize rock and pyrrhotite rejection, thereby maximizing concentrate grade, that is, the %Ni in the concentrate.

Total particle residence times in the above flotation cells are typically between one and one and a half hours (Kerr *et al.*, 2003).

15.6. FLOTATION PRODUCTS

15.6.1. Concentrate

The product of the pentlandite flotation circuit contains approximately 60% water, which has to be removed before the concentrate is smelted.

TABLE 15.5 Details of Three Industrial Cleaner Flotation Plants

Concentrator	Kambalda, Australia	Raglan, Canada	Thompson, Canada
Feed	Flash flotation froth cyclone undersize	Rougher flotation froth and cleaner–scavenger froth	Rougher flotation froth
Feed size			70% Passing 150 μm
Number of cells	5	2	20
Cell volume (each cell)	5 m^3 cubic	160 m^3 rectangular	3 m^3 cubic
Cell type	OK 5, mechanical	Minnovex 2 m × 6 m × 13 m, column	Denver DR 30, mechanical
% Solids in feed			40
pH	Natural ~9	Natural ~9	10
Collector	Sodium ethyl xanthate	Potassium ethyl xanthate	Potassium amyl xanthate
Frother	Polyfroth H 20	None	Methyl isobutyl carbinol
Other	Guar gum to depress talc	Carboxy methyl cellulose to depress talc	
Froth destination	Concentrate thickener	Concentrate thickener	Pentlandite/ chalcopyrite separation
Underflow destination	Fine rougher flotation	Cleaner recycle or ball mill grinding then cleaner–scavengers	Scavenger cleaners then waste tailings

Dewatering is done by settling the mineral particles in large (15–30 m diameter) thickeners or settlers (Appendix E). The solids settle under the influence of gravity to the bottom of the settler from where they are moved to a central bottom discharge by a slowly rotating rake.

Faster settling is obtained by adding organic flocculants (for example, polyacrylamides) at a dosage of 0.002–0.004 kg/tonne of ore (Ferrera, Arinaitwe, & Pawlik, 2009) to the feed. These create large 'flocs' of fine particles that settle faster than the individual fine particles.

TABLE 15.6 Details of Three Industrial Scavenger Flotation Plants

Concentrator	Kambalda, Australia	Raglan, Canada	Thompson, Canada
Feed	Fine rougher underflow	Ground cleaner underflow	Reground rougher underflow
Feed size			
Number of cells	6	4	90
Cell volume, each cell	30 m^3	30 m^3	3 m^3 cubic
Cell type	OK 28	OK 28	Denver DR 30
% Solids in feed			~40
pH	Natural ~9	Natural ~9	10
Collector	Sodium ethyl xanthate	Potassium ethyl xanthate	Potassium ethyl xanthate
Frother	Polyfroth H 20	None	Methyl isobutyl carbinol
Other	Guar gum to depress talc	Carboxy methyl cellulose to depress talc	None
Froth destination	Cleaner flotation	Cleaner flotation	Cleaner flotation
Underflow destination	waste tailings	Waste tailings	Waste tailings

The underflow from the thickeners still contains between 30% and 40% water. This is lowered to 10% or 15% water in rotary vacuum filters and then to 8% water in pressure filters (Larox, 2010).

Lastly, the filtered concentrate is dried to about 0.2% moisture in steam-heated driers, hydrocarbon-fired fluid bed driers or fluid bed roasters before it is fed to a smelting furnace.

15.6.2. Tailings

About 85% of the feed to the concentrator reports to the flotation tailings. The tailings are mostly stored in large dams near the mine property. Water is usually reclaimed from the dams and recycled to the concentrator (Liechti, Mayhew, Kerr, & Truskoski, 2000).

Recycling water minimizes the consumption of water and the possibility of contamination of the environment by concentrator effluents. Because the pH of the tailing water is close to that required for flash and rougher–scavenger flotation, recycling of tailings water also minimizes the consumption of CaO and Na_2CO_3.

Rock tailings are often used as backfill in underground mines (Wyshynski, Easton, Wood, & Farr, 2000). They are dewatered to approximately 30% water and combined with cement as mine backfill.

Pyrrhotite tailings are often stored under water to minimize oxidation and accidental sulfuric acid generation (Liechti *et al.*, 2000; Wyshynski *et al.*, 2000).

15.7. OPERATION AND CONTROL

Modern flotation plants are well equipped with sensors and automated control systems. The objectives of this equipment are to maintain continuous steady state operation while meeting the objectives of maximizing the recovery of nickel and the concentrate grade.

Measured flotation variables are as follows:

(a) particle size after grinding and re-grinding (Outotec, 2010a);
(b) %Ni, %solids, pH and mass flow rates of the process streams (Outotec, 2010b);
(c) froth thickness in the flotation cells (a thick froth layer allows more time for gangue rock and pyrrhotite to drain than does a thin froth layer); and,
(d) impeller speeds and air input rates.

Adjustments made on the basis of these measurements are as follows:

(a) flow rates of water into hydrocyclone feed sumps of the mill to control flotation feed size;
(b) addition rates of flotation reagent (collector, frother and modifier) and water throughout the flotation plant; and,
(c) froth thickness, by adjusting the concentration of the frother and levels of the slurry in the flotation cell (Elmahdy & Finch, 2009).

Details of the flotation measurements and the controller adjustments that are made in the flotation plant are given in Table 15.7.

15.8. RECENT DEVELOPMENTS

Important recent developments in pentlandite flotation are as follows:

(a) quantitative electron microscope analysis of ores and concentrator process streams; and,
(b) flash flotation.

TABLE 15.7 Flotation Measurements and Their Use in Flotation Control and Optimization

Sensing instrument	Purpose	Type of device	Use in automatic control
Process stream particle size analyzer	Senses particle sizes after grind and re-grind mills	Laser beam diffraction (Outotec, 2010a)	Controls water addition rates to hydrocyclone feed (which controls particle recycle to ball mills and final grind size)
In-stream X-ray chemical analyzer	Senses nickel content of solids in process streams	X-ray fluorescence analysis of automatically taken process stream samples (Outotec, 2010b)	Controls collector, frother, depressant and water addition rates throughout the plant. Adjusts valves in flotation cells to vary froth thickness
Slurry level sensor	Indicates pulp surface location	Float level, hydrostatic pressure, conductivity	Adjusts underflow valves to maintain prescribed froth layer thicknesses
Slurry mass flow meter and % solids in slurry meter	Measures mass and volumetric flow rates of process streams	Magnetic induction, Doppler effects, ultrasonic energy loss	Measures flows in flotation circuit, permits optimization of recycle flow rates

15.8.1. Electron Microscope Analysis of Ores and Process Streams

Technologies for quantifying the mineralogical composition of individual ore particles were developed during the 1990s. The heart of the technology is the scanning electron microscope coupled with energy dispersive X-ray analysis (SEM-EDX).

This technology permits the evaluation of an assemblage of particles (for example, flotation feed) to determine the size and shape of each particle and provides a mineral map of each particle that gives the minerals present, their quantity and their association with other minerals (Dai *et al.*, 2009; Ford *et al.*, 2009).

SEM-EDX may be used to answer questions such as:

(a) What fraction of the pentlandite in a sample is present as liberated pentlandite particles?
(b) What fraction of the pentlandite in a sample is present as grains ‘locked’ (hidden) inside rock and pyrrhotite particles?

(c) How much more of the pentlandite in a sample would be liberated if the particle size of the sample was decreased from 70% passing 45 μm to 80% passing 45 μm?

Such analysis of samples collected from process streams under different operating conditions is used to optimize plant operation.

It seems unlikely that any new concentrator project will be approved without quantitative SEM analysis of the ore to predict its likely behavior during concentration (Dai *et al.*, 2009).

15.9. SUMMARY

Froth flotation of pentlandite ores makes concentrates containing 5%–15% Ni suitable for smelting and refining to high-purity nickel from ground ore containing 1%–3% Ni.

This is achieved by

(a) causing ground pentlandite particles (~50 μm) to become attached to small (~1 mm diameter) air bubbles rising in a water–ore particle slurry;
(b) capturing these particles in a short-lived froth that overflows into a concentrate collection tank; and,
(c) leaving gangue rock and pyrrhotite in the slurry.

The process is made selective for pentlandite by using flotation reagents, called collectors, which make the pentlandite hydrophobic while leaving the gangue rock and pyrrhotite hydrophilic.

Xanthate is the universal collector for pentlandite flotation. It is assisted by modifiers, such as hydroxyl ions, which prevent other minerals, for example, pyrrhotite and talc, from floating with the pentlandite.

Chalcopyrite and pentlandite float together under the conditions described in this chapter. They are often separated by selective flotation of chalcopyrite, which is discussed in the next chapter.

REFERENCES

Boldt, J. R., & Queneau, P. (1967). *The winning of nickel.* (p. 39). Longmans.

Comeau, G., Hardie, C., Pelletier, S., & Rhabani, R. (2008). Commissioning and testing of thickener feed de-aeration cyclones at the Xstrata nickel Raglan operation. In *Paper 15 in proceedings of the 40th annual meeting of the Canadian mineral processors* (pp. 237–253). CIM.

Dai, Z., Bos, J. A., Quinn, P., et al. (2009). Flowsheet development for Thompson ultramafic low-grade nickel ores. In C. O. Gomez, J. E. Nesset & S. R. Rao (Eds.), *Advances in mineral processing science and technology: Proceedings of the 7th UBC-MCGill-UA international symposium on fundamentals of mineral processing* (pp. 217–227). CIM.

Doucet, J., Price, C., Barrette, R., & Lawson, V. (2009). Evaluating the effect of operational changes at Vale Inco's Clarabelle mill. In C. O. Gomez, J. E. Nesset & S. R. Rao (Eds.), *Advances in mineral processing science and technology: Proceedings of the 7th UBC-MCGill-UA international symposium on fundamentals of mineral processing* (pp. 337–347). CIM.

Elmahdy, A. M., & Finch, J. A. (2009). Effect of frother blends on hydrodynamic properties. In C. O. Gomez, J. E. Nesset & S. R. Rao (Eds.), *Advances in mineral processing science and technology: Proceedings of the 7th UBC-MCGill-UA international symposium on fundamentals of mineral processing* (pp. 125–133). CIM.

Ferrera, V., Arinaitwe, E., & Pawlik, M. (2009). A role of flocculant conformation in the flocculation process. In C. O. Gomez, J. E. Nesset & S. R. Rao (Eds.), *Advances in mineral processing science and technology: Proceedings of the 7th UBC-MCGill-UA international symposium on fundamentals of mineral processing* (pp. 397–408). CIM.

Finch, J. A., Cilliers, J., & Yianatos, J. (2007). Column flotation. In M. C. Fuerstenau, G. Jameson & R. H. Yoon (Eds.), *Froth flotation, A century of innovation* (pp. 681–755). SME.

Ford, F., Lee, A., Davis, C., et al. (2009). Predicting Clarabelle mill recoveries using Mineral Liberation Analyzer (MLA) grade-recovery curves. In C. Hamilton, B. Hart & P. J. Whittaker (Eds.), *Mineralogy: Proceeding of the 48th annual conference of metallurgists of CIM* (pp. 31–41). CIM.

Fuerstenau, M. C., Jameson, G. & Yoon, R.-H. (Eds.). (2007). *Froth flotation, A century of innovation.* SME.

Gomez, C. O., Nesset, J. E. & Rao, S. R. (Eds.). (2009). *Advances in mineral processing science and technology: Proceedings of the 7th UBC-MCGill-UA international symposium on fundamentals of mineral processing.* CIM.

Gorain, B. K., Oravainen, H., Allenius, H., et al. (2007). Mechanical froth flotation cells. In M. C. Fuerstenau, G. Jameson & R. H. Yoon (Eds.), *Froth flotation, A century of innovation* (pp. 637–680). SME.

Hagihara, H. (1952). Mono- and multilayer adsorption of aqueous xanthate on galena surfaces. *Journal of Physical Chemistry, 56*, 606–621.

Hernandez-Aguilar, J. R. (2009). Gas dispersion studies at Highland Valley Copper. In C. O. Gomez, J. E. Nesset & S. R. Rao (Eds.), *Advances in mineral processing science and technology: Proceedings of the 7th UBC-MCGill-UA international symposium on fundamentals of mineral processing* (pp. 349–359). CIM.

Kerr, A. (2002). An overview of recent developments in flotation technology and plant practice for nickel ores. In A. L. Mular, D. N. Halbe & D. J. Barratt (Eds.), *Mineral processing, plant design, practice and control proceedings, Vol. 1* (pp. 1142–1158). SME.

Knights, B. D. H., & Bryson, M. A. W. (2009). Current challenges in PGM flotation of South African ores. In C. O. Gomez, J. E. Nesset & S. R. Rao (Eds.), *Advances in minerals processing* (pp. 285–396). CIM.

Larox. (2010). Larox mineral concentrates [Company brochure].

Liechti, D., Mayhew, M., Kerr, A., & Truskoski, J. (2000). Inco Limited, Ontario Division – Clarabelle mill. In B. Damjanovic & J. R. Goode (Eds.), *Canadian milling practice, Special Vol. 48* (pp. 97–101). CIM.

Nagaraj, D. R., & Ravishankar, S. A. (2007). Flotation reagents – A critical overview from an industry perspective. In M. C. Fuerstenau, G. Jameson & R. H. Yoon (Eds.), *Froth flotation, A century of innovation* (pp. 375–423). SME.

Outotec. (2010a). Grinding control solutions [Company brochure].

Outotec. (2010b). On-line elemental analyzers [Company brochure].

Warder, J., & McQuie, J. (2005). The role of flash flotation in reducing overgrinding of nickel at WMC's Leinster nickel operation. In G. J. Johnson (Ed.), *Centenary of flotation symposium* (pp. 931–936). AusIMM.

Warner, A. E. M., Diaz, C. M., Dalvi, A. D., et al. (2007). JOM world nonferrous smelter survey part IV: Nickel: Sulfide. *JOM, 59*, 58–72.

Wiese, J. G., Becker, M., Bradshaw, D. J., & Harris, P. J. (2007). Interpreting the role of reagents in the flotation of platinum-bearing Merensky ores. *Journal of the Southern African Institute of Mining and Metallurgy, 107*, 29–36.

Wyshynski, M., Easton, D., Wood, D., & Farr, I. (2000). Inco Limited, Manitoba Division – Thompson mill. In B. Damjanovic & J. R. Goode (Eds.), *Canadian milling practice, Special Vol. 48* (pp. 101–103). CIM.

SUGGESTED READING

Folinsbee, J. (Ed.). (2007). *Proceedings of the 39th annual meeting of the Canadian mineral processors*. CIM.

Fuerstenau, M. C., Jameson, G., & Yoon, R.-H. (Eds.). (2007). *Froth flotation: A century of innovation*. SME.

Advances in mineral processing science and technology. (2009). In C. O. Gomez, J. E. Nesset & S. R. Rao (Eds.), *Proceedings of the 7th UBC-MCGill-UA international symposium on fundamentals of mineral processing*. CIM.

Kerr, A., Bouchard, A., Truskoski, J., et al. (2003). The "Mill Redesign Project" at Inco's Clarabelle Mill. *CIM Bulletin, 96*, 58–66.

Malhotra, D., Taylor, P., Spiller, E., & LeVier, M. (Eds.). (2009). *Recent advances in mineral processing plant design*. SME.

Chapter 16

Separation of Chalcopyrite from Pentlandite by Flotation

All nickel sulfide ores contain chalcopyrite, $CuFeS_2$ and pentlandite, $(Ni,Fe)_9S_8$. During 'bulk' flotation, which is done at pH of approximately 9, the chalcopyrite floats with the pentlandite.

If the quantity of chalcopyrite is small, it is usually left in the chalcopyrite–pentlandite concentrate. The contained copper is then recovered during downstream smelting and refining of the nickel.

However, if the quantity is large, the chalcopyrite is usually floated from the bulk concentrate at a pH value of approximately 12 to form a separate copper concentrate that is ready for smelting and refining to high-purity copper. A flowsheet of the chalcopyrite flotation circuit is given in Figure 16.1.

Chalcopyrite flotation is usually done when the mass ratio of Cu:Ni in the ore is greater than 0.3, as shown in Table 16.1. At these concentrations, it is more economical to separate copper and nickel physically (by flotation) than it is to separate them chemically (by smelting and refining).

16.1. CHAPTER OBJECTIVES

The objectives of this chapter are

(a) to describe the principles of floating chalcopyrite from mixed chalcopyrite–pentlandite concentrates; and,
(b) to indicate how these principles are applied industrially to make separate nickel and copper concentrates.

16.2. SEPARATION OF CHALCOPYRITE AND PENTLANDITE

The universal strategy for the separation of copper and nickel is to float chalcopyrite from mixed pentlandite–chalcopyrite concentrate by deactivating the pentlandite, so that it does not float, then floating the chalcopyrite.

The pentlandite is deactivated by adding lime, CaO, to the slurry so that the pH increases to 12. The increased concentration of the OH^- ions displaces xanthate ions from the pentlandite surfaces, causing them to become hydrophilic and hence unfloatable.

Extractive Metallurgy of Nickel, Cobalt and Platinum-Group Metals. DOI: 10.1016/B978-0-08-096809-4.10016-4

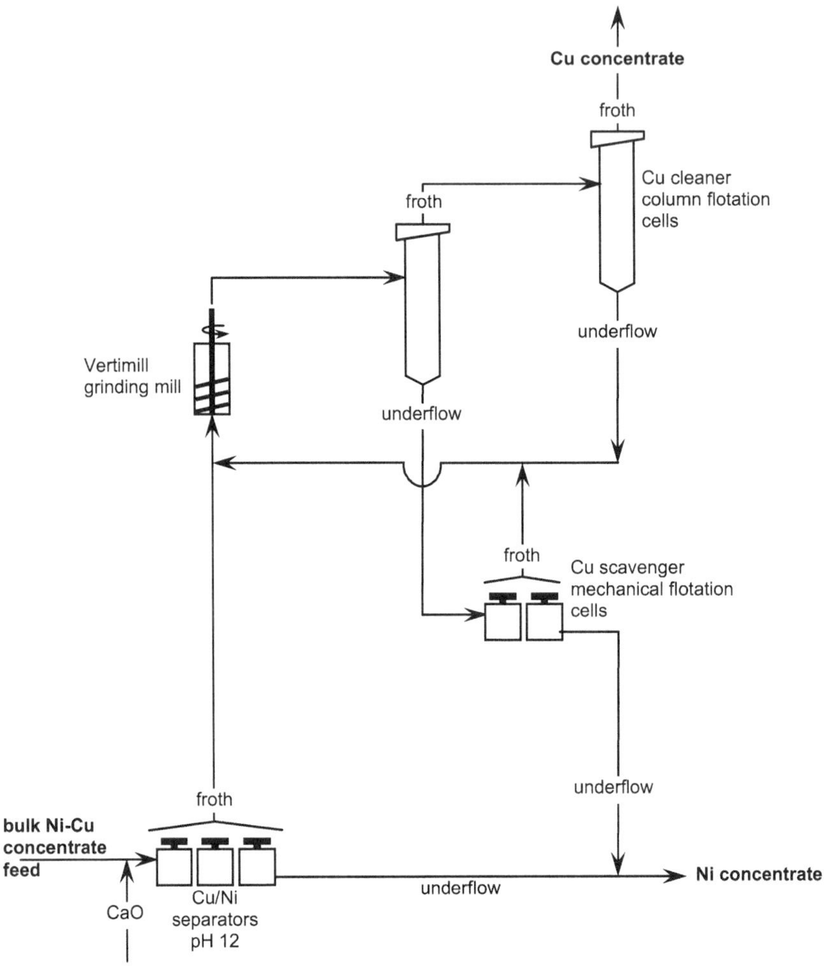

FIGURE 16.1 Schematic flowsheet of copper–nickel (chalcopyrite–pentlandite) separation circuit. Chalcopyrite is floated from the bulk concentrate – giving a chalcopyrite concentrate froth and a pentlandite concentrate underflow. The principal reagent is CaO, added as slaked lime, $Ca(OH)_2$. Re-grinding may not be required (Xu & Wells, 2005). The nickel concentrate can also be cleaned (not shown) to produce a high-grade nickel concentrate for sale and a lower nickel grade 'middlings' concentrate for the mining company's smelter.

The xanthate ions continue to bond with the chalcopyrite at this pH. As a result, the chalcopyrite particles remain hydrophobic and readily attach to rising bubbles in a flotation cell.

16.2.1. Pyrrhotite and Gangue

Pyrrhotite and gangue rock in the bulk concentrate report to the pentlandite concentrate. The gangue rock is hydrophilic, as is pyrrhotite at high pH values.

TABLE 16.1 Treatment of 10 Pentlandite/Chalcopyrite Ores

Concentrator	Ni in feed ore, %	Cu in feed ore, %	Cu:Ni ratio	Products and destinations
Kambalda, W. Australia (BHP Billiton)	2.25	0.21	0.1	Ni bulk concentrate (13% Ni) to company's flash furnace smelter
Leinster, W. Australia (BHP Billiton)				Ni bulk concentrate (86% Ni recovery) to company's flash furnace smelter
Mt. Keith, W. Australia (BHP Billiton)	0.58	<0.1		Ni bulk concentrate (19% Ni) to company's flash furnace smelter (Kerr, 2002)
Selebi Phikwe, Botswana (Norilsk)	0.7	0.7	1	Bulk concentrate (5% Ni, 3% Cu) to company's flash furnace smelter
Thompson, Manitoba (Vale Limited)	2.61, 1.74	0.16, 0.11	0.05	Ni concentrate (12% Ni, 0.2% Cu) to company's electric furnace smelter.
				Middling Cu concentrate (6% Ni, 6% Cu) to company's flash furnace smelter
Clarabelle, Sudbury, Ontario (Vale Limited)	1–2	0.9–2.5	1.1	Ni concentrate to company's flash furnace smelter, Cu concentrate to market
Voisey's Bay, Labrador (Vale Limited)	3	2	0.66	Ni concentrate to flash or electric furnace smelter, Cu concentrate to market
Strathcona, Sudbury, Ontario (Xstrata)	1.8	1.9	1.1	Ni concentrate (12% Ni, 3% Cu) to company's electric furnace Ni smelter; Cu concentrate (31% Cu, 0.4% Ni) to company's Noranda furnace smelter
Raglan, North Quebec (Xstrata)	2.8	0.8	0.25	Bulk concentrate (17% Ni, 4% Cu) to company's electric furnace smelter
Nadezda, Siberia, (Norilsk)				Bulk concentrate (12% Ni, 5% Cu) to company's flash furnace smelter

Ore whose Cu:Ni ratio is >0.3 is usually made into separate nickel and copper concentrates. Ore whose Cu:Ni ratio is <0.3 is usually left as a single 'bulk' concentrate.

As a result, pyrrhotite and gangue rock do not float and hence report to the underflow, which forms the pentlandite concentrate.

16.2.2. Objective: High-Grade (%Cu) Concentrate

The above discussion indicates that the dominant Cu:Ni separation strategy is to produce a high-grade chalcopyrite concentrate free of pentlandite, pyrrhotite and gangue rock and a relatively low-grade pentlandite concentrate, containing pyrrhotite and rock.

The nickel content of copper concentrates must be kept at a low level (less than 0.5% Ni) because nickel is costly to deal with in electrolytic copper refineries. The nickel enters the copper sulfate solution from which it must be removed to avoid contaminating the copper cathode product, which must be greater than 99.99% pure.

16.3. INDUSTRIAL PRACTICE

A schematic diagram of an industrial flotation circuit that separates copper and nickel is shown in Figure 16.1. The principal objective is to produce separate nickel and copper concentrates. A slurry of the bulk concentrate enters the flowsheet at the lower left. Cleaned copper concentrate froth leaves upper right while nickel concentrate underflow leaves lower right.

The operating conditions for several industrial concentrators are given in Table 16.2.

As shown in Figure 16.1, copper concentrate froth from the copper–nickel separator cells is cleaned in two column cells before being dewatered and smelted. A schematic diagram of a column flotation cell is shown in Figure 16.2. The combined underflow from the separator and copper scavenger cells is nickel concentrate. It is dried and then smelted in a nickel smelter.

16.3.1. Reagents

The principal reagent in the circuit shown in Figure 16.1 is lime, CaO. The addition of lime increases the concentration of OH^- ions in the slurry, thereby depressing pentlandite and pyrrhotite particles, and separating them from the copper concentrate. Dextrin and sodium cyanide are also occasionally used while triethylenetetramine has been suggested but not used (Xu & Wells, 2005).

Xanthate collectors and frothers are not usually needed. There is usually enough remaining in the slurry feed.

16.4. GRINDING

The feed to copper–nickel separator flotation may contain a significant quantity of unliberated chalcopyrite–pentlandite particles. In this case, the floated chalcopyrite froth may be reground to liberate the chalcopyrite before cleaner

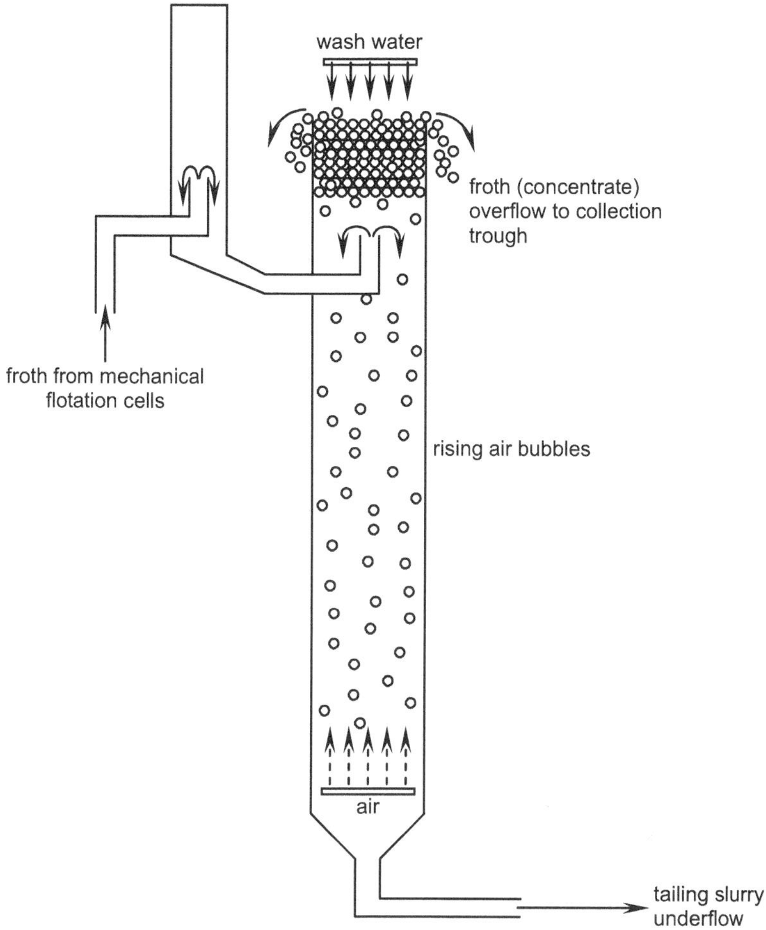

FIGURE 16.2 Sketch of column flotation cell. Air bubbles carry chalcopyrite particles up to the top of the column where they overflow in concentrate froth. Water spraying on top of the froth layer washes weakly held pentlandite, pyrrhotite and gangue from the froth back into the slurry and eventually out in the underflow tailings slurry. Column cells are typically 1–4 m in diameter and 9–14 m in height.

flotation, as shown in Figure 16.1. This, however, is seldom done (Doucet, Price, Barrette, & Lawson, 2009; Holmes *et al.*, 2000).

16.5. SUMMARY

Nickel sulfide ores contain pentlandite, $(Ni,Fe)_9S_8$, and chalcopyrite, $CuFeS_2$. These minerals are always recovered together in a metal-rich bulk concentrate by the following steps:

(a) crushing/grinding; and,

TABLE 16.2 Details of Three Chalcopyrite/Pentlandite Separator Flotation Plants

Location	Thompson, Manitoba	Timmins, Ontario (now shut)	Voisey's Bay, Labrador
Feed to pentlandite/ chalcopyrite separator	Cleaned Ni rougher concentrate	Cleaned Ni rougher concentrate	Cleaned Ni rougher concentrate
% Ni	12	10	~17
% Cu	1	5	~12
Tonnes feed per day	1900	250	800
Particle size	70% passing 150 μm	80% passing 50 μm	80% passing 150 μm
% Solids in slurry	40		
Reground before feeding	No	No	No
Flotation conditions			
Aeration	No	No	Yes
Pentlandite/pyrrhotite depressant	CaO (3 kg per tonne of separator feed)	CaO, dextrin	CaO
pH	~12 (CaO saturated)	~12	~12
Collector	Residual potassium amyl xanthate	Residual potassium isobutyl xanthate	Added Flex 31
Frother	Residual methyl isobutyl carbinol	Added methyl isobutyl carbinol	Residual Dowfroth 250

Flotation cell type	Mechanical 2.8 m^3	Column	Mechanical 38 m^3
Products			
Ni concentrate (underflow)			
Tonnes per day	1750	210	~500
% Ni	12	12	Not available
% Cu	0.2	1	1
Destination	Company Ni smelter	Company Ni smelter	Company Ni smelters
Cu concentrate (froth)			
Tonnes per day	150	40	~300
% Ni	6	0.4	0.6
% Cu	6	30	32
Destination	Company flash smelter	Company Noranda process smelter	Market
Reference	Wyshynski, Easton, Wood, & Farr, 2000	Charland *et al.*, 2006	Wells *et al.*, 2007

All receive cleaned bulk pentlandite–chalcopyrite concentrate and float chalcopyrite particles from pentlandite, gangue and pyrrhotite particles.

(b) froth flotation of the ground pentlandite and chalcopyrite particles at pH ~9, which separates them from the pyrrhotite and gangue rock particles in the ore.

This bulk concentrate is sometimes separated into individual copper and nickel concentrates by froth flotation of chalcopyrite from pentlandite at a pH of about 12. Lime, CaO, is used to adjust the pH of the feed slurry.

Separation of the bulk concentrate into a nickel concentrate and a copper concentrate is usually performed when the Cu:Ni mass ratio of the bulk concentrate is greater than 0.3.

These concentrates are then dried, smelted and refined in separate copper and nickel smelters and refineries. This is discussed in the next chapter.

REFERENCES

Charland, A., Kormos, L., Whittaker, P., et al. (2006). *A case study for integrated use of automated mineralogy in plant optimization: the Falconbridge Montcalm concentrator, In Automated Mineralogy, 06*. MEI.

Doucet, J., Price, C., Barrette, R., & Lawson, V. (2009). Evaluating the effect of operational changes at Vale Inco's Clarabelle mill. In C. O. Gomez, J. E. Nesset & S. R. Rao (Eds.), *Advances in mineral processing science and technology: Proceedings of the 7th UBC-MCGill-UA international symposium on fundamentals of mineral processing* (pp. 337–347). CIM.

Holmes, W., MacNamara, D., Marrs, G., et al. (2000). Falconbridge Limited – Strathcona mill. In B. Damjanovic & J. R. Goode (Eds.), *Canadian milling practice, Special, Vol. 48* (pp. 95–97). CIM.

Wells, P., Langlois, P., Barrett, J., et al. (2007). Flowsheet development, commissioning and start-up of the Voisey's Bay Mill. In J. Folinsbee (Ed.), *Proceedings of the 39th annual meeting of the Canadian mineral processors* (pp. 3–18). CIM.

Xu, M., & Wells, P. F. (2005). Laboratory and miniplant studies on Cu/Ni separation. In J. Donald & R. Schonewille (Eds.), *Nickel and cobalt 2005, challenges in extraction and production* (pp. 347–358). CIM.

Wyshynski, M., Easton, D., Wood, D., & Farr, I. (2000). Inco Limited, Manitoba Division – Thompson mill. In B. Damjanovic & J. R. Goode (Eds.), *Canadian milling practice, Special, Vol. 48* (pp. 101–103). CIM.

SUGGESTED READING

Kerr, A. (2002). An overview of recent developments in flotation technology and plant practice for nickel ores. In A. L. Mular, D. N. Halbe & D. J. Barratt (Eds.), *Mineral processing, plant design, practice and control proceedings, Vol. 1* (pp. 1142–1158). SME.

Xu, M., & Wells, P. F. (2005). Laboratory and miniplant studies on Cu/Ni separation. In J. Donald & R. Schonewille (Eds.), *Nickel and cobalt 2005, challenges in extraction and production* (pp. 347–358). CIM.

Chapter 17

Smelting of Nickel Sulfide Concentrates by Roasting and Electric Furnace Smelting

Production of a nickel matte suitable for refining to high-purity nickel from nickel sulfide concentrates occurs in two steps:

(a) smelting (oxidizing and melting) the concentrates to make a molten sulfide matte; and,
(b) converting (oxidizing) more of the iron and sulfur from this molten matte.

About half of the global production of primary nickel makes use of these processes (Warner *et al.*, 2007). The product of sulfide smelting and converting is a molten matte that is nickel rich (40%–75% Ni). The content of iron (1%–5% Fe) and sulfur (21%–24% S) are relatively low. Copper and cobalt may also be present. This matte is suitable for hydro- and pyro-metallurgical refining to high-purity nickel and other products.

Smelting produces nickel-enriched matte by:

(a) removing sulfur from the nickel concentrate by oxidizing it to sulfur dioxide;
(b) removing iron from the concentrate by oxidizing it to iron oxide and fluxing the iron oxide with silica to form a molten iron silicate slag at 1300°C; and,
(c) removing gangue rock from the concentrate by dissolving it in the molten silicate slag.

Two industrial methods are used for smelting nickel sulfides: (i) roasting followed by smelting in an electric furnace; or, (ii) flash smelting. These different routes for smelting are illustrated in Figure 17.1.

Roasting and smelting in an electric furnace are described in this chapter, flash smelting is discussed in Chapter 18 and converting is discussed in Chapter 19.

The objectives of this chapter are to describe:

(a) principles of the roasting and smelting of nickel sulfide concentrates;
(b) industrial practice of fluidized-bed roasting and electric furnace smelting;

Extractive Metallurgy of Nickel, Cobalt and Platinum-Group Metals. DOI: 10.1016/B978-0-08-096809-4.10017-6

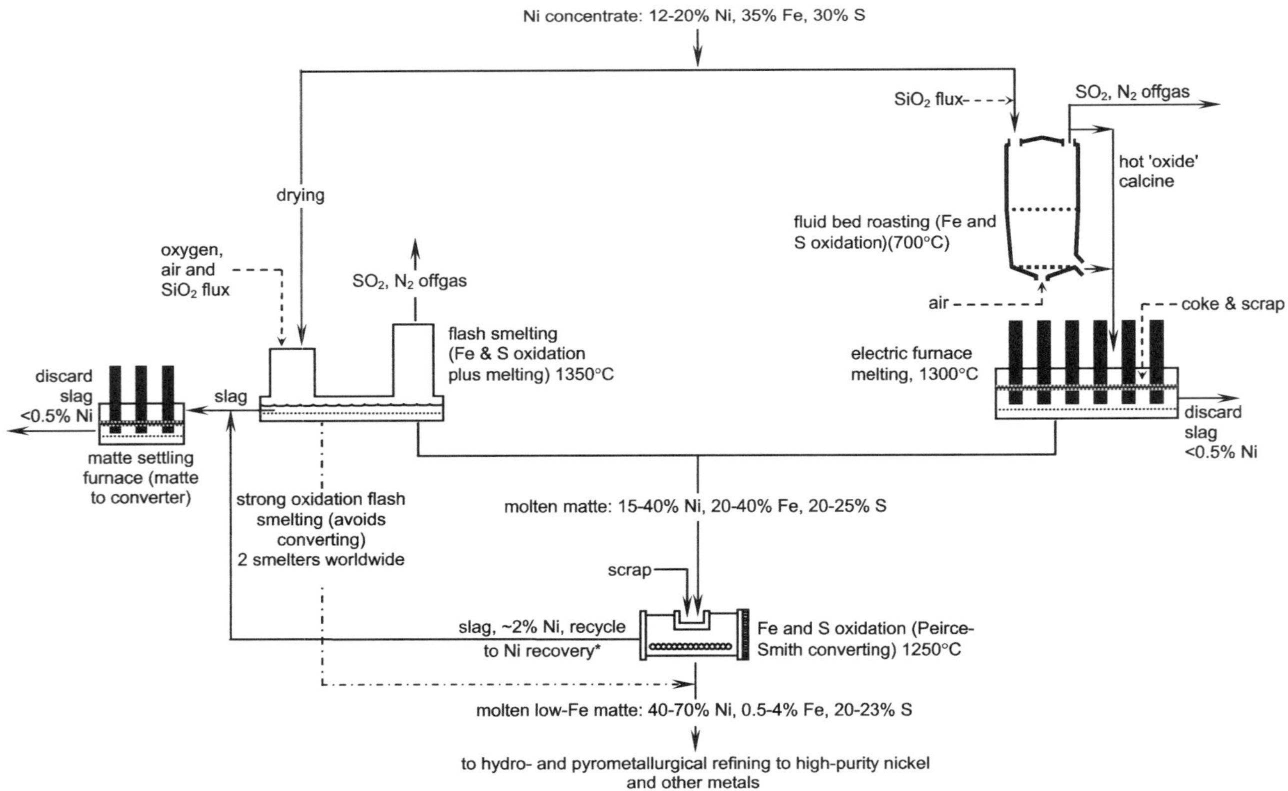

FIGURE 17.1 Alternative ways of producing nickel-rich, low-iron matte from nickel sulfide concentrates. Flash smelting is on the left, roasting/electric furnace smelting on the right. Roasting and smelting are continuous. Peirce–Smith converting is batch. Other metals in the concentrate feed (copper, cobalt, silver, gold, platinum-group elements) mostly accompany nickel through smelting and converting. They are recovered during refining of the converter matte. *Converter slag is usually sent to an electric furnace for nickel matte recovery (shown on the left). It is occasionally recycled back to the electric *smelting* furnace.

(c) metal recoveries and energy requirements; and,
(d) control and optimization.

Industrial details for the operation of roaster and electric smelting furnaces are given in Tables 17.1 and 17.2.

TABLE 17.1 Details of Two Nickel Smelter Fluidized-Bed Roasters*

Smelter	Falconbridge, Canada	Thompson, Canada
Start-up date	1978	1961
Roaster type	Fluidized bed	Fluidized bed
Number		
Inside dimensions, m		
Bed diameter	5.6	5.5
Freeboard diameter	8	6.4
Height above tuyeres		6.5
Nominal capacity, each roaster tonnes of dry solids/h	40	55
Number of tuyeres		256
Distance apart, m		0.2
Air pressure below tuyeres bar		0.3
Feed	Concentrate–water slurry	Concentrate–water slurry
% Concentrate	70	90
% Water	30	10
% Feed sulfur oxidized to SO_2 in roaster	70	40
SO_2 destination	Sulfuric acid plant	Atmosphere
Oxygen utilization, %		~100
Temperatures		
Bed	760	600
Discharge calcine	760	580
Offgas		530
Offgas		
Volume% SO_2	11–13	25
Discharge rate, Nm^3/h	40 000	48 000
Offgas handling system	Cyclones, electrostatic precipitators, scrubbing, dewatering, acid plant	Cyclones to balloon flue to chimney stack
Product		
% Reporting to cyclones		75
% Overflowing the fluidized bed		25

They oxidize concentrate in preparation for electric furnace smelting (Schonewille *et al.*, 2005).
**Warner et al., 2007*

TABLE 17.2 Production Details of Three Electric Smelting Furnaces Treating Roasted Nickel Concentrate*

Smelter	Falconbridge, Canada	Thompson, Canada	Norilsk, Russia
Start-up date	1979	1961	
Feed material	Roasted (oxidized) concentrate	Roasted (oxidized) concentrate	Roasted (oxidized) concentrate
Number of furnaces	1	2	3
Size, outside, m			
Hearth, *w* × *l* × *h*	9 × 30 × 3	11 × 32 × 6	10 × 27 × 5
Wall cooling system	Water-cooled copper plates and fingers	Coolers on walls and around tapholes	Water-cooled copper elements
Electrodes	6 in line	6 in line	6 in line
Diameter, m	1.4	1.2	
Material	Solidified paste (Söderberg)	Solidified paste (Söderberg)	
Paste consumption, kg/tonne of dry solid feed		3.5	2–3
Converter slag additions	No	Yes	No
Feed details, tonnes/h			
Dry solid charge, mainly roasted concentrate	80	65 each furnace	75 each furnace
Approximate composition	12% Ni, 4.5% Cu, 0.5% Co, 31% Fe, 28% S	14%Ni, 0.3% Cu, 0.3% Co, 37% Fe, 29% S	5% Ni, 2.5% Cu, 0.2% Co, 43% Fe, 7% S (roasted)
Molten converter slag	None		
Reductant added	Coke (~0.04 tonnes/tonne of calcine)		
Operating details			
Power consumption, kWh/tonne of dry solid feed	440	470	515
Average operating power, MW	40	16	
Average in furnace current, A	38 000	17 000	50 000
Average electrode to electrode voltage, V	1050	320	500
Production details			
Matte, tonnes/h	~30	~30 each furnace	
Matte composition	36% Ni, 11% Cu, 1% Co, 33% Fe, 17% S	30% Ni, 1% Cu, 1% Co, 37% Fe, 27% S	13% Ni, 8% Cu, 0.7% Co, 53% Fe, 23% S

TABLE 17.2 Production Details of Three Electric Smelting Furnaces Treating Roasted Nickel Concentrate*—cont'd

Smelter	Falconbridge, Canada	Thompson, Canada	Norilsk, Russia
Slag, tonnes/h	~50	~35 (each furnace)	
Destination	Discard	Granulation and discard	Discard
Slag composition	35% Fe, 35% SiO_2	0.4% Ni, 0.2% Co, 37% Fe, 35% SiO_2	0.1% Ni, 0.1% Co, 33% Fe, 37% SiO_2
Matte from converter slag recovery method	Cylindrical slag cleaning furnace	Recycle to smelting furnace	Electric slag settling furnace
Offgas, Nm^3/h			
Vol% SO_2 leaving furnace	1000	3300	70
Matte/slag temperature, °C	1265/1310	1190/1310	1200/1300

Electric smelting efficiently recovers nickel, cobalt, copper and precious metals but it requires considerable electrical energy.
**Warner et al., 2007*

17.1. PRINCIPLES OF ROASTING AND SMELTING

About a quarter of nickel sulfide smelting is done by roasting and smelting (Makinen & Ahokainen, 2009). It entails two operations:

(a) partially oxidizing the nickel concentrate in a fluidized-bed roaster to form a calcine; and,
(b) melting the calcine plus silica flux in an electric furnace to form a nickel-rich molten matte and a nickel-lean molten slag.

The fluidized-bed roaster is illustrated in Figure 17.2 and the electric furnace is illustrated in Figure 17.3.

Matte smelting in an electric furnace is also used for laterite concentrates (see Chapter 8) and for extracting platinum-group metals from sulfide concentrates (see Chapter 35).

The products from smelting are (i) molten nickel-rich sulfide matte; (ii) molten nickel-lean silicate slag; and, (iii) offgas bearing sulfur dioxide. The molten matte is transferred to the converting process for further oxidation of iron and sulfur; the molten slag is discarded and the sulfur dioxide is converted to sulfuric acid.

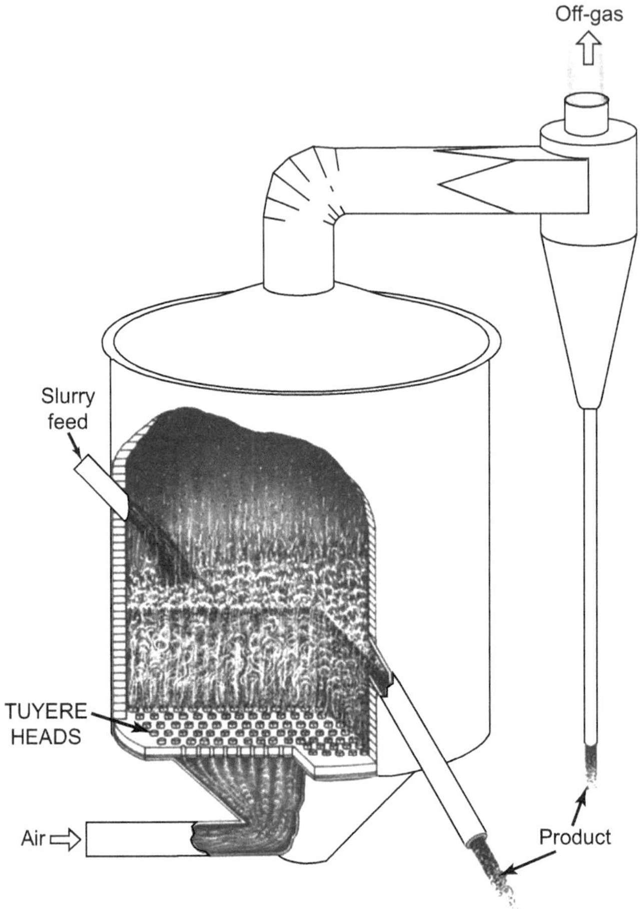

FIGURE 17.2 Fluidized-bed roaster for partially oxidizing 50–100 μm pentlandite concentrate particles. A slurry of concentrate, silica flux and water is fed to the top of the roaster. Air is blown upwards through tuyeres at the bottom. The result is a fluidized air-concentrate particle bed where oxygen and concentrate particles react quickly and uniformly. The product is hot partially oxidized calcine particles ready for electric furnace smelting to nickel-rich molten sulfide matte. This roaster is ~5 m in diameter and 6 m in height. It roasts ~50 tonnes/h concentrate. Industrial details are given in Table 17.1. *Source: Drawing from Boldt and Queneau (1967), courtesy of Vale.*

17.2. CHEMISTRY OF ROASTING

Roasting nickel concentrates is the *partial oxidation* of concentrates with air. It is described by the following reactions:

$$\underset{\text{pentlandite in concentrate}}{Ni_{4.5}Fe_{4.5}S_8(s)} + \underset{\text{in air}}{12.5O_2(g)} \xrightarrow{700^\circ C} \underset{\text{in calcine}}{4.5NiO(s) + 4.5FeO(s)} + \underset{\text{in } N_2\text{-}SO_2 \text{ offgas}}{8SO_2(g)} \tag{17.1}$$

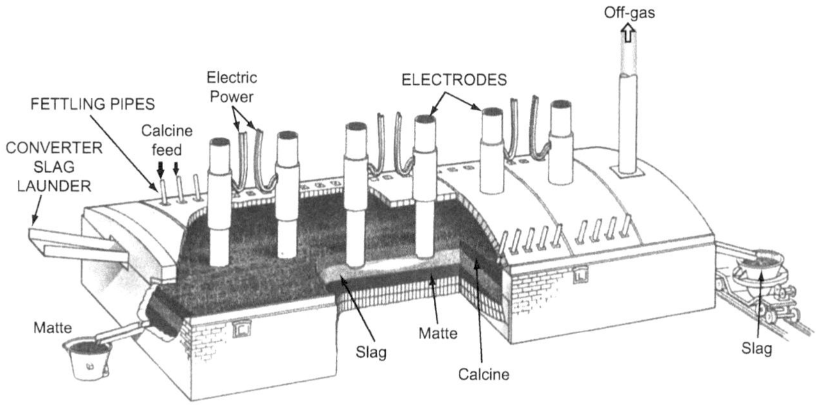

FIGURE 17.3 Electric furnace for continuously smelting roasted nickel concentrates to nickel-rich molten sulfide matte at 1300°C. The furnace is heated by passing electrical current between large suspended carbon electrodes through molten slag and matte. Operational details are given Table 17.2. *Source: Drawing from Boldt and Queneau (1967), courtesy of Vale.*

$$\underset{\text{pyrrhotite in concentrate}}{Fe_8S_9(s)} + \underset{\text{in air}}{13O_2(g)} \xrightarrow{700^\circ C} \underset{\text{in calcine}}{8FeO(s)} + \underset{\text{with } N_2 \text{ in offgas}}{9SO_2(g)} \qquad (17.2)$$

Both reactions remove sulfur from the concentrate, thereby increasing the concentration of nickel in the calcine product. Also, both reactions are exothermic, so that they heat up the incoming concentrate and air.

An important feature of the roasting of nickel concentrates is that the air supply to the roaster is restricted, so that the concentrate is only partially oxidized. The smallest concentrate particles are probably completely oxidized. The larger concentrate particles, however, consist of an oxidized rim with an unoxidized core consisting of sulfide mineral.

These oxidized and unoxidized sulfides are smelted in the electric furnace to form (i) a molten sulfide matte that is rich in nickel; and, (ii) a silicate slag that is lean in nickel.

17.2.1. Industrial Roasting

Modern industrial nickel concentrate roasting is done in fluidized-bed roasters (Adham, Lee, & Sarvinis, 2005; Adham *et al.*, 2010). Details of industrial roasting are given in Table 17.1.

These roasters use the heat generated from the exothermic reactions, given by Equations (17.1) and (17.2), to heat the incoming concentrate to approximately 700°C.

Fluidized-bed roasters also produce a gas product that is concentrated in sulfur dioxide (10%–25% by volume SO_2), suitable for capture as sulfuric acid.

(The capture of sulfur dioxide and the production of sulfuric acid are discussed in Chapter 20.)

17.2.2. Roasting Temperature

The operating temperature of the roaster is chosen to give rapid oxidation while producing gas that is cool enough to go directly to a cyclone. Temperatures between 600°C and 760°C met these constraints. These temperatures also promote oxide formation rather than sulfate formation (Schonewille, Boissoneault, Ducharme, & Chenier, 2005).

17.2.3. Temperature Control

The temperature of the roaster is controlled by adjusting the water content of the concentrate feed. Water evaporates endothermically in the roaster so that increasing the water content of the concentrate cools the roaster and *vice versa.*

17.2.4. Extent of Oxidation

The extent of oxidation in the roaster is controlled by varying the ratio of the input rate of the air to the input rate of the concentrate. Increasing this ratio gives more extensive oxidation and *vice versa.*

The air-to-concentrate ratio is varied by adjusting the input rate of the feed while maintaining the input rate of the air. The input rate of the air is maintained to keep the bed fluidized.

The extent of oxidation is typically between 40 and 70% of the sulfur fed to the roaster. This range produces a calcine that, when subsequently smelted, produces a matte that contains approximately 30% Ni.

17.2.5. Choice of Extent of Oxidation

A roaster is the most efficient sulfur dioxide collection unit in the smelter. Those smelters with sulfuric acid plants tend, therefore, to do most of the oxidation of sulfur in the roaster. Thus, the Falconbridge smelter does more than 90% of the oxidation of sulfur in the roaster and the remainder in the converters. More sulfur oxidation than this is undesirable, because it can lead to a metal-rich, low-sulfur matte that has a high melting point in the downstream electric furnace (Schonewille *et al.*, 2005).

17.2.6. Input Rate of Air

The input rate of air to the fluidized-bed roasters is typically 40 000 Nm^3/h (Warner *et al.*, 2007). This gives a stable fluidized bed with about three quarters of the calcine being carried to the top of the roaster and one quarter overflowing the bed.

17.3. ELECTRIC FURNACE SMELTING

The smelting of nickel matte in an electric furnace entails

(a) charging warm (~300°C) roaster calcine into a hot (~1300°C) electrically heated furnace;
(b) melting the calcine with a silica flux to form nickel-rich molten sulfide matte and molten silicate slag; and,
(c) periodically tapping molten matte and slag separately through low and high tapholes.

A schematic diagram of an electric furnace is shown in Figure 17.3.

The molten slag, containing less than 0.5% Ni, is discarded. The molten matte is transferred to a converting furnace where iron and sulfur are further oxidized to give the final product from the smelter. This matte is suitable for refining to high-purity nickel and other metals.

17.3.1. Calcine Electrical Energy Requirement

Heating and melting the calcine charge to the furnace requires about 500 kWh of energy/tonne of charge (Warner *et al.*, 2007). Most of this energy is supplied electrically, and the remainder is supplied by oxidizing a small amount of carbon (coke) in the furnace charge.

17.3.2. Objectives of Smelting

The main objectives of smelting in an electric furnace are to melt the calcine feed and provide conditions whereby the melted roaster calcine separates cleanly into a molten matte that is rich in nickel and a molten slag that is lean in nickel.

Efficient separation is encouraged by minimizing the oxidation of nickel in the furnace. This is achieved by avoiding ingress of air and by adding metallurgical coke (carbon).

Carbon, from the coke in the feed and from the electrodes, reacts with any air that may leak into the furnace, ensuring that the desired reducing conditions are maintained. This reaction can be represented as:

$$\underset{\substack{\text{in electrodes}\\\text{and coke}}}{2C(s)} + \underset{\text{in air}}{O_2(g)} \xrightarrow{1300^\circ C} \underset{\substack{\text{in offgas}\\\text{with } N_2}}{2CO(g)} \qquad (17.3)$$

The reducing environment in the furnace ensures that the Ni mostly reports to the sulfide matte.

17.3.3. Separating Matte and Slag

Upon melting, the sulfides and oxides in the calcine and the silica flux separate into two immiscible liquids: the molten sulfide matte and the silica-saturated iron silicate slag (see also Chapter 8). At steady state, therefore, the furnace contains the following three layers:

(a) hot, solid roaster calcine at the top;
(b) molten slag in the middle (specific gravity 2.7–3.3; Jones, 2005); and,
(c) molten matte at the bottom (specific gravity 4.8–5.3; Jones, 2005).

As the calcine melts at the upper slag surface, the oxides join the slag and the sulfide droplets sink through the slag to the matte layer. This rain of sulfide droplets through the slag provides an ideal arrangement for attaining conditions that are close to equilibrium in the furnace.

17.3.4. Sulfidation of Nickel Oxide

The calcine from the roaster contains nickel oxide, as shown in Equation (17.1). This nickel oxide is sulfidized during smelting so that the contained nickel mainly reports to the matte product rather than to the slag.

The sulfidation takes place in the smelting furnace slag by the following reaction:

$$\mathrm{NiO(slag)} + \mathrm{FeS}(\ell) \xrightarrow{1300^{\circ}\mathrm{C}} \mathrm{NiS}(\ell) + \mathrm{FeO(slag)} \qquad (17.4)$$

NiO(slag)	FeS(ℓ)	NiS(ℓ)	FeO(slag)
dissolved from calcine into slag	partially oxidized pyrrhotite droplet descending through slag to matte layer	in Fe–Ni–S droplet descending to to matte layer	dissolved in slag

The free energy for this reaction, ΔG_R°, is approximately –40 kJ/mol of NiO, which indicates that the right-hand side is favored. This is confirmed by industrial electric furnace data (Warner *et al.*, 2007).

The partition ratio of nickel between matte and slag is given by the following expression:

$$\frac{\text{Ni concentration (\%) in slag}}{\text{Ni concentration (\%) in matte}}$$

The partition ratio in industrial furnaces is approximately 0.01. This is equivalent to a recovery of nickel to the matte of approximately 98% in the electric furnace (taking matte and slag masses into account).

17.3.5. Recovery of Copper and Cobalt to Matte

Pentlandite concentrates contain some copper, mainly as chalcopyrite, $CuFeS_2$ and cobalt, which is dissolved in the pentlandite. The recovery of copper to the matte phase in the electric furnace is similar to the recovery of nickel, that is, about 98%.

The recovery of the cobalt to the matte phase is, however, much lower. Typical values range between 50% and 80%. This is because cobalt oxidizes more readily and hence dissolves more extensively in the slag than nickel and copper (Grimsey, 1993; Matousek, 1993).

Nevertheless, one roasting and smelting operation specializes in recycling cobalt waste materials. This smelter recovers about 80% of the cobalt in its feed by operating the electric furnace under strongly reducing conditions (that is, with about 0.04 tonnes of coke per tonne of calcine) even though this requires matte tapping at a higher temperature of 1250°C (Schonewille *et al.*, 2005).

This smelter also avoids recycling of highly oxidized converter slags to its electric smelting furnace, which assists in maintaining the required reducing conditions.

17.3.6. Recovery of Silver, Gold and Platinum-Group Elements to Matte

Silver, gold and platinum-group metals do not oxidize during roasting and smelting. These metals report almost completely to the matte product of smelting furnace. The partition ratio is given by the following expression:

$$\frac{\text{metal concentration (\%) in slag}}{\text{metal concentration (\%) in matte}}.$$

Industrial and pilot plant data suggest, for example, that the partition ratio for platinum in electric furnaces is 0.001–0.01. This is equivalent to 98% or 99% recovery in the matte, which is somewhat higher than the range of 94%–98% estimated by Jones (2005).

17.4. INDUSTRIAL ELECTRIC FURNACES

The smelting of nickel calcines in an electric furnace is mostly done in rectangular furnaces with three pairs of self-baking electrodes in line, as shown in Figure 17.3. They are similar in construction to rectangular furnaces used for smelting laterites and the rectangular furnaces used for smelting mattes in the production of platinum-group elements.

All electric furnaces heat the charge by passing electricity from electrode to electrode through molten slag and matte (Sheng, Irons, & Tisdale, 1997; Warner *et al.*, 2007).

17.4.1. Construction Details

The electric furnace is constructed on 2.5-cm-thick steel base plates seated on concrete or brick piers. The bottom of the furnace is cooled by natural convection of air beneath the furnace. Air can also be blown against the furnace bottom with fans if additional cooling is required.

The walls of the furnace are 60% MgO magnesia–chrome brick, cooled with water-cooled copper plates and fingers (Schonewille *et al.*, 2005). The roof is a sprung arch of high-duty fireclay (silica–alumina) brick.

The furnace structure is kept under compression with a set of solid vertical beams (called '*buckstays*') and horizontal spring-loaded tie rods. The pressure on the tie rods is kept between 3 and 10 bar during heating and operation to contain refractory expansion. Some expansion is offset by inserting 'expansion papers' between refractory bricks during construction. Refractory temperatures are monitored by thermocouples during start-up and continuous operation. Infrared radiation 'thermal scans' are also done.

Electric furnaces are operated with a thick (1.5 m) layer of slag. This permits the electrodes to be moved up and down for precise power and temperature control.

The electrodes are self-baking Söderberg electrodes. These electrodes are formed by adding a 10% volatile carbon paste into cylindrical steel casings at the top of the electrodes. These casings, made of steel that is about 3 mm thick, are about 1 m in diameter and are finned inside. Heat from the furnace and from electrode resistance melts the paste and vaporizes the volatile components of the paste, so that a 'baked' carbon electrode forms.

The submerged tips of the electrodes are slowly oxidized to carbon monoxide, CO, by reaction with the slag bath. This is countered by lowering the electrodes through a hydraulic (non-flammable fluid) slipping-collar mechanism and by adding new paste on the top. New sections of steel casing are periodically welded to the top of the existing electrode casing for this purpose (while the power is on). Typical electrode consumption is 2 or 3 kg of paste per tonne of calcine (Warner *et al.*, 2007). This rate of consumption is equivalent to about 0.3 m of electrode per day. The electrodes are about 15 m high from submerged tip to paste addition level.

Electrode paste, casing materials and operating procedures are chosen carefully to avoid electrode breakage in the furnace (Ord, Schofield, & Tan, 1995).

Power inputs to matte smelting furnaces are about 100 kW/m^2 of hearth area. Electrode current densities are 10–30 kA/m^2 of electrode cross-sectional area. Power inputs and currents greater than these tend to cause excessive inductive turbulence in the liquids and overheating of the electrodes.

17.4.2. Covered Bath

Calcine is charged at a rate that keeps the molten slag covered with calcine so that the furnace has a 'black top'. This minimizes heat loss, avoids overheating of the furnace roof and prevents accidental coke combustion (Tisdale & Ransom, 1997).

17.4.3. Furnace Start-Up

Industrial electric furnaces are started by (Stober *et al.*, 2009):

(a) building a start-up bed of steel rails and slag pieces between the electrodes of each electrode pair (Schonewille *et al.*, 2005);
(b) heating the furnace to approximately 1000°C with natural gas burners around the hearth – at about 10°C/h;
(c) initiating electricity flow between electrodes through the start-up bed;
(d) charging about 800 tonnes of granulated solid slag into the furnace, melting it and bringing it up to temperature electrically;
(e) starting calcine production in the fluidized-bed roaster; and,
(f) charging fluxed calcine to the furnace and beginning matte production, increasing applied power as calcine addition rate increases.

This start-up procedure takes about a week.

17.4.4. Steady-State Operation

Operating the furnace continuously at steady state entails:

(a) charging fluxed calcine and coke at their prescribed rates;
(b) removing the offgas;
(c) maintaining prescribed temperatures through the furnace by controlling electric energy input rate (power); and,
(d) intermittently tapping molten matte and slag from the furnace to maintain constant matte and slag levels in the furnace.

The matte is tapped into ladles and transferred to converting furnace. The slag is tapped into ladles or granulated then discarded.

17.4.5. Smelting Rate, Power, Voltage

The rate at which electrical energy must be supplied to the electric furnace (that is, the power requirement of the furnace) is almost directly proportional to the rate at which calcine is being smelted, as given by the following formula:

$$\begin{aligned}&\text{Electrical energy supply rate requirement (J/s)} = \text{power requirement (W)}\\&= Q\text{, calcine smelting energy requirement (J/tonne of calcine)}\\&\quad\times \text{smelting rate (tonnes of calcine/s)}\end{aligned} \tag{17.5}$$

where Q is approximately 500 kW h/tonne of calcine (Warner *et al.*, 2007).

This electrical power is provided by applying voltage between the electrodes of each electrode pair. It is related to applied voltage and the resistance of the slag is related to current flow by the following equation:[1]

$$\text{Electrical energy supply rate requirement (J/s, W)} = \text{power (W)}$$

$$= \frac{(\text{voltage applied between electrodes})^2\ (\text{V}^2)}{\text{resistance to current flow between electrodes (ohms)}} \qquad (17.6)$$

In combination, these two equations show that an increased rate of charging calcine to the furnace must be accompanied by an increase in applied electrode-to-electrode voltage. The calcine will otherwise not melt at the rate at which it is being charged to the furnace. An excessive applied voltage will, on the other hand, overheat the furnace.

Applied voltage is altered by changing the transformer load taps of the electrode pairs.

17.4.6. Control

Once the applied power has been roughly selected by means of transformer voltage taps, it is then controlled to a setpoint by automatically raising and lowering the electrodes. The control is based on two factors:

(a) the low electrical resistivity of molten matte and the high electrical resistivity of slag; and,
(b) the existence of parallel current paths between electrodes.

The parallel current paths are electrode–slag–electrode and electrode–slag–matte–slag–electrode. Raising the electrodes favors the slag path, which increases the resistance between electrodes and decreases applied power, given by Equation (17.6). Lowering the electrodes favors current flow through the low-resistance matte, increasing the applied power.

Furnace temperature is also controlled by raising and lowering the electrodes. Temperatures below the setpoint are increased by lowering the electrodes, which increases the power. Temperatures above the setpoint are decreased by raising the electrodes, which decreases the power.

Matte and slag temperatures can also be adjusted somewhat independently. Matte temperature is increased by lowering the electrodes while slag temperature is increased by raising the electrodes. The transformer voltage taps may have to be changed during these electrode movements to keep the overall applied power at its requisite level.

1. More precisely, this equation should be $P = (V^2/R) \times \phi$ where ϕ is the electric furnace power factor ~0.9 where power factor is (power to the furnace)/(VA at the transformer).

17.4.7. Furnace Shutdown

An electric furnace is shut down by:

(a) raising matte and slag temperatures by ~100°C to ensure that slag and matte are fully molten and fluid;
(b) tapping matte from the furnace (power on);
(c) tapping slag from the furnace (power on then power off); and,
(d) allowing the furnace to cool at its natural rate.

This is done for only a long planned maintenance or a complete furnace rebuild (usually after about 10 years).

17.5. SUMMARY

Roasting and electric furnace smelting account for about a quarter of nickel sulfide smelting. The other three quarters are done by flash smelting.

The main advantage of roasting and smelting is the high recovery of nickel, copper, cobalt and by-product precious metals. Its main disadvantage is its large consumption of electricity.

Other advantages of electric smelting are (i) its ability to attain and control high slag temperatures, which is critical when the calcine feed contains considerable MgO, which has a high melting point; and, (ii) its ability to efficiently smelt metal-rich scrap and other recycle materials.

REFERENCES

Adham, K., Buchholz, T., Kokourine, A., et al. (2010). Design of copper-cobalt sulphating roasters for Katanga Mining Limited in D.R. Congo. In J. Harre (Ed.), *Copper 2010 Proceedings: Vol. 2. Pyrometallurgy I* (pp. 587–600). GDMB.

Adham, K., Lee, C., & Sarvinis, J. (2005). Fluidized bed technology applications for nickel extraction. In J. Donald & R. Schonewille (Eds.), *Nickel and Cobalt 2005, Challenges in Extraction and Production* (pp. 3–17). CIM.

Boldt, J. R., & Queneau, P. (1967). *The winning of nickel.* (pp. 228–234, 247–249, 263–267)Longmans.

Grimsey, E. J. (1993). Metal recovery in nickel smelting. In R. G. Reddy & R. N. Weizenbach (Eds.) *Proceedings of the Paul E. Queneau International Symposium: Extractive Metallurgy of Copper, Nickel and Cobalt: Vol. I. Fundamental Aspects* (pp. 1239–1251). TMS.

Jones, R. T. (2005). An overview of Southern African PGM (platinum group metal) smelting. In J. Donald & R. Schonewille (Eds.) *Nickel and Cobalt 2005, Challenges in Extraction and Production* (pp. 147–178). CIM.

Makinen, T., & Ahokainen, T. (2009). 50 years of nickel flash smelting – still going strong. In J. Liu, J. Peacey & M. Barati, et al. (Eds.), *Pyrometallurgy of Nickel and Cobalt 2009, Proceedings of the International Symposium* (pp. 209–220). CIM.

Matousek, J. W. (1993). The behavior of cobalt in pyrometallurgical processes. In R. G. Reddy & R. N. Weizenbach (Eds.), *Proceedings of the Paul E. Queneau International Symposium: Extractive Metallurgy of Copper, Nickel and Cobalt: Vol. I. Fundamental Aspects* (pp. 129–141). TMS.

Ord, R. J., Schofield, J. G., & Tan, C. G. (1995). Improved performance of Soderberg electrodes. *CIM Bulletin, 88*, 97–101.

Schonewille, R., Boissoneault, M., Ducharme, D., & Chenier, J. (2005). Update on Falconbridge's Sudbury nickel smelter. In J. Donald & R. Schonewille (Eds.) *Nickel and Cobalt 2005, Challenges in Extraction and Production* (pp. 479–498). CIM.

Sheng, Y. Y., Irons, G. A., & Tisdale, D. G. (1997). Power, fluid flow and temperature distributions in electric smelting of nickel matte. In C. Diaz, I. Holubec & C. G. Tan (Eds.) *Nickel/Cobalt 97: Vol. III. Pyrometallurgical Operations, the Environment and Vessel Integrity in Nonferrous Smelting and Converting* (pp. 45–66). CIM.

Stober, F., Jastrzebski, M., Walker, C., et al. (2009). Start-up and ramp-up of metallurgical furnaces to design production rate. In J. Liu, J. Peacey & M. Barati et al. (Eds.) *Pyrometallurgy of Nickel and Cobalt 2009, Proceedings of the International Symposium* (pp. 487–507). CIM.

Tisdale, D. G., & Ransom, C. G. (1997). Adapting to one furnace at Falconbridge. In C. Diaz, I. Holubec & C. G. Tan (Eds.) *Nickel/Cobalt 97: Vol. III. Pyrometallurgical Operations, the Environment and Vessel Integrity in Nonferrous Smelting and Converting* (pp. 35–43). CIM.

Warner, A. E. M., Diaz, C. M., Dalvi, A. D., et al. (2007). JOM world nonferrous smelter survey part IV: nickel sulfide. *JOM, 59*, 58–72.

SUGGESTED READING

Diaz, C., Holubec, I., & Tan, C. G. (Eds.). (1997). *Nickel/Cobalt 97: Vol. III: Pyrometallurgical Operations, the Environment and Vessel Integrity in Nonferrous Smelting and Converting*. CIM.

Donald, J., & Schonewille, R. (Eds.). (2005). *Nickel and Cobalt 2005, Challenges in Extraction and Production*. CIM.

Liu, J., Peacey, J., & Barati, M., et al. (Eds.). (2009). *Pyrometallurgy of Nickel and Cobalt 2009: Proceedings of the International Symposium*. CIM.

Mahant, R. P., McKague, A. L., Norman, G. E., & Michelutti, R. E. (1984). The development and operation of the smelting and environmental improvement process: Falconbridge Nickel Mines – Sudbury Operations. *CIM Bulletin, 77*, 79–98.

McMcKague, A. L., Norman, G. E., & Jackson, J. F. (1980). Falconbridge Nickel Mines Limited new smelting process. *CIM Bulletin, 73*, 132–141.

Chapter 18

Flash Smelting of Nickel Sulfide Concentrates

Roasting and electric furnace smelting of nickel sulfides were discussed in Chapter 17. An alternative process is flash smelting. A schematic diagram of a flash smelting furnace is shown in Figure 18.1.

Flash smelting combines roasting and melting in one furnace. In flash smelting, most of the heat required comes from oxidizing the iron and sulfur in the concentrate feed. It requires very little fuel or electricity. Flash smelting accounts for about three quarters of the nickel sulfide that is processed by smelting (Makinen & Ahokainen, 2009).

Flash smelting entails continuously blowing oxygen, air, sulfide concentrate and silica flux into a 1300°C furnace. The products of flash smelting are

(a) molten sulfide matte, considerably richer in nickel than the concentrate feed;
(b) molten slag that is lean in nickel; and,
(c) hot, dust-laden offgas containing 20%–50% SO_2 by volume.

Details of the operation of the furnaces are given in Tables 18.1 and 18.2.

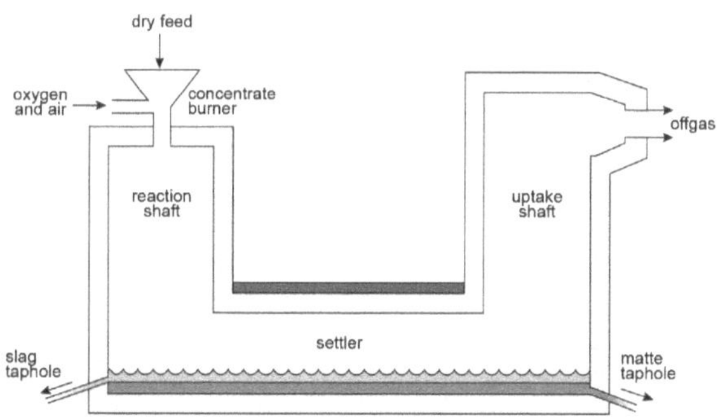

FIGURE 18.1 The Outotec-type flash furnace for smelting pentlandite concentrate, which contains 15% Ni, to molten Ni–Fe–S matte, which contains 40% Ni. A furnace is typically 25 m long and 8 m wide. The capacity of such a furnace is typically between 1000 and 2000 tonnes/day of dried concentrate. *Source: Drawing courtesy CM Solutions (Pty) Ltd.*

Extractive Metallurgy of Nickel, Cobalt and Platinum-Group Metals. DOI: 10.1016/B978-0-08-096809-4.10018-8

TABLE 18.1 Dimensions and Production Details of Five Outotec Nickel Concentrate Flash Smelting Furnaces

Smelter	Kalgoorlie, Australia (BHP Billiton)	Selebi Phikwe, Botswana (BCL Ltd.)	Gansu, China (Jinchuan Group Ltd.)	Norilsk, Russia (Nadezda Smelter)	Harjavalta, Finland (Boliden)
Start-up date	1972	1973	1992	1981	1959
Ni-in-matte production, tonnes/year	100 000	27 000	65 000	140 000 (two furnaces)	38 000
Size, inside brick, m					
Hearth, $w \times l \times h$	8 × 37 × 3.5	8 × 22 × 4	7 × 32 × 3.5	10 × 31 × 6	7 × 19 × 2.6
Reaction shaft					
Diameter	7	8	6	8	4.6
Height above settler roof	6	11	6	9	7.6
Concentrate burners	4	4	4	1	1
Gas uptake					
Diameter	3.5 x 8	5	3		
Height above settler roof	7	17	7		

Furnace operating days per year		356	330	330	330
Campaign life, years	10	9	8	5	10
Feed details, tonne/h					
New dry concentrate	140	120	50	150	45
Composition	15% Ni, 0.3% Cu, 0.4% Co, 34% Fe, 32% S	5% Ni, 4% Cu, 0.2% Co, 43% Fe, 31% S	9% Ni, 4% Cu, 0.2% Co, 38% Fe, 27% S	12% Ni, 5% Cu, 0.4% Co, 42% Fe. 33% S	15% Ni, 0.8% Cu, 0.4% Co, 30% Fe, 29% S
Blast details					
Temperature, °C	500	260 (steam heated)			ambient
Volume% O_2	35	31	42	70	75
Flowrate, Nm^3/h	85 000	150 000	33 000	34 000	7000
Oxygen supply, tonne/h	25	40	12	20	4
Production details					
Matte, tonne/h	25	25		55	5
Matte composition	47% Ni, 1.5% Cu, 0.8% Co, 20% Fe, 27% S	17% Ni, 15% Cu, 0.4% Co, 33% Fe, 25% S	29% Ni, 15% Cu, 0.6% Co, 29% Fe, 23% S	32% Ni, 15% Cu, 0.8% Co, 23% Fe, 27% S	65% Ni, 5% Cu, 0.7% Co, 5% Fe, 22% S

(*Continued*)

TABLE 18.1 Dimensions and Production Details of Five Outotec Nickel Concentrate Flash Smelting Furnaces—cont'd

Smelter	Kalgoorlie, Australia (BHP Billiton)	Selebi Phikwe, Botswana (BCL Ltd.)	Gansu, China (Jinchuan Group Ltd.)	Norilsk, Russia (Nadezda Smelter)	Harjavalta, Finland (Boliden)
Slag, tonne/h	70	110	50	130	20
Slag composition	0.7% Ni, 0.1% Cu, 0.2% Co, 40% Fe, 33% SiO_2	1.5% Ni, 1.3% Cu, 0.1% Co, 40% Fe, 28% SiO_2	0.2% Ni, 0.2% Cu, 0.1% Co, 41% Fe, 36% SiO_2	0.6% Ni 0.3% Cu, 0.2% Co, 40% Fe, 34% SiO_2	4% Ni, 0.3% Cu, 0.5% Co, 38% Fe, 29% SiO_2
Matte from slag recovery method	Electric furnace appendage on flash furnace	Electric furnaces (2)	Electric furnaces (2)	Electric furnaces (4)	Electric furnace (1)
Matte from converter slag recovery method	Recycle to flash furnace	Electric furnaces (2)	Electric furnaces (2)	Electric furnaces (4)	No converting

Offgas, Nm^3/h		100 000	60 000 (at precipitators	56 000	16 000
Vol% SO_2 leaving furnace		7.2	8 (at H_2SO_4 plant)	30	30
SO_2 destination	H_2SO_4 plant	Atmosphere	H_2SO_4 plant	Atmosphere	H_2SO_4 plant
Matte/slag/offgas temp., °C	1170/1300/1400	1160/1240/1400	1320/1380/1380	1150/1250/1150	1360/1400/1400
Fuel inputs					
Hydrocarbon fuel burnt per hour	1.5 m^3 oil	6 tonnes coal	1.3 m^3 oil + 0.8 tonnes coal	6000 Nm^3 gas/h (18 settler burners)	1 tonne heavy oil
Electrical power input	6.25 MW, 6 electrodes	0	5 MW, 6 electrodes		

The furnace in the last column avoids converting by smelting nickel concentrate directly to very rich nickel matte (Section 19.12).
The data are mostly from Warner *et al.*, 2007, who also kindly provided additional information. The furnace in the last column avoids converting by direct smelting to very rich additional information. The Kalgoorlie and Gansu furnaces have electrically heated appendages.

TABLE 18.2 Dimensions and Production Details of the World's Two Inco-Type Nickel Sulfide Smelting Furnaces, Sudbury, Canada

Smelter	Sudbury, Canada, Vale
Start-up date	1993
Ni-in-matte production, tonnes/year	133 000
Size, inside brick, m	2 Furnaces
Hearth, *w* × *l* × *h*	8 × 30 × 9 (extensively water-cooled throughout)
Concentrate burners	4 Horizontal
Gas uptake (square)	
Dimensions	4 × 4 (in center of furnace roof)
Height above settler roof	15 With top circular gas exit hole
Tapholes	
Matte	2 On each side wall, total of 4
Slag	1 On end wall
Converter slag additions	Via chute through water-cooled door on end wall (opposite slag taphole)
Feed details, tonnes/h	
New dry concentrate	125
Composition	10% Ni, 12% Cu, 0.3% Co, 39% Fe, 34% S (2007)
Operating details	
Campaign life, years	2–3
Operating days per year	307
Blast details	
Temperature, °C	Ambient
Volume% O_2	96
Flowrate, Nm^3/h	14 000
Oxygen supply, tonnes/h	11
Production details	
Matte, tonnes/h	65
Matte composition	23% Ni, 25% Cu, 0.6% Co, 24% Fe, 26% S (2007)
Slag, tonnes/h	120
Slag composition	0.5% Ni, 0.5% Cu, 0.2% Co, 43% Fe, 36% SiO_2
Matte from slag recovery method	discard
Matte from converter slag recovery method	Recycle to flash furnace
Offgas, Nm^3/h	26 000
Vol% SO_2 leaving furnace	55 (dry)
Matte/slag/offgas temp., °C	1210/1280/1350
Fuel inputs	
Hydrocarbon fuel burnt/h	340 Nm^3 natural gas and 1.4 tonnes of coke (on slag surface)

The furnaces use oxygen (rather than air) to oxidize iron and sulfur. This minimizes (i) the need for hydrocarbon fuel; and, (ii) the amount of offgas that has to be cooled, cleaned and made into sulfuric acid. The furnaces also treat recycle molten converter slag for metal rich matte recovery. The data updated from Warner *et al.* (2007).

The molten matte and slag are immiscible (see Chapter 8). They form separate layers in the furnace – matte below and slag above. The matte is tapped and transferred to converting, which is discussed in Chapter 19.

The slag is tapped separately then usually sent to an electric settling furnace for recovery of entrained matte, as shown in Figure 17.1, and then it is discarded. The electric settling furnace is also called a '*slag-cleaning furnace*'. Additional details of slag-cleaning furnaces are given in Appendix E.

Dust is removed from the offgas, which is then sent to a sulfuric acid plant where sulfur dioxide is converted to sulfuric acid (see Chapter 20).

18.1. OBJECTIVE OF THE PROCESS – NICKEL ENRICHMENT

The principal objective of flash smelting is to produce molten matte that is richer in nickel (and other metals) than the original concentrate. The nickel enrichment is defined as follows:

$$\frac{\text{Ni concentration in matte, \%}}{\text{Ni concentration in concentrate, \%}}$$

Typical values for the nickel enrichment in a flash furnace are between 2.5 and 5 (Warner *et al.*, 2007). The enrichment of nickel is obtained by

(a) oxidizing most of the S in the concentrate and removing it as SO_2 gas;
(b) oxidizing most of the Fe in the concentrate and removing it in molten slag as an iron silicate; and,
(c) removing the gangue rock in the concentrate by dissolving it in the molten slag.

The principle of the process is that iron and sulfur oxidize more readily than nickel.

18.2. ADVANTAGES AND DISADVANTAGES

There are two main advantages of flash smelting over roasting and smelting:

(a) flash smelting requires much less energy than electric smelting because most of the required energy is generated by the roasting reactions:

$$\underset{\text{pentlandite in concentrate}}{Ni_{4.5}Fe_{4.5}S_8(s)} + \underset{\text{in oxygen+air}}{5.75O_2(g)}$$

$$\xrightarrow{1300^\circ C} \underset{\text{in matte}}{4.5NiS(\ell)} + \underset{\text{in silicate slag}}{4.5FeO(\ell)} + \underset{\text{in offgas with } N_2}{3.5SO_2(g)} \quad (18.1)$$

$$\Delta H_R = -1200 \text{ kJ/mol of } Ni_{4.5}Fe_{4.5}S_8$$

$$\underset{\text{pyrrhotite in concentrate}}{Fe_8S_9(s)} + \underset{\text{in oxygen +air}}{13O_2(g)} \xrightarrow{1300^\circ C} \underset{\text{in silicate slag}}{8FeO(\ell)} + \underset{\text{in offgas with } N_2}{9SO_2(g)} \quad (18.2)$$

$$\Delta H_R = -2500 \text{ kJ/mol of } Fe_8S_9$$

(b) flash smelting avoids the emission of weak sulfur dioxide gas (Schonewille, Boissoneault, Ducharme, & Chenier, 2005).

The main disadvantage of flash smelting is that more nickel is lost to slag than in roasting and smelting, because the conditions in the flash furnace are more oxidizing. As a result, a subsequent step for the recovery of nickel from the slag using an electric settling furnace is usually required. This slag-cleaning step could be a separate furnace or it could be included as an appendage to the flash furnace. The slag-cleaning appendage to a flash furnace is shown in Figure 18.2.

18.3. EXTENT OF OXIDATION

The reactions that occur in flash smelting are similar to those of roasting and smelting: the partial oxidation of concentrate and the melting of the oxidation products.

The objective of flash smelting is to oxidize iron and sulfur but not nickel, copper and cobalt. This objective is achieved by limiting the rate at which oxygen is supplied to the furnace.

In practice, the feed rate of the concentrate is chosen to meet production objectives. The input rate of oxygen (in both air and oxygen) is then adjusted to obtain the desired degree of oxidation. This is often represented by the nickel content in the matte, which increases with the increasing extent of oxidation of iron and sulfur and *vice versa*.

The nickel content of the matte is an important parameter for subsequent converting and refining. Industrially, it varies 20%–60% (Warner *et al.*, 2007).

18.4. CHEMISTRY

The concentrate burners of flash furnaces create a dispersion of concentrate, oxygen and nitrogen in the hot furnace. Most of the oxidation of the concentrate takes place in this dispersion.

Microscopic analysis of the oxidized particle indicates that this oxidation produces (i) fully oxidized small particles; and, (ii) partially oxidized large particles with oxide rims and unoxidized sulfide cores (Jorgensen, 1997).

These particles fall into the molten slag layer where they melt and react toward equilibrium by reactions such as:

$$\underset{\substack{\text{dissolved}\\ \text{over-oxidized}\\ \text{particles in slag}}}{NiO(\ell)} + \underset{\substack{\text{in partially oxidized}\\ \text{pyrrhotite droplets}\\ \text{descending through}\\ \text{slag to the matte layer}}}{FeS(\ell)} \xrightarrow{1300^\circ C} \underset{\substack{\text{in Fe, Ni, S droplets}\\ \text{descending to}\\ \text{matte layer}}}{NiS(\ell)} + \underset{\substack{\text{dissolved}\\ \text{in slag}}}{FeO(\ell)} \qquad (18.3)$$

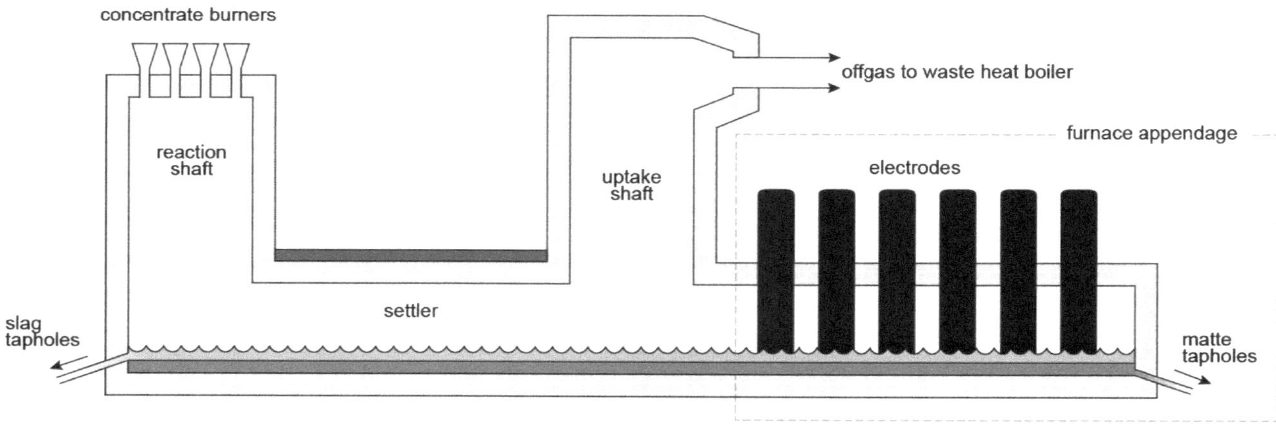

FIGURE 18.2 The Outotec-type flash furnace with electrically heated appendage, Kalgoorlie, Australia. The appendage is used to heat the slag and settle nickel-rich matte from it before it is tapped from the furnace. The furnace in Gansu, China, is of a similar design. *Source: Drawing courtesy CM Solutions (Pty) Ltd.*

The Gibbs free energy for this reaction, ΔG_R°, is approximately –40 kJ/mol of NiO, which indicates that the reaction tends to go to the right-hand side of Equation (18.3).

This is confirmed by industrial data for the operation of flash furnaces (Warner *et al.*, 2007). The partition ratio of nickel between matte and slag is defined as follows:

$$\frac{\text{Ni concentration (\%) in slag}}{\text{Ni concentration (\%) in matte}}$$

The values for the partition ratio of nickel range between 0.01 and 0.09. This suggests that the right-hand side of the reaction given in Equation (18.3) is favored. However, the partition ratio for nickel smelting in a flash furnace is somewhat worse than that from smelting in an electric furnace. As a result, slags from flash furnaces are almost always settled or cleaned in electric furnaces before they are discarded.

The total recovery of nickel from the flash furnace, including the settling or cleaning of the slag, is typically 95%. This is somewhat lower than the 98% obtained by roasting and smelting (see Chapter 17).

18.4.1. Behavior of Other Metals

Pentlandite concentrates contain copper and cobalt. The recovery of copper to the matte in a flash furnace is between 80% and 93%, whereas the recovery of cobalt is between 26% and 70% (Warner *et al.*, 2007). Both are significantly lower than those achieved by roasting and smelting, because the conditions in the flash furnace are more oxidizing than those in roasting and smelting.

18.4.2. Precious Metals

Platinum-group elements are recovered efficiently to matte in flash smelting (Makinen, Fagerlund, Anjal, & Rosenback, 2005). This suggests that the recovery of platinum-group elements to the matte should be at least as good as the recovery of nickel, that is, approximately 95%.

18.5. INDUSTRIAL FLASH SMELTING

Flash smelting comes in two types: Outotec flash smelting and Inco flash smelting (Davenport, Jones, King, & Partelpoeg, 2010).

Outotec-type flash smelting, which is shown in Figures 18.1 and 18.2, oxidizes the concentrate by mixing dried concentrate and flux with *oxygen-enriched air* and blowing the mixture *downwards* into a hot furnace.

Inco flash smelting mixes dried concentrate and flux with *oxygen* and blows it *horizontally* into a hot furnace.

The Outotec and Inco furnaces are similar in most other respects.

There are seven Outotec-type nickel smelting furnaces worldwide. These furnaces account for about 70% of the production of nickel by flash smelting. There are two Inco-type flash furnaces at one smelter. These furnaces account for about 30% of the production of nickel by flash smelting.

18.6. OUTOTEC-TYPE FLASH FURNACE

The Outotec-type flash furnaces for nickel sulfides vary considerably in size and shape; however, they all have the following five features:

(a) between one and four concentrate burners that mix the pentlandite concentrate, the silica flux and the oxygen-air blast at the burner tip and blow the mixture downwards into a hot (1300°C) furnace;
(b) a reaction shaft, where most of the reactions between oxygen and concentrate take place and where most of the oxidized concentrate melts;
(c) a settler, where molten oxidized concentrate drops collect and settle as molten matte and slag layers;
(d) low and high tapholes, which are water cooled, for separately tapping molten matte and molten slag from the furnace; and,
(e) an uptake for removing the offgas bearing the sulfur dioxide.

These features of Outotec-type flash furnaces are shown in Figure 18.1. The smelting is continuous except for matte and slag tapping, which are intermittent.

18.6.1. Construction Details

The interior of an Outotec-type flash furnace consists of direct-bonded bricks. These bricks are high-purity magnesia-chrome bricks. The external bricks are cooled in three different ways. Water-cooled copper jackets are used in high temperature regions of the walls. Air-cooled steel plates or water spray-cooled steel plates are used elsewhere (Makinen & Ahokainen, 2009).

18.6.2. Concentrate Burner

Dried concentrate, flux and air/oxygen blast are combined in the concentrate burner and blown into the reaction shaft. A schematic diagram of a concentrate burner is shown in Figure 18.3.

Recent concentrate burners consist of an annulus through which the air–oxygen blast is blown into the reaction shaft, a central pipe through which the concentrate and the flux falls into the reaction shaft, and a distributor cone at the burner tip which blows air and oxygen horizontally through the descending solid feed.

Special attention is paid to obtaining a uniform distribution of concentrate particles and blast down the reaction shaft. This uniform distribution is obtained

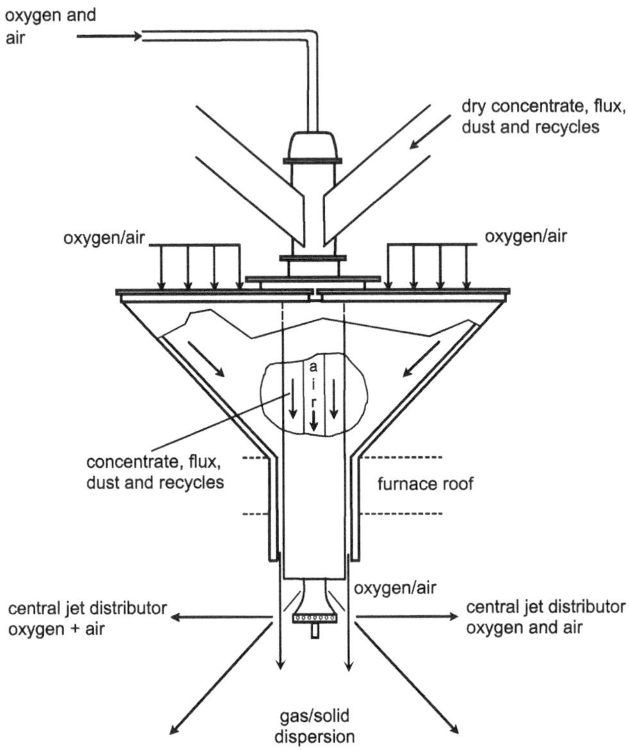

FIGURE 18.3 The concentrate burner used in the Outotec-type furnaces. Note the central jet distributor at the bottom of the burner. The main goal of the burner is to create a uniform dispersion of concentrate and gas around the burner. This type of burner can smelt up to 200 tonnes/h of feed. The feed consists mainly of (i) dried pentlandite concentrate, about 50–100 μm; (ii) silica flux, about 1 mm; and, (iii) recycle dust and other particulate recycle materials. Burner design is discussed by Nagai, Kawanaka, Yamamoto, & Sasai, 2010.

by introducing blast and solids vertically and uniformly into quadrants around the burner and by blowing the particles outward with central jet distributor.

Ports for viewing and cleaning the burner tip are also provided.

18.6.3. Matte and Slag Tapholes

Molten matte and slag are tapped through separate tapholes. Each taphole is a single hole, typically between 0.06 and 0.08 m in diameter. The taphole consists of a water-cooled copper 'chill block' that is imbedded in the furnace walls. The tapholes are plugged with moist fireclay, which is solidified by the heat of the furnace when the clay is pushed into the hole. The tapholes are opened by chipping out the clay and by melting it out with steel oxygen lances.

Matte is tapped down a refractory-lined steel or copper chute into cast steel ladles for transport to converting.

Slag is tapped down a water-cooled copper chute into an electric matte-from-slag settling furnace.

Both withdrawals are only partial. Reservoirs of matte and slag that are each about 0.5 m deep are retained in the furnace.

Tapping of matte is moved around between six matte tapholes in the furnace walls. This washes out solid buildups on the furnace floor by providing matte flow over the entire hearth.

18.6.4. Electrically Heated Appendage to the Outotec-Type Flash Furnace

A schematic diagram of the Outotec-type flash furnace in Kalgoorlie, Australia, is shown in Figure 18.2. This diagram shows the electric furnace appendage. The furnace in Gansu, China, has a similar design. The appendage is designed to independently heat the slag so that it can be easily tapped from the furnace and settle nickel-rich matte from the slag before the slag is tapped. This eliminates the need for a separate slag-cleaning furnace.

18.7. INCO-TYPE FLASH FURNACE

The design of the two Inco-type flash furnaces in Sudbury, Canada, is shown in Figures 18.4 and 18.5. Their design is similar to the Outotec-type furnaces but they differ in the following aspects:

(a) they use oxygen blast (rather than air/oxygen blast);
(b) they are fed horizontally (two burners at each end); and,
(c) they discard their slag without an external matte-from-slag settling furnace.

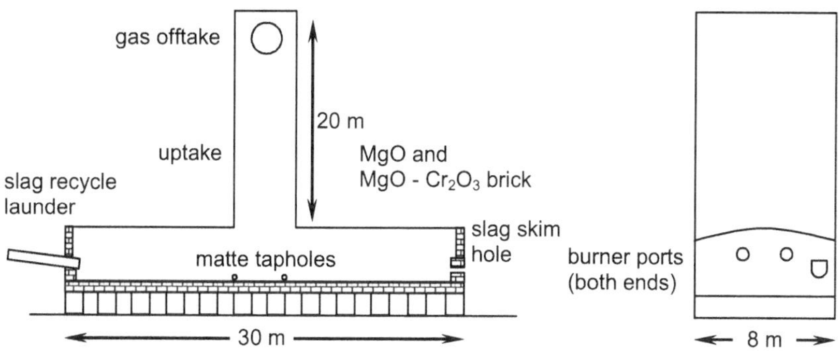

FIGURE 18.4 The Inco-type flash furnace for smelting pentlandite concentrate to nickel-rich molten matte. The four horizontal burners and the tall central uptake are notable. The oxidant is industrial oxygen. The furnace smelts about 2000 tonnes of dried concentrate/day.

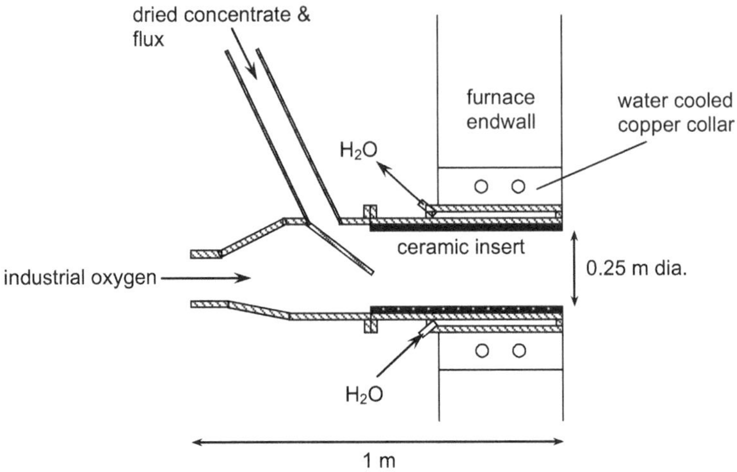

FIGURE 18.5 The concentrate burner for the Inco-type flash furnace. The down-coming concentrate and the horizontal industrial oxygen input are notable. Recent burners have a flexible concentrate/flux downcomer.

The construction and operation of the Inco-type flash furnace are described in more detail by Davenport *et al.* (2010) and Donald and Scholey (2005).

18.8. PERIPHERAL EQUIPMENT

Flash furnaces are surrounded by the following peripheral equipment:

(a) concentrate blending equipment;
(b) concentrate dryers (sometimes steam heated, Chen & Mansikkaviita, 2009; Talja, Chen, & Mansikkaviita, 2010);
(c) flash furnace feed bins and feed system;
(d) oxygen plant;
(e) blast preheater (optional);
(f) waste-heat boiler;
(g) dust-recovery and recycle system;
(h) gas-cleaning system;
(i) sulfuric acid plant (Chapter 20); and,
(j) electric matte-from-slag settling furnace (slag-cleaning furnace) (Appendix F).

The peripheral equipment is described by Davenport *et al.* (2010).

18.9. OPERATION AND CONTROL OF THE FLASH FURNACE

Steady-state flash furnace operation entails the following actions:

(a) feeding particulate solids and blast at constant rates;

(b) withdrawing sulfur dioxide-rich gas through the gas uptake at a constant rate;
(c) tapping matte from the furnace on a scheduled basis or as-needed by the converters; and,
(d) tapping slag on a scheduled basis or when it reaches a prescribed level in the furnace.

18.9.1. Control

The operation must meet the following metallurgical criteria for the control of the furnace:

(a) the grade (that is, % Ni) of the matte;
(b) the concentration of silica (% SiO_2) of the slag; and,
(c) the temperature of the slag.

At the same time, a protective coating of solidified slag on the inside walls of the furnace, called the freeze lining, must be maintained.

Control of the Matte Grade (% Ni)

The target for the matte grade is obtained by changing the ratio of the feed rate of oxygen in the blast to the feed rate of concentrate. This changes the extent of oxidation of iron and sulfur, so that target concentration of nickel in the matte is met.

In practice, this ratio is changed by adjusting the rate at which oxygen and air enter the furnace while the feed rate of concentrate is maintained.

Control of the Silica in the Slag (or % SiO_2)

The iron oxide formed by oxidation of the concentrate is fluxed with silica to form liquid slag. The amount of silica that is added is such that the slag is sufficiently fluid, making it easy to tap. A fluid slag allows for a clean separation of the matte and the slag.

The values of the mass ratio of silica to iron vary between 0.7 and 1.0. This ratio is adjusted by changing the rate at which flux is fed to the solids feed dryer.

Temperature Control

The temperatures of the matte and the slag are measured as these molten materials flow from the furnace. Disposable thermocouple probes and optical pyrometers are used. The temperatures of the matte and the slag are adjusted by changing the following factors:

(a) the rate at which nitrogen, mainly in air, enters the furnace since it acts as a 'coolant';

(b) the rate at which smelter reverts, which are low in iron and sulfur, are recycled to the furnace; and,
(c) the rate of combustion of hydrocarbon fuel.

The temperature of the slag can be adjusted somewhat independently of matte temperature by (i) adjusting combustion rate of fuel in the settler burner; and, (ii) adjusting the electrical power in furnaces with electrically heated appendages.

The temperature of the slag in a nickel flash furnace is typically 1250°C–1350°C. The higher values in this temperature range are generally for slags that are high in MgO because these types of slags have higher melting points. The source of the MgO is the concentrate feed to the smelter.

Reaction Shaft and Hearth Control

Long campaign life for the flash furnace requires the controlled deposition of slag that is rich in magnetite on the walls and hearth of the furnace. Deposition of the magnetite slag is encouraged by highly oxidizing conditions in the furnace, low operating temperatures (specifically cool walls) and low silica content in slag.

It is discouraged by reversing these conditions and by adding coke to the furnace.

18.10. APPRAISAL

Nickel flash smelting is a high-productivity smelting process that

(a) efficiently uses the heat from sulfide mineral oxidation to heat and melt the furnace products; and,
(b) efficiently captures the sulfur dioxide byproduct.

However, flash smelting is an oxidizing process and, as a result, significant amounts of nickel and other metals are lost to slag. Flash smelters are almost always accompanied either by an auxiliary electric settling furnace for the recovery of nickel from the slag or by an electrically heated appendage.

Flash smelting has also had difficulties smelting concentrates that are high in magnesium, because these concentrates produce slags that have melting points near the upper limit of the operating temperature of flash furnaces, that is, near 1300°C.

These difficulties have not, however, prevented flash smelting from becoming the dominant method of smelting nickel sulfide.

18.11. RECENT TRENDS

One Outotec flash furnace makes matte that is so low in iron that it eliminates the need for converting. This restricts the emission of sulfur dioxide to one continuous source, which greatly simplifies the capture of sulfur dioxide. This furnace is discussed in Chapter 19.

18.12. SUMMARY

Flash smelting oxidizes and melts pentlandite concentrates. The principal product is a molten Ni–Fe–S matte that is considerably richer in nickel (20%–60% Ni) than the concentrate feed (12%–20% Ni).

This matte product is transferred to converting (Chapter 19) and from there to the production of high-purity metal (Chapters 21–27).

Flash smelting entails blowing a dispersion of dry concentrate, flux and oxygen/air into a furnace at 1300°C. It produces a furnace matte that is rich in nickel by the following steps:

(a) oxidizing much of the sulfur in the concentrate and removing it as sulfur dioxide gas;
(b) oxidizing much of the iron in the concentrate and removing it in molten iron-silicate slag (with silica flux); and,
(c) removing the gangue rock in the concentrate by dissolving it in the molten slag.

The main advantage of flash smelting over roasting/electric furnace melting is its smaller energy requirement, due to its efficient use of heat from the oxidation of sulfur and iron.

The main disadvantage of flash smelting is the greater loss of nickel and other metals to slag – which almost always requires subsequent settling of matte from slag in an purpose-built electric settling furnace.

REFERENCES

Chen, S., & Mansikkaviita, H. (2009). Steam dryer plant for a variety of non-ferrous metal concentrates. In J. Liu, J. Peacey & M. Barati, et al. (Eds.), *Pyrometallurgy of nickel and cobalt 2009, proceedings of the international symposium* (pp. 621–626). CIM.

Davenport, W. G., Jones, D. M., King, M. J., & Partelpoeg, E. H. (2010). *Flash smelting: Analysis, control and optimization* (2nd ed.). Wiley.

Donald, J. R., & Scholey, K. (2005). An overview of Copper Cliff's operations. In J. Donald & R. Schonewille (Eds.), *Nickel and cobalt 2005, challenges in extraction and production* (pp. 457–477). CIM.

Jorgensen, F. R. A. (1997). Dust formation during the combustion of nickel concentrate. In C. Levac & R. Berryman (Eds.), *Nickel/cobalt 97: Vol. II. Pyrometallurgy fundamentals and process development* (pp. 323–336). CIM.

Makinen, T., & Ahokainen, T. (2009). 50 Years of nickel flash smelting – Still going strong. In J. Liu, J. Peacey & M. Barati, et al. (Eds.), *Pyrometallurgy of nickel and cobalt 2009, proceedings of the international symposium* (pp. 209–220). CIM.

Makinen, T., Fagerlund, K., Anjal, Y., & Rosenback, L. (2005). State of the art in nickel smelting: Direct Outokumpu nickel technology. In J. Donald & R. Schonewille (Eds.), *Nickel and cobalt 2005, challenges in extraction and production* (pp. 71–89). CIM.

Nagai, K., Kawanaka, K., Yamamoto, K., & Sasai, S. (2010). Development of Sumitomo concentrate burner. In J. Harre (Ed.), *Copper 2010 proceedings: Vol. 3. Pyrometallurgy II* (pp. 1025–1034). GDMB.

Schonewille, R., Boissoneault, M., Ducharme, D., & Chenier, J. (2005). Update on Falconbridge's Sudbury nickel smelter. In J. Donald & R. Schonewille (Eds.), *Nickel and cobalt 2005, challenges in extraction and production* (pp. 479–498). CIM.

Talja, J., Chen, S., & Mansikkaviita, H. (2010). Kumera technology for copper smelters. In J. Harre (Ed.), *Copper 2010 proceedings: Vol. 3. Pyrometallurgy II* (pp. 1183–1197). GDMB.

Warner, A. E. M., Diaz, C. M., Dalvi, A. D., et al. (2007). JOM world nonferrous smelter survey part IV: Nickel: Sulfide. *JOM, 59*, 58–72.

SUGGESTED READING

Davenport, W. G., Jones, D. M., King, M. J., & Partelpoeg, E. H. (2010). *Flash smelting: Analysis, control and optimization* (2nd ed.). Wiley.

Makinen, T., & Ahokainen, T. (2009). 50 Years of nickel flash smelting – Still going strong. In J. Liu, J. Peacey & M. Barati, et al. (Eds.), *Pyrometallurgy of nickel and cobalt 2009, proceedings of the international symposium* (pp. 209–220). CIM.

Makinen, T., Fagerlund, K., Anjal, Y., & Rosenback, L. (2005). State of the art in nickel smelting: Direct Outokumpu nickel technology. In J. Donald & R. Schonewille (Eds.), *Nickel and cobalt 2005, challenges in extraction and production* (pp. 71–89). CIM.

Chapter 19

Converting – Final Oxidation of Iron From Molten Matte

The final product from a nickel smelter is a sulfide matte that is low in iron (Figure 13.1). This matte is produced by oxidizing iron in the molten smelting furnace matte with air (sometimes air enriched with oxygen) mostly using Peirce-Smith converters (see Figures. 8.2 and 8.3).

This chapter examines the converting of nickel smelter matte. The objectives of the chapter are to describe:

(a) the advantages of low-iron matte for making high-purity metals;
(b) the principles and chemistry of the converting process;
(c) the behavior of other valuable elements during converting, that is, copper, cobalt, silver, gold and platinum-group metals;
(d) industrial Peirce-Smith converting; and,
(e) alternatives to Peirce-Smith converting.

Details of the industrial operation of Peirce-Smith converters are given in Table 19.1.

19.1. STARTING AND FINISHING COMPOSITIONS

The matte product from electric and flash furnace smelting typically contains 15%–40% Ni, 20%–40% Fe, and 20%–25% S. The matte product from converting typically contains 40%–70% Ni, 0.5%–4% Fe, 21%–24% S, and the remainder is Cu and Co (Warner *et al.*, 2007).

These compositions indicate that matte converting is mainly an iron-oxidation process. It is an efficient way of removing most of the iron in the matte before the matte is sent for hydrometallurgical or vapometallurgical refining. These refining operations are discussed in Chapters 23–27.

Converting removes iron as iron silicate in the molten slag. The slag is a dense, unreactive and readily disposable waste product.

Extractive Metallurgy of Nickel, Cobalt and Platinum-Group Metals. DOI: 10.1016/B978-0-08-096809-4.10019-X

TABLE 19.1 Production Details of Five Nickel Matte Peirce-Smith Converting Operations*

Smelter	Kalgoorlie, Australia	Falconbridge, Canada	Sudbury, Canada	Jinchuan, China	Norilsk, Russia
Start-up date	1972	1930	1911	1992	1981
Ni-in-matte production, tonnes per year	100 000	70 000	130 000	70 000	40 000
Number of converters	3	3	5	3	4
Converter details					
Outside diameter × length, m	3.6 × 7.3	(a) two @ 4 × 9 (b) one @ 4 × 15	(a) three @ 4 × 13.7 (b) two @ 4 × 10.7	6 × 8.2	4 × 9
Number of tuyeres					
Total	28	(a) 42 (b) 6	(a) 51 (b) 42	34	52
Tuyere diameter, cm	6.4	(a) 5 (b) 3.2	(a) 5.1 (b) 5.1	4.8	5
Usual blast rate per converter, Nm^3/h	19 000	(a) 30 000 (b) 6500	35 000	20 000	36 000
Usual volume % O_2 in blast	21	(a) 21 (b) 33–43	24–27	21	21
Production details					
Input molten matte					
Composition	47% Ni, 1.5% Cu, 0.8% Co, 20% Fe, 27% S	35% Ni, 11% Cu, 1% Co, 33% Fe, 17% S	22% Ni, 25% Cu, 0.6% Co, 24% Fe, 26% S	29% Ni, 15% Cu, 0.6% Co, 29% Fe, 23% S	13% Ni, 8% Cu, 0.7% Co, 53% Fe, 23% S
Source	Outokumpu flash furnace with	Electric furnace	Inco flash furnace	Outokumpu flash furnace with	Electric furnace

	electric appendage			electric appendage	
Flux, % SiO_2	99		~95	95	
Diameter, mm	<25		10–40	40	
Solid reverts	10% of matte feed			25–30% of matte feed	
Product molten matte					
Composition	66% Ni, 2% Cu, 0.9% Co 4.2% Fe, 24% S	56% Ni, 18% Cu, 2% Co 2% Fe, 21% S	35% Ni, 42% Cu, 0.4% Co 0.5% Fe, 22% S	48% Ni, 25% Cu, 0.8% Co 4.2% Fe, 22% S	35% Ni, 32% Cu, 0.7% Co 3.2% Fe, 23% S
Matte pouring temperature, °C	1280	1250	1020	1250	1200
Destination	H_2O granulated then leached	H_2O granulated then leached	Slow cooled, milled	Slow cooled, milled	Slow cooled, milled
Product slag					
Composition	21% SiO_2, 55% Fe, 32% Fe_3O_4	21% SiO_2, 48% Fe	26% SiO_2, 51% Fe, 23% Fe_3O_4	26% SiO_2, 48% Fe 16% Fe3O_4	18% SiO_2, 55% Fe
Slag pouring temperature, °C	1280		1230	1280	
Slag destination	Recycle to flash furnace	Fuel fired slag cleaning furnace	Recycle to flash furnace	Electric slag cleaning furnaces	Electric slag cleaning furnaces
Offgas, Nm^3/h	100 000				
Volume% SO_2	4		4	3	2
Destination	40% to H_2SO_4 plant, 60% to atmosphere	Atmosphere	Atmosphere	Sulfuric acid plant	Atmosphere

Warner et al. (2007)

19.2. CHEMISTRY OF CONVERTING

Converting removes iron from matte by reaction with oxygen. The reaction is given as follows:

$$\underset{\substack{\text{in molten smelting}\\ \text{furnace matte}}}{\overset{1200^\circ\text{C}}{2\text{Fe}(\ell)}} + \underset{\substack{\text{in injected air}\\ \text{or air + oxygen}}}{\overset{25^\circ\text{C}}{\text{O}_2(g)}} + \underset{\substack{\text{in crushed rock}\\ \text{or sand}}}{\overset{25^\circ\text{C}}{\text{SiO}_2(s)}} \rightarrow \underset{\substack{\text{molten silicate}\\ \text{slag}}}{\overset{1275^\circ\text{C}}{\text{Fe}_2\text{SiO}_4(\ell)}} \quad (19.1)$$

Some sulfur is also oxidized during converting. The reaction is given as follows:

$$\underset{\substack{\text{in molten smelting}\\ \text{furnace matte}}}{\overset{1200^\circ\text{C}}{\text{S}(\ell)}} + \underset{\substack{\text{in injected air}\\ \text{or air + oxygen}}}{\overset{25^\circ\text{C}}{\text{O}_2(g)}} \rightarrow \underset{\text{in SO}_2/\text{N}_2\text{ offgas}}{\overset{\sim 1275^\circ\text{C}}{\text{SO}_2(g)}} \quad (19.2)$$

The oxidation of sulfur, given in Equation (19.2), becomes more prevalent as the iron content of the matte decreases.

The oxidation of nickel (and copper, silver, gold, and platinum-group elements) under these conditions is thermodynamically unfavorable. Consequently, these elements mostly remain in the matte.

The heat generated by the reactions given in Equations (19.1) and (19.2) is sufficient to heat the incoming blast and flux and to keep the matte and slag hot and molten. In fact, with an oxygen-enriched blast, the reactions produce sufficient heat to melt nickel-rich scrap and recycled smelter solids (referred to as *reverts*).

Molten slag, at a temperature of 1275°C, is typically poured from the converter several times during a converting cycle. The slag that is removed is replaced with fresh furnace matte. In this way, nickel and other metals gradually build up in the converter until there is enough (approximately 60 tonnes of nickel as matte) to do a final iron 'end-point' blow.

19.3. PRINCIPLES OF CONVERTING

There are two basic principles of converting:

(a) iron oxidizes preferentially to nickel (as well as copper, gold, silver and platinum-group elements); and,
(b) the slag produced by the oxidation of iron is immiscible with the unoxidized matte.

The products of converting are a layer of molten slag and a layer of molten matte. The specific gravity of the slag is about 3.0 and that of the matte is about 5.0 (Jones, 2005). Because the specific gravity of the slag is less than that of the matte, the slag floats on the matte.

The slag is rich in iron and lean in nickel; the matte is rich in nickel and lean in iron.

The slag is poured from the converter and almost always sent to an electric settling furnace (also called a slag-cleaning furnace) for the additional recovery of nickel from the slag. This recovery is necessary because the slag inevitably contains matte droplets entrained by the turbulent Peirce-Smith converting process.

The matte is then poured from the converter and transferred to metal production. The pouring of the matte is shown in Figure 8.3.

19.4. BEHAVIOR OF OTHER METALS

Copper, silver, gold and platinum-group elements that are present in the feed to the smelter are not oxidized by oxygen during converting. They remain almost entirely in the matte.

Cobalt remains mostly in the matte, but some may oxidize and dissolve in the slag. Its behavior is between that of iron and nickel. Industrially, 30%–50% of the cobalt in the feed to the smelter is lost in slag (Warner *et al.*, 2007), mostly in converter slag.

Some of the oxidized cobalt is recovered by operating the electric settling furnace under reducing conditions by adding a carbon reductant.

19.5. CHOICE OF FINAL IRON CONTENT

The first step in making high-purity metal from converter matte is often leaching. The objective of removing iron before leaching is to minimize the production of iron residue in the leach plant. Converter matte destined for leaching typically contains 1%–4% Fe.

Other processes require even less iron, for example, matte solidification/flotation requires 0.5% Fe, and matte anode electrorefining requires 0.6% Fe, because iron tends to report to the final nickel products (Donald & Scholey, 2005). (Matte leaching is described in Chapter 24, matte solidification/flotation is described in Chapter 21 and anode electrorefining is described in Appendix F.)

19.6. END-POINT DETERMINATION

The converting sequence ends when the following two criteria are met:

(a) the converter contains the specified amount of nickel in the matte, often after several molten feed matte additions; and,
(b) the iron has been 'blown down' to its specified target concentration.

The converter operator ascertains when this end point has been reached by taking a matte sample through one of tuyeres in the converter and measuring content of iron in the sample by X-ray fluorescence analysis.

In one case, the end point is approached slowly by blowing the converter with industrial nitrogen that has a low concentration of oxygen (Donald & Scholey, 2005).

Once the target end point has been reached, the converter is rolled into its pouring position and the slag and the matte are poured into separate ladles.

The molten slag is usually transferred by crane to an electric settling furnace for matte recovery.

The molten matte is transferred by crane to either (i) a water granulation plant; (ii) a casting slow-cooling plant; or, (iii) a matte anode-casting plant (Donald & Scholey, 2005; Dunkley, Norval, Jones, & Chennells, 2008; Jacobs, 2006; Marcuson & Diaz, 2005; Schonewille, Boissoneault, Ducharme, & Chenier, 2005).

Granulation prepares the matte for leaching. Slow cooling prepares the matte for flotation into a concentrate rich in nickel, copper and platinum-group metals for subsequent metal production. Matte anode-casting prepares the matte for electrorefining.

19.7. CAPTURE OF SULFUR DIOXIDE

Nearly all of the converting of nickel matte is performed in Peirce-Smith converters. This is due to the simplicity and high chemical efficiency of this type of converter. This process suffers, however, from the following drawbacks:

(a) the converter emits sulfur dioxide into the environment during charging and skimming and at the interface between the hood and the converter during blowing; and,
(b) the production of sulfur dioxide is interrupted during charging and skimming, which gives a discontinuous flow of sulfur dioxide to the sulfuric acid plant.

These drawbacks of Peirce-Smith converters have led to the following developments:

(a) continuous converting; and,
(b) flash smelting directly to the final matte (Makinen & Ahokainen, 2009; Makinen, Fagerlund, Anjala, & Rosenback, 2005).

Both of these developments avoid batch Peirce-Smith converting and the problems associated with the capture of sulfur dioxide.

Continuous converting has been implemented at one nickel sulfide (and platinum) smelter, that is, at Anglo American Platinum's Waterval smelter in South Africa. The continuous converter is described in Chapter 35.

Direct-to-final-matte flash smelting (DON Process) is being used in two smelters, described in Section 19.12.

19.8. TUYERES AND OXYGEN ENRICHMENT

The tuyeres in Peirce-Smith converters are carbon, steel or stainless steel pipes that are embedded in the converter refractory, as shown in Figure 8.2. They are about 0.05 m in diameter. They are joined to a gas distribution 'bustle pipe', which is affixed along the length of the converter and connected through a rotatable seal to a blast supply flue.

The blast air is pressurized by electric or steam driven blowers. Industrial oxygen (when used) is added to the supply flue just before it connects to the converter.

Steady flow of blast requires periodic clearing (or '*punching*') of the tuyeres to remove matte accretions that build up at their tips. Punching is done by ramming a steel bar completely through the tuyere holes. It is usually done with a Gaspe mobile carriage puncher that runs on rails behind the converter. The puncher is sometimes automatically positioned and operated.

19.8.1. Limitation on the Rate of Blast

The input rate of blast per tuyere is about 12 Nm^3/min at 1.2 bar. This gives tuyere tip velocities of 80–120 m/s. Blowing rates above 17 Nm^3/min per tuyere cause slopping of matte and slag from the converter. High blowing rates without slopping are favored by deep submergence of the tuyere in the matte.

The feed rate of oxygen can be increased without increasing the input rate of the blast by enriching the blast with oxygen.

19.8.2. Oxygen Enrichment

Nickel sulfide smelters are gradually adopting oxygen enrichment of their blast, especially the nickel smelters in Canada. There are a number of advantages of oxygen enrichment, including the following:

(a) the rate of reaction is increased for a given total blast input rate;
(b) the concentration of sulfur dioxide in the offgas is increased, making gas handling and sulfuric acid manufacture cheaper; and,
(c) the amount of nitrogen, which must be heated, entering the converter is diminished.

Reducing the amount of nitrogen permits (i) rapid heating of matte and slag; and, (ii) melting of valuable solids, such as nickel-rich scrap and reverts.

The only disadvantage of an oxygen-enriched blast is that it gives a high reaction temperature at the tuyere tips. This leads to rapid refractory erosion around the tip. Blowing at high velocity, which promotes tubular accretion formation and pushes the reaction zone out and away from the tuyere tips, discourages this erosion.

On balance, the advantages of oxygen enrichment outweigh the disadvantages. This balance is particularly in favor of oxygen enrichment for those smelters that wish to maximize (i) the rate of converting, especially if converting is a production bottleneck; and, (ii) the melting of nickel-rich scrap and reverts.

The present upper practical limit of oxygen enrichment appears to be about 30% O_2 (by volume). Above this level, erosion of the refractory lining becomes excessive. This is because strong tubular accretions do not form in front of the tuyeres above 30% O_2, causing the reactions between oxygen and matte to take place flush with the tuyeres and refractory.

This limitation of about 30% O_2 can only be overcome by using nitrogen-shrouded tuyeres, which are discussed in Section 19.9.

19.8.3. Maximizing Converter Productivity

The production rate of a converter, in tonnes of product matte per day, is maximized by:

(a) blowing converter blast at its maximum rate (including avoidance of tuyere blockages);
(b) enriching the blast to its maximum feasible oxygen level;
(c) maximizing oxygen utilization by keeping the tuyere tips deeply submerged in the matte; and,
(d) maximizing converter campaign life.

19.9. NITROGEN-SHROUDED BLAST INJECTION

Oxygen enrichment above about 30% O_2 can be achieved by shrouding the air-oxygen blast with nitrogen 'coolant'. This system of blast is shown in Figure 19.1.

Nitrogen-shrouded blast is used in the Falconbridge smelter in Canada (Kapusta, Stickling, & Tai, 2005; Navarra & Kapusta, 2009). So far, this type of injection is used only at this one smelter.

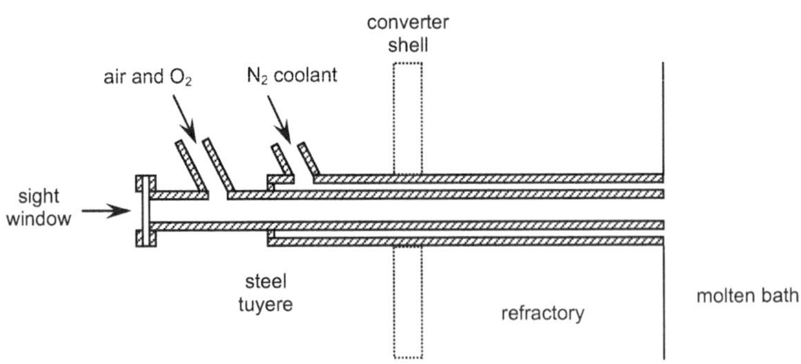

FIGURE 19.1 Shrouded tuyere for injecting oxygen-enriched air (~40 volume% O_2) into a Peirce-Smith converter. Oxygen-enriched air is blown through the center pipe. Nitrogen is blown through the annulus as a coolant.

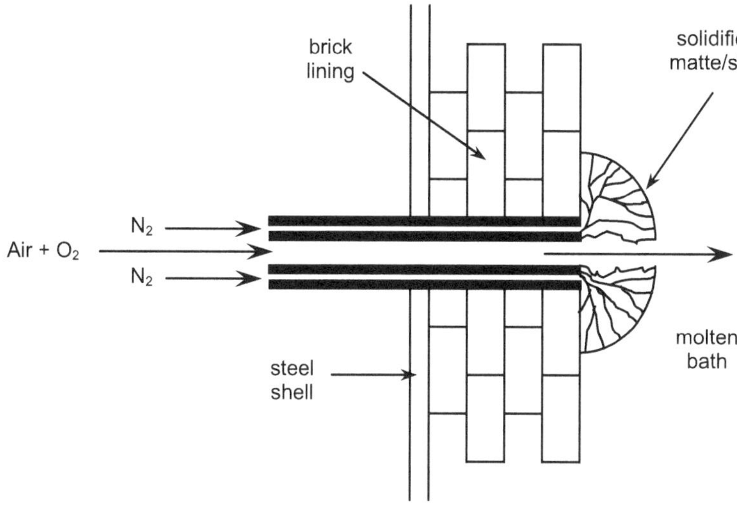

FIGURE 19.2 Sketch of accretion growth in front of nitrogen-shrouded oxygen-enriched air tuyere. The accretion at the tip of the tuyere protects the adjacent refractory from high-temperature wear.

The objectives of the process are:

(a) to oxidize iron and sulfur with up to 43% O_2 (by volume) in blast thereby increasing the productivity of the converter and its ability to melt nickel-rich solids;
(b) to eliminate the need to punch the tuyeres of the converter; and,
(c) to minimize refractory wear, to lengthen converter campaign life and lower refractory replacement costs.

The blast is blown into the converter at about 6 bar. The nitrogen is blown in at about 10 bar. These pressures prevent accretions from bridging across the tuyere tips, thereby eliminating the need for punching. The build up of accretion at the tuyere tip is shown in Figure 19.2.

19.10. CONVERTER CONTROL

19.10.1. Temperature Control

All the heat for maintaining the converter liquids at their specified temperatures is provided by the reactions (Equations 19.1 and 19.2). As a result, the temperature of the converter is readily controlled in two ways:

(a) by raising or lowering oxygen-enrichment level, which raises or lowers the rate at which nitrogen enters the converter and cools the matte (at a constant total blast flow rate); and,
(b) by adjusting the rate of addition of scrap and reverts.

Temperature Measurement

The temperature of liquid in the converter is measured by means of (i) an optical pyrometer sighted downwards through the converter mouth; or, (ii) a two-wavelength optical pyrometer telescope sighted through a tuyere. The tuyere pyrometer appears to be more satisfactory because it sights directly on the matte rather than through a dust-laden atmosphere.

Choice of Temperature

Converting temperature is chosen to give (i) rapid slag formation; and, (ii) fluid slag with a minimum of entrained matte. A slag temperature of about 1275°C discourages tuyere blockage without causing excessive refractory wear.

19.10.2. Control of Slag and Flux

The primary objective of creating a slag in the converter is to liquefy newly formed iron oxides so that they can be poured from the converter. Silica flux is added for this purpose.

A common indicator of the composition of slag is the ratio of the concentration of silica (% SiO_2) in slag to the concentration of iron (% Fe) in slag. Industrially, this value is typically between 0.45 and 0.5.

Flux is added through chutes above the converter and/or by a flux gun through an end wall. It is added at a rate that matches the rate of oxidation of iron after an initial delay of several minutes during which the converter heats up. The flux is commonly crushed silicate rock that is between 0.01 and 0.05 m in diameter.

The following factors increase the rate of reaction between oxygen, matte and flux to form liquid slag:

(a) high operating temperature;
(b) steady input of small, evenly sized particles of silica flux (Schonewille, O'Connell, & Toguri, 1993);
(c) deep tuyere placement in the matte (to avoid over-oxidizing the slag);
(d) vigorous mixing, which the Peirce-Smith converter provides; and,
(e) reactive flux.

Schonewille *et al.* (1993) reported that the most reactive fluxes are those with a high percentage of quartz (rather than tridymite or feldspar).

19.10.3. Control of the Matte Composition

The main objective of converting is to produce matte that is low in iron. Some smelters also control the sulfur content of their final matte. The Thompson, Canada smelter, for example, produces low-iron, low-sulfur converter matte, which has a composition of about 0.6% Fe and 18.7% S. This low content of sulfur gives stronger matte anodes (Marcuson & Diaz, 2005).

The main variable for controlling the final sulfur content of converter matte is temperature. A high converting temperature favors rapid removal of sulfur because it thermodynamically favors the reaction given in Equation (19.2) (Donald *et al.*, 2005; Kellogg, 1987). This allows sulfur content in the matte to be adjusted while iron is blown down to its specified level.

19.11. ALTERNATIVES TO PEIRCE-SMITH CONVERTING

Peirce-Smith converting accounts for about 95% of the converting of nickel matte. This is due to its simplicity and high chemical efficiency. However, it is a poor collector of SO_2 and it is a batch process, which requires considerable materials handling.

These deficiencies are remedied by the following developments:

(a) continuous converting (Donald *et al.*, 2005); and,
(b) flash smelting directly to a low-iron matte (Makinen *et al.*, 2005; Makinen & Ahokainen, 2009).

Continuous converting is described in Chapter 35.

19.12. DIRECT TO LOW-IRON MATTE FLASH SMELTING

The Outokumpu-type flash smelter in Harjavalta, Finland, avoids Peirce-Smith converting by producing a matte that has a low iron content (about 5% Fe) in the flash smelting furnace. This arrangement avoids the problems of the capture of sulfur dioxide of the Peirce-Smith converter. A schematic diagram of the flowsheet of the process is illustrated in Figure 19.3. Industrial data for the Harjavalta operation are given in Table 18.1.

19.12.1. Process Description

The basic strategy of the process, referred to as the DON Process, is to increase iron oxidation in the flash furnace by increasing the

$$\frac{O_2\text{-in-blast input rate}}{\text{concentrate feed rate}}$$

ratio until its product matte contains only ~5% Fe.

The temperature of the furnace is controlled by adjusting the feed rates of nitrogen, fuel and solid reverts. The composition of the slag is controlled by adjusting feed rate of flux.

There are three products from this flash furnace:

(a) the molten matte, containing 5% Fe and 65% Ni, which is granulated (and then leached prior to metal production);

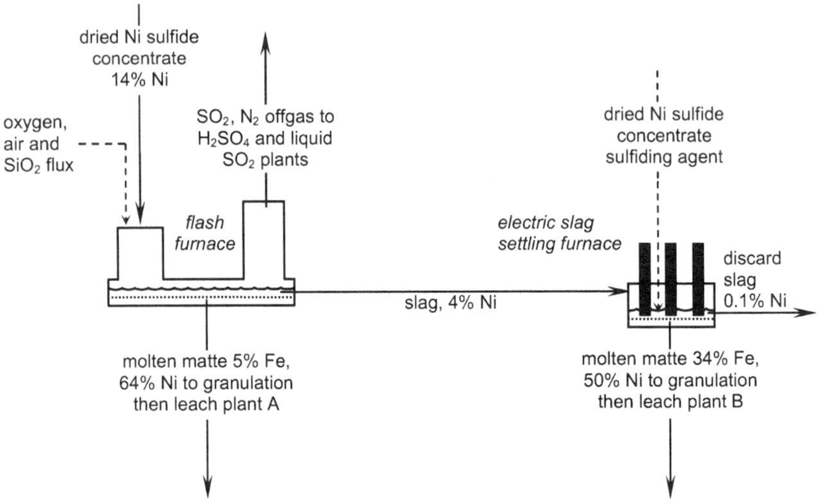

FIGURE 19.3 Schematic flowsheet for flash smelting directly to 5% Fe matte (Makinen *et al.*, 2005). The single continuous sulfur dioxide offgas stream and the high and low Fe matte products are noteworthy. The two matte products are leached separately (Makinen & Taskinen, 2008).

(b) the slag, containing about 4% Ni, which is sent to an electric furnace for nickel-matte recovery; and,
(c) the offgas, containing about 30% SO_2 and 70% N_2 gas by volume, which is sent to sulfuric acid and liquid sulfur dioxide plants.

Overall, the capture of sulfur dioxide is greater than 99%.

19.12.2. Subsequent Recovery of Nickel in Electric Furnace Matte

A unique aspect of the process is that it produces a second high-iron matte from the slag-cleaning furnace. This matte is also granulated and leached for recovery of nickel in a separate leach plant that has been designed specifically for this high-iron matte.

This, of course, increases the complexity of the leach plant, a price that is paid for the efficient capture of sulfur dioxide. The leaching plant is described in Chapters 23 and 24.

19.13. SUMMARY

In the nickel industry, converting is the production of a low-iron matte, containing 0.5%–4% Fe, from a furnace matte that contains between 20% and 40% Fe.

The iron in the furnace matte is most often removed from the matte into the slag by oxidation with air or oxygen-enriched air in Peirce-Smith converters.

Nickel and other metals are not oxidized during this oxidation. These metals remain in the matte phase and are recovered during downstream processing.

Peirce-Smith converting is a batch process that is chemically efficient but a poor collector of sulfur dioxide from inadvertent sulfur oxidation.

An alternative to Peirce-Smith converting is flash furnace direct-to-low-iron matte production. It avoids the sulfur dioxide collection problem but it requires specialized technology for the recovery of nickel from the slag.

The product of converting, the low-iron matte, is an excellent starting point for making nickel (and other metals) by leaching and other metal-production processes.

REFERENCES

Donald, J. R., & Scholey, K. (2005). An overview of Inco's Copper Cliff operations. In J. Donald & R. Schonewille (Eds.), *Nickel and cobalt 2005 challenges in extraction and production* (pp. 457–477). CIM.

Donald, J. R., Warner, A. E. M., Dalvi, A. D., et al. (2005). Risks and opportunities in continuous converting for nickel at Inco. In A. Ross, T. Warner & K. Scholey (Eds.), *Converter and fire refining practices* (pp. 295–307). TMS.

Dunkley, J. J., Norval, D., Jones, R. T., & Chennells, P. (2008). Water atomization of PGM-containing intermediate alloys. In M. Rogers (Ed.), *Third international platinum conference 'Platinum in Transformation'* (pp. 155–159). SAIMM.

Jacobs, M. (2006). Process description and abbreviated history of Anglo Platinum's Waterval smelter. In R. Jones (Ed.), *Southern African pyrometallurgy 2006* (pp. 17–28). SAIMM.

Jones, R. T. (2005). An overview of Southern African PGM (platinum group metal) smelting. In J. Donald & R. Schonewille (Eds.), *Nickel and cobalt 2005, challenges in extraction and production* (pp. 147–178). CIM.

Kapusta, J. P. T., Stickling, H., & Tai, W. (2005). High oxygen shrouded injection at Falconbridge: Five years of operation. In A. Ross, T. Warner & K. Scholey (Eds.), *Converter and fire refining practices* (pp. 47–60). TMS.

Kellogg, H. H. (1987). Thermochemistry of nickel matte converting. *Canadian Metallurgical Quarterly, 26*, 285–298.

Makinen, T., & Ahokainen, T. (2009). 50 Years of nickel flash smelting – Still going strong. In J. Liu, J. Peacey & M. Barati, et al. (Eds.), *Pyrometallurgy of nickel and cobalt 2009, proceedings of the international symposium* (pp. 209–220). CIM.

Makinen, T., Fagerlund, K., Anjala, Y., & Rosenback, L. (2005). Outokumpu's technologies for efficient and environmentally sound nickel production. In J. Donald & R. Schonewille (Eds.), *Nickel and cobalt 2005, challenges in extraction and production* (pp. 71–89). CIM.

Makinen, T., & Taskinen, P. (2008). State of the art in nickel smelting: Direct Outokumpu nickel technology. *Mineral Processing and Extractive Metallurgy, 117*, 86–94.

Marcuson, S. W., & Diaz, C. M. (2005). The changing Canadian nickel smelting landscape – Late 19th century to early 21st century. In J. Donald & R. Schonewille (Eds.), *Nickel and cobalt 2005 challenges in extraction and production* (pp. 179–208). CIM.

Navarra, A., & Kapusta, J. P. T. (2009). Decision-making software development for incremental; Improvement of nickel matte conversion. In J. Liu, J. Peacey & M. Barati, et al. (Eds.), *Pyrometallurgy of nickel and cobalt 2009, proceedings of the international symposium* (pp. 611–619). CIM.

Schonewille, R., Boissoneault, M., Ducharme, D., & Chenier, J. (2005). Update on Falconbridge's Sudbury nickel smelter. In J. Donald & R. Schonewille (Eds.), *Nickel and cobalt 2005, challenges in extraction and production* (pp. 479–498). CIM.

Schonewille, R. H., O'Connell, G. J., & Toguri, J. M. (1993). A quantitative method for silica flux evaluation. *Metallurgical Transactions B, 24B*, 63–73.

Warner, A. E. M., Diaz, C. M., Dalvi, A. D., et al. (2007). JOM world nonferrous smelter survey part IV: Nickel Sulfide. *JOM, 59*, 58–72.

SUGGESTED READING

Boldt, J. R., & Queneau, P. (1967). *The winning of nickel.* Toronto: Longmans Canada Ltd.

Donald, J. & Schonewille, R. (Eds.), (2005). *Nickel and cobalt 2005, challenges in extraction and production.* CIM.

Kapusta, J. & Warner, T. (Eds.), (2009). *International Peirce-Smith converting centennial.* Wiley.

Liu, J., Peacey, J. & Barati, M.et al. (Eds.), (2009). *Pyrometallurgy of nickel and cobalt 2009, proceedings of the international symposium.* CIM.

Ross, A. G., Warner, T. & Scholey, K. (Eds.), (2005). *Converter and fire refining practices.* TMS.

Warner, A. E. M., Diaz, C. M., Dalvi, A. D., et al. (2007). JOM world nonferrous smelter survey part IV: Nickel: Sulfide. *JOM, 59*, 58–72.

Chapter 20

Sulfur Dioxide Capture in Sulfuric Acid and Other Products

About half the global production of primary nickel is extracted from sulfide minerals. Sulfur in some form is an inevitable by-product of this extraction. The most common form is as sulfur dioxide gas from roasting, smelting and converting.

Sulfur dioxide can be harmful to fauna and flora. Every effort should, therefore, be made to capture it, usually as sulfuric acid, which has many uses including fertilizer production and ore leaching (Davenport & King, 2006).

This chapter describes the following:

(a) offgases from nickel extraction;
(b) production of sulfuric acid from these gases; and,
(c) other sulfur dioxide-based by-products.

20.1. NICKEL EXTRACTION OFFGASES

Nickel sulfide roasting and flash smelting produce continuous steams of offgas that contain 10%–12% SO_2 (by volume) after dilution with air. These gases are ideal for making sulfuric acid.

The flow of offgases from an electric furnace is continuous, but the concentration of sulfur dioxide is too weak (~1% by volume) for efficient manufacture of sulfuric acid. These gases are usually vented to the atmosphere.

Peirce-Smith converter gases are discontinuous and difficult to capture. They are also mostly vented to the atmosphere.

20.2. PRODUCTION OF SULFURIC ACID FROM ROASTER AND FLASH FURNACE OFFGASES

Production of sulfuric acid from roaster and flash furnace gases entails the following:

(a) cooling, cleaning and drying the gases;

Extractive Metallurgy of Nickel, Cobalt and Platinum-Group Metals. DOI: 10.1016/B978-0-08-096809-4.10020-6

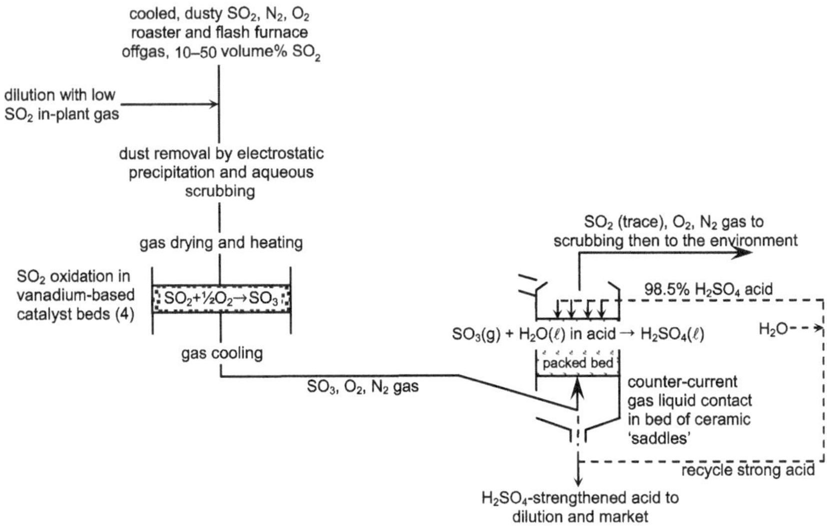

FIGURE 20.1 Simplified flowsheet for capturing sulfur dioxide as sulfuric acid.[a] The oxidation of sulfur dioxide to sulfur trioxide and then absorption of the sulfur trioxide into sulfuric acid are shown.

(b) catalytically oxidizing the sulfur dioxide to sulfur trioxide with oxygen in the gases or with additional air; and,
(c) making sulfuric acid by reacting the sulfur trioxide produced in step (b) with the water contained in 98.5% H_2SO_4.

A simplified flowsheet for the manufacture of sulfuric acid is shown in Figure 20.1.

20.3. GAS COOLING, CLEANING, AND DRYING

Cooling, cleaning and drying of smelter offgas are detailed by Davenport and King (2006) and King (2010). These operations entail the following:

(a) cooling in heat recovery boilers or water spray coolers;
(b) de-dusting in electrostatic precipitators;
(c) removing the remaining dust (and some gaseous impurities, such as halides) by water scrubbing;
(d) condensing water vapor from the water-scrubbed gas by cooling the gas to about 30°C; and,
(e) dehydrating the gas with strong sulfuric acid.

a. Several slightly different H_2SO_4-making process also exist (King, 2010). None are being used for nickel roaster and flash furnace gases.

20.3.1. Condensation

The gas entering catalytic converters for the oxidation of sulfur dioxide must be bone dry; otherwise, the water vapor and sulfur trioxide produced by the oxidation reaction will inadvertently form corrosive sulfuric acid in the flues and equipment.

Scrubbing saturates the gas with water vapor. Most of this is condensed from the gas by cooling the gas to about 30°C, which leaves about 5% H_2O in the exit gas from the condenser (Davenport & King, 2006).

20.3.2. Dehydration

Final dehydration of the gas is done by contact with strong sulfuric acid. The reaction is as follows:

$$\underset{\substack{\text{in partially}\\ \text{dried gas}}}{\overset{\sim 30^\circ\text{C}}{H_2O(g)}} + \underset{\substack{\text{in 96\% } H_2SO_4,\\ 4\%\ H_2O\\ \text{sulfuric acid}}}{\overset{\sim 60^\circ\text{C}}{H_2SO_4(\ell)}} \longrightarrow \underset{\substack{\text{slightly weakened acid}\\ \Delta H_R^o \approx -80\ \text{kJ/mol}\ H_2O}}{\overset{\sim 80^\circ\text{C}}{(H_2SO_4(\ell) + H_2O(\ell))}} \quad (20.1)$$

This reaction lowers the water content in the gas to approximately 50 mg/Nm^3. This is low enough to prevent accidental downstream production of sulfuric acid.

20.4. OXIDATION OF SULFUR DIOXIDE TO SULFUR TRIOXIDE

Sulfur dioxide and water do not react to form sulfuric acid at a rate fast enough for an industrial process (Davenport & King, 2006). Sulfur dioxide must be first oxidized to sulfur trioxide, which then reacts with water. The oxidation reaction is given as follows:

$$\underset{\substack{\text{in dried sulfur}\\ \text{dioxide feed gas}}}{SO_2(g)} + \underset{\text{in feed gas}}{0.5O_2(g)} \xrightarrow{\substack{\text{vanadium based}\\ \text{catalyst}}} \underset{\text{in product gas}}{SO_3(g)} \quad (20.2)$$

This oxidation reaction is almost always performed in a catalytic reactor at a temperature of between 400°C and 625°C. The reactor has a diameter of about 10 m, and the vanadium-based catalyst is laid in 1-m-thick beds inside the reactor. Without the catalyst, the oxidation reaction given in Equation (20.2) is too slow for the process to be profitable.

The optimum concentration of sulfur dioxide in the feed gas to the catalyst bed is between 10% and 12% by volume. The optimum volumetric ratio of O_2:SO_2 is 1:1. At this ratio, the oxidation of sulfur dioxide to sulfur trioxide goes almost to completion. Note that this ratio is twice the stoichiometric requirement of oxygen in the oxidation reaction given in Equation (20.2).

FIGURE 20.2 Photograph of catalyst pieces used in the oxidation of sulfur dioxide. The outside diameter of the largest piece is 20 mm. The pieces are used in ~1 m thick, ~10-m-diameter beds in stainless-steel towers. Rings, star rings and pellets are shown. The star rings maximize catalyst area and minimize resistance to gas flow. At temperatures of 420°C–620°C, a molten layer of vanadium–alkali metal sulfate/pyrosulfate covers the porous solid silica substrates. *Photograph courtesy of Haldor Topsoe A/S.*

20.5. CATALYST FOR THE OXIDATION OF SULFUR DIOXIDE

The exact compositions of commercial catalysts for the oxidation of sulfur dioxide are proprietary. However, they are known to contain about 5%–9% V_2O_5, 15%–20% potassium sulfate/pyrosulfate (SO_4/S_2O_7), 2%–5% sodium sulfate/pyrosulfate, 0%–15% cesium sulfate/pyrosulfate and 55%–70% porous silica substrate, which is made from diatomaceous earth.

The catalysts become active when they melt at approximately 400°C. They decompose and become inactive above about 625°C. At their operating temperature, they consist of a molten layer of vanadium, potassium, sodium, cesium, sulfur and oxygen on a solid silicate substrate. For this reason, the catalysts are referred to as *supported liquid phase catalysts*. Catalyst pieces are shown in Figure 20.2.

20.5.1. Catalyst Beds

The oxidation of sulfur dioxide to sulfur trioxide is mostly carried out in four (occasionally three) separate catalyst beds, each about 1 m thick, with gas cooling between the beds. A catalyst bed is shown in Figure 20.3, and the inter-stage cooling is shown in Figure 20.4. The intermediate cooling keeps the gas temperature in the operating range of the catalyst, which is from 400°C–625°C; otherwise, the large amount of heat produced by the exothermic reaction would overheat and chemically degrade the catalyst (Davenport & King, 2006).

20.5.2. Composition of Feed Gas

This problem of degradation at high temperature also explains why the catalyst feed gases mostly contain less than 12% SO_2 by volume. More SO_2 than this (and less N_2) will cause overheating of even a thin catalyst bed.

20.5.3. Efficiency of the Oxidation of Sulfur Dioxide

Four-bed converters like that shown in Figure 20.4 oxidize about 99.5% of the sulfur dioxide in the feed gas. The remaining unoxidized sulfur dioxide is

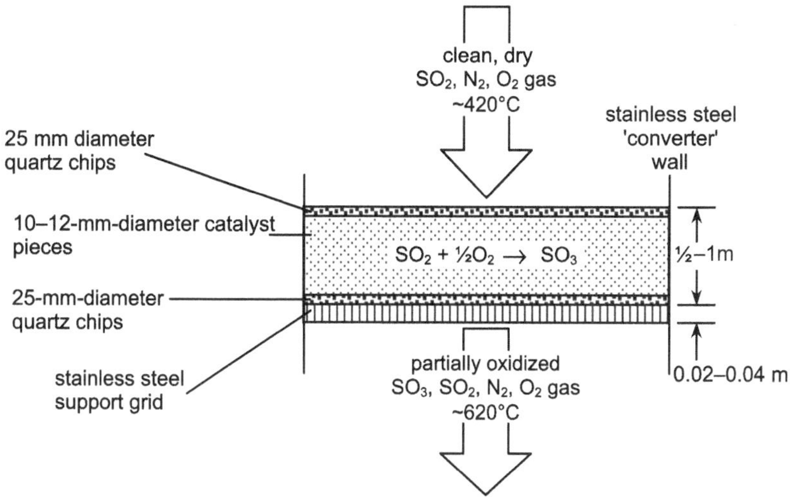

FIGURE 20.3 Bed of catalyst pieces for oxidizing sulfur dioxide to sulfur trioxide. It is circular with a diameter of about 10 m. Industrial oxidation is usually done in a 'converter' of four such beds. Downward gas flows are about 25 $Nm^3/min/m^2$ of converter cross-sectional area. The top layer of quartz chips holds the catalyst in place, and the bottom layer prevents the catalyst from sticking to the stainless-steel support grid.

often removed by absorption in a basic solution of sodium oxide (Ishiyama & Hirai, 1997; King, 2010; Tokyo Nickel, 2006). Other basic solutions can also be used.

20.6. MAKING ACID FROM SULFUR TRIOXIDE

Sulfuric acid is made from sulfur trioxide by reacting exit gas from the catalytic oxidation with the water in 98.5% sulfuric acid, as shown in Figure 20.5. The reaction produces strengthened sulfuric acid:

$$\underset{\substack{\text{in } SO_3,\ N_2,\ O_2\\ \text{catalytic oxidation}\\ \text{exit gas}}}{SO_3(g)} + \underset{\substack{\text{in 98.5\% } H_2SO_4\\ \text{sulfuric acid}\\ \text{solution}}}{H_2O(\ell)} \xrightarrow{80-110^\circ C} \underset{\substack{\text{in strengthened}\\ \text{acid}\\ \Delta H^\circ = -130\ \text{kJ/mol of } SO_2}}{H_2SO_4(\ell)} \quad (20.3)$$

20.6.1. Industrial Sulfuric Acid-Making

The acid-making reaction, Equation (20.3), is carried out industrially by blowing cooled catalyst-bed exit gas up around a bed of ceramic saddles. At the same time, 98.5% sulfuric acid is trickled counter-currently down around the 'saddles'.

Typical industrial acid-making parameters are as follows:

(a) tower diameter 7–9 m;
(b) saddle sizes 0.05–0.08 m;
(c) saddle packing height 2–4 m;

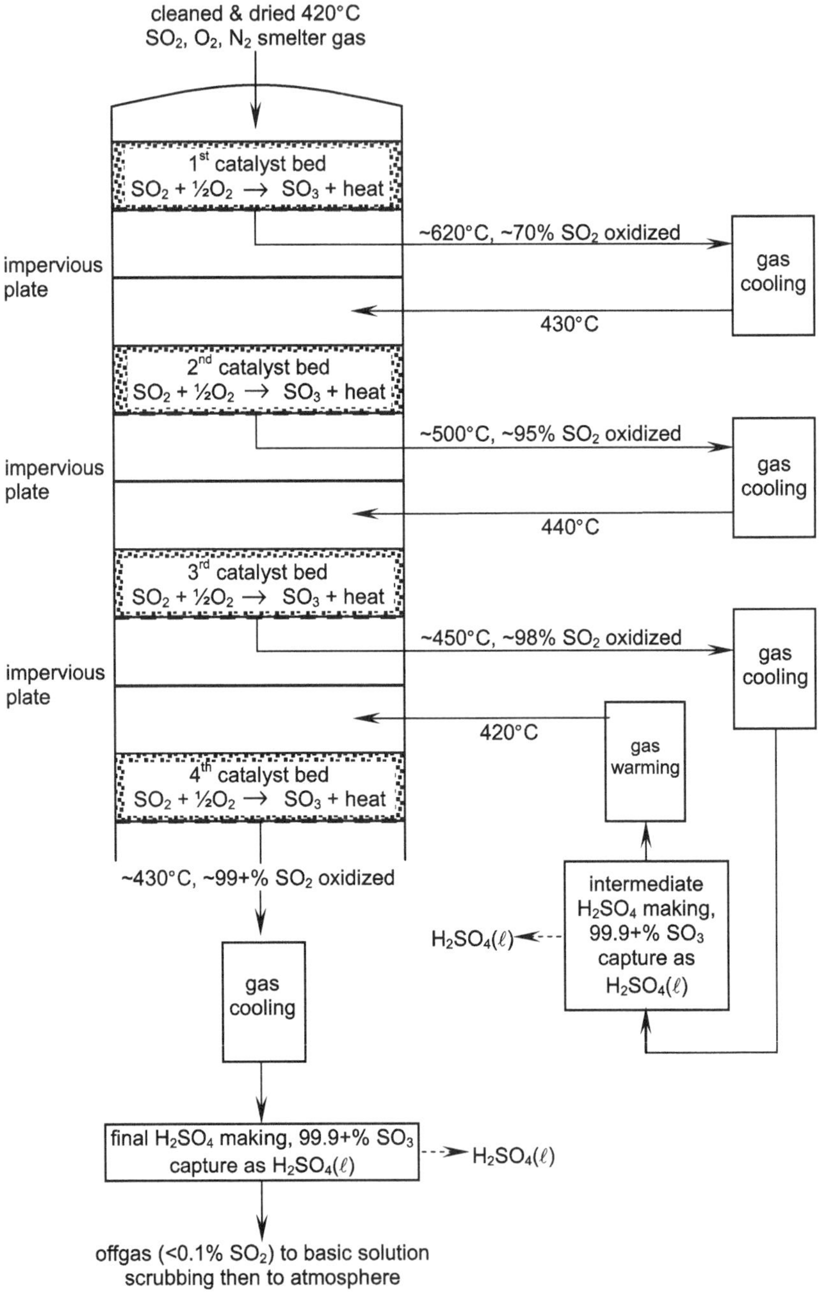

FIGURE 20.4 Schematic of 3-1 double-contact sulfuric acid plant, which oxidizes greater than 99% of the incoming sulfur dioxide and captures 99.9% of the resulting sulfur trioxide as sulfuric acid. Gas cooling between the catalyst beds avoids overheating and catalyst degradation. Note the increase in the overall oxidation of sulfur dioxide as the gas moved through each catalyst bed.

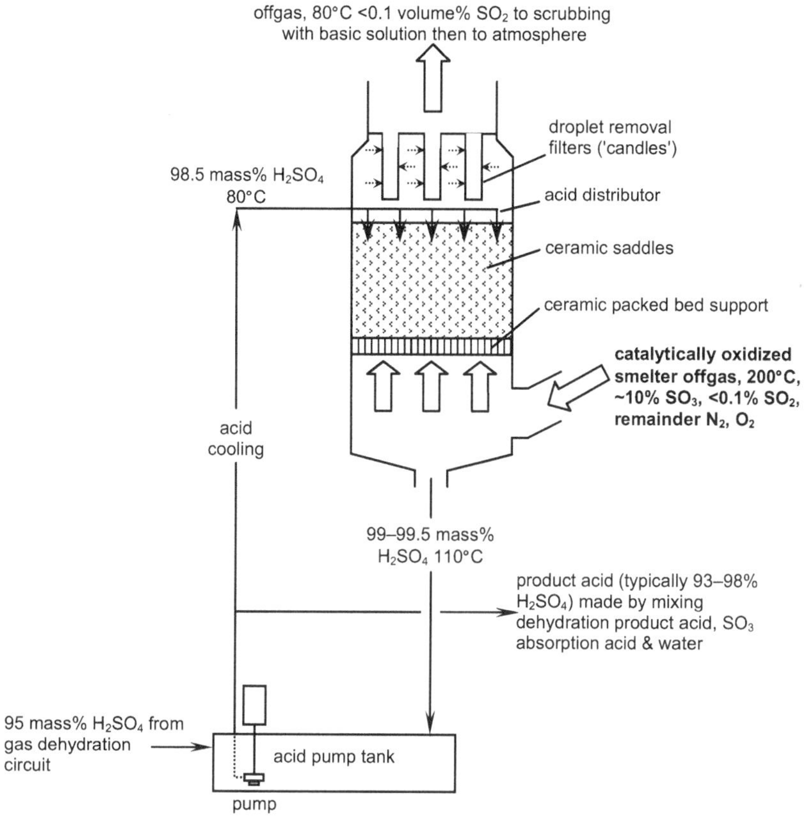

FIGURE 20.5 Sketch of a sulfuric acid-making tower. The feed is gas rich in sulfur trioxide (~10% SO_3) from catalytic oxidation of sulfur trioxide (center right). The product is 99%–99.5% sulfuric acid (bottom). The commercial products (typically 93%–98% H_2SO_4 acid) are made by mixing acid from dehydration and acid from the absorption tower with water.

(d) gas residence time in packing 2–4 s; and,
(e) acid residence time in packing 300–500 s.

At these conditions, the acid-making reaction, given in Equation (20.3), is greater than 99% complete.

20.6.2. Acid-Making Tower Product

The industrial product of acid making is 99%–99.5% sulfuric acid with 1%–0.5% water.

20.6.3. Acid Mist

The exit gas from the acid-making tower contains a fine spray of 10–250-μm-acid droplets. These droplets are removed from the gas to prevent accidental downstream corrosion. Circular fabric filters inside the acid-making tower are used. The collected acid is returned to the product acid stream or sent to impurity removal.

20.7. DOUBLE-CONTACT ACID-MAKING

A schematic flowsheet of a double-contact acid-making plant is shown in Figure 20.5. It is the most common flowsheet used for capture of sulfur dioxide. The feed is clean, dry smelter offgas. It passes through the following equipment:

(a) a catalytic converter containing three sequential catalyst beds with inter-stage cooling;
(b) an sulfur trioxide absorption, or acid-making, tower[1];
(c) a fourth catalyst bed and gas cooler; and,
(d) a second acid-making tower.

The gas contacts sulfuric acid twice, hence the name 'double-contact'. It gives extremely efficient oxidation of sulfur dioxide and acid production (Davenport & King, 2006).

20.8. ACID PLANT PRODUCTS

A double-contact sulfuric acid plant has the following three acid product streams:

(a) weakened acid from dehydration; and,
(b) two strengthened acid streams from the two acid-making towers.

These are combined by automatically blending to give various acid products. Water is added when necessary. Sulfuric acid is mostly sold in grades of 93%–98% sulfuric acid according to market demand. The main product in cold climates is ~94% sulfuric acid because of its low freezing point (−35°C). A small amount of oleum (sulfur trioxide dissolved in sulfuric acid) is also made and sold (BASF, 2009).

Sulfuric acid is shipped in stainless-steel trucks, steel rail-tank cars and double-hulled steel barges and ships. Great care is taken to avoid spillage.

20.8.1. Production of Liquid Sulfur Dioxide

An alternative method to the production of sulfuric acid for the capture of sulfur dioxide is the production of liquid sulfur dioxide. It is made by cooling and compressing clean smelter offgas that has a high concentration of sulfur

---

1. Single-contact acid-making finishes after the first acid-making tower in Figure 20.4. The efficiency of single-contact acid-making is considerably lower than that of double-contact acid-making. New single-contact installations are rare.

dioxide. Production of liquid sulfur dioxide is mainly used in pulp and paper bleaching (Innovation Group, 2009). Its production is only 1% or 2% of the production of sulfuric acid.

20.8.2. Gas Scrubbing Products

Several commercial by-products are made from scrubbing the exit gas of an acid plant. The most prominent of these are gypsum ($CaSO_4 \cdot 2H_2O$) and sodium bisulfite ($NaHSO_3$).

20.9. ENVIRONMENTAL PERFORMANCE OF THE NICKEL INDUSTRY

The capture of sulfur dioxide during the roasting, smelting and converting of nickel sulfide concentrates varies from greater than 99% to 0% (Makinen & Ahokainen, 2009).

The worst performers are those in remote areas where there are no markets for sulfuric acid and where acid transportation costs would far exceed sulfuric acid revenues. A potential remedy might be to reduce the sulfur dioxide to elemental sulfur, which then could be stored in underground nickel sulfide mines. However, this is energy-intensive and expensive (Rameshni & Santo, 2010).

20.10. MAKING SULFURIC ACID FOR THE LEACHING OF NICKEL LATERITE

The leaching of nickel laterites requires about 0.3 tonnes of sulfuric acid per tonne of concentrate (see Chapter 11). This is equivalent to about 25 tonnes of sulfuric acid per tonne of product nickel from a laterite ore that contains about 1.3% Ni.

Sumitomo produces sulfuric acid in the copper smelter at Niihama, Japan, and transports it to the Coral Bay leach plant. All other leach plants make their sulfuric acid on site from imported elemental sulfur. This acid is made in following two steps:

(a) burning molten elemental sulfur with dried air[2] to make sulfur dioxide:

$$\underset{\substack{\text{melted with}\\ \text{pressurized steam}}}{\overset{140^{\circ}\text{C}}{S(\ell)}} + \underset{\substack{\text{in filtered}\\ \text{then dried air}}}{O_2(g)} \longrightarrow \underset{\substack{\text{in 11 volume\% } SO_2\text{, 10 vol\%}\\ O_2\text{, 79 vol\% } N_2 \text{ gas}}}{\overset{1100^{\circ}\text{C}}{SO_2(g)}} \tag{20.4}$$

(b) making sulfuric acid from the product gas as described above.

Industrial aspects of the process are described by Davenport & King (2006).

2. The air is dried much as described in Section 20.3.2.

The leach plant at Tenke Fungurume, Democratic Republic of the Congo, also burns sulfur to make sulfuric acid for leaching copper–cobalt ores.

20.11. SUMMARY

Sulfur dioxide gas is produced by all nickel sulfide smelters. It is often made into sulfuric acid, which is a useful industrial chemical.

The acid is made by the following steps:

(a) cleaning and drying the smelter gas;
(b) catalytically oxidizing the sulfur dioxide to sulfur trioxide; and,
(c) reacting the resulting sulfur trioxide with the water contained in concentrated sulfuric acid.

The result is strengthened sulfuric acid, which serves as the basis for the range of sulfuric acid products from the acid plant, which usually has concentrations that range between 93% and 98% H_2SO_4.

Modern acid plants remove 99.5% of the sulfur dioxide in the offgas from the smelter. Most of the remainder can be removed by scrubbing the acid plant exit gas with basic solutions. This scrubbing also produces useful by-products, such as gypsum, $CaSO_4 \cdot 2H_2O$ and sodium bisulfite ($NaHSO_3$).

REFERENCES

BASF (2009). Oleum (Company Brochure).

Davenport, W. G., & King, M. J. (2006). *Sulfuric acid manufacture: Analysis, control and optimization*. Elsevier.

Innovation Group (2004). Sulfur dioxide.

Ishiyama, H., & Hirai, Y. (1997). Fluid bed roasting at Tokyo nickel company. In C. Diaz, I. Holubec & C. G. Tan (Eds.), *Proceedings of the nickel/cobalt 97 international symposium, Vol. 3. Pyrometallurgical operations, the environment and vessel integrity in nonferrous smelting and converting* (pp. 133–140). CIM.

King, M. J. (2010). Recent developments and future directions in sulphuric acid manufacture. In N. L. Piret (organizer), Course notes from sulphuric acid production technologies short course at copper 2010 conference. Hamburg.

Makinen, T., & Ahokainen, T. (2009). 50 years of nickel flash smelting – still going strong. In J. Liu, J. Peacey & M. Barati et al, (Eds.), *Pyrometallurgy of nickel and cobalt 2009, proceedings of the international symposium* (pp. 209–220). CIM.

Rameshni, M., & Santo, S. (2010). Production of elemental sulphur from SO_2. In J. Harre (Ed.), *Copper 2010 proceedings. Downstream fabrication, application and new products sustainable development/health, safety and environmental control, Vol. 1* (pp. 477–495). GDMB.

Tokyo Nickel. (2006). *Inco TNC Limited.* (Company Handout).

SUGGESTED READING

Davenport, W. G., & King, M. J. (2005). *Sulfuric acid manufacture: Analysis, control and optimization*. Elsevier.

Schlesinger, M. E., King, M. J., Sole, K. C., & Davenport, W. G. (2011). *Extractive metallurgy of copper* (5th ed.). Elsevier.

Louie, D. K. (2008). *Handbook of sulphuric acid manufacturing*. DKL Engineering Inc.

Warner, A. E. M., Diaz, C. M., Dalvi, A. D., et al. (2007). JOM world nonferrous smelter survey part IV: Nickel sulfide. *Journal of Metals, 59*, 58–72.

Chapter 21

Slow Cooling and Solidification of Converter Matte

The molten converter matte, discussed in Chapter 19, contains nickel, sulfur and a small amount of iron. It also contains copper, cobalt, silver, gold and platinum-group elements to a greater or lesser extent.

These elements cannot be separated while the matte is molten. They are separated by solidifying the matte and treating it by two different techniques as follows:

(a) vapometallurgical refining (Chapter 22); and,
(b) hydrometallurgical refining (Chapters 23 through 27).

However, about half of the global production of converter matte from sulfide smelting undergoes a preliminary treatment before these processes (Warner *et al.*, 2007). The following steps are used:

(a) between 12 and 25 tonnes of molten matte are poured into large molds;
(b) this matte is slowly solidified and cooled in covered molds, which results in the formation of large individual grains of heazlewoodite (Ni_3S_2), chalcocite (Cu_2S) and metallic alloy (see Figures 21.1 and 21.2);
(c) the solid matte is crushed and ground, which liberates the grains of the different minerals from each other; and,
(d) the ground ore is separated into alloy, copper–sulphide and nickel–sulphide concentrates by magnetic separation and froth flotation.

A schematic diagram of the process is shown in Figure 21.3. Operating details are given in Tables 21.1 and 21.2.

The advantages of this preliminary separation are that it simplifies the subsequent metal extraction and minimizes the in-plant residence times.

21.1. SOLIDIFICATION AND SLOW COOLING PROCESS

The equilibrium phase diagram for the Ni–Cu–S system is shown in Figure 21.2. This phase diagram indicates that the products of solidifying the molten matte, then cooling it to ambient temperature, are nearly pure Ni_3S_2 and Cu_2S.

Extractive Metallurgy of Nickel, Cobalt and Platinum-Group Metals. DOI: 10.1016/B978-0-08-096809-4.10021-8

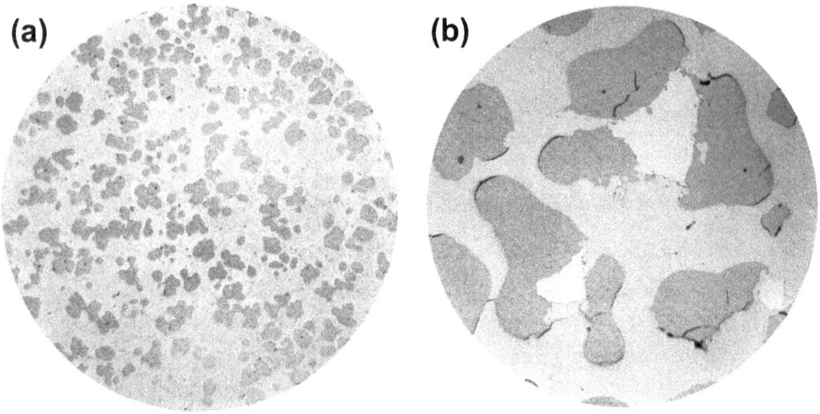

FIGURE 21.1 Photomicrograph (×100) of (a) quenched and (b) slowly cooled nickel-copper converter matte, showing that the quenched matte is much more finely grained than the slow-cooled matte. The grain size of the quenched matte is about 10 μm while that of the slow-cooled matte is about 100 μm. The dark grey grains are Cu_2S; the light grey grains are Ni-Cu alloy. The matrix is Ni_3S_2. Grinding the solidified matte to a mean particle size of 70 μm liberates the three phases so that they can be separated from one another by magnetic and froth flotation processing. *Source: Boldt and Queneau, 1967, courtesy of Vale.*

The equilibrium conditions are closely approximated industrially by solidification and slow cooling of the converter matte.

Slow cooling encourages the precipitated grains of Ni_3S_2 and Cu_2S to grow to a size of approximately 100 μm in size, as shown in Figure 21.1.

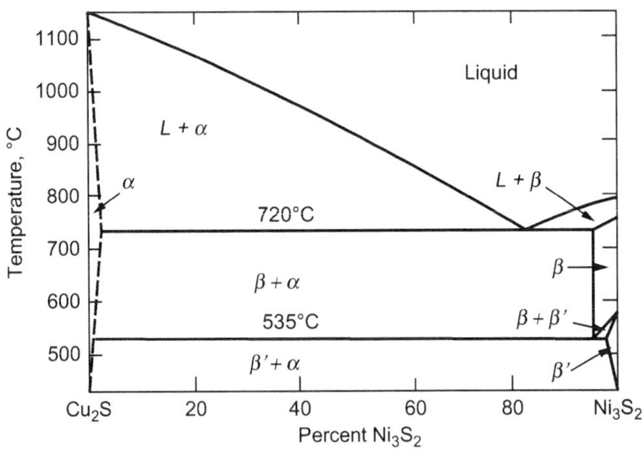

FIGURE 21.2 Equilibrium Cu_2S–Ni_3S_2 phase diagram. The complete miscibility of Cu_2S and Ni_3S_2 above the Cu_2S melting point (~1130°C) and the almost complete immiscibility of Cu_2S and Ni_3S_2 below 450°C are notable. The latter causes molten matte to separate into two individual phases during slow cooling.

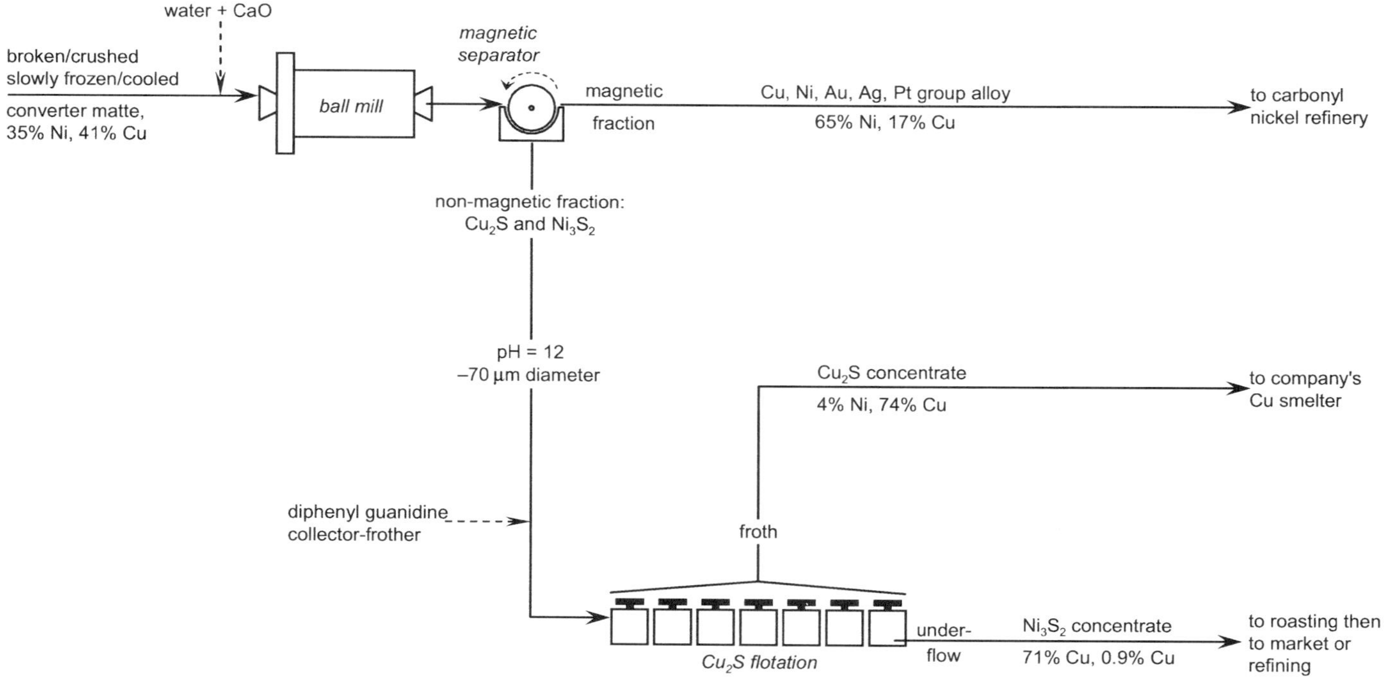

FIGURE 21.3 Simplified flowsheet for separating slowly solidified and cooled converter matte into (i) magnetic alloy concentrate; (ii) copper-rich sulphide concentrate; and, (iii) nickel-rich sulphide concentrate (Tables 21.1 and 21.2). Concentrate cleaning circuits are not shown. Donald and Scholey (2005) and Dorigo, Yalcin, Duncan, & Montgomery, (2009) give more details. Wet magnetic separation is done with a magnetic drum, which rotates and lifts magnetic particles up and out of a slurry bath. The magnetic particles overflow, and the non-magnetic particles underflow.

TABLE 21.1 Details of the Sudbury Matte Separation Plant (Vale) for Making Magnetic Alloy, Copper Sulphide and Nickel Sulphide Concentrates from Solidified Low Sulfur (21% S) Converter Matte

Location	Sudbury, Canada
Startup date	1948
Nickel production all products	~140 000 tonnes per year
Casting and slow solidification/cooling	
Source of matte	Peirce-Smith converters
Matte casting temperature	980°C
Molds	
Number	220
Material	Castable fire clay, coated with lime wash (imbedded in floor)
Size	4 × 2.4 × 0.6 m deep with 45° walls
Matte mass per mold	23 tonnes
Cover	Refractory-lined steel
Casting method	Molten matte poured from ladle via overhead crane
Cooling duration	Three days with cover on to ~480°C followed by 1 day with cover off to 200°C
Breaking, crushing and grinding	
Ingot transport	Via overhead crane using 'scissor' tongs.
Ingot breaking	With overhead chisel-tipped hammers (pile drivers)
Crushing	Jaw crusher (to ~1 m) then two open-circuit gyratory crushers in series to ~0.01 m
Grinding	Rod mill (open circuit) then ball mills (closed circuit)
Size control	Cyclone classifiers
Final matte	
Particle size	99% < 70 μm diameter
Destination	Magnetic and froth flotation separation of alloy, Cu sulphide, and Ni sulphide concentrates (Table 21.2)

This table gives matte casting, slow cooling, crushing and grinding details (Donald & Scholey, 2005). Similar procedures are followed at the Waterval Smelter, South Africa (Anglo American Platinum) but with round molds (Jacobs, 2006).

TABLE 21.2 Details of Magnetic and Flotation Separations at Vale's Sudbury Matte Separation Plant*

Location	Sudbury, Canada
Startup date	1948
Nickel production all products	~140 000 tonnes per year
Particle separations	
Isolation of magnetic metallic concentrate	Magnetic separation, consisting of the following steps: (a) magnetic separation of magnetic metal particles from non-magnetic sulphide particles; (b) ball mill regrind; and, (c) second magnetic separation (cleaning)
Product	Alloy concentrate: 65% Ni: 17% Cu with high concentrations of Ag, Au and Pt-group elements
Destination	To top blown rotary converting (Appendix G) and carbonyl refining (Chapter 22)
Isolation of Ni_3S_2 concentrate, low in Cu_2S	Cu_2S is floated from non-magnetic matte fraction by froth flotation using CaO for pH control (pH > 12) and diphenyl guanidine collector/frother Underflow Ni_3S_2 slurry is reground and has (i) Cu_2S removed in a column flotation cell; and, (ii) alloy removed in a drum magnetic separator
Product	Ni_3S_2 concentrate: 71% Ni and 0.9% Cu
Destination	To NiO production by oxidation roasting. The NiO is sold or sent to Clydach, Wales for reduction, sulfur activation and ambient pressure carbonyl nickel refining (Chapter 22)
Isolation of Cu_2S concentrate	
Product	Cu_2S concentrate: 4% Ni and 74% Cu
Destination	Company's copper smelter
Middlings concentrate	Underflow from grinding and cleaning Cu_2S concentrate
Product	Ni_3S_2 middlings concentrate: 66% Ni and 6.6% Cu (not shown in Figure. 21.3)
Destination	Roasted then sent to Sudbury nickel refinery for high-pressure carbonyl nickel refining

The Anglo American Platinum (South Africa) Magnetic Concentrator Plant uses only magnetic separation (that is, no flotation) to produce a magnetic concentrate rich in platinum-group elements (~30% platinum-group metals) and a non-magnetic bulk sulphide concentrate (Jacobs, 2006).
**Donald & Scholey, 2005*

The driving force for the grain growth is the minimization of the interfacial energy.

The nickel converter matte purposely contains slightly less sulfur than is required to form stoichiometric Ni_3S_2 and Cu_2S (Donald & Scholey, 2005). The low sulfur content is achieved by converting at a high temperature.

This deficiency of sulfur causes a third phase to form during solidification and slow cooling. This third phase is a magnetic alloy that contains nickel, copper and most of the silver, gold and platinum-group elements in the feed matte.

21.2. INDUSTRIAL MATTE CASTING, SOLIDIFICATION AND SLOW COOLING

Between 12 and 25 tonnes of molten converter matte, at about 1000°C, are cast into large refractory-lined steel depressions, or molds, at floor level. This is shown in Figure 21.4. A steel loop is placed into the solidifying matte, and the mold is covered with a steel lid. The matte is allowed to cool for 3 days. After 3 days, the lid is removed, and the matte is cooled for another day. The solid casting is lifted out of the mold using the steel loop and then sent to breaking, crushing, grinding and mineral separation.

21.2.1. Matte Structure

The solidified matte contains grains of three major components: (i) a magnetic nickel–copper alloy (containing most of the silver, gold and platinum-group elements); (ii) chalcocite (Cu_2S); and, (iii) heazlewoodite (Ni_3S_2). A small amount of other minerals, for example, bornite (Cu_5FeS_4) may also be present.

As mentioned earlier, the grains are quite large, approximately 100 μm in size (Donald & Scholey, 2005). Grinding the matte to a mean particle size of 70 μm liberates the minerals.

21.2.2. Isolation of the Liberated Grains

The ground matte is separated into three concentrates. Two major methods are used for this purpose:

(a) magnetic separation of the precious metals-rich alloy particles into a magnetic alloy concentrate; and,
(b) flotation separation of the remaining sulphides particles into copper-rich and nickel-rich sulphide concentrates (this separation is not always done; Jacobs, 2006).

FIGURE 21.4 Casting of low-iron, sulfur-deficient matte into floor-level molds in preparation for slow cooling. Slow cooling causes the matte to separate into ~100 μm Cu_2S, Ni_3S_2 and alloy grains, which are separated by comminution then magnetic separation and froth flotation. The alloy grains contain most of the silver, gold and platinum-group elements in the matte. The newly-filled molds will be covered with the lid (left). *Photograph courtesy of Anglo American Platinum.*

21.3. CONCENTRATE DESTINATIONS

Matte concentrate destinations vary from smelter to smelter.

The most straightforward arrangement is that at the Waterval, South Africa, smelter (Anglo American Platinum). A magnetic concentrate that is rich in platinum-group elements and a non-magnetic concentrate are made. The magnetic fraction is upgraded by leaching and then sent to a precious metals refinery. The non-magnetic fraction is sent to a nickel–copper refinery (Jacobs, 2006). This arrangement maximizes the recovery of the metals and minimizes the in-plant residence time.

The Sudbury, Canada, smelter (Vale) makes three major concentrates:

(a) a magnetic alloy concentrate, which is sent directly to its nickel refinery;
(b) a copper-rich sulphide flotation concentrate, which is smelted in a copper smelter; and,
(c) an nickel-rich sulphide concentrate, which is oxidation roasted to nickel oxide.

Residues from the nickel refinery are treated for copper, cobalt and precious metals extraction.

The nickel oxide from the oxidation roasting is either sold or it is refined to high-purity nickel in the company's carbonyl refineries in Sudbury, Canada, and Clydach, Wales. Carbonyl refining is described in Chapter 22.

21.4. SUMMARY

The converter matte contains nickel, iron and sulfur. It may also contain copper, cobalt, silver, gold and platinum-group elements to a greater or lesser extent. These elements are ultimately separated and made into metal either hydrometallurgically or vapometallurgically.

Often, however, the elements are given a preliminary separation by (i) slow cooling and solidification of the molten matte; (ii) ingot breaking, crushing and grinding; (iii) magnetic separation; and, (iv) froth flotation.

This treatment results in the following concentrates:

(a) an alloy concentrate that is rich in precious metals;
(b) a copper sulphide concentrate; and,
(c) a nickel sulphide concentrate.

These concentrates are then transferred to specialist metal refineries for the recovery of metal.

REFERENCES

Boldt, J. R., & Queneau, P. (1967). *The winning of nickel.* (pp. 280–283). Longmans.

Donald, J. R., & Scholey, K. (2005). An overview of Copper Cliff's operations. In J. Donald & R. Schonewille (Eds.), *Nickel and cobalt 2005, challenges in extraction and production* (pp. 457–477). CIM.

Dorigo, U. A., Yalcin, T., Duncan, D., & Montgomery, J. (2009). Front line planning and scheduling at Vale Inco's Ni-Cu matte processing plant. In J. Liu, J. Peacey & M. Barati, et al. (Eds.), *Pyrometallurgy of nickel and cobalt 2009, proceedings of the international symposium* (pp. 391–399). CIM.

Jacobs, M. (2006). Process description and abbreviated history of Anglo Platinum's Waterval smelter. In R. Jones (Ed.), *Southern African pyrometallurgy 2006* (pp. 17–28). SAIMM.

Warner, A. E. M., Diaz, C. M., Dalvi, A. D., et al. (2007). JOM world nonferrous smelter survey, part IV: Nickel sulfide. *Journal of Metals, 59*, 58–72.

SUGGESTED READING

Donald, J. R., & Scholey, K. (2005). An overview of Copper Cliff's operations. In J. Donald & R. Schonewille (Eds.), *Nickel and cobalt 2005, challenges in extraction and production* (pp. 457–477). CIM.

Marcuson, S. W., & Diaz, C. M. (2005). The changing Canadian nickel smelting landscape – late 19th Century to early 21st Century. In J. Donald & R. Schonewille (Eds.), *Nickel and cobalt 2005, challenges in extraction and production* (pp. 179–207). CIM.

Chapter 22

Carbonyl Refining of Impure Nickel Metal

Most nickel converter matte is made into high-purity nickel by leaching, solution purification and electrowinning or hydrogen reduction. However, about a quarter of the converter matte is made into high-purity nickel by gaseous carbonyl refining. This process is used in Canada, China, Russia, and Wales.

A schematic diagram of the flowsheet of the process is shown in Figure 22.1. In its simplest form, it consists of oxidizing the matte, reducing

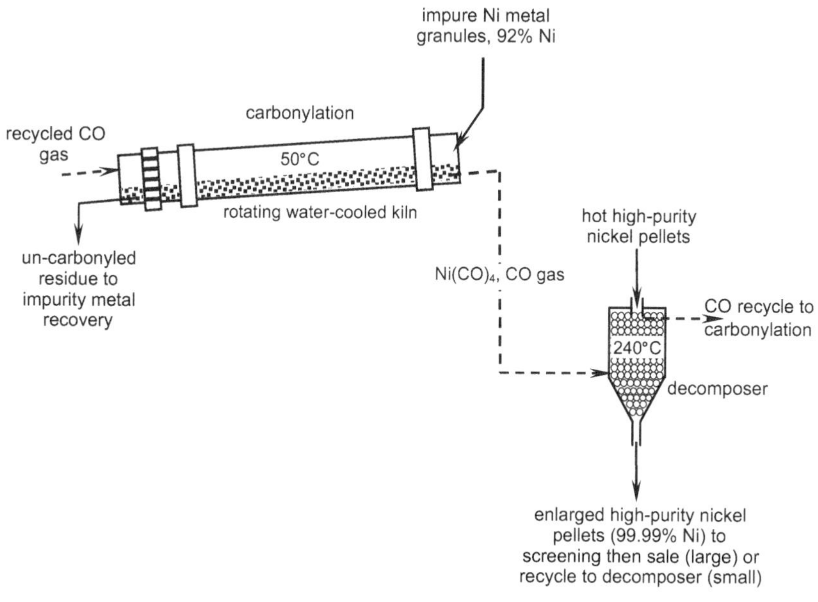

FIGURE 22.1 Schematic flowsheet for ambient pressure carbonyl refining (as practiced at Clydach, Wales). The cool carbonylation kiln and hot pellet decomposer are noteworthy. The Clydach refinery is designed to produce ~40 000 tonnes of refined nickel per year. (The slope of the kiln is exaggerated – it is actually ~1° from horizontal.) Nickel carbonyl and carbon monoxide are both extremely toxic and must not be allowed to escape into the workplace.

Extractive Metallurgy of Nickel, Cobalt and Platinum-Group Metals. DOI: 10.1016/B978-0-08-096809-4.10022-X

the oxide product to impure nickel metal (not shown) and then undertaking the following steps:

(a) continuously reacting the impure nickel metal with carbon monoxide gas (~50°C, ambient pressure) to form gaseous nickel carbonyl, $Ni(CO)_4$, in a rotating kiln;
(b) continuously removing the nickel carbonyl gas and the unreacted residue from the kiln;
(c) transferring the residue to a metals recovery plant;
(d) continuously decomposing the nickel carbonyl gas to carbon monoxide and high-purity nickel by contacting it with hot (~240°C) high-purity nickel pellets;
(e) continuously removing carbon monoxide and enlarged nickel pellets from the decomposer;
(f) continuously cooling the carbon monoxide and recycling it to the carbonylation kiln; and,
(g) sending the enlarged high-purity nickel pellets to market.

22.1. CHEMISTRY OF THE PROCESS

The formation of gaseous nickel carbonyl, step (a) above, occurs by the carbonylation reaction:

$$\underset{\substack{\text{92\% Ni in impure}\\ \text{nickel granules}\\ \text{remainder Cu, Co,}\\ \text{Fe \& precious metals}}}{Ni(s)} + \underset{\substack{\text{in recycle gas from}\\ \text{decomposers, 99.5\% CO}\\ \text{remainder, } CO_2,\ N_2}}{4CO(g)} \xrightarrow{\substack{50^\circ\text{C, ambient}\\ \text{pressure}}} \underset{\text{in } Ni(CO)_4\text{, CO gas}}{Ni(CO)_4(g)} \tag{22.1}$$

The equilibrium constant for this reaction at 50°C and ambient pressure is given by the following expression:

$$K_E^{50^\circ C} \approx \frac{X_{Ni(CO)_4}}{X_{CO}^4} \approx 4 \times 10^4 \tag{22.2}$$

where X is the mole fraction of the gaseous species.

Similarly, the decomposition of the gaseous $Ni(CO)_4$, step (d), is described by the following expression:

$$\underset{\text{from carbonylation kiln}}{Ni(CO)_4(g)} \xrightarrow{\substack{240^\circ\text{C, ambient}\\ \text{pressure}}} \underset{\substack{\text{high-purity nickel}\\ \text{deposited on hot}\\ \text{nickel pellet seed}}}{Ni(s)} + \underset{\substack{\text{cooled and}\\ \text{recycled to}\\ \text{carbonylation}\\ \text{kiln}}}{4CO(g)} \tag{22.3}$$

The equilibrium constant for this reaction at 240°C and ambient pressure is given by the following expression:

$$K_E^{240^\circ C} \approx \frac{X_{CO}^4}{X_{Ni(CO)_4}} \approx 7 \times 10^4 \tag{22.4}$$

(Refer to Appendix H for thermodynamic calculations.)

The equilibrium constants for the reactions given in Equations (22.1) and (22.3) indicate the following:

(a) at equilibrium at 50°C and ambient pressure, the product gas of the carbonylation will mostly be nickel carbonyl; and,
(b) at equilibrium at 240°C and ambient pressure, the exit gas from the nickel carbonyl decomposer will mostly be carbon monoxide.

The first principle of nickel carbonyl refining is that nickel tends to form carbonyl gas, while other metals do not. This gasification allows nickel to be separated from the other metals and impurities.

The second principle of nickel carbonyl refining is that while nickel can be carbonyled to form nickel carbonyl gas at a low temperature, ~50°C, the resulting nickel carbonyl can be decomposed to nickel metal at a high temperature, ~240°C.

22.2. INDUSTRIAL AMBIENT PRESSURE CARBONYLATION

The operating details for carbonylation kilns are given in Table 22.1. The process is continuous, performed at 50°C in rotating kilns. The carbonylation reaction, Equation (22.1), is highly exothermic, generating about 160 kJ/mol of $Ni(CO)_4$. As a result, the rotating kilns are water cooled to maintain the temperature of the kiln at 50°C.

22.2.1. Kiln Complexity

A carbonylation kiln has three inputs: (i) impure nickel alloy; (ii) carbon monoxide gas; and, (iii) cooling water. All of these input flows must be fed to the kiln while it rotates.

There are also three outputs: (i) uncarbonyled residue; (ii) gas consisting of nickel carbonyl and carbon monoxide; and, (iii) warm cooling water.

In addition, nickel carbonyl gas is extremely toxic. This means that absolutely no gas can be allowed to escape the kiln. This requires perfect seals between the rotating kilns and their static charging and discharging devices.

22.2.2. Sulfidation of the Feed

As can be seen from the operating details given in Table 22.1, the feed to the kiln is sulfided by adding carbonyl sulfide gas, COS, to the input

TABLE 22.1 Operating Details of Industrial Ambient Pressure Carbonylation Kilns (2) at Clydach, Wales

Refinery	Vale, Clydach, Wales
Startup date	1902 (many modifications since)
Feed material	Roaster calcine (impure nickel oxide) from Canada
Nickel production rate, t/a (all products)	40 000
Carbonylation kilns	
Purpose	Volatilize nickel as $Ni(CO)_4(g)$ from reduced NiO alloy while leaving impurities behind in solid form
Reaction	$Ni(s) + 4CO(g) \rightarrow Ni(CO)_4(g)$
Number of kilns in refinery	2
Length × diameter, m	50 × 4
Slope	1° from horizontal
Rotational speed, rotations per minute	0.2–0.5
Construction material	0.05-m thick steel
Operating pressure	Ambient (~1 bar)
Operating temperature	50°C
Cooling system (to offset exothermic reaction)	Water flowing through 0.1 m diameter steel pipes throughout the kiln (water temperature rise, ~4°C)
Operating details (continuous process)	
Feed solids	Sulfidized [with COS(g)] impure nickel metal, which has been reduced from refinery's impure NiO feed
Composition	~92% Ni, remainder Cu, Co, Fe, precious metals
Particle diameter	0.2–0.5 mm
Feed rate, tonnes/h	2.5–4.4
Feed temperature	Ambient
Solids residence time, h	~100
Input gas	
Source	Recycle CO from decomposers plus CO from cracked natural gas
Composition, volume%	80%–95% CO (remainder CO_2 and N_2)
Flowrate, Nm^3/h	~10 000
Input temperature	Ambient
Gas residence time, h	~0.06
Gas product	
Composition, volume%	15% $Ni(CO)_4$, 85% CO
Destination	De-dusting then on to decomposers for high-purity nickel production
Solid product	Un-carbonyled residue
Composition	40%–50% Ni plus impurity metals
Destination	Recycle to Sudbury nickel refinery for metal recovery
Ni carbonylation efficiency	~90% (remainder is recovered in Canada)

The process is counter-current and continuous, Figure 22.1. It makes high-purity nickel carbonyl gas from sulfidized impure nickel metal. The nickel carbonyl gas is sent to decomposers for high-purity nickel pellet and powder production.

gas of the kiln. This addition of carbonyl sulfide increases the rate of carbonylation.

Without sulfidation, a coating of metallic copper forms on the surface of partially de-nickeled alloy particles. Because carbon monoxide diffuses slowly through this copper coating, the rate of carbonylation is unacceptably slow.

22.3. DECOMPOSITION OF NICKEL CARBONYL

The decomposition of nickel carbonyl is done industrially either by passing the exit gas from the carbonylation kiln up through a descending packed bed of hot (~240°C) high-purity nickel pellets (Figure 22.1) or by thermally shocking the gas at about 300°C in vertical furnaces that are electrically heated.

Operating details of these different methods of decomposition are given in Tables 22.3 and 22.4.

22.3.1. Pellets

Nickel deposits on the hot pellets as they descend. Sufficiently large pellets, about 1 cm in diameter, are removed by screening and sent to market. They are used mainly for making high-nickel and iron-base alloys (Vale, 2010a). Undersize pellets fall through the screens. They are re-heated and recycled to the decomposer for further growth. Small nickel spheres, approximately 0.05 cm in diameter, are added as replacement seed if needed.

22.3.2. Powder

The thermal shocking of the nickel carbonyl gas causes the precipitation of small, high-purity powder with a particle size of between 2 and 10 μm. The powder is sold for use in powder metallurgy, batteries, fuel cells, pigments, conductive polymers, chemicals and many other applications (Vale, 2010b). The morphology, density and size of the powder are varied by altering the temperature of the feed gas and the temperature profile of the furnace, and by chemical additives (Donald & Scholey, 2005).

22.4. HIGH-PRESSURE CARBONYL REFINING

The nickel refinery in Sudbury, Canada, (Vale) operates the carbonylation reactor at a pressure of 70 bar (Donald & Scholey, 2005; Inco, 2006; Mroczynski, 2009; Wiseman, Bale, Chapman, & Martin, 1988).

The effect of pressure is to push the carbonylation reaction, Equation (22.1), to the right. Increased pressure also increases the rate of diffusion of carbon monoxide through the coating of metallic copper and other impurities that form on the surface of the partially reacted feed particles. The high-pressure carbonylation reactor is shown in Figures 22.2 and 22.3, and the operating details are given in Table 22.2.

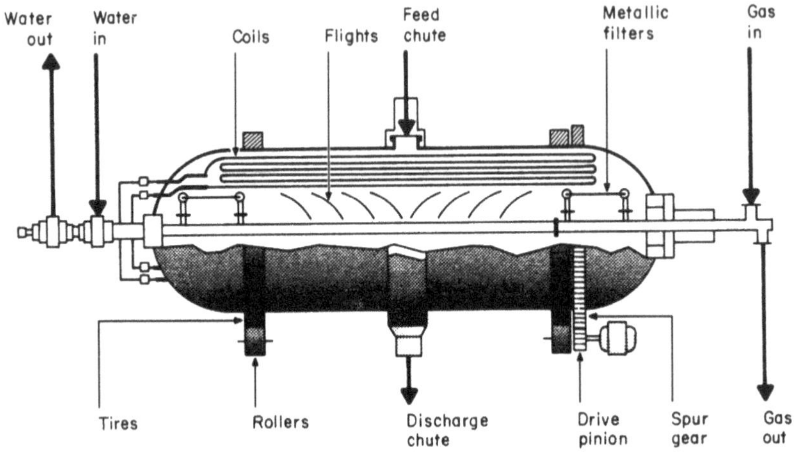

FIGURE 22.2 High-pressure (~70 bar, 170°C) carbonylation reactor at Vale's Sudbury nickel refinery. It is 4 m diameter × 13 m long and rotates at ~1 rpm. Its position in the carbonyl refining flowsheet is shown in Figure 22.3. Operating details are given in Table 22.2. The operation is started by charging a 150 tonne batch of impure nickel metal. Rotation is then begun and carbon monoxide is continuously passed through the tumbling charge, producing nickel carbonyl gas slightly contaminated with iron carbonyl. The operation is stopped when about 98% of the feed nickel has been gasified. The reactor is then purged with nitrogen, the residual solids are removed and the process is started again.

High-pressure carbonylation is performed as a batch process rather than as continuous process. This avoids moving solids in and out of a rotating reactor, which minimizes the risk of escape of toxic nickel carbonyl gas into the workplace.

22.4.1. Iron Carbonylation

High-pressure carbonylation is very efficient for the removal of nickel from the solid phase to the gas phase, but it also produces iron carbonyl gas, $Fe(CO)_5(g)$.

This means that the carbonylation feed should be low in iron, ~2%, and that the nickel carbonyl be separated from iron carbonyl.

The separation of iron carbonyl from nickel carbonyl is accomplished by condensing the carbonyl gas and then selectively distilling nickel carbonyl. The distillation is possible because the normal boiling point of nickel carbonyl is 43°C, whereas the normal boiling point of iron carbonyl is 103°C.

High-purity nickel pellets and powders are then produced from the purified nickel carbonyl, as described in Section 22.3.

22.4.2. Production of Ferronickel Pellets

A liquid residue, consisting of nickel carbonyl and iron carbonyl, is the bottom product in the distillation column. This liquid is removed from the distillation column, vaporized and decomposed into ferronickel pellets in the same manner

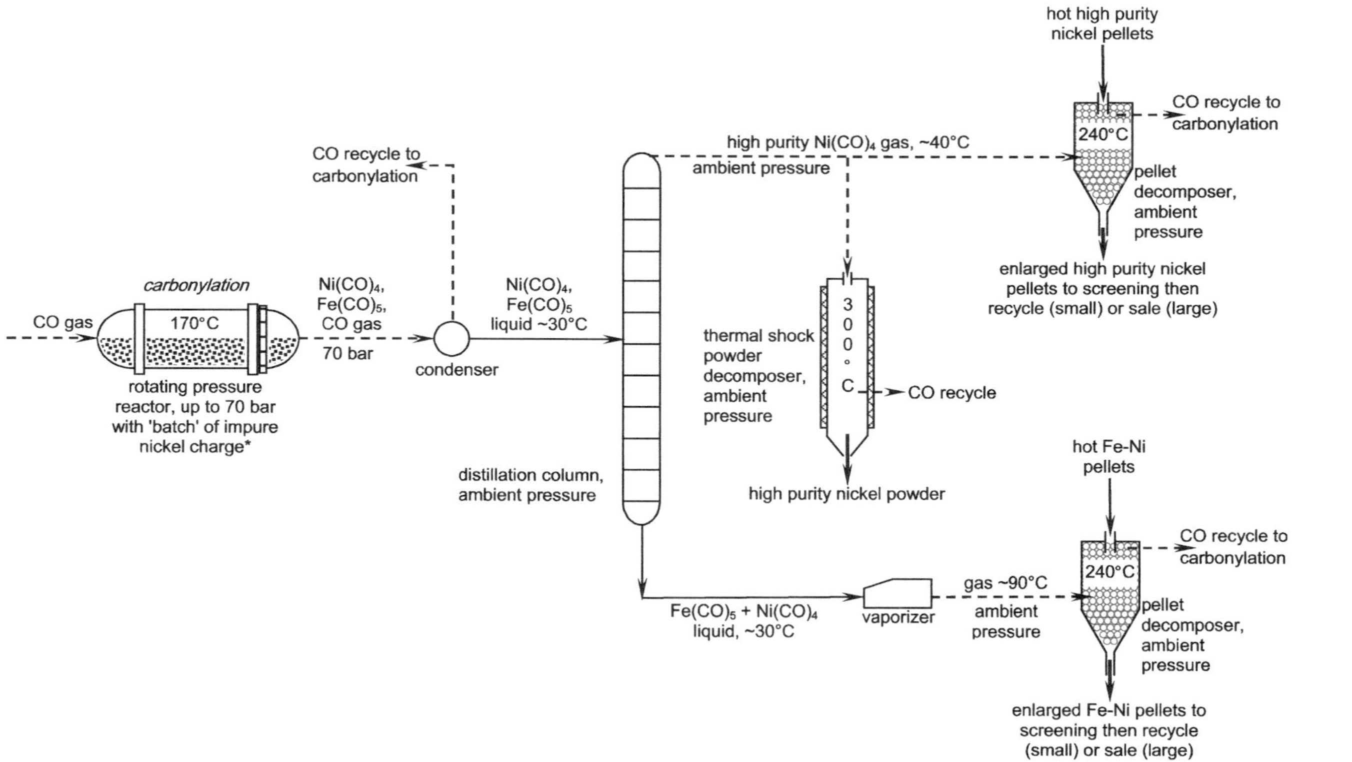

FIGURE 22.3 Schematic flowsheet for high-pressure nickel carbonyl refining (Donald & Scholey, 2005). The rotating reactor, condenser, distillation column, nickel powder-making decomposer and nickel and ferronickel pellet decomposers are notable. Carbonyl-liquid storage tanks are not shown. *The residue remaining after a carbonylation batch is sent to leaching for recovery of copper, cobalt, silver, gold and platinum-group metals. Sulfur is also added in the feed to maintain a high reaction rate. The efficiency, that is, the conversion of nickel from the solid phase to the gas phase, is approximately 98%.

TABLE 22.2 Operating Details of Industrial High-Pressure Carbonylation Reactors in Sudbury, Canada*

Refinery	Vale Ltd., Sudbury, Canada
Startup date	1972
Nickel production all products	65 000 tonnes per year
Pressure reactor details	
Number in refinery	3 (2 in use at any given time)
Length × diameter	13 × 4 m
Wall thickness	0.05 m
Wall material	Carbon steel
Rotational speed	~1 rpm
Operating temperature	~170°C
Operating pressure	~70 bar
Compressors	one 1600 kW, 70 bar compressor per reactor
Gas flow through reactor	CO in and $Ni(CO)_4$-bearing gas out through micro-metallic filters, flow periodically reversed
Reactor cooling and gas/solid mixing	Water-cooled pipe lifters rotating with reactor and cascading solid feed through CO gas
Operating details	
Feed material	Coarse, impure Ni granules from reduction and water granulation like coarse black sand, <1 cm diameter
Quantity per reactor batch	150 tonnes, charged through top chute
Composition, mass%	80% Ni; 12% Cu; 2% Co; 2% Fe; 3% S; 1% other including Ag, Au and Pt-group elements (the S is necessary for a rapid reaction rate)
Input carbon monoxide	
Purity	99.5% CO (CO_2 and N_2 impurities)
Source	Recycle from $Ni(CO)_4$ decomposers plus CO from natural gas cracking
Reaction time per batch	42 hours, start to start
Carbonylation efficiency	Ni: 97.5%, Fe: 30%
Output gas	
Composition	Proprietary
Destination	Liquefaction, distillation, decomposition of pure $Ni(CO)_4$ to nickel metal
Residue	
Mass per batch	~30 tonnes
Approximate composition, mass%	9% Ni; 57% Cu; 9% Co; 6% Fe; 15% S; 4.4% other, including Ag, Au and Pt-group elements
Removal procedure	Stop rotation, flush reactor with inert gas, empty through bottom chute
Destination	Slurry, grind, leach, metal production (Sabau & Bech, 2007)

The solid feed and residue are charged and emptied batch wise. Input and output gas flows are continuous during each batch. The process operates at ~70 bar in a rotating reactor, Figure 22.2. Note the sulfur content of the feed.

**Donald & Scholey 2005; Inco, 2006; Wiseman et al., 1988.*

TABLE 22.3 Details of Industrial Pellet-Making Decomposers

Refinery	Vale: Clydach, Wales	Vale: Sudbury, Canada
Startup date	1902 (many modifications since)	1972
Nickel production all products, tonnes per year	40 000	70 000
Nickel pellet production (remainder powder)	~35 000	~60 000
$Ni(CO)_4(g)$ decomposer pelletmakers		
Number in refinery	18	17
Decomposition furnaces, height × diameter, m	3 × 1.2	3 × 1.2
Nickel production rate, per decomposer	7 tonnes per day	10 tonnes per day
Decomposition temperature, °C	240	240
Source of heat (to offset endothermic decomposition reaction)	Hot descending recycle undersize pellets	Hot descending recycle undersize pellets
Decomposition pressure	Ambient	Ambient
Feed gas	De-dusted gas from carbonylation kilns	Gas from top of liquid carbonyl distillation column
Composition	15 volume% $Ni(CO)_4$, remainder CO	32 volume% $Ni(CO)_4$, remainder CO (850 g/Nm^3 Ni)
Input rate, each decomposer, Nm^3 per minute	~14	~8
Exit gas		
Composition, volume%	99.9% CO, 0.1% $Ni(CO)_4(g)$	99.9% CO, 0.1% $Ni(CO)_4(g)$
Destination	Carbonylation kilns (2)	Carbonylation reactors (3)
Pellets		
Seed	0.5 mm diameter purchased or self-made spheres	0.5 mm diameter purchased or self-made spheres
Growth time in decomposers, days	~90	~90
Product size	0.8–2 cm diameter pellets	0.8–2 cm diameter pellets
Size-control system	Slotted screens for undersize to fall through to recycle system	Slotted screens for undersize to fall through to recycle system
Composition, % Ni	99.99	99.99
Other pellet products		85% Fe, 15% Ni pellets made from vaporized distillation 'bottoms' liquid (1% of total Ni production)

The pellets grow during nickel carbonyl decomposition. They are screened as they leave the decomposer. Undersize pellets are heated and returned to the decomposer. Oversize pellets are cooled and sent to market. They are replaced by fine high-purity nickel 'seed' spheres.

TABLE 22.4 Details of Industrial Powder-Making Decomposers

Refinery	Vale: Clydach, Wales	Vale: Sudbury, Canada
Startup date	1902 (many modifications since)	1972
Nickel production all products, tonnes per year	~40 000	~70 000
Nickel powder production, tonnes per year (remainder, pellets)	3000–8000, depending on customer demand	10 000–15 000, depending on customer demand
Nickel carbonyl decomposer pelletmakers		
Number	5	10
Decomposition furnaces, height × diameter, m	10 × 2	10 × 2
Production rate per decomposer, t/d Ni	3	5
Decomposition temperature at gas outlet, °C	~300	~300
Heat (to offset endothermic decomposition reaction) provided by:	Electric wall heaters arranged vertically in four independent zones	Electric wall heaters arranged vertically in four independent zones
Feed gas	De-dusted gas from carbonylation kilns	$Ni(CO)_4$ gas vaporized from liquid $Ni(CO)_4$ tank atop the distillation column
Composition, volume%	~12%–16% $Ni(CO)_4$, remainder CO	35%–45% $Ni(CO)_4$, remainder CO
Input rate, each decomposer, Nm^3/min	~6	~4
Exit gas		
Composition, volume%	99.9% CO, 0.1% $Ni(CO)_4(g)$	99.9% CO, 0.1% $Ni(CO)_4$
Destination	Recycle to carbonylation kilns	Recycle to decomposition reactors
Powder product		
Approximate diameter, μm	2–10	2–10
Size and morphology control system	Gas input rate, wall temperature profile (four zones) and chemical additives	Gas input rate, wall temperature profile (four zones) and chemical additives
Composition, % Ni	>99.5	>99.5

The sizes and morphologies of the powder particles are altered by adjusting the temperature profile, gas feed rate and chemical additives.

that pure nickel pellets are made. These ferronickel pellets typically contain 65% Fe and 35% Ni (Vale, 2010c).

22.5. APPRAISAL

Carbonyl refining is an efficient method for producing high-purity nickel pellets and powders (99.99% Ni). It is nearly continuous and requires little labor.

The disadvantages seem to be the following:

(a) the use of extremely toxic and potentially explosive gases; and,
(b) the need for a hydrometallurgical plant for recovering un-carbonyled nickel and by-product metals (Sabau & Bech, 2007).

22.6. SUMMARY

About 25% of nickel sulfide converter matte is refined to high-purity nickel (99.99% Ni) by the following steps:

(a) oxidizing the matte to impure nickel oxide;
(b) reducing the oxide product to impure nickel alloy;
(c) selectively producing nickel carbonyl gas, $Ni(CO)_4$, from the alloy; and,
(d) decomposing the carbonyl gas to high-purity nickel, 99.99% nickel.

The carbonylation is done continuously at ambient pressure or batch wise at about 70 bar.

The products of the process are high-purity nickel pellets and powders. The pellets are mostly used for making high-nickel and iron-base alloys. The powders are used for powder metallurgy, batteries, fuel cells and many other applications.

REFERENCES

Donald, J. R., & Scholey, K. (2005). An overview of Copper Cliff's operations. In J. Donald & R. Schonewille (Eds.), *Nickel and cobalt 2005, challenges in extraction and production* (pp. 457–477). CIM.

Inco Limited (2006). *Copper cliff nickel refinery overview.* (Company Brochure).

Mroczynski, S. A. (2009). TBRC slag flux control at the Copper Cliff nickel refinery. In J. Liu, J. Peacey & M. Barati (Eds.), *Pyrometallurgy of nickel and cobalt 2009, proceedings of the international symposium* (pp. 293–304). CIM.

Sabau, M., & Bech, K. (2007). Status and improvement plans in Inco's electrowinning tankhouse. In G. E. Houlachi, J. D. Edwards & T. G. Robinson (Eds.), *Copper 07-Cobre 07 proceedings of the sixth international conference, Vol. 5, Electrorefining and electrowinning* (pp. 439–450). CIM.

Vale (2010a). *Nickel pellets* (Company Brochure).

Vale (2010b). *Nickel powders* (Company Brochure).

Vale (2010c). *Ferronickel pellets* (Company Brochure).

Wiseman, L. G., Bale, R. A., Chapman, E. T., & Martin, B. (1988). Inco's Copper Cliff nickel refinery. In G. P. Tyroler & C. A. Landolt (Eds.), *Extractive metallurgy of nickel and cobalt* (pp. 373–390). TMS.

SUGGESTED READING

Boldt, J. R., & Queneau, P. (1967). *The winning of nickel.* (pp. 280–283). Longmans.

Marcuson, S. W., & Diaz, C. M. (2005). The changing Canadian nickel smelting landscape – late 19th century to early 21st century. In J. Donald & R. Schonewille (Eds.), *Nickel and cobalt 2005 challenges in extraction and production* (pp. 179–207). CIM.

Queneau, P., O'Neill, C. E., Illis, A., & Warner, J. S. (1969). Some novel aspects of the pyrometallurgy and vapometallurgy of nickel. *Journal of Metals, 21*, 35–45.

Tyroler, P. M., Sanmiya, T. S., & Hodkin, E. W. (1988). Hydrometallurgical processing of Inco's pressure carbonyl residue. In G. P. Tyroler & C. A. Landolt (Eds.), *Extractive metallurgy of nickel and cobalt* (pp. 391–401). TMS.

Wender, I., & Pino, P. (1968 & 1977). *Organic synthesis via metal carbonyls, Vols. 1 & 2*. John Wiley and Sons.

Wiseman, L. G., Bale, R. A., Chapman, E. T., & Martin, B. (1988). Inco's Copper Cliff nickel refinery. In G. P. Tyroler & C. A. Landolt (Eds.), *Extractive metallurgy of nickel and cobalt* (pp. 373–390). TMS.

Chapter 23

Hydrometallurgical Production of High-Purity Nickel and Cobalt

Two different nickel sulfide intermediates must be refined to pure nickel and cobalt products:

(a) sulfide matte from nickel sulfide smelting and converting (Chapter 19); and,
(b) sulfide precipitates from nickel laterite leaching (Chapter 12).

The methods used to refine these mattes and sulfide intermediates are shown in Table 23.1. These data indicate that hydrometallurgy is the principal method for making high-purity nickel from these mattes and sulfide intermediates.[1]

The objective of this chapter is to describe the overall hydrometallurgical processes used to refine mattes and sulfide intermediates.

The sulfide precipitates from laterite leaching operations have little or no copper, which simplifies the refinery. Those refineries treating this type of material are described first. Mattes from smelting operations usually contain significant amounts of copper and these processes are described later in the chapter.

The emphasis of the descriptions in this chapter is on the integration of the overall process. Detailed descriptions of leaching, solvent extraction (Chapter 24), solvent extraction (Chapter 25), electrowinning (Chapter 26) and hydrogen reduction (Chapter 27) follow this chapter.[2]

---

1. Other methods are (a) oxidation/reduction roasting, Chapter 8; (b) oxidation/reduction/carbonylation, Chapter 19 and (c) aqueous electrorefining, Appendix H.

2. If the reader is not familiar with the principles of leaching, solvent extraction, electrowinning and hydrogen reduction, it may be more profitable to read those chapters first before returning to this chapter.

Extractive Metallurgy of Nickel, Cobalt and Platinum-Group Metals. DOI: 10.1016/B978-0-08-096809-4.10023-1

TABLE 23.1 Details of Matte and Intermediate Sulfide Refineries

Company: location	Refining process	Estimated nickel production, tonnes/year
BHP Billiton: Kwinana, Australia	Ammonia-air leach, H_2 reduction	65 000
Minara: Murrin Murrin, Australia	Oxygen-sulfuric acid leach, H_2 reduction	40 000
Vale: Sudbury, Canada	Carbonyl refining	70 000
Vale: Thompson, Canada	Matte anode electrorefining	55 000
Sherritt: Fort Saskatchewan, Canada	Ammonia-air leach, H_2 reduction	35 000
Jinchuan: China	Matte anode electrorefining and carbonyl refining	1 20 000
Jilin: China[a]	Oxygen-sulfuric acid leach, metal anode refining and carbonyl refining	20 000
Norilsk: Harjavalta, Finland	Oxygen-sulfuric acid leach, electrowinning and H_2 reduction	60 000
Eramet: Le Havre, France	Chlorine leach, electrowinning	15 000
Sumitomo: Niihama, Japan	Chlorine leach, electrowinning	40 000
Sherritt: Ambatovy, Madagascar	Oxygen-sulfuric acid leach, H_2 reduction	60 000 (in start-up phase)
Xstrata: Kristiansand, Norway	Chlorine leach, electrowinning	90 000
Norilsk: Monchegorsk, Russia	Carbonyl refining	35 000
Norilsk: Norilsk, Russia	Electrorefining of 7% S anodes	40 000
Anglo-American Platinum: South Africa	Oxygen-sulfuric acid leach, electrowinning	22 000
Impala Platinum: South Africa	Oxygen-sulfuric acid leach, H_2 reduction	13 000
Lonmin Platinum: South Africa	Oxygen-sulfuric acid leach, nickel sulfate product	3 000
Northam Platinum: South Africa	Oxygen-sulfuric acid leach, nickel sulfate product	2 000
Norilsk: Stillwater, U.S.A.	Oxygen-sulfuric acid leach, nickel sulfate product	100
Vale: Clydach, Wales	Carbonyl refining	40 000

Annual nickel productions are estimated from published sources. Hydrometallurgy (including electrorefining) accounts for about three quarters while carbonyl refining about a quarter of the refining production.
[a]*Donghe (2009)*.

23.1. REFINING OF SULFIDE PRECIPITATES FROM LATERITE LEACHING OPERATIONS

Sulfide precipitates are refined to nickel and cobalt at the following refineries: (i) Fort Saskatchewan, Canada (Sherritt); (ii) Murrin Murrin, Australia (Minara Resources); and, (iii) Niihama, Japan (Sumitomo).

These processes are described in the following three sections.

23.1.1. Sherritt Process, Fort Saskatchewan, Canada

Nickel is leached from laterite ores and precipitated as a sulfide at Moa Bay, Cuba. This sulfide precipitate is transported to Fort Saskatchewan, Canada, for processing to nickel metal (Budac, Kofluk & Belton, 2009; Cordingley & Krentz, 2005).

The process used at Fort Saskatchewan is shown in Figure 23.1. This process is a modification of the original Sherritt process. The changes have been necessitated by the change in the feed, from nickel concentrates and mattes to mixed nickel–cobalt precipitates from Moa Bay.

The mixed nickel–cobalt sulfide precipitate, which contains 55% Ni, 5% Co and 35% S, is transported to Fort Saskatchewan in 1 m^3 polypropylene bags. The contents are crushed and screened, and re-pulped using process solution and concentrated ammonium sulfate solution from the feed to the ammonium sulfate crystallizer.

The mixed nickel–cobalt sulfide is leached with ammonia and air at about 120°C. The extraction of nickel and cobalt are high, exceeding 98% for cobalt and 99% for nickel. The objectives of the leaching stage are to extract the nickel and cobalt into solution, to oxidize the cobalt so that it is present as cobaltic hexammine, to precipitate iron as $Fe_2O_3 \cdot H_2O$ and to oxidize the sulfur to sulfate.

The leach solution is thickened and filtered using lamella thickeners, pressure leaf filters and a Larox pressure filter.

The solution from leaching contains 85 g/L Ni, 10–13 g/L Co, 0.5 g/L Cu, 3 g/L Zn, 130 g/L NH_3 and 180 g/L $(NH_4)_2SO_4$. This solution is cooled to below 35°C. Ammonia is added in the form of a 72% aqueous solution and anhydrous ammonia. This addition results in the precipitation of a complex salt of cobaltic hexammine sulfate, nickel hexammine sulfate and ammonium sulfate: $[Co(NH_3)_6]_3[SO_4]_2 \cdot 2Ni(NH_3)_6SO_4 \cdot (NH_4)_2SO4 \cdot xH_2O$. The slurry is filtered, and typically the filtrate contains less than 0.5 g/L Co.

The complex cobaltic hexammine salt is filtered out and re-pulped with water to selectively dissolve nickel hexammine sulfate. The filtrate is sent to ammonia stripping while the solids are sent to the cobalt circuit. The salt is redissolved and reprecipitated so that the ratio of Co:Ni is greater than 10 000:1. The salt is dissolved in ammonium sulfate solution so that the concentration of cobalt is 50 g/L. These steps are all performed in flat or conical

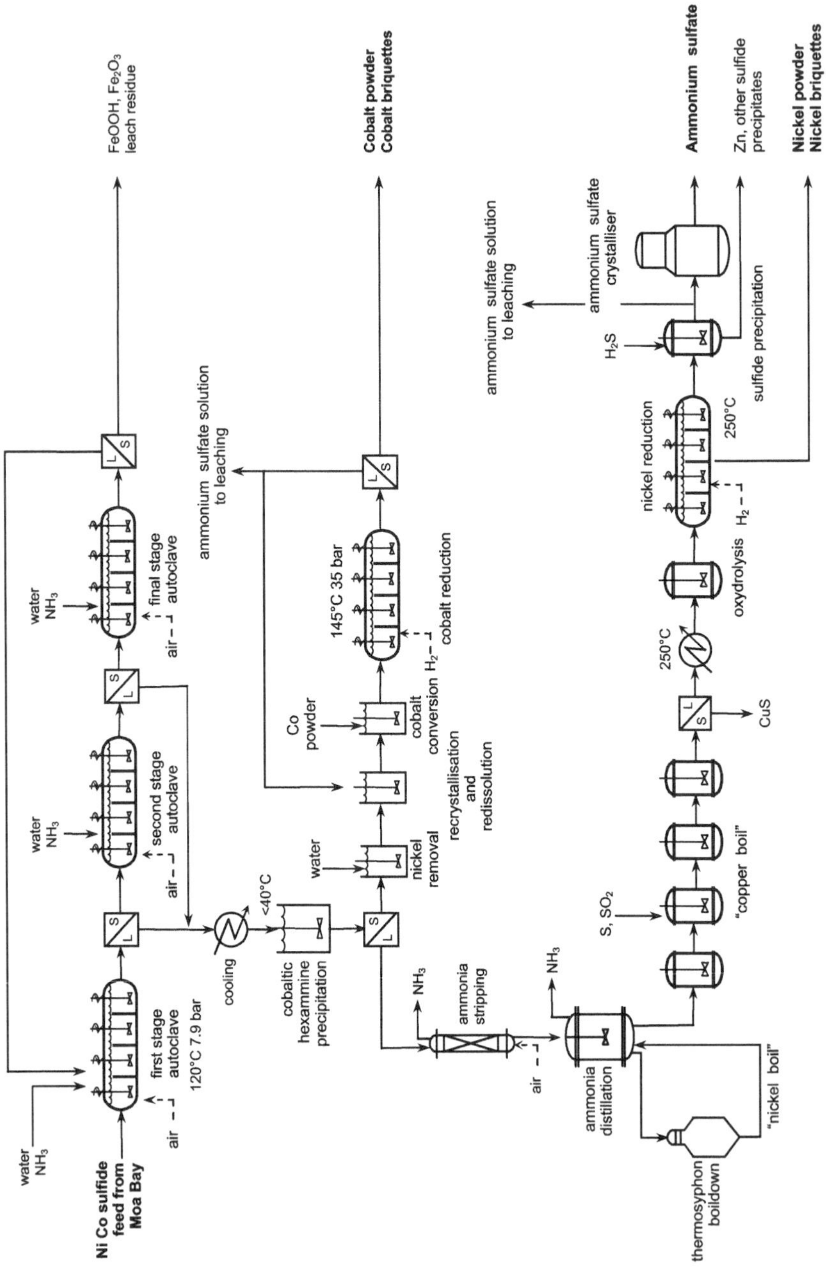

FIGURE 23.1 The Sherrit (Corefco) refinery at Fort Saskatchewan, Canada.

bottomed tanks at atmospheric pressure. The final salt precipitation step is performed in a Swenson cooling crystallizer.

The mole ratio of ammonia to cobalt is adjusted to 2.4:1 and the liquor is then transferred to the cobalt reduction circuit. Cobalt powder is produced by hydrogen reduction at 35 bar and 175°C. The cobalt powder is briquetted and sintered in hydrogen. The briquettes typically contain 99.9% Co and less than 0.002% S and less than 0.002% C.

The solution from the cobaltic hexammine precipitation is treated to reduce the concentration of ammonia in two steps: stripping with air and distillation. The distillation consists of a packed bed column and a thermosyphon reboiler, called the 'nickel boil'. These two steps reduce the mole ratio of NH_3:Ni from 8:1 to 5:1.

This solution passes through four reactors known as the 'copper boil' where copper is removed. Elemental sulfur and sulfur dioxide are added to the second reactor, which results in the precipitation of copper as copper sulfide. This precipitate is filtered from solution.

Sulfuric acid is added to the copper removal reboiler to adjust the NH_3:Ni ratio to 2:1. The nickel boil and copper boil result in a 15%–20% reduction in volume, so that the concentration of the solution is typically 70–75 g/L Ni, 3 g/L Co, 35–40 g/L NH_3, 320–340 g/L $(NH_4)_2SO_4$ and less than 1 mg/L Cu.

The solution, however, contains small amounts of sulfamate, $NH_2SO_3^-$, and unsaturated sulfur anions, such as thiosulfate. These anions are oxidized in a continuous oxidation and hydrolysis step called 'oxydrolysis'. Air is injected into the solution in a column reactor at 250°C. Ammonium sulfate is added to the solution. Thiosulfate and sulfamate are oxidized to ammonium sulfate.

The solution is transferred to hydrogen reduction. Nickel powder is precipitated at 250°C and 35 bar.

The nickel powder is briquetted and sintered in hydrogen. The briquettes typically contain 98.8% Ni, 0.1% Co and less than 0.01% S and less than 0.005% C.

The solution from nickel hydrogen reduction is crystallized to produce ammonium sulfate, which is sold as fertilizer.

Thus, the Sherritt refinery produces nickel and cobalt powder or briquettes, ammonium sulfate and hydrated iron oxide from mixed sulfide precipitates.

23.1.2. Minara Resources Process, Murrin Murrin, Australia

The Minara Resources plant at Murrin Murrin treats a limonitic ore that is predominantly smectite.

A simplified flowsheet of the refinery operation is shown in Figure 23.2 (Sole & Cole, 2002). The Murrin Murrin operation produces the same products from the same type of feed as the process at Fort Saskatchewan. However, it does this in a substantially different manner.

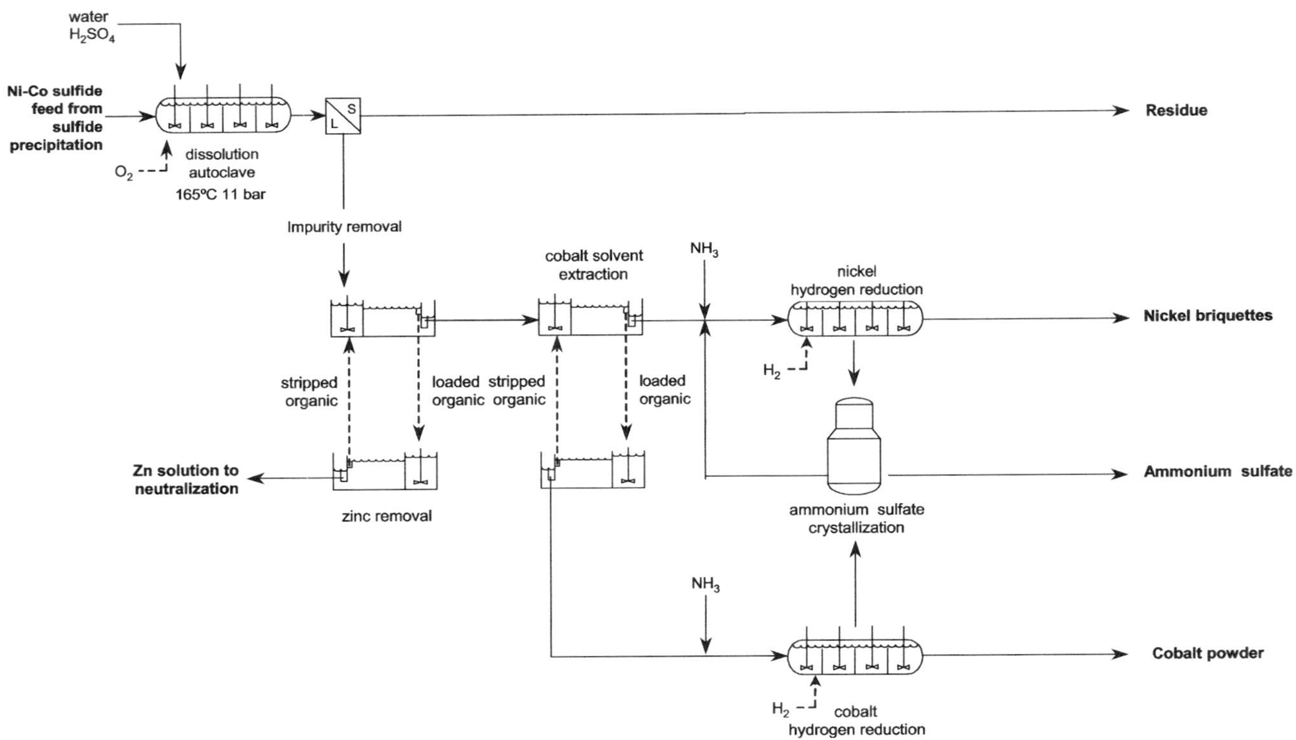

FIGURE 23.2 Simplified flowsheet of the Murrin Murrin refinery. Both the nickel- and cobalt-containing solutions are ammoniated. The mole ratio of NH_3:Ni is about 5:1 and of NH_3:Co is about 2.4:1. These are similar values used in the Sherritt operations. Nickel and cobalt are produced from these solutions by hydrogen reduction.

The laterite ore at Murrin Murrin is leached at high temperature using sulfuric acid. The dissolved nickel and cobalt are precipitated using hydrogen sulfide at 95°C. The partial pressure of hydrogen sulfide is 1 bar. This precipitate is the feed to the refinery.

The first step in the refinery is the re-dissolution of the mixed sulfide precipitate. This re-dissolution is done in an autoclave at 165°C and 11 bar using oxygen gas (Sole & Cole, 2002). The solution from re-dissolution contains between 87 and 95 g/L Ni and about 8 g/L Co (Mayze, 1999).

The high concentration of nickel in solution has a major impact on the size of the solvent extraction circuit (Mayze, 1999). The solvent-extraction circuit at Murrin Murrin is much smaller than those installed at Cawse and Bulong (both now closed) where the concentration of nickel in solution was only 12 and 3 g/L, respectively.

The solution is purified by solvent extraction to remove zinc at low pH using a combination of CYANEX 272 and tributyl phosphate as the extractants (Barnard, 2010). CYANEX 272 is a commercial extractant containing *bis*(2,4,4-trimethylpentyl)phosphinic acid. The pH is controlled by the addition of ammonium hydroxide. There are two extraction and three stripping stages. Zinc is stripped from the loaded organic with dilute sulfuric acid.

Cobalt is extracted from the nickel solution using the same extractants, CYANEX 272 and tributyl phosphate, at a pH higher than zinc extraction. The pH is controlled with ammonium hydroxide. Any nickel that is co-extracted with the cobalt is removed by scrubbing the loaded organic with a cobalt sulfate solution. This scrub solution is the strip solution that has been diluted. Scrubbing reduces the Co:Ni ratio to less than 1000:1. Cobalt is stripped from the loaded organic using a solution of sulfuric acid (Sole & Cole, 2002).

Thus, the Murrin Murrin refinery produces nickel and cobalt powder or briquettes, ammonium sulfate and hydrated iron oxide from mixed sulfide precipitates.

23.1.3. Matte Chlorine Leach Electrowinning Process, Niihama, Japan

The Sumitomo plant at Niihama, Japan, treats mixed sulfide precipitate from Coral Bay in the Philippines. The flowsheet for the process, known as the matte chlorine leach electrowinning (MCLE) process, is shown in Figure 23.3 (Higuchi, Ozaki, Sugiura, & Kemori, 2006; Sole & Cole, 2002).

Like the operations at Fort Saskatchewan, the operations at Niihama have been recently modified. In the case of Niihama, the modifications are to enable it to treat mixed sulfide precipitates from Coral Bay (Higuchi *et al.*, 2006).

The mixed sulfide precipitate from Coral Bay contains 51% Ni, 4.5% Co, 1.7% Fe and 31% S. The Niihama refinery also treats mattes that contain less cobalt and more copper. For example, Higuchi *et al.* (2006) reported on the

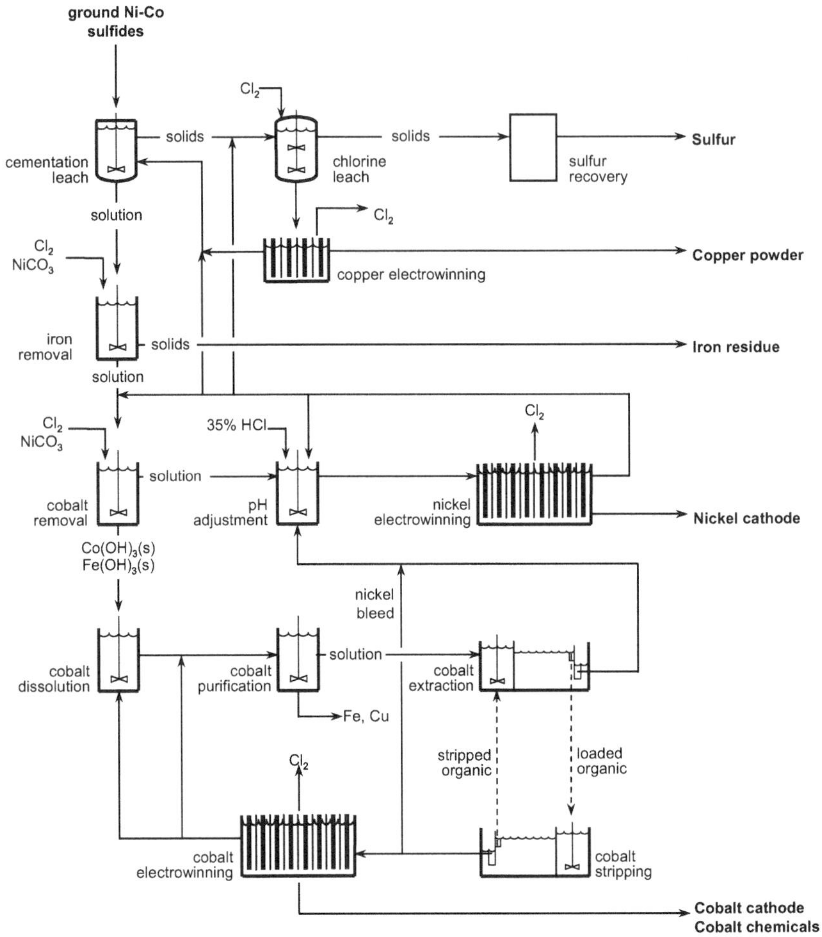

FIGURE 23.3 A schematic diagram of the MCLE process, Niihama, Japan (Sumitomo) (Higuchi *et al.*, 2006).

treatment of mattes that typically contain 68%–78% Ni, 0.6%–0.9% Co, 0.1%–2.5% Cu, 0.5%–4.5% Fe and 20%–24% S.

Matte that is received is ground to a mean particle size of about 44 μm. The ground matte is mixed with sulfide precipitate from Coral Bay and pulped with solution from copper electrowinning. This slurry forms the feed to the cementation circuit.

The objective of the cementation circuit is to remove copper from solution. The reaction involves the dissolution of NiS and the precipitation of Cu_2S. About 20% of the nickel dissolves in the cementation circuit.

The solids and solution from the cementation circuit are separated. The solids are transferred to the chlorine leach and the solution to iron removal.

The remaining 80% of the nickel that is not dissolved in the cementation circuit is dissolved with hydrochloric acid using chlorine gas. There are three brick-lined reactors in series. The reaction temperature is between 105°C and 110°C. The redox potential of solution is controlled to between 400 and 530 mV (vs. Ag/AgCl). Air is also injected into the last reactor. Any metallic components in the matte react with oxygen and consume acid so that the final pH is between 0.5 and 1.0.

The leach slurry is filtered. The residue, which consists of unleached copper sulfides and sulfur, is transferred to sulfur removal (see Chapter 24 for chemical reactions). The solution is pumped to copper electrowinning.

Some of the copper is dissolved in the chlorine leaching vessels. As a result, copper must be removed from the circuit. This is done by electrowinning. Copper powder is deposited at the cathode, automatically shaken off the cathodes and collected from the bottom of the cells. The solution from electrowinning is returned to the cementation circuit.

The solution from the cementation circuit is pumped to the iron removal circuit. Some of the solution is pumped directly to cobalt removal. Iron is precipitated as ferric hydroxide, $Fe(OH)_3$, by the addition of chlorine gas and nickel carbonate, $NiCO_3$. Chlorine oxidizes the ferrous ions to ferric ions, and the nickel carbonate increases the pH so that ferric hydroxide precipitates. The redox potential is controlled at 1000 mV (vs. Ag/AgCl) by adjusting the rate at which chlorine is added. The pH is maintained at a value of 2 by adjusting the rate at which nickel carbonate is added. The temperature of the operation is 60°C. The iron precipitate is transferred to the Shisaka smelter.

The solutions from cementation and iron removal are pumped to cobalt removal. In this operation, chlorine gas is used to oxidize Co^{2+} to Co^{3+} at a redox potential of 1000 mV (vs. Ag/AgCl). Nickel carbonate is added to adjust the pH to between 4 and 5. As a result of these additions, cobalt and iron precipitate as $Co(OH)_3$ and $Fe(OH)_3$, respectively. Arsenic is removed by co-precipitation with the $Fe(OH)_3$.

The solution from cobalt removal is treated to remove zinc. Then, the pH is adjusted to about 1.5 by the addition of concentrated hydrochloric acid. This solution, which contains 70 g/L Ni, 85 g/L Cl^- and less than 0.001 g/L of Cu, Fe or Co, is then fed to nickel electrowinning.

Nickel is electrowon from the solution at a temperature of about 60°C at a current density of 260 A/m^2. The anode is placed in an anode bag so that the chlorine that is evolved can be collected efficiently. The solution leaving electrowinning is returned to pH adjustment, cobalt removal and iron removal.

The final part of the circuit is the 'chemical section' where cobalt is produced. In the first step of the chemical section, cobalt and iron hydroxides from cobalt removal are dissolved in sulfuric acid. Sodium sulfite is used in the dissolution to reduce Co^{3+} to Co^{2+}.

The pH of the solution is increased using NaOH so that iron re-precipitates as ferric hydroxide. The re-precipitate is filtered from the solution, which is pumped to copper removal.

Copper is removed from the solution by precipitation with hydrogen sulfide. The precipitate is filtered from the solution, which is pumped to solvent extraction.

The solvent extraction occurs in two steps: (i) both cobalt and nickel are extracted; and then, (ii) cobalt and nickel are separated.

Cobalt and nickel are extracted from the solution using Versatic 10 (Sole & Cole, 2002), which is a highly branched carboxylic acid extractant. The raffinate is returned to hydroxide dissolution. Nickel and cobalt are stripped from solution using hydrochloric acid. The solution is pumped next to nickel–cobalt separation.

Cobalt and nickel are separated using solvent extraction with tri-*n*-octyl amine as the extractant. Nickel remains in the raffinate, while cobalt is extracted. The raffinate is transferred to the nickel electrowinning circuit. The loaded organic is stripped using hydrochloric acid to give a concentration of 60 g/L Co. This solution is pumped to cobalt electrowinning. Cobalt electrowinning is similar to nickel electrowinning. Other cobalt chemicals are also produced in the chemical section.

Thus, the process produces nickel, cobalt and copper metals, elemental sulfur and iron precipitate in a chloride environment.

23.2. REFINING OF NICKEL MATTES FROM SMELTING OPERATIONS

Nickel–copper mattes are refined to nickel and cobalt at the following refineries:

Nikkelverk, Kristiansand, Norway (Xstrata); and, Harjavalta, Finland (Norilsk).

Other operations that treat nickel–copper mattes form part of the platinum industry. As a result, these processes are described in detail in Chapter 36.

The processing of nickel–copper mattes is significantly different from the processing of mixed sulfide precipitates. This difference arises from the need to separate nickel from copper in the refinery.

The Nikkelverk and Harjavalta processes are described in the following two sections.

23.2.1. Nikkelverk Process, Kristiansand, Norway

The Nikkelverk refinery at Kristiansand (Xstrata) is a large integrated refinery that produces nickel, copper, sulfuric acid and platinum-group metals (Dotterud, Peek, Stenstad, & Ramsdal, 2009; Stensholt, Zachariasen, & Lund, 1986; Stensholt, Zachariansen, Lund, & Thornhill, 1988; Stensholt *et al.*, 2001). It treats converter matte from Sudbury, Canada, and BCL, Botswana. The chemical composition of these feeds is given in Table 23.2, which reveal

TABLE 23.2 Composition of the Converter Mattes that Are Fed to the Nikkelverk Refinery*

Source of converter matte	Ni, %	Cu, %	Fe, %	S, %	Co, %
Sudbury	54	18	2.5	23	2.2
BCL	38	36	2.5	22	0.9

Stensholt et al., 2001

the significant amounts of copper in the matte. The nickel is present as Ni_3S_2 and the copper as Cu_2S.

A schematic diagram of the flowsheet is shown in Figure 23.4.

The blended converter matte is re-ground to 100% less than 150 μm in a dry ball mill. The mill is swept with air to reduce operator exposure to Ni_3S_2, which

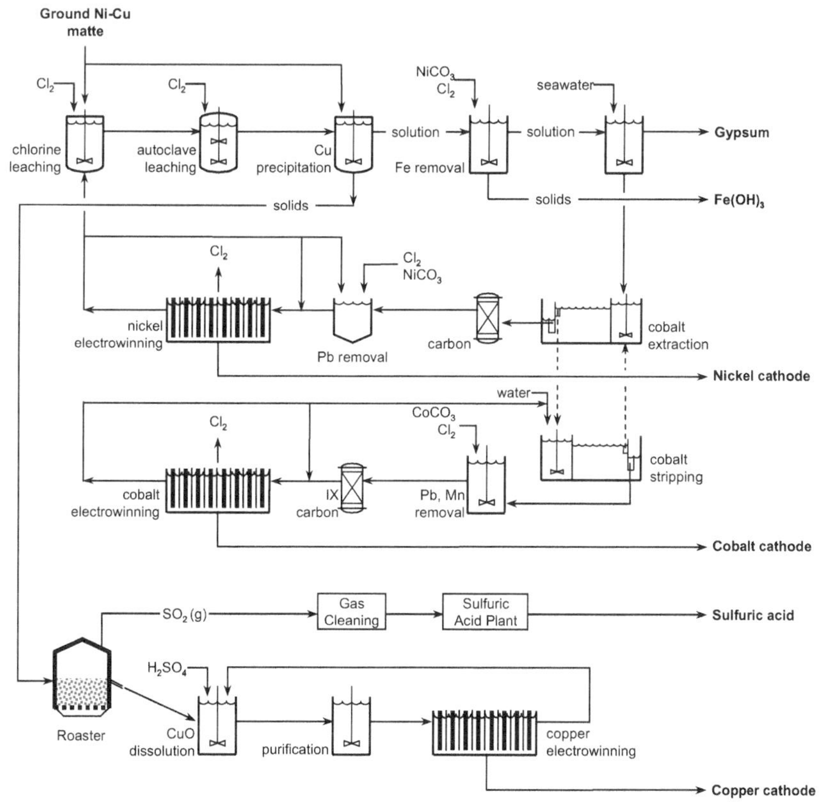

FIGURE 23.4 A schematic diagram of the process flowsheet used by Nikkelverk (Kristiansand, Norway).

is sometimes referred to as nickel sub-sulfide. Nickel sub-sulfide has been implicated as a carcinogen.

The milled matte is pulped with solution from nickel electrowinning and fed to five leaching tanks. Chlorine gas, recompressed from nickel electrowinning, is injected under the impeller of the tank agitator. The reaction is exothermic and is maintained at the boiling point. The solution contains about 230 g/L Ni and about 50 g/L Cu. The principal reactions are the dissolution of Ni_3S_2 and Cu_2S to yield nickel and copper in solution, and elemental sulfur in the residue.

The slurry is pumped to two autoclaves, which operate at a temperature of 150°C.

The residue from autoclave leaching consists mainly of copper sulfide, elemental sulfur, and unreacted nickel sulfide. The solution contains about 230 g/L Ni and about 7 g/L Cu. The higher temperature of the autoclave encourages the following reaction:[3]

$$\mathrm{NiS}(s) + 2\mathrm{CuCl}(aq) \xrightarrow{150^\circ\mathrm{C}} \mathrm{NiCl_2}(aq) + \mathrm{Cu_2S}(s) \qquad (23.1)$$

Reactions of this form are the predominant method for separating nickel and copper in a nickel–copper refinery.

The slurry is transferred to copper precipitation, where fresh matte is added to reduce the concentration of copper in solution to less than 0.5 g/L by reactions that are similar to that shown in Equation (23.1).

This slurry is washed and filtered so that the chloride content of the residue is less than 0.1%. This residue, which contains 5% Ni, 50% Cu, 40% S, 1% Co and 1% Fe, is transferred to the copper roaster. The solution, which after washing contains 220 g/L Ni, 11 g/L Co, 7 g/L Fe and 0.5 g/L Cu, is pumped to iron removal.

Iron is removed from the solution by oxidizing all the Fe^{2+} to Fe^{3+} using chlorine and raising the pH using nickel carbonate, $NiCO_3$. Iron precipitates as ferric hydroxide, $Fe(OH)_3$, which is filtered from solution and disposed of underground. It typically contains 43% Fe and 1.5% Ni. The solution is cooled and filtered to remove gypsum. This solution is transferred to the cobalt removal section.

Cobalt is removed from solution by solvent extraction using 17% tri-iso-octylamine in an aromatic diluent. Four stages of extraction are used to reduce the concentration of cobalt from 11 g/L to about 0.001 g/L Co. The loaded organic is scrubbed in two mixer settlers with solution from cobalt electrowinning to remove entrained aqueous solution.

Cobalt is stripped from the loaded organic using water and solution from cobalt electrowinning. The concentration of cobalt in the solution from stripping is about 90 g/L. This solution is purified using ion exchange, to remove

3. Reactions of this form are sometimes called metathesis reactions or ion-exchange reactions.

zinc and trace amounts of copper. The solution also passes through carbon columns before it is pumped to cobalt electrowinning.

Cobalt is recovered from solution by electrowinning. The feed solution contains about 50 g/L Co at a pH of between 1.4 and 1.5. The solution from cobalt electrowinning is recycled to the cobalt purification section.

The raffinate from cobalt removal contains 220 g/L Ni, 0.1 g/L Co, 0.25 g/L Pb, 0.2 g/L Mn and 2 g/L HCl. Entrained organic is removed using carbon columns. This solution is diluted with solution from nickel electrowinning so that the concentration of nickel is about 85 g/L and then transferred to lead and manganese removal.

Lead and manganese are removed from solution by precipitation with nickel carbonate and chlorine gas. This treatment reduces the concentrations of lead and manganese to less than 0.02 mg/L and less than 0.05 mg/L, respectively. Other impurities, such as cobalt, iron, copper and arsenic are also removed. This purified solution is transferred to the nickel tankhouse.

The solution from lead and manganese removal is diluted using nickel electrowinning solution to 65 g/L before being fed to nickel electrowinning. Nickel is plated at the cathode, and chlorine gas is evolved at the anode. Chlorine is collected and used in leaching. The solution from electrowinning contains about 54 g/L Ni and is recycled to leaching.

The residue from copper precipitation is re-pulped in water and transferred to the roasting plant. The roaster is a fluidized bed reactor that operates at a temperature of about 875°C. The feed gas is air that has been enriched with oxygen so that it contains 40% O_2. The residue is roasted to completion to form a calcine of CuO and SO_2 gases.

The gas from the roaster is cleaned to remove dust and selenium using electrostatic precipitators and venturi scrubbers. The cleaned gas is sent to a double-absorption sulfuric acid plant.

Calcine from the roaster is transferred to copper leaching, where it is mixed with solution from copper electrowinning and with fresh acid. The sulfuric acid dissolves the copper calcine so that the concentration of copper in solution is about 85 g/L. This solution is fed to the copper tankhouse after purification and dilution with return solution from copper electrowinning.

The concentration of copper in the feed to the tankhouse is 60 g/L. Copper is plated over 7 days onto copper starter sheets. The solution leaving copper electrowinning contains 50 g/L Cu and 95 g/L H_2SO_4. This solution is recycled to copper leaching and purification. A bleed is withdrawn from this stream and returned to the chlorine leach to remove nickel from the copper circuit.

The Nikkelverk operation at Kristiansand recovers precious metals from the residue from copper leaching.

Thus, the Nikkleverk refinery produces nickel, copper and cobalt metals and sulfuric acid from the base metals refinery. In addition, Platinum-group metals are produced.

23.2.2. Harjavalta Operations, Harjavalta, Finland

The Norilsk refinery at Harjavalta, Finland, treats nickel matte from the Boliden smelter at Harjavalta. The smelter produces two products from the direct Outokumpu Nickel (DON) process: (i) a flash furnace (FF) matte; and, (ii) an electric furnace (EF) matte. The compositions of these two mattes are shown in Table 23.3.

The main minerals in the matte are heazlewoodite (Ni_3S_2), chalcocite (Cu_2S) and a nickel–copper alloy.

A simplified flowsheet of the process is shown in Figure 23.5. The process description that follows is based on the information provided by Knuutila, Hultholm, Saxen, & Rosenback (1997), Latva-Kokko (2006) and Makinen, Fagerlund, Anjala, & Rosenback (2005).

The FF matte is dissolved in four steps: copper removal, atmospheric leaching, nickel pressure leaching and copper pressure leaching.

The first step in leaching the FF matte is copper removal in which copper is precipitated as $Cu_3SO_4(OH)_4$ in five agitated tanks connected in series. Oxygen is injected below in the impeller of the agitator. About 25% of the nickel in the matte feed dissolves in this step. The reaction consumes oxygen and acid, which results in an increase in the pH to between 5.5 and 6.0, which is required for the precipitation of copper.

Solid–liquid separation after copper removal is done in a thickener. The overflow from the thickener is filtered in a filter press and transferred to the solvent-extraction circuit. The solids are transferred to nickel atmospheric leaching.

The second step in leaching the FF matte is nickel atmospheric leaching, in which Ni_3S_2 is leached by oxygen at pH values in the range of 3.0–3.5. Four agitated tanks connected in series are used. The extraction of nickel is about 40% after nickel atmospheric leaching.

The third step in leaching the FF matte is nickel pressure leaching. A five-compartment horizontal autoclave is used. Little or no oxygen is added.

TABLE 23.3 Composition of the FF and EF Mattes from the Harjavalta Smelter*

Matte	Ni, %	Cu, %	Fe, %	S, %
FF	65	5	4	21
EF	50	6	30	7

**Knuutila et al., 1997*

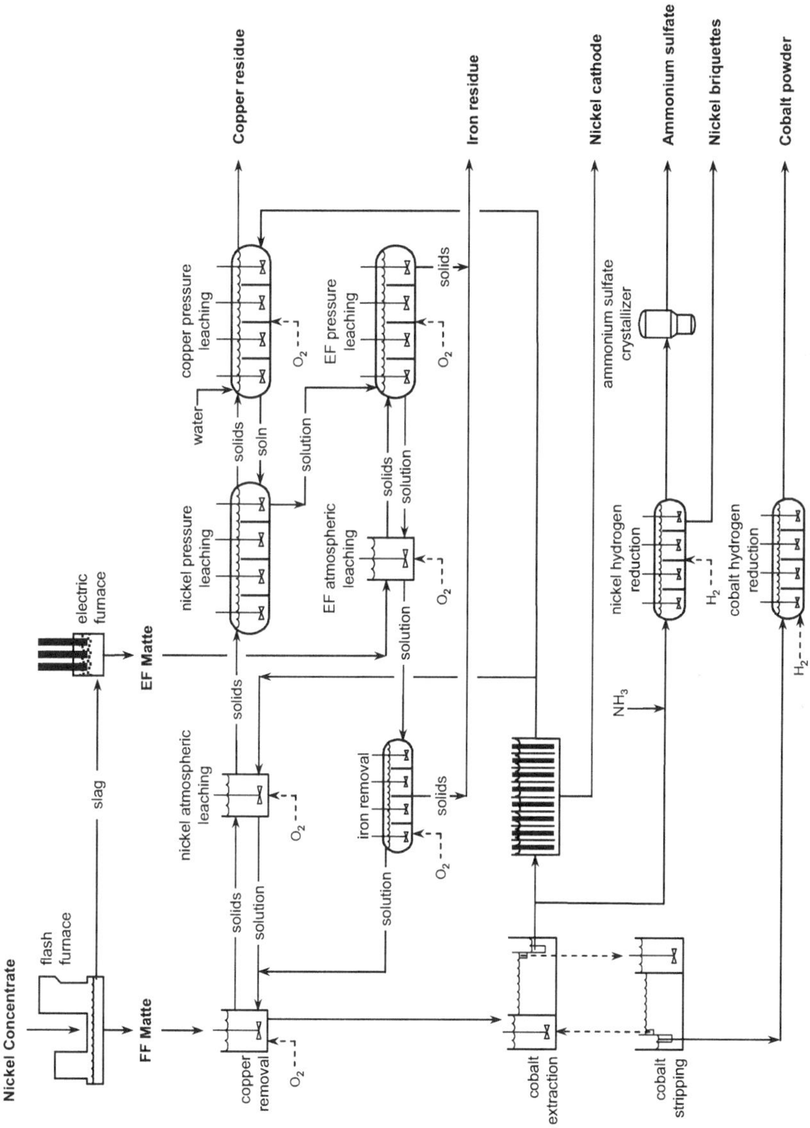

FIGURE 23.5 A schematic diagram of the process flowsheet of the refinery at Harjavalta, Finland.

Instead, leaching is effected by copper ions, by metathesis reactions similar to the following reaction:

$$6NiS(s) + 9CuSO_4(aq) + 4H_2O(\ell) \xrightarrow{150^\circ C} Cu_9S_5(s) + 6NiSO_4(aq) + 4H_2SO_4(aq) \quad (23.2)$$

The fourth, and last, stage of the leaching of the FF matte is copper pressure leaching. A seven-compartment, titanium-lined autoclave is used. Oxygen is injected into each compartment. Copper and nickel are leached at elevated temperature by sulfuric acid and oxygen. The primary purpose of this leaching stage is to produce sufficient copper to drive the reaction given in Equation (23.2).

The EF matte is dissolved in two steps: atmospheric leaching and pressure leaching. The purpose of the leaching of the EF matte is to dissolve all of the nickel, copper and cobalt, and to precipitate the iron as goethite or hematite.

The first step in the leaching of the EF matte is atmospheric leaching, which occurs in five agitated tanks in series. Oxygen is injected into each reactor. The pH at the end of the series of reactors is about 2.5. Because the EF matte is mostly metallic (the sulfur content is low), there is the possibility of the formation of hydrogen. This is minimized by adequate supply of oxygen. All of the metallic components of the EF matte are leached in atmospheric leaching.

The second step in the leaching of the EF matte is pressure leaching, which occurs in an autoclave at elevated temperature. Sulfuric acid is added and oxygen is injected into each compartment of the autoclave. The objective of pressure leaching is to dissolve the sulfide components in the EF matte and to precipitate iron, usually as jarosite, $RFe_3(SO_4)_2(OH)_6$ where R represents H_3O^+, NH_4^+, Na^+ or K^+.

Solution from the EF atmospheric leaching is transferred to iron removal, which occurs in an autoclave at a temperature of about 120°C. This iron removal step removes residual iron, since the majority of the iron is precipitated as jarosite in EF pressure leaching.

Solution from copper removal, containing about 130 g/L Ni and 1 g/L Co, is filtered and cooled in heat exchangers to below 50°C. This solution is pumped to solvent extraction, where cobalt is removed in four extraction stages. The extractant is CYANEX 272. The raffinate contains 130 g/L Ni and 0.01 g/L Co. Impurities, such as zinc, copper and manganese, are also loaded onto the organic. The raffinate is pumped to nickel electrowinning and nickel hydrogen reduction.

Cobalt is stripped from the loaded organic in four stages. The loaded organic is washed in one stage. Zinc and iron that are loaded onto the organic are then scrubbed from the loaded organic in a single stage.

The solution from cobalt stripping contains 110 g/L Co, 30 mg/L Cu and 23 mg/L Mn. This solution is pumped to cobalt electrowinning.

Nickel is plated onto starter sheets in the nickel tankhouse. A diaphragm-type cell is used. A slight hydrostatic head in the cathode compartment ensures that flow is from the cathode compartment to the anode compartment, so as to minimize hydrogen evolution at the cathode. Oxygen is evolved at the anode. Cathodes are harvested after about 7 days. Cathodes are cut into squares and packed for delivery.

The solution leaving nickel electrowinning contains about 100 g/L Ni and about 50 g/L sulfuric acid. This solution is recycled to nickel atmospheric leaching.

Part of the raffinate is sent to nickel reduction rather than to nickel electrowinning. The solution is ammoniated and pumped into the reduction autoclave. Nickel is reduced to metal by standard Sherritt technology in five autoclaves each with a volume of 40 m^3. The nickel powder is dried, briquetted, sintered in a hydrogen atmosphere and packed for delivery.

The solution leaving hydrogen reduction is sent to the ammonium sulfate crystallizer. Ammonium sulfate is the main outlet of sulfur from the plant.

Thus, the Harjavalta refinery produces nickel in the form of cathode, nickel powder or briquettes, cobalt metal and ammonium sulfate.

23.2.3. Other Refining Options

There are several other options that may be considered. Three different, but related, flowsheets for processing nickel–copper mattes in the platinum industry are described in Chapter 36. These processes use Sherritt technology for leaching, and produce nickel sulfate by crystallization, nickel powder for hydrogen reduction or nickel cathode by electrowinning. The options for sulfur removal are either as ammonium sulfate or sodium sulfate.

Another process that is worth investigation is that of the Yabulu refinery at Townsville, Queensland, Australia. In that process, solvent extraction is carried out in ammoniacal solutions to separate nickel and cobalt and then nickel is precipitated as nickel carbonate. The carbonate is calcined and reduced in hydrogen.

23.3. APPRAISAL

The five processes that have been presented demonstrate the variety of refining options. Sulfate, chloride and ammonium sulfate solutions are used to produce nickel cathode and nickel powder by either electrowinning or hydrogen reduction. Sulfur is removed either as sulfate, in the form of ammonium sulfate, or as elemental sulfur. Nickel and cobalt may be separated by fractional crystallization or by solvent extraction. Impurities, such as iron, zinc and manganese, can be removed by precipitation or ion exchange.

Many of the refineries have had to adapt to new feedstocks, resulting in modifications to the original operation. All seem to have done so successfully.

23.4. SUMMARY

Two types of sulfide intermediates are refined hydrometallurgically: mixed sulfide precipitates and converter matte. The mixed sulfides are significantly easier to treat because the levels of copper in the feed material are much lower than those in converter matte.

Three different chemistries are used: ammonia, sulfate and chloride. The Sherritt refinery in Fort Saskatchewan, Canada, is based on ammonia chemistry. Nickel and cobalt are separated by crystallization. This refinery used hydrogen reduction to produce nickel and cobalt briquettes. Sulfur is removed as ammonium sulfate.

The Minara Resources refinery at Murrin Murrin, Australia, uses similar chemistry and produces the same products as the Sherritt refinery. However, nickel and cobalt are separated by solvent extraction rather than by crystallization.

The Sumitomo refinery at Niihama, Japan, uses chloride chemistry to produce nickel and cobalt by electrowinning. Sulfur is removed as elemental sulfur. Nickel and cobalt are separated by precipitation.

The Nikkelverk refinery at Kristiansand, Norway, uses chloride chemistry to refine nickel–copper matte. The nickel and copper are separated by leaching chemistry, and nickel and cobalt are separated by solvent extraction. Nickel and cobalt are produced by electrowinning.

The Norilsk refinery at Harjavalta, Finland, uses sulfate chemistry to refine mattes that are relatively low in copper. Copper and nickel are separated by leaching chemistry, while nickel and cobalt are separated by solvent extraction. Nickel is produced both by electrowinning and by hydrogen reduction. Cobalt is produced by hydrogen reduction. Sulfur is removed as ammonium sulfate.

REFERENCES

Barnard, K. (2010). Indentification and characterization of a CYANEX 272 degradation product formed in the Murrin Murrin solvent extraction circuit. *Hydrometallurgy, 103*, 190–195.

Budac, J. J., Kofluk, R., & Belton, D. (2009). Reductive leach process for improved recovery of nickel and cobalt in the Sherritt hexamine leach process. In J. J. Budac, R. Fraser & I. Mihaylov, et al. (Eds.), *Hydrometallurgy of nickel and cobalt 2009* (pp. 77–85). CIM.

Cordingley, P. D., & Krentz, R. (2005). Corefco refinery – Review of operations. In J. Donald & R. Schonewille (Eds.), *Nickel and cobalt 2005 challenges in extraction and production* (pp. 407–425). CIM.

Donghe, P. (2009). Technology development in Jinlin Nickel Industry Co. Ltd. In J. Liu, J. Peacey & M. Barati, et al. (Eds.), *Pyrometallurgy of nickel and cobalt 2009* (pp. 131–135). CIM.

Dotterud, O. M., Peek, E. M. L., Stenstad, O., & Ramsdal, P. O. (2009). Iron control and tailings disposal in the Xstrata chlorine leach process. In J. J. Budac, R. Fraser & I. Mihaylov, et al. (Eds.), *Hydrometallurgy of nickel and cobalt 2009* (pp. 321–333). CIM.

Higuchi, H., Ozaki, Y., Sugiura, T., & Kemori, N. (2006). Iron removal from the MCLE [matte chlorine leach electrowin] circuit. In J. E. Dutrizac & P. A. Riveros (Eds.), *Iron control technologies* (pp. 403–413). CIM.

Knuutila, K., Hultholm, S. E., Saxen, B., & Rosenback, L. (1997). *New nickel process increasing production at Outokumpu Harjavalta Metals Oy, Finland. ALTA 1997 nickel and cobalt pressure leaching and hydrometallurgy forum.* ALTA Metallurgical Services.

Latva-Kokko, M. J. (2006). Iron removal as part of the nickel matte leaching process. In J. E. Dutrizac & P. A. Riveros (Eds.), *Iron control technologies* (pp. 391–401). CIM.

Makinen, T., Fagerlund, K., Anjala, Y., & Rosenback, L. (2005). Outokumpu's technologies for efficient and environmentally sound nickel production. In J. Donald & R. Schonewille (Eds.), *Nickel and cobalt 2005: Challenges in extraction and production* (pp. 71–89). CIM.

Mayze, R. (1999). *An engineering comparison of the three treatment flowsheets in WA nickel laterite projects. Alta 1999 nickel and cobalt pressure leaching and hydrometallurgy forum.* ALTA Metallurgical Services.

Sole, K. C., & Cole, P. M. (2002). Purification of nickel by solvent extraction. In Y. Marcus, A. K. SenGupta & J. A. Marinsky (Eds.), *Ion exchange and solvent extraction* (pp. 143–195). Marcel Dekker Inc.

Stensholt, E. O., Dotterud, O. M., Henriksen, E. E., et al. (2001). Development and plant practice of the Falconbridge chlorine leach process. *CIM Bulletin, 94*, 101–144.

Stensholt, E. O., Zachariasen, H., & Lund, J. H. (1986). Falconbridge chlorine leach process. Section C. *Transactions of the Institution of Mining and Metallurgy, 5*, C10–C16.

Stensholt, E. O., Zachariansen, H., Lund, J. H., & Thornhill, P. G. (1988). Recent improvements in the Falconbridge Nikkelverk nickel refinery. In G. P. Tyroler & C. A. Landolt (Eds.), *Extractive metallurgy of nickel and cobalt* (pp. 403–413). TMS.

SUGGESTED READING

Budac, J. J., Fraser, R., & Mihaylov, I., et al. (Eds.), (2009). *Hydrometallurgy of nickel and cobalt 2009*. CIM.

Cooper, W. C., & Mihaylov, I. (Eds.), (1997). *Nickel/cobalt 97: Vol. I. Hydrometallurgy and refining of nickel and cobalt*. CIM.

Donald, J., & Schonewille, R. (Eds.), (2005). *Nickel and cobalt 2005 challenges in extraction and production*. CIM.

Egedahl, R. D., & Collins, M. J. (2009). Vital status of Sherritt nickel refinery workers (1954–2003). In J. J. Budac, R. Fraser & I. Mihaylov, et al. (Eds.), *Hydrometallurgy of nickel and cobalt 2009* (pp. 689–699). CIM.

Palmer, J., Malone, J., & Loth, D. (2005). WMCR – Kalgoorlie Nickel Smelter operations overview 1972–2005. In J. Donald, & R. Schonewille (Eds.), *Nickel and cobalt 2005: Challenges in extraction and production* (pp. 441–455). CIM.

Stevens, D., Bishop, G., Singhal, A., et al. (2009). Operation of the pressure oxidative leach process for Voisey's Bay nickel concentrate at Vale Inco's hydromet demonstration plant. In J. J. Budac, R. Fraser & I. Mihaylov, et al. (Eds.), *Hydrometallurgy of nickel and cobalt 2009* (pp. 3–16). CIM.

Chapter 24

Leaching of Nickel Sulfide Mattes and Precipitates

The first step in the hydrometallurgical treatment of nickel intermediates, such as matte and mixed sulfide precipitates, is to dissolve the nickel from the sulfide intermediate into an aqueous solution. Three different chemical systems are used:

(a) leaching in chloride solutions using chlorine;
(b) leaching in ammonium sulfate solutions using air and ammonia; and,
(c) leaching in sulfuric acid using oxygen.

The objectives of this chapter are to describe the leaching of mattes and sulfide intermediates.

24.1. CHLORINE LEACHING

Approximately 160 000 tonnes of nickel and 10 000 tonnes of cobalt per year are leached in chloride solutions using chlorine.

The process for leaching matte using chlorine at Nikkelverk (Kristiansand, Norway) is shown in Figure 24.1 (Stensholt, Zachariasen, & Lund, 1986; Stensholt, Zachariansen, Lund, & Thornhill, 1988; Stensholt *et al.*, 2001). Operating details for Nikkleverk are given in Table 24.1. The leaching of the matte entails the following steps:

(a) atmospheric leaching followed by leaching at high temperature (~150°C) in an autoclave of finely ground matte particles with recycle solution containing nickel and copper chloride using chlorine gas as the oxidant; and,
(b) precipitation of copper sulfide, CuS, from the cooled slurry by controlled addition of more matte.

The compositions of the inputs and outputs to the leaching plant are given in Table 24.1.

Extractive Metallurgy of Nickel, Cobalt and Platinum-Group Metals. DOI: 10.1016/B978-0-08-096809-4.10024-3

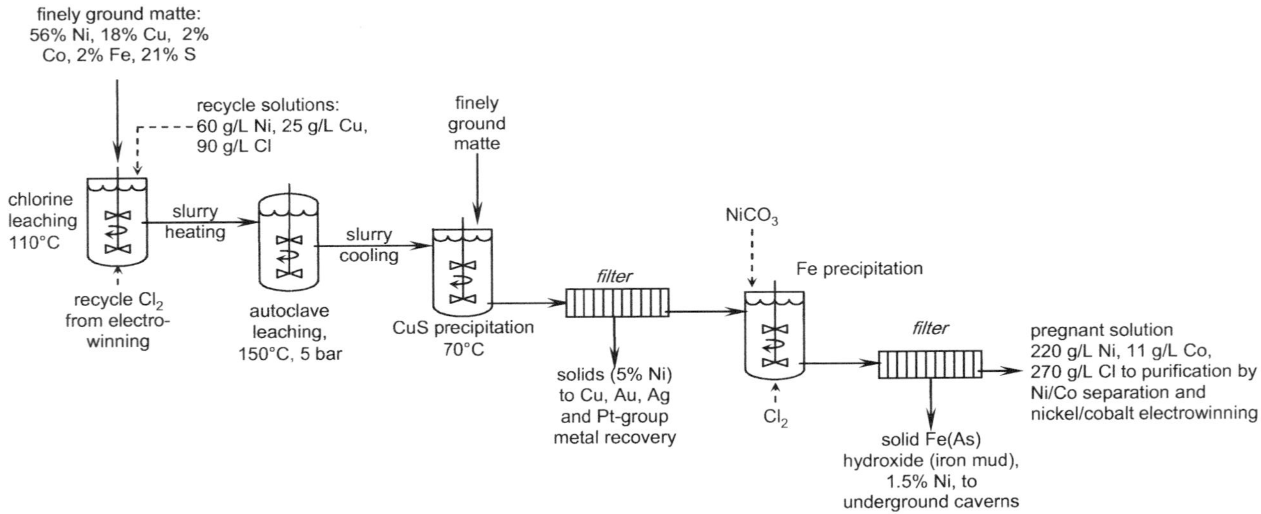

FIGURE 24.1 Schematic flowsheet of the chlorine leaching of nickel matte (Kristiansand, Norway). The principal feed is finely ground nickel–copper sulfide matte. The principal products are pregnant solution of nickel and cobalt, and solids containing copper, silver, gold and platinum sulfide (Dotterud *et al.*, 2009; Stensholt *et al.*, 1986, 1988, 2001). The solution goes to nickel and cobalt separation then electrowinning. The solids go to roasting, leaching and copper, silver, gold and platinum-group metals recovery. The two products of leaching are (a) pregnant leach solution, containing 220 g/L Ni and 11 g/L Co, which is transferred to iron precipitation, solution purification, nickel and cobalt separation and electrowinning of high-purity nickel and cobalt; (b) copper sulfide precipitate, which is sent to a copper/precious metal plant for copper, sulfuric acid and precious metals production (Dotterud *et al.*, 2009; Stensholt *et al.*, 1986, 1988, 2001).

TABLE 24.1 Compositions of Nikkelverk Chlorine Leaching Plant Inputs and Outputs*

	Contents					
Input or Output	Ni	Cu	Co	Fe	S	Cl
Falconbridge matte, %	55	18	2	2	21	
Botswana matte, %	40	34	1	2	17	
Custom feed (high Fe), %	3	30	25	5–30		
Custom feed (low Fe), %	15	15	20	3		
Lixiviant, g/L tonne[a]	60	25	0	1		90
Leach residue and precipitate, %[b]	5	50	1	1	40	0.2
Iron mud, %[b]	1.5			43		
Leach solution, g/L[b]	220	0.5	11			270

The lixiviant contains recycled nickel electrowinning anolyte (which provides nickel in solution), copper residue leach solution (which provides copper in solution) and several low-impurity effluent solutions (Stensholt *et al.*, 2001). Sumitomo's Niihama and Eramet's Le Havre-Sandouville plants also use chlorine leaching (Eramet, 2010; Higuchi, Ozaki, Sugiura, & Kemori, 2006; Sumitomo, 2007).
[a]*Stensholt et al., 1986, 1988.*
[b]*Stensholt et al., 2001.*
**Dotterud et al., 2009.*

24.1.1. Chemistry

The main reactions in the leaching tanks are as follows:

$$\underset{\substack{\text{in recycle chloride}\\ \text{solutions, 25 g/L Cu}^+}}{6CuCl(aq)} + \underset{\substack{\text{recycled from Ni}\\ \text{and Co electrowinning}}}{3Cl_2(g)} \xrightarrow{\sim 110^\circ C} \underset{\text{in aqueous solution}}{6CuCl_2(aq)} \quad (24.1)$$

$$\underset{\substack{\text{in finely}\\ \text{ground}\\ \text{50\% Ni}\\ \text{matte}}}{Ni_3S_2(s)} + 6CuCl_2(aq) \xrightarrow{110^\circ C} \underset{\substack{\text{for nickel}\\ \text{production}\\ \text{by}\\ \text{electrowinning}}}{3NiCl_2(aq)} + 6CuCl(aq) + \underset{\substack{\text{tocopper}\\ \text{roaster}}}{2S(\ell)} \quad (24.2)$$

$$\underset{\substack{\text{in finely}\\ \text{ground}\\ \text{50\% Ni}\\ \text{matte}}}{Ni_3S_2(s)} + 2CuCl_2(aq) \xrightarrow{110^\circ C} \underset{\substack{\text{for nickel}\\ \text{production}\\ \text{by}\\ \text{electrowinning}}}{NiCl_2(aq)} + 2CuCl(aq) + 2NiS(s) \quad (24.3)$$

The solution leaving leaching contains 230 g/L Ni and 50 g/L Cu.

The reaction in the leaching autoclave is the metathesis or exchange reaction:

$$\underset{\text{from Equation (24.3)}}{NiS(s)} + 2CuCl(aq) \xrightarrow{150^\circ C} \underset{\text{for nickel production by electrowinning}}{NiCl_2(aq)} + \underset{\text{to copper roaster}}{Cu_2S(s)} \qquad (24.4)$$

The solution leaving the autoclave contains 230 g/L Ni and 7 g/L Cu. The decrease in the concentration of copper over the autoclave, from 50 g/L–7 g/L, indicates that the reaction given in Equation (24.4) essentially effects the separation of copper and nickel in the refinery. However, high temperatures, about 150°C, are required for this reaction to occur at an appreciable rate.

The slurry from these reactions is cooled. Ground matte is added, which results in the precipitation of Cu^+ as Cu_2S. The principal precipitation reactions are as follows:

$$2CuCl(aq) + S(\ell) + \underset{\text{in matte addition}}{Ni_3S_2(s)} \xrightarrow{70^\circ C} \underset{\text{for nickel electrowinning}}{NiCl_2(aq)} + 2NiS(s) + Cu_2S(s) \qquad (24.5)$$

$$\underset{\text{in matte feed}}{2Cu_2S(s)} + 2S(\ell) \xrightarrow{70^\circ C} \underset{\text{precipitation product to copper/precious metals plant}}{4CuS(s)} \qquad (24.6)$$

The concentration of copper in solution leaving the copper precipitation tanks contains 0.5 g/L Cu.

About 90% of the matte is leached in the chlorine leaching tanks and the autoclave [Equations (24.1)–(24.4)]. The remainder is used to precipitate copper sulfide, CuS [Equations (24.5) and (24.6)].

The resulting slurry is filtered to give:

(a) a filter cake of leaching residue and precipitation solids, which are washed and transferred to copper roasting for the eventual recovery of copper, silver, gold and platinum-group metals; and,
(b) a solution, rich in nickel, which is purified and transferred to cobalt removal by solvent extraction and then to electrowinning of high-purity nickel.

24.1.2. Industrial Operation

The leaching operation is continuous. There are five leaching tanks operated at about 110°C, followed by two autoclaves in series operated at about 150°C to drive the metathesis reaction [Equation (22.4)]. These leaching operations are followed by two precipitation tanks in series operated at about 70°C.

About 500 tonnes of matte are leached per day. As mentioned earlier, about 90% of the matte reacts in the leaching tanks and autoclaves and the remainder reacts in the copper sulfide precipitation tanks. The leaching efficiency of nickel is approximately 98%; most of the remaining nickel is, however, recovered by recycle from the copper recovery circuit. The overall recovery of nickel is 99%, which is similar to that obtained by the Niihama chlorine leaching plant (Sumitomo, 2007).

The plant is highly instrumented with on-line chemical analyzers, pH meters and redox potential meters. The redox measurements are especially important for automatically adjusting matte feed rates into the leaching and precipitation tanks (Dotterud, Peek, Stenstad, & Ramsdal, 2009).

24.2. OXYGEN–AMMONIA LEACHING

Approximately 100 000 tonnes/year of nickel and about 5000 tonnes/year of cobalt are produced by leaching in ammonium sulfate solutions using ammonia and oxygen.

Leaching of sulfides using oxygen and ammonia was first installed in 1954 in Fort Saskatchewan, Canada. The same autoclaves are still in operation. Ammonia leaching is also used at Kwinana, Australia. The discussion here is largely based on the Fort Saskatchewan plant.

24.2.1. Fort Saskatchewan Operation

The Sherritt leaching process at Fort Saskatchewan is shown in Figure 24.2 (Budac, Kofluk & Belton, 2009; Cordingley & Krentz, 2005; Kerfoot & Cordingley, 1997). The process entails the following steps:

(a) leaching the mixed nickel–cobalt sulfide precipitate in ammonium sulfate using ammonia and oxygen at about 95°C–120°C; and,
(b) cooling and filtering the resulting slurry into a solution that is rich in nickel (85 g/L Ni) and a solid residue that is lean in nickel (about 4% Ni).

The residue is washed and discarded. The solution is pumped to nickel and cobalt separation, impurity removal and finally to hydrogen reduction where high-purity nickel and cobalt is produced (Chapters 27).

The compositions of the inputs and outputs to the Sherritt leaching plant are given in Table 24.2 and the operational details are given in Table 24.3.

24.2.2. Leaching Chemistry

The main nickel and cobalt leaching reactions are given as follows:

$$\underset{\text{in sulphide feed slurry}}{NiS(s)} + \underset{\text{in injected air}}{2O_2(g)} + \underset{\text{in injected 72\% } NH_3 \text{ aqueous solution}}{6NH_3(\ell)} \xrightarrow{\sim 105-120^\circ C} \underset{\text{nickelous hexammine sulphate solution}}{Ni(NH_3)_6 \cdot SO_4(aq)} \tag{24.7}$$

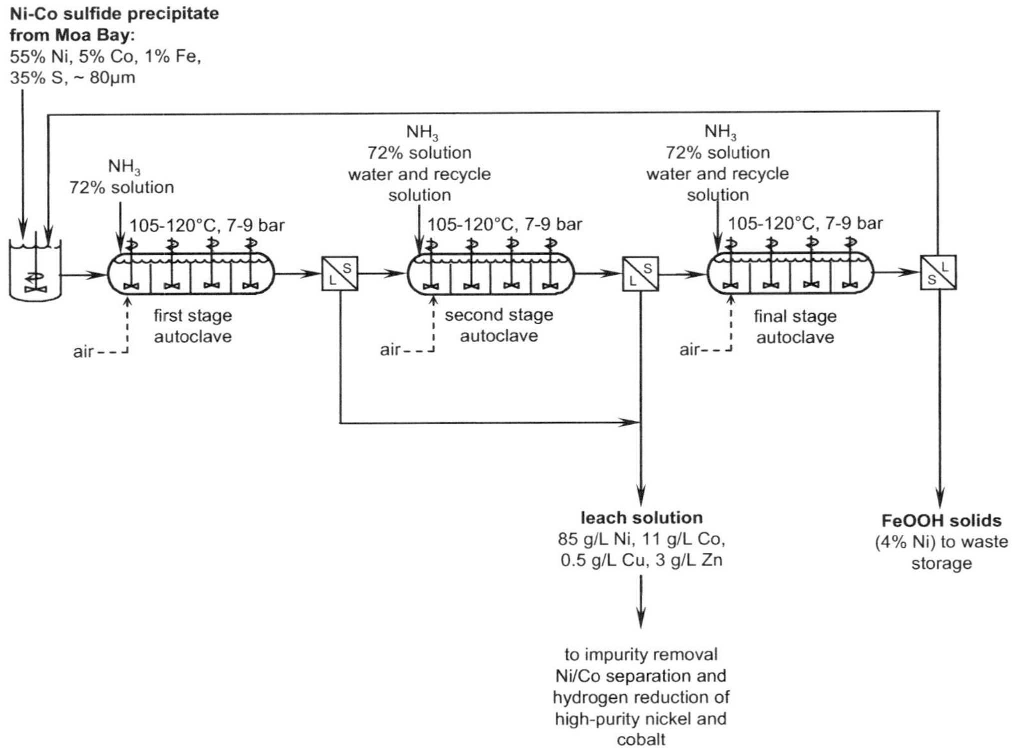

FIGURE 24.2 Schematic flowsheet for high temperature ammonia–air leaching of Ni–Co–S intermediate precipitates. Nickel loss is minimized by starting with low-iron feed, which minimizes production of iron precipitate. All processes are continuous. There are eight autoclaves in total. Production of the feed is described in Chapters 10–12. Hydrogen reduction of nickel and cobalt is described in Chapter 27.

TABLE 24.2 Compositions of Fort Saskatchewan Air–Ammonia Leaching Process Inputs and Outputs*

	Composition								
Solid or solution	Ni	Co	Cu	Zn	Fe	S	NH_3	$(NH_4)_2SO_4$	SiO_2
Ni–Co sulfide precipitate from Moa Bay laterite leach, %	55	5	0.2	0.9	0.8	35			
Leach solution, g/L	85	11	0.5	3			130	180	
Solid unleached residue and precipitated $Fe_2O_3 \cdot H_2O$, %	4	0.5			25				20

Cordingley & Krentz, 2005; Kerfoot & Cordingley, 1997.

$$\underset{\text{in feed slurry}}{2CoS(s)} + \underset{\text{in air}}{4.5O_2(g)} + \underset{\substack{\text{in 72\% aqueous}\\ NH_3\text{ solution}}}{10NH_3(\ell)} + \underset{\text{in input solution}}{(NH_4)_2SO_4(aq)} \xrightarrow{\sim 105-120^\circ C} \underset{\substack{\text{cobaltic hexammine}\\ \text{sulphate solution}}}{[Co(NH_3)_6]_2(SO_4)_3(aq)} + H_2O(\ell) \quad (24.8)$$

There are two objectives for the leaching section: (i) to maximize the leaching efficiency of nickel and cobalt; and, (ii) to maximize the oxidation and complexation of cobalt in solution as cobaltic hexamine, $Co(NH_3)_6^{3+}$.

The leaching efficiencies for nickel and cobalt are greater than 99% and 97%, respectively (Kofluk & Freeman, 2006). The proportion of the total cobalt present as cobaltic hexamine is about 92%.

Iron sulfide, FeS, dissolves in a similar manner to cobalt sulfide. The ferric ammine complex is, however, unstable and rapidly decomposes to form a solid $Fe_2O_3 \cdot H_2O$ precipitate. As a result, iron is removed from the solution. Since copper and zinc form stable ammine complexes, they are removed later by precipitation as sulfides with hydrogen sulfide.

24.2.3. Leaching Products

After solid/liquid separation (Cordingley & Krentz, 2005), the two products of leaching are:

(a) the unleached solids and precipitated $Fe_2O_3 \cdot H_2O(s)$, which are then washed and discarded in a tailings pond; and,
(b) the solution containing nickel and cobalt as hexamine sulfates.

The composition of the solution is given in Table 24.3. This solution is transferred to nickel–cobalt separation, where cobalt is removed by the

TABLE 24.3 Operational Details of the Fort Saskatchewan Air–Ammonia Leaching Process*

Feed	**Ni–Co sulfide precipitate from Moa Bay, Cuba**
Rate, tonnes/day	up to 180
Particle size	80% <70 μm
Composition	See Table 24.2
Autoclaves	
Number	8
Installation date	1954
Diameter × length, m	3.5 × 13
Working volume, m^3	75
Construction material	Mild steel with 316 stainless steel interior lining
Compartments, each autoclave	4
Connection between compartments	Adjustable weirs
Agitators	One shaft per compartment two impellers per shaft
Air and ammonia solution injection	Below stirrers, where needed to form hexammines
Operating temperature, °C	105–120
Operating pressure, bar	7.5–9
O_2 utilization, %	~55
Extractions (Kofluck & Freeman, 2006)	
Ni%	>99
Co%	98
Solution concentrations	
Ni, g/L	80–85
Co, g/L	10–13
Cu, g/L	0, 5
NH_3,g/L	130
$(NH_4)_2SO_4$, g/L	180

Cordingley & Krentz, 2005; Kerfoot & Cordingley, 1997.

formation of cobaltic hexamine salt, and finally to hydrogen reduction, where high-purity nickel powder is produced. The cobalt hexamine salt is dissolved and then processed by hydrogen reduction to high-purity cobalt powder.

24.3. LEACHING BY SULFURIC ACID SOLUTIONS USING OXYGEN

The following sulfide intermediates and mattes are leached in sulfuric acid using oxygen:

(a) flash furnace and electric furnace mattes at Harjavalta, Finland (Latva-Kokko, 2006);

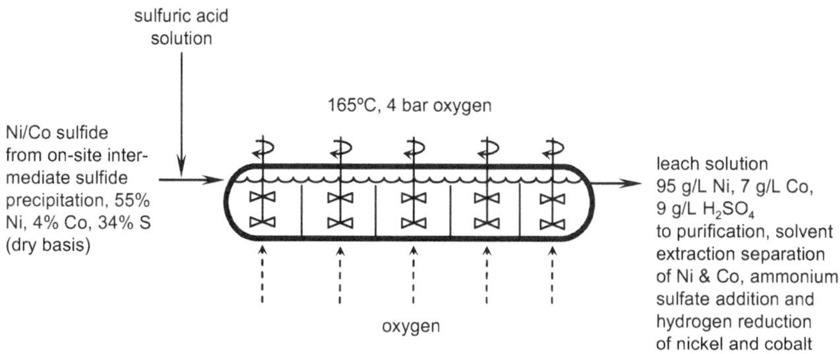

FIGURE 24.3 Sulfuric acid–oxygen leaching of mixed nickel–cobalt sulfide precipitate at Ambatovy, Madagascar (Collins, Buban, Holloway, Masters, & Raudsepp, 2009). The feed is a mixed sulfide precipitate obtained after the leaching of laterite ore (Chapters 10 and 11).

(b) converter mattes at most platinum-group metal refineries; and,
(c) mixed nickel–cobalt sulfide precipitates from several laterite leaching plants (Murrin Murrin, Australia, and Ambatovy, Madagascar).

A schematic diagram of an autoclave for the pressure leaching of sulfides in sulfuric acid using oxygen gas is shown in Figure 24.3.

24.3.1. Leaching of Mixed Nickel and Cobalt Sulfides

The main reactions for leaching of sulfide intermediates arising from the processing of laterites in sulfuric acid using oxygen are as follows:

$$\underset{\substack{\text{in intermediate}\\\text{precipitate}}}{NiS(s)} + \underset{\text{in injected oxygen}}{2O_2(g)} \xrightarrow{\substack{\text{in aqueous sulfuric}\\\text{acid solution}}} \underset{\text{sulfate solution}}{NiSO_4(aq)} \tag{24.9}$$

$$\underset{\substack{\text{in intermediate}\\\text{precipitate}}}{CoS(s)} + \underset{\text{in injected oxygen}}{2O_2(g)} \xrightarrow{\substack{\text{in aqueous sulfuric}\\\text{acid solution}}} \underset{\text{sulfate solution}}{CoSO_4(aq)} \tag{24.10}$$

The leaching efficiencies of nickel and cobalt are 98% and 99%, respectively. The leaching operation is continuous. The composition of the leaching solution typically contains 95 g/L Ni, 7 g/L Co, 0.3 g/L Fe and 9 g/L H_2SO_4.

Iron and other impurities are removed by neutralization with aqueous ammonia solution and sulfidation. Separate solutions containing nickel and cobalt are then produced by solvent extraction. High-purity nickel and cobalt are produced from these solutions by hydrogen reduction.

24.3.2. Matte Leaching

The leaching of matte is more complicated than the leaching of sulfide precipitate because the matte usually contains considerable amounts of copper. Details are provided by Bryson, Hofirek, Collins, Stiksma, and Berezowsky (2008), Corbett (2009), Latva-Kokko (2006), and Makinen, Fagerlund, Anjala, and Rosenback (2005).

This description is based mainly on the Harjavalta process. However, nickel–copper matte is also leached in the platinum industry. These processes are described in Chapter 36.

The flowsheet of the Harjavalta process was shown in Figure 23.5.

The Harjavalta refinery treats two mattes: the flash furnace matte and the electric furnace matte. The furnace matte contains Ni_3S_2, Cu_2S and an Ni–Cu alloy. Leaching of the flash furnace matte occurs in four stages.

In the copper removal stage, matte reacts with acid and oxygen:

$$\underset{\text{in feed matte}}{Ni_3S_2(s)} + H_2SO_4(aq) + \underset{\substack{\text{gas supplied}\\ \text{to reactors}}}{0.5O_2(g)} \xrightarrow{85^\circ C} \underset{\substack{\text{purified solution}\\ \text{to cobalt SX}}}{NiSO_4(aq)} + \underset{\substack{\text{undissolved solids to}\\ \text{nickel atmospheric leaching}}}{2NiS(s)} + H_2O(\ell) \quad (24.11)$$

The consumption of acid results in an increase in pH, so that copper precipitates:

$$\underset{\substack{\text{in solution}\\ \text{from leaching}}}{3CuSO_4(s)} + 4H_2O(\ell) \xrightarrow{85^\circ C} \underset{\text{to nickel atmospheric leaching}}{CuSO_4 \cdot 2Cu(OH)_2(s)} + 2H_2SO_4(aq) \quad (24.12)$$

In nickel atmospheric leaching, the remaining portion of the Ni_3S_2 dissolves by the reaction:

$$\underset{\text{in feed matte}}{Ni_3S_2(s)} + H_2SO_4(aq) + \underset{\substack{\text{gas supplied}\\ \text{to reactors}}}{0.5O_2(g)} \xrightarrow{85^\circ C} \underset{\substack{\text{purified solution}\\ \text{to copper removal}}}{NiSO_4(aq)} + \underset{\substack{\text{undissolved solids to}\\ \text{nickel pressure leaching}}}{2NiS(s)} + H_2O(\ell) \quad (24.13)$$

In the pressure leaching stage, nickel sulfide, which is formed in copper removal and nickel atmospheric leaching, reacts with copper in solution:

$$\underset{\text{from nickel atmospheric leaching}}{6NiS(s)} + \underset{\text{from copper pressure leaching}}{9CuSO_4(aq)} + 4H_2O(\ell) \xrightarrow{150°C} \underset{\text{to copper pressure leaching}}{Cu_9S_5(s)}$$

$$+ \underset{\text{to nickel atmospheric leaching}}{6NiSO_4(aq)} + 4H_2SO_4(aq) \quad (24.14)$$

The final stage of the leaching train is copper pressure leaching, which provides copper in solution for reaction in nickel pressure leaching by the following reactions:

$$\underset{\text{from nickel pressure leaching}}{Cu_{1.8}S(s)} + 0.8H_2SO_4(aq) + \underset{\text{oxygen injected into autoclave}}{0.4O_2(g)} \xrightarrow{150°C} \underset{\text{to nickel pressure leaching}}{CuS(s)}$$

$$+ 0.8CuSO_4(aq) + 0.8H_2O(\ell) \quad (24.15)$$

$$\underset{\text{from nickel pressure leaching}}{CuS(s)} + \underset{\text{oxygen injected into autoclave}}{2O_2(g)} \xrightarrow{150°C} \underset{\text{to nickel pressure leaching}}{CuSO_4(aq)} \quad (24.16)$$

Copper is not completely dissolved in copper pressure leaching so the residue provides an exit for copper from the leaching circuit. At other operations, such as Anglo American Platinum (see Chapter 36), copper is completely dissolved but only a portion of the exit stream is returned to the nickel pressure leaching. The chemistry of matte leaching in sulfate solutions in nickel–copper refineries is also described in Chapter 36.

24.4. APPRAISAL

Pressure leaching is a well-established, efficient first step in hydrometallurgical production of nickel and cobalt.

Chlorine, air–ammonia and oxygen–sulfuric acid leaching are comparable in their efficiencies. Existing leaching plants of each type will continue to be used.

Oxygen–sulfuric acid leaching seems to have the edge for future projects, because of cheaper reagents and materials of construction and the avoidance of corrosive gases.

24.5. SUMMARY

The leaching of sulfide precipitates and mattes in chlorine, air–ammonia and oxygen–sulfuric acid was described in this chapter. All of these chemistries successfully dissolve nickel and cobalt, precipitate and reject nickel- and cobalt-lean iron compounds, principally hematite and iron hydroxides, and produce leach solutions that are suitable for producing high-purity nickel and cobalt.

Some also make metal-rich residue (from matte), suitable for producing copper, silver, gold and platinum-group metals.

New intermediate sulfide leaching plants are likely to be based on oxygen–sulfuric acid leaching (Masters, Barta, Stiksma, & Collins, 2006). The reagents are cheap, toxic gases are avoided and downstream iron-removal and metal recovery methods are well established.

REFERENCES

Bryson, L. J., Hofirek, Z., Collins, M. J., Stiksma, J., & Berezowsky, R. M. (2008). New matte leaching developments at Anglo Platinum's base metal refinery. In C. A. Young, C. G. Anderson & Y. Choi (Eds.), *Hydrometallurgy 2008* (pp. 570–589). SME.

Budac, J. J., Kofluk, R., & Belton, D. (2009). Reductive leach process for improved recovery of nickel and cobalt in the Sherritt hexamine leach process. In J. J. Budac, R. Fraser & I. Mihaylov, et al. (Eds.), *Hydrometallurgy of nickel and cobalt 2009* (pp. 77–85). CIM.

Collins, M. J., Buban, K. R., Holloway, P. C., Masters, I. M., & Raudsepp, R. (2009). Ambatovy laterite ore preparation plant and high pressure acid leach pilot plant operation. In J. J. Budac, R. Fraser & I. Mihaylov, et al. (Eds.), *Hydrometallurgy of Nickel and Cobalt 2009* (pp. 499–510). CIM.

Corbett, A. (2009). *An overview of the base metal removal plant at Northam Platinum Limited. In Southern African hydrometallurgy conference*. SAIMM.

Cordingley, P. D., & Krentz, R. (2005). Corefco refinery – Review of operations. In J. Donald & R. Schonewille (Eds.), *Nickel and cobalt 2005 challenges in extraction and production* (pp. 407–425). CIM.

Dotterud, O. M., Peek, E. M. L., Stenstad, O., & Ramsdal, P. O. (2009). Iron control and tailings disposal in the Xstrata chlorine leach process. In J. J. Budac, R. Fraser & I. Mihaylov, et al. (Eds.), *Hydrometallurgy of nickel and cobalt 2009* (pp. 321–333). CIM.

Eramet. (2010). *Le Havre-Sandouville Refinery* [Company brochure].

Higuchi, H., Ozaki, Y., Sugiura, T., & Kemori, N. (2006). Iron removal from the MCLE [matte chlorine leach electrowin] circuit. In J. E. Dutrizac & P. A. Riveros (Eds.), *Iron control technologies* (pp. 403–413). CIM.

Kerfoot, D. G. E., & Cordingley, P. D. (1997). The acid pressure leach process for nickel and cobalt laterite. Part II: Review of operations at Fort Saskatchewan. In W. C. Cooper & I. Mihaylov (Eds.), *Nickel/cobalt 97: Vol. I. Hydrometallurgy and refining of nickel and cobalt* (pp. 355–369). CIM.

Kofluk, R. P., & Freeman, G. K. W. (2006). Iron control in the Moa Bay Operation. In J. E. Dutrizac & P. A. Riveros (Eds.), *Iron control technologies* (pp. 573–589). CIM.

Latva-Kokko, M. J. (2006). Iron removal as part of the nickel matte leaching process. In J. E. Dutrizac & P. A. Riveros (Eds.), *Iron control technologies* (pp. 391–401). CIM.

Makinen, T., Fagerlund, K., Anjala, Y., & Rosenback, L. (2005). Outokumpu's technologies for efficient and environmentally sound nickel production. In J. Donald & R. Schonewille (Eds.), *Nickel and cobalt 2005: Challenges in extraction and production* (pp. 71–89). CIM.

Masters, L. M., Barta, L. A., Stiksma, J., & Collins, M. J. (2006). Iron control in processes developed at Dynatec. In J. E. Dutrizac & P. A. Riveros (Eds.), *Iron control technologies* (pp. 653–671). CIM.

Stensholt, E. O., Dotterud, O. M., Henriksen, E. E., et al. (2001). Development and plant practice of the Falconbridge chlorine leach process. *CIM Bulletin, 94*, 101–144.

Stensholt, E. O., Zachariasen, H., & Lund, J. H. (1986). Falconbridge chlorine leach process. *Transactions of the Institution of Mining and Metallurgy, Section C, 5*, C10–C16.

Stensholt, E. O., Zachariansen, H., Lund, J. H., & Thornhill, P. G. (1988). Recent improvements in the Falconbridge Nikkelverk nickel refinery. In G. P. Tyroler & C. A. Landolt (Eds.), *Extractive metallurgy of nickel and cobalt* (pp. 403–413). TMS.

Sumitomo. (2007). Outline of MCLE [matte chlorine leach electrowin] process. Notes received during visit to MCLE plant, May 29, 2007.

SUGGESTED READING

Budac, J. J., Fraser, R., & Mihaylov, I. et al. (Eds.), (2009). *Hydrometallurgy of nickel and cobalt 2009*. CIM.

Cooper, W. C., & Mihaylov, I. (Eds.), (1997). *Nickel/cobalt 97: Vol. I. Hydrometallurgy and refining of nickel and cobalt*. CIM.

Donald, J., & Schonewille, R. (Eds.), (2005). *Nickel and cobalt 2005 challenges in extraction and production*. CIM.

Egedahl, R. D., & Collins, M. J. (2009). Vital status of Sherritt nickel refinery workers (1954-2003). In J. J. Budac, R. Fraser & I. Mihaylov, et al. (Eds.), *Hydrometallurgy of nickel and cobalt 2009* (pp. 689–699). CIM.

Roux, J. O., Du Toit, M., & Shklaz, D. (2009). Novel redesign of a pressure leach autoclave by a South African platinum producer. In Retrieved from *Proceedings of The Fifth Base Metals Conference, SAIMM*. http://www.basemetals.org.za/Kasane2009/Papers/index.htm.

Chapter 25

Separation of Nickel and Cobalt by Solvent Extraction

All nickel ores contain cobalt, to a greater or lesser extent. Thus, all of the leach solutions contain nickel and cobalt. Industrial examples of leach solutions are given in Table 25.1.

High-purity nickel and cobalt are made from these solutions by electrowinning or hydrogen reduction (described in Chapters 26 and 27). Although there is a degree of separation of nickel and cobalt in hydrogen reduction, this is not complete. In addition, the standard reduction potentials for depositing nickel and cobalt from aqueous solutions are very close, –0.25 V

TABLE 25.1 Compositions of Solutions from the Leaching of Sulfide Intermediates

Company and Location	Type of Material Leached	Type of Leaching Chemistry	Concentration in Leach Solution, g/L	
			Ni	Co
Norilsk, Harjavalta, Finland	Flash furnace matte	Oxygen–sulfuric acid	120	1
Sherritt, Ambatovy, Madagascar	Mixed nickel-cobalt sulfide	Oxygen-sulfuric acid	90	7
Xstrata Nikkelverk, Kristiansand, Norway	Nickel–copper matte	Chlorine–hydrochloric acid	220	11
Sherritt, Fort Saskatchewan, Canada	Mixed nickel–cobalt sulfide	Air–ammonia	85[a]	12[a]

[a] *To fractional crystallization, not solvent extraction (Cordingley & Krentz, 2005).*

Extractive Metallurgy of Nickel, Cobalt and Platinum-Group Metals. DOI: 10.1016/B978-0-08-096809-4.10025-5

and –0.28 V, respectively. This means that cobalt will deposit during nickel electrowinning.

As a result, nickel and cobalt in solution must be separated into individual aqueous solutions containing either nickel or cobalt. There are three techniques that are currently used industrially for achieving this objective: (i) precipitation of the cobaltic hexamine salt; (ii) precipitation of cobaltic hydroxide using nickelic hydroxide; and, (iii) solvent extraction. Solvent extraction is currently favored, although a universal set of reagents has not been established.

25.1. CHAPTER OBJECTIVES

Solvent extraction of nickel and cobalt is practiced in three different solution environments: (i) chloride; (ii) sulfate; and, (iii) ammoniacal solutions. Generally, cobalt is extracted selectively from the solution in chloride and sulfate chemistry, leaving nickel in solution. In ammoniacal solutions, however, nickel is extracted selectively, leaving cobalt in solution. Some plants extract both nickel and cobalt and then separate them (for example, Niihama).

The objectives of this chapter are to describe the principles and the industrial practice of solvent extraction of solutions containing nickel and cobalt. The organic extractants that are used to separate cobalt from nickel, the reasons for their selectivity and how these extractants are applied industrially are also described.

25.2. PRINCIPLES OF SOLVENT EXTRACTION

Solvent extraction occurs in at least two steps: extraction and stripping.

The extraction step involves the following sub-processes:

(a) mixing the leaching solution, which contains both nickel and cobalt, with a liquid organic that contains an extractant that, under particular operating conditions, will extract cobalt (but not nickel) from solution into the organic liquid; and,
(b) separating the organic phase, which is loaded with cobalt, from the aqueous solution, which is depleted in cobalt, by gravity.

The specific gravity of the organic phase is about 0.85, whereas the specific gravity of the aqueous solution is about 1.1. As a result, the organic phase floats on top of the aqueous phase, which allows the two phases to be separated in a settler. The aqueous solution from the extraction stage is known as the '*raffinate*', and the organic from the extraction stage is known as the '*loaded organic*'.

The raffinate, which contains mainly nickel, is sent for further processing either by nickel electrowinning or hydrogen reduction.

The loaded organic is sent to the stripping section, which involves the following sub-processes:

(a) mixing the loaded organic phase with an aqueous solution that is low in cobalt – causing cobalt to 'strip' from the loaded organic phase into the aqueous phase; and,
(b) separating the organic and aqueous phases by gravity.

The aqueous solution leaving stripping, which is enriched in cobalt, is transferred to cobalt electrowinning or hydrogen reduction, while the organic phase is recycled to the mixing tank of the extraction step.

The solvent extraction of cobalt from nickel is shown in Figure 25.1. The equipment that is generally used is a mixer-settler, shown in Figure 25.2. The contacting of the organic phase with the aqueous phase is done in the mixer and the separation of the phases is done in the settler. There are two mixers shown in Figure 25.2. In the first mixer, the agitator creates a pumping action with sufficient head to move the fluids through the mixer-settler. Also shown in the first mixer is a false bottom, which creates a region for suction of the pumping impeller that is common to both the aqueous and organic phases. The flow of the emulsion of organic and aqueous phases to the settler must be distributed evenly across the width of the settler to ensure optimal use of the settling area. Picket fences are often placed in the settler to distribute the flow evenly across the width of the settler.

In practice, extraction and stripping are each done in several mixer-settlers connected in a counter-current manner. Inadvertently, extracted nickel may be *scrubbed* from the cobalt-loaded organic. Physically entrained droplets of aqueous phase may be *washed* from the organic phase. Entrained organic in the aqueous phase may be settled in aftersettlers.

25.3. CHLORIDE SOLVENT EXTRACTION

Solvent extraction in chloride solutions is applied at the following operations (Dotterud, Peek, Stenstad, & Ramsdal *et al.*, 2009; Eramet, 2010; Higuchi, Ozaki, Sugiura, & Kemori, 2006): (i) Xstrata Nikkelverk, Kristiansand, Norway; (ii) Sumitomo Metal Mining, Niihama, Japan; and, (iii) Eramet, Le-Havre-Sandouville, France.

Chloride solvent extraction accounts for about 150 000 tonnes per year of nickel and 10 000 tonnes per year of cobalt.

This section is based on the Nikkelverk plant, which is described in several publications (Dotterud *et al.*, 2009; Stensholt, Zachariasen, & Lund, 1986; Stensholt, Zachariansen, Lund, & Thornhill, 1988; Stensholt *et al.*, 2001).

In the Nikkelverk operation, cobalt is extracted from solution using tri-isooctyl amine. The extractant is available commercially as Alamine 308 (Cognis, 2010) and several other brands. The concentration of the tri-isooctyl amine is 17% by volume in an aromatic diluent.

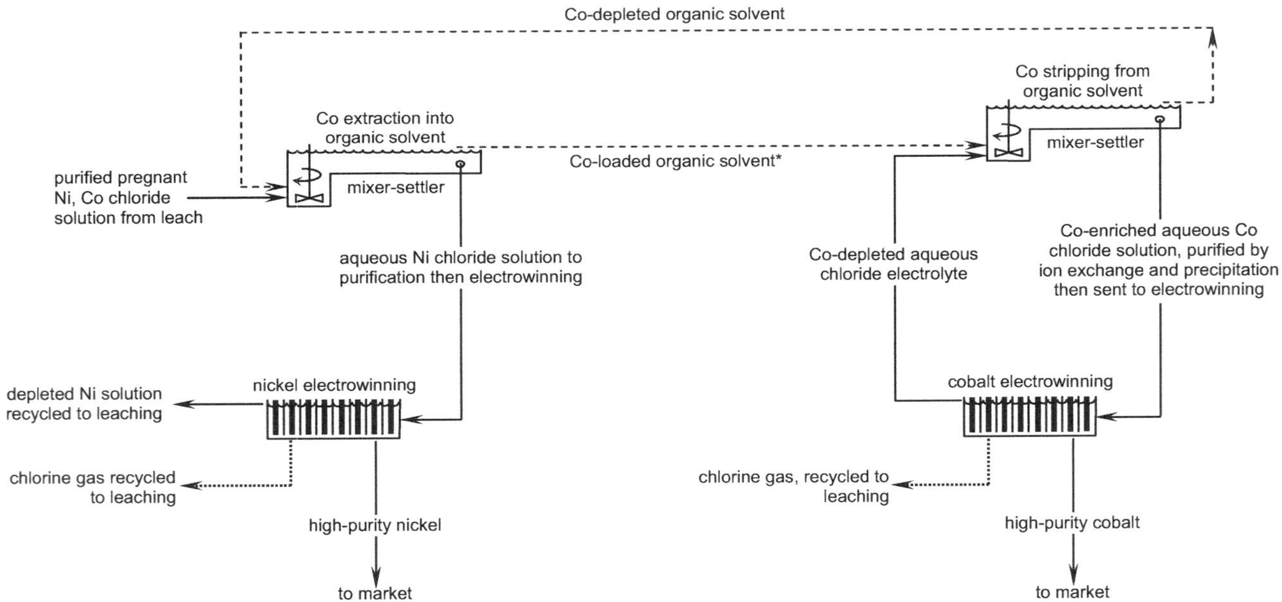

FIGURE 25.1 Simplified solvent extraction of cobalt/nickel separation flowsheet including metal electrowinning. Cobalt extraction and stripping are both done in several mixer-settlers arranged in counter-current flow. This flowsheet is for chloride leach pregnant leach solution and chloride electrowinning aqueous electrolytes. Solvent extraction in sulfate solutions is similar but (i) the aqueous solutions are both sulfates; and, (ii) the oxygen produced during electrowinning is not recycled to leach. *Loaded organic is washed or scrubbed to remove inadvertently entrained or co-extracted nickel.

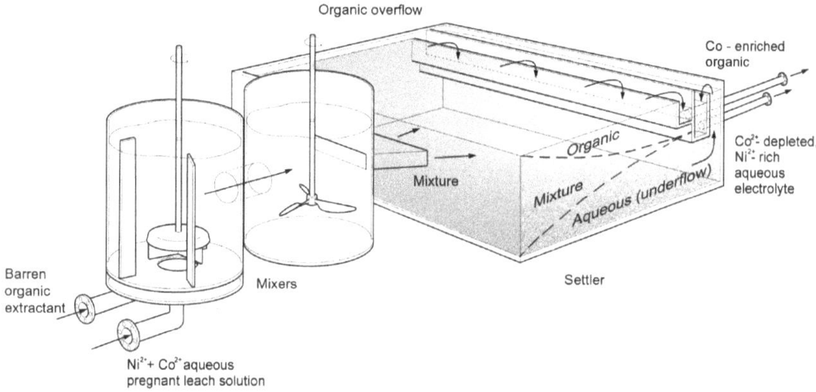

FIGURE 25.2 A solvent extraction mixer-settler. The two mixing compartments, the large settler and the organic overflow/aqueous underflow system are notable. Flow is distributed evenly in the settler by picket fences (not shown). The feed is leach solution containing nickel and cobalt. The products are a cobalt-rich, nickel-lean organic phase and a nickel-rich, cobalt-lean aqueous phase. The aqueous phase goes to nickel production. The organic phase goes to washing then stripping of the cobalt into aqueous electrolyte then cobalt production. Spiral and pulse mixers are also used (Mihaylov, Grinbaum, Ilan, & Efraim, 2009; Paatero *et al.*, 2009).

25.3.1. Cobalt Extraction

Tri-isooctyl amine extracts cobalt from chloride leaching solution by the following reaction (Sole & Cole, 2002):

$$\underset{\substack{\text{organic extractant}\\ \text{17\% by volume}\\ \text{in aromatic diluent}}}{2R_3N(\ell)} + \underset{\substack{\text{strong chloride aqueous}\\ \text{pregnant leach solution}\\ \text{270 g/L Cl, 11 g/L Co,}\\ \text{220 g/L Ni}}}{CoCl_2(aq)} + 2HCl(aq) \xrightarrow{30^\circ C} \underset{\substack{\text{Co-loaded organic}\\ \text{extractant in aromatic}\\ \text{diluent, } \sim 6 \text{ g/L Co}}}{(R_3NH)_2CoCl_4(\ell)} \tag{25.1}$$

where R_3N represents the tri-isooctyl amine (R = alkyl group, N = nitrogen).

At high concentrations of chloride ions in solution, cobalt forms tetrahedral $CoCl_4^{2-}$ ions, which readily dissolve in protonated amines (Sole, 2008a). In contrast, nickel forms neutral $NiCl_2$ complexes that are not extracted by amines.

Loading of the organic extractant with cobalt is favored by a high concentration of hydrochloric acid in the aqueous phase.

The Nikkelverk plant uses four mixer-settlers in a counter-current arrangement to extract cobalt from the leach solution. This multistage counter-current arrangement ensures efficient cobalt extraction.

After the organic loading step, the organic and aqueous phases are separated by gravity. The raffinate, depleted in cobalt, typically has the following concentration: 220 g/L Ni, 0.1 g/L Co and 2 g/L HCl. This solution is purified, diluted and sent to nickel electrowinning, while the organic phase, enriched in cobalt, is sent to washing.

25.3.2. Washing to Remove Nickel from the Organic Phase

A small amount of nickel is entrained as nickel-bearing solution in the organic phase. This entrained nickel must be removed from the cobalt-rich organic to prevent contamination of cobalt. The nickel is removed by washing the organic with water that has been slightly acidified with hydrochloric acid in small mixer-settlers. The nickel is recovered by recycling the wash water to the extraction circuit.

25.3.3. Stripping Cobalt

The washed organic phase, which is loaded with cobalt, is pumped to the strip mixer-settlers, where it is contacted with an aqueous solution that has a low concentration of chloride ions, usually the return solution from electrowinning. It is the difference in the concentration of chloride that determines whether the cobalt is loaded or stripped.

The stripping reaction is the reverse of the extraction reaction (Equation 25.1). The stripping reaction is given as follows:

$$\underset{\substack{\text{Co-loaded organic}\\ \text{extractant in aromatic}\\ \text{diluent, } \sim 6\text{ g/L Co}}}{(R_3NH)_2CoCl_4(\ell)} \xrightarrow{30^\circ C} \underset{\substack{\text{Co-depleted}\\ \text{organic extractant}}}{2R_3N(\ell)} + \underset{\substack{Co^{2+}\text{-enriched (90 g/L Co),}\\ \text{low } Ni^{2+}\ (\sim 0.01\text{ g/L Ni})\\ \text{and low HCl aqueous solution}}}{CoCl_2(aq) + 2HCl(aq)} \tag{25.2}$$

The organic phase is depleted in cobalt and the aqueous phase is enriched in cobalt. As mentioned before, these two phases are separated by gravity in the mixer-settler. Nikkelverk uses two mixer-settlers for stripping in a counter-current arrangement. The loaded organic is stripped with a mixture of cobalt solution and water. The depleted organic phase is returned to the extraction stage, while the aqueous solution is purified then sent to cobalt electrowinning.

25.3.4. Summary

High concentrations of chloride ions result in the extraction of cobalt from the aqueous phase while low concentrations of chloride ions result in the stripping of cobalt from the organic phase.

Consequently, Xstrata Nikkelverk produces the following two solutions from the leaching solution that contains 220 g/L Ni and 11 g/L Co (Stensholt *et al.*, 2001): (i) nickel-rich raffinate solution containing about 60 g/L Ni and 0.03 g/L Co; and, (ii) cobalt-rich solution containing about 0.01 g/L Ni and 50 g/L Co.

25.4. SOLVENT EXTRACTION IN SULFATE SOLUTIONS

Solvent extraction of cobalt from sulfate solutions is practiced at the following operations: (i) Minara Resources, Murrin Murrin, Australia; (ii) Norilsk Nickel, Harjavalta, Finland; and, (iii) Sherritt, Toamasina, Madagascar.

Solvent extraction in sulfate accounts for about 140 000 tonnes per year of nickel and 15 000 tonnes per year of cobalt.

The extractant used is bis(2,4,4-trimethylpentyl)phosphonic acid, which is commercially available as CYANEX 272, LIX 272 and Ionquest 290 (Sole, 2008b). The concentration of the extractant is typically 10%– 25% by volume in an aliphatic diluent.

The aqueous solution is purified to remove zinc and copper impurities prior to solvent extraction so as to prevent their co-extraction when using CYANEX 272 or LIX 272 (Rickelton & Nucciarone, 1997).

The extraction reaction is given as follows:

$$\underset{\substack{\text{organic extractant}\\ 10\%-25\%\text{ by volume}}}{2H_2A_2(\ell)} + \underset{\text{purified solution}}{CoSO_4(aq)} \xrightarrow{50^\circ C} \underset{\text{Co-loaded organic}}{CoA_2\cdot H_2A_2(\ell)} + \underset{\substack{\text{low-}H_2SO_4\\ (\text{pH } 3-5)\text{ solution}}}{H_2SO_4(aq)} \tag{25.3}$$

where A represents the 2,4,4-trimethylpentyl phosphonic conjugate base.

The stripping reaction is the reverse of the extraction reaction:

$$\underset{\substack{\text{Co-loaded organic}\\ \text{in aliphatic diluent}}}{CoA_2\cdot H_2A_2(\ell)} + \underset{\substack{\text{strong } H_2SO_4\\ \text{solution}}}{H_2SO_4(aq)} \xrightarrow{50^\circ C} \underset{\text{Co-depleted extractant}}{2H_2A_2(\ell)} + \underset{\text{Co-enriched solution}}{CoSO_4(aq)} \tag{25.4}$$

Extraction is favored at the low concentrations of sulfuric acid, about 0.001 g/L, in the aqueous solution to the extraction stage, while stripping is favored by high concentrations of sulfuric acid, about 180 g/L, in the aqueous solution to the strip.

Extraction and stripping are each carried out in several mixer-settlers. Murrin Murrin, for example, uses four stages of extraction and three stages of stripping.

The maximum separation factor of CYANEX 272 [bis(2,4,4-trimethylpentyl) phosphinic acid] for Co:Ni is 6700:1 (Cytec, 2010). This means that up to 6700 times more cobalt than nickel will be extracted into CYANEX 272 from an aqueous sulfate solution containing equal amounts of nickel and cobalt (Rickelton & Nucciarone, 1997).

Sole (2008b) indicates that the preferential extraction of cobalt is due to the simple structure and hydrophobicity of the organic complex of cobalt.

25.5. SOLVENT EXTRACTION IN AMMONIACAL SOLUTIONS

Solvent extraction of cobalt from ammoniacal solutions is practiced at Queensland Nickel, Australia. This refinery at Yabulu, near Townsville, Australia, uses LIX-84-I to extract nickel from an ammoniacal solution (MacKenzie & Virnig, 2004; Sole & Cole, 2002). The reaction is:

$$\underset{\substack{\text{leaching solution}\\ \text{9 g/L Ni,}\\ \text{0.3 g/L Co}}}{Ni(NH_3)_4^{2+}(aq)} + \underset{\substack{\text{ketoxime}\\ \text{extractant,}\\ \text{30\% by volume}}}{2RH(\ell)} \rightarrow \underset{\text{loaded organic}}{R_2Ni(\ell)} + \underset{\substack{\text{Ni-depleted}\\ \text{raffinate}}}{2NH_3(aq) + 2NH_4^+(aq)} \tag{25.5}$$

Solvent extraction from ammoniacal solutions is different from chloride and sulfate solutions because nickel is loaded preferentially, not cobalt. However, in order for this extraction to be effective, cobalt must be present as Co^{3+} prior to nickel solvent extraction.

The extraction occurs in three stages of counter-current mixer-settlers with advance organic:aqueous ratio of 1.5:1. The loaded organic is stripped with a solution of 270 g/L ammonia in three stages using an organic:aqueous ratio of 15:1. The strip solution is contains 75–80 g/L Ni, with low levels of impurities.

The cobalt is precipitated from the nickel extraction raffinate as a sulfide. This sulfide is redissolved. The impurities, mainly iron and zinc, are removed from the solution using CYANEX 272. The cobalt and nickel are extracted from this solution using D2EPHA and stripped into a solution containing ammonia and ammonium carbonate. The cobalt is oxidized to Co^{3+}, and nickel, manganese and copper are removed from the strip solution using LIX 84-I. The raffinate from this extraction is the final purified cobalt solution.

There are two disadvantages to the process chemistry used at Queensland Nickel. The first disadvantage is that the Co^{3+} oxidizes the extractant to form a ketone, which means that the costs of replenishing extractant are higher than at other facilities. The second disadvantage is that the high ammonia concentrations cause the extractant to form an imine. The extractant is regenerated using hydroxylamine at Queensland Nickel.

25.6. SOLVENT EXTRACTION IN SULFATE AND CHLORIDE SOLUTIONS

Vale has implemented a novel process at Goro in New Caledonia (Bacon *et al.*, 2005; Sole & Cole, 2002). Laterite ore is leached at 270°C and 56 bar. The residence time in the leaching autoclave is about 30 min. The discharge slurry is washed in the counter-current decantation circuit. The solution from leaching is partially neutralized to remove aluminum, chromium, silicon, copper and iron. Copper is further removed using ion exchange. This purified solution is transferred to solvent extraction.

Nickel and cobalt are extracted from the purified solution, along with zinc, using bis(2,4,4-trimethylpentyl)dithiophosphinic acid, available as CYANEX 301. Calcium, magnesium and manganese are not extracted. The loaded organic is stripped from solution using concentrated hydrochloric acid. Pulsed columns are used. The stripped solution contains nickel, cobalt and zinc. Zinc is removed from solution using ion exchange, and the purified solution pumped to cobalt solvent extraction.

Cobalt is separated from the nickel using Alamine 336. Cobalt is stripped from the loaded organic using a dilute solution. The raffinate contains mainly nickel chloride and the strip solution contains cobalt chloride.

The nickel chloride solution is treated by pyrohydrolysis to form nickel oxide powder and to regenerate the hydrochloric acid needed for stripping the nickel and cobalt from the CYANEX 301.

The cobalt is recovered from solution by precipitation with sodium carbonate to product cobalt carbonate.

25.7. DILUENTS

The organic extractants are always dissolved in low viscosity, low density organic diluents. This combination gives organic liquids that are (i) readily pumped and mixed with aqueous liquids; (ii) immiscible with aqueous solutions; and, (iii) easily separated from aqueous liquids by gravity.

Typical specific gravities for the organic phase are about 0.85, whereas those for the aqueous solution are about 1.1. Typical viscosities of the organic phase are about 3 cP (0.003 kg/ms).

The diluents used for tri-isooctyl amine (that is, Alamine 308) are typically aromatics (Stensholt *et al.*, 2001), which inhibit the formation of a second immiscible organic phase.

The diluents used for CYANEX 272 and LIX 272 are typically aliphatics, for example, Shell MSB 210. Aliphatics have the advantages of a low vapor pressure, high combustion flash point temperature and a high resistance to oxidation. Unlike amines, phosphonic acids tend not to form a second organic phase.

25.8. WASHING AND SCRUBBING THE ORGANIC

Washing is the physical removal of aqueous solution that is entrained in the organic phase. The Nikkelverk refinery at Kristiansand in Norway washes the loaded organic phase with water that is slightly acidified with hydrochloric acid in two small mixer-settlers. Washing removes entrained nickel-rich aqueous solution from the organic.

Scrubbing is the chemical removal of nickel that has co-extracted with the cobalt. Nickel can, for example, be scrubbed from the LIX 272 or CYANEX

272 organic phase with an aqueous solution containing 30 g/L Co at a pH of 3.7 and a temperature of 50°C (Cytec, 2010).

Both washing and scrubbing improve the purity of the cobalt product, such that it is greater than 99.9% Co.

25.9. IMPURITY REMOVAL

Solvent extraction does not remove impurities, for example, iron, arsenic, lead, manganese and calcium. The impurities must be removed either before or after solvent extraction.

The Nikkelverk refinery, for example:

(a) precipitates iron, arsenic and calcium with chlorine and nickel carbonate before solvent extraction;
(b) precipitates lead and manganese from nickel electrolyte after solvent extraction by adding recycle electrolyte, chlorine and nickel carbonate and removes entrained organic solvent with activated carbon; and,
(c) purifies product cobalt electrolyte after solvent extraction by precipitation of lead and manganese using chlorine and cobalt carbonate, removes zinc using ion exchange and removes entrained organic using activated carbon (Stensholt *et al.*, 2001).

25.10. APPRAISAL

Solvent extraction is an efficient way of separating aqueous nickel and cobalt solutions into:

(a) an nickel-rich, cobalt-lean solution, ready for high-purity nickel production; and,
(b) a cobalt-rich, nickel-lean solution, ready for high-purity cobalt production.

Impurities have to be eliminated before or after cobalt–nickel solvent extraction.

The success of solvent extraction has led to increased adoption of hydrometallurgy in the nickel and cobalt extraction industry.

As mentioned in Chapter 10, a new plant, at Goro in New Caledonia, uses solvent extraction to extract nickel and cobalt directly from the solutions of laterite leaching operations (Okita, Singhal, & Perraud, 2006). Unfortunately, the solvent extraction columns have failed mechanically, and the plant currently produces mixed hydroxide for refining elsewhere.

25.11. SUMMARY

Nickel ores contain cobalt, so solutions from leaching nickel mattes and sulfide precipitates also contain cobalt. Solvent extraction is used to extract cobalt from the leaching solution into an organic extractant, leaving a nickel-rich, cobalt-lean electrolyte. The cobalt is then stripped from the organic extractant to produce a cobalt-rich, nickel-lean aqueous solution.

Commercial extractants for the solvent extraction of nickel and cobalt are available from numerous suppliers.

REFERENCES

Bacon, W. G., Colton, D. F., Krause, E., Mihaylov, I. O., Singhai, A., & Duterque, J. P. (2005). *Development of the Goro nickel process. In ALTA 2005 nickel/cobalt forum.* ALTA Metallurgical Services.

Cognis. (2010). *Alamine 308 Technical brochure.* Cognis Corporation.

Cordingley, P. D., & Krentz, R. (2005). Corefco refinery – Review of operations. In J. Donald & R. Schonewille (Eds.), *Nickel and cobalt 2005 challenges in extraction and production* (pp. 407–425). CIM.

Cytec. (2010). *CYANEX 272 extractant, Technical brochure.* Cytec Corporation.

Dotterud, O. M., Peek, E. M. L., Stenstad, O., & Ramsdal, P. O. (2009). Iron control and tailings disposal in the Xstrata chlorine leach process. In J. J. Budac, R. Fraser, I. Mihaylov, V. G. Papangelakis & D. J. Robinson (Eds.), *Hydrometallurgy of nickel and cobalt 2009* (pp. 321–333). CIM.

Eramet. (2010). Le Havre-Sandouville Refinery. *Company brochure.*

Higuchi, H., Ozaki, Y., Sugiura, T., & Kemori, N. (2006). Iron removal from the MCLE [matte chlorine leach electrowin] circuit. In J. E. Dutrizac & P. A. Riveros (Eds.), *Iron control technologies* (pp. 403–413). CIM.

MacKenzie, J. M. W., & Virnig, M. J. (2004). Solvent extraction technology for the extraction of nickel using LIX 84-INS. An update and circuit comparisons. In W. P. Imrie & D. M. Lane (Eds.), *International laterite symposium* (pp. 457–475). TMS.

Mihaylov, I., Grinbaum, B., Ilan, Y., & Efraim, A. (2009). Opportunities for nickel-cobalt extraction and separation using CYANEX 301. In J. J. Budac, R. Fraser, I. Mihavylov, V. G. Papangelakis & D. J. Robinson (Eds.), *Hydrometallurgy of nickel and cobalt 2009* (pp. 383–391). CIM.

Okita, Y., Singhal, A., & Perraud, J.-J. (2006). Iron control in the Goro nickel process. In J. E. Dutrizac & P. A. Riveros (Eds.), *Iron control technologies* (pp. 635–651). CIM.

Paatero, E., Nyman, B., Laital, H., et al. (2009). Extraction of nickel with pre-neutralized organic acids using Outotec mixer settler technology. In J. J. Budac, R. Fraser, I. Mihavylov, V. G. Papangelakis & D. J. Robinson (Eds.), *Hydrometallurgy of nickel and cobalt 2009* (pp. 211–220). CIM.

Rickelton, W. A., & Nucciarone, D. (1997). The treatment of cobalt/nickel solutions using CYANEX extractants. In W. C. Cooper & I. Mihaylov (Eds.), *Nickel/Cobalt 97, Vol. I, Hydrometallurgy and refining of nickel and cobalt* (pp. 275–292). CIM.

Sole, K. (2008a). *Solvent extraction in base metal hydrometallurgy.* Johannesburg: Short course at Anglo Research. July 9, 2008.

Sole, K. (2008b). Solvent extraction in hydrometallurgical processing and purification of metals. In M. Aguilar & J. L. Cortina (Eds.), *Solvent extraction and liquid membranes, fundamentals and application in new materials* (pp. 141–200). CRC Press.

Sole, K. C., & Cole, P. M. (2002). Purification of nickel by solvent extraction. In Y. Marcus, A. K. SenGupta & J. A. Marinsky (Eds.), *Ion exchange and solvent extraction* (pp. 143–195). Marcel Dekker Inc.

Stensholt, E. O., Dotterud, O. M., Henriksen, E. E., et al. (2001). Development and plant practice of the Falconbridge chlorine leach process. *CIM Bulletin, 94*, 101–104.

Stensholt, E. O., Zachariasen, H., & Lund, J. H. (1986). Falconbridge chlorine leach process. *Trans. Inst. Min. Met. C, 5*, C10–C16.

Stensholt, E. O., Zachariansen, H., Lund, J. H., & Thornhill, P. G. (1988). Recent improvements in the Falconbridge Nikkelverk nickel refinery. In G. P. Tyroler & C. A. Landolt (Eds.), *Extractive metallurgy of nickel and cobalt* (pp. 403–413). TMS.

SUGGESTED READING

Flett, D. S. (2005). Solvent extraction in hydrometallurgy: The role of organophosphorus extractants. *Journal of Organometallic Chemistry, 690*, 2426–2438.

Makinen, T., Fagerlund, K., Anjala, Y., & Rosenback, L. (2005). Outokumpu's technologies for efficient and environmentally sound nickel production. In J. Donald & R. Schonewille (Eds.), *Nickel and cobalt 2005 challenges in extraction and production* (pp. 71–89). CIM.

Matsumoto, N., Matsumoto, S., Nakagawa, H., & Sugita, I. (2009). Solvent extraction technology for nickel and cobalt separation with crowding reaction. In J. J. Budac, R. Fraser, I. Mihaylov, V. G. Papangelakis & D. J. Robinson et al. (Eds.), *Hydrometallurgy of nickel and cobalt 2009* (pp. 283–293). CIM.

Paatero, E., Nyman, B., Laitala, H., et al. (2009). Extraction of nickel with pre-neutralized organic acids using Outotec mixer settler technology. In J. J. Budac, R. Fraser, I. Mihaylov, V. G. Papangelakis & D. J. Robinson (Eds.), *Hydrometallurgy of nickel and cobalt 2009* (pp. 211–220). CIM.

Chapter 26

Electrowinning of Nickel from Purified Nickel Solutions

Leaching followed by solvent extraction produces a purified solution of nickel chloride or sulfate. This chapter discusses the electrowinning of high-purity nickel from these solutions.

Nickel electrowinning entails:

(a) immersing metal cathodes and anodes in an aqueous nickel electrolyte; and,
(b) applying a DC electrical potential between these anodes and cathodes.

These actions cause the following processes to occur:

(a) electrons are generated at the anode by the anodic reaction, which is the evolution of either chlorine or oxygen;
(b) these electrons are driven by the applied potential though the external circuit to the cathode;
(c) electrons are accepted at the cathode by the cathodic reactions, which are the deposition of high-purity nickel and the evolution of hydrogen; and,
(d) current flows by convective and diffusive transport of ions through the solution (electrolyte) between the cathode and the anode, thus completing the electrical circuit.

Details for the operation of industrial nickel tankhouses are given in Tables 26.1 and 26.2.

26.1. OBJECTIVES OF THIS CHAPTER

This chapter describes the electrowinning of high-purity nickel (>99.9% Ni) from purified solutions obtained from leaching and solvent extraction. Specifically, the objectives of the chapter are the following:

(a) to discuss the chemical principles of electrowinning; and,
(b) to describe how these principles are applied industrially.

About 200 000 tonnes of nickel are electrowon per year. Other methods for producing high-purity nickel are carbonyl refining of impure metallic nickel (Chapter 22), hydrogen reduction (Chapter 27) and electrorefining of impure

Extractive Metallurgy of Nickel, Cobalt and Platinum-Group Metals. DOI: 10.1016/B978-0-08-096809-4.10026-7

TABLE 26.1 Industrial Electrowinning of Nickel from Chloride Solution*

Location	Niihama, Japan	Kristiansand, Norway
Cathode production, tonnes per year	30 000	90 000
Chloride Electrolyte		
Composition, g/L		
Ni	95	60 (in), 54 (out)
Cl		60
Co	5	
Cu	0.0002	
Fe	<0.0001	
Pb	0.0002	
Zn	<0.0001	
Temperature, °C	60 (in cells)	60
pH	2	1.4–2.0
Flow rate in and out of each cell, m^3/min	0.035	0.07
Anodes		
Material	Ru oxide-coated Ti mesh	Ru oxide-coated Ti mesh
Depth × width × thickness, m	1 × 0.8 × 0.002	1 × 0.8 × 0.002
Anode Compartments		
Construction materials		
Frame	Polypropylene	NA
Width, m	0.02	
Membrane	Polyester	
Cathodes		
Material	Nickel starter sheets	Nickel starter sheets
Depth × width × thickness, m	1 × 0.8 × 0.001	1.3 × 0.7 × 0.001
Center to center distance apart, m	0.11	0.13

TABLE 26.1 Industrial Electrowinning of Nickel from Chloride Solution*—cont'd

Location	Niihama, Japan	Kristiansand, Norway
Number per cell	52	51 anodes, 52 cathodes
Plating time, days	7	7
Cells		
Number	200	430
Materials	Concrete with fiber-reinforced polymer inserts + 32 polymer concrete	Reinforced monolithic concrete with fiber-reinforced polymer inserts
Inside length × width × height, m	6 × 0.9 × 1.1	7 × 0.8 × 1.6
Electrical		
Cell voltage, V	3	
Cell current, A	23 000	24 000
Cathode current density, A/m^2	233–279	260
Current efficiency, %	99.5	98–99
Cathode Deposit		
Composition, %		
Ni	99.99	>99.98 and 99.99
C		<0.002
Co	0.003	<0.0002
Cu	0.0002–0.0005	<0.0001
Fe	<0.0001	<0.001
Pb	0.0003	<0.0002
Zn	<0.0001	<0.0002

The feed is high-purity nickel chloride solution. The product is high-purity sheets of electrowon nickel, ~99.99% Ni.

**Niihama, 2007; Stensholt, Zachariasen, & Lund, 1986; Stensholt, Zachariansen, Lund, & Thornhill, 1988 with updates*

TABLE 26.2 Industrial Electrowinning of Nickel from Sulfate Solution

Location	Harjavalta, Finland, 1997	Rustenburg, South Africa, 2009	Rustenburg, new tankhouse, 2011
Cathode production, tonnes per year	18 000	21 500	33 000
Sulfate Electrolyte			
Composition, g/L			
Ni	130 in, 65 out	80 in, 50 out	80 in, 50 out
Addition agents	Na lauryl sulfate	Boric acid 6–10 g/L	Boric acid 6–10 g/L
Temperature, °C	65	60–65	60–65
pH	3.5–3.8	3.5	3.5
Flow rate in and out of each cell, m^3/h	0.12–0.15		
Anodes			
Material	Rolled chemical lead	Lead alloy (0.6% Sn and 0.05% Sr)	Lead alloy (0.6% Sn and 0.05% Sr)
Depth × width × thickness, m	0.95 × 0.95 × 0.01		
Cathodes			
Material	Nickel starter sheets	Nickel starter sheets	Titanium cathodes
Depth × width × thickness, m	1 × 1 × 0.01		
Center to center distance apart, m	0.13	0.16	
Number per cell	50 anodes and 49 cathodes	41 anodes and 40 cathodes	49 anodes and 48 cathodes
Plating time, days	7 + 2 days making starter sheets on Ti blanks	6 + 2 days making starter sheets on Ti blanks	10–14 days
Cathode Compartment (Construction materials)		Yes	Yes

TABLE 26.2 Industrial Electrowinning of Nickel from Sulfate Solution—cont'd

Location	Harjavalta, Finland, 1997	Rustenburg, South Africa, 2009	Rustenburg, new tankhouse, 2011
Frame	PVC coated 2 cm steel rod	Oregon pine	PVC
Thickness, m	0.02		
Membrane	Polyester	Woven terylene	Woven terylene
Cells			
Number	126 including 18 starter sheet cells		208
Materials	Reinforced monolithic concrete with PVC inserts	Precast concrete	Polymer concrete
Inside length × width × height, m	6 × 1.2 × 1.2	6.6 × 1.2 × 1.2	
Electrical			
Cell potential, V	3.9	3.6–3.9	3.6–3.9
Cell current, kA	22	15	18
Cathode current density, A/m^2	200–230	220	220–260
Current efficiency, %	93	97	>95
Cathode Deposit %			
Ni	>99.85		
Co	<0.1		
Cu	<0.02		
Fe	<0.02		
Pb	<0.001		
S	<0.001		
Zn	<0.001		

The feed is high-purity nickel sulfate solution. The product is high-purity sheets of electrowon nickel, >99.8% Ni.

metallic nickel and metalized nickel matte (Appendix F). New projects seem to be favoring the use of electrowinning and hydrogen reduction (Collins *et al.*, 2009; Stevens *et al.*, 2009).

26.2. ELECTROWINNING NICKEL FROM CHLORIDE ELECTROLYTE

Nickel is electrowon from chloride electrolyte at the following operations: (i) Xstrata Nikkelverk, Kristiansand, Norway (Dotterud, Peek, Stenstad, & Ramsdal, 2009); (ii) Sumitomo nickel refinery, Niihama, Japan (Kobayashi & Imamura, 2009); and, (iii) Eramet, Le Havre-Sandouville, France (Eramet, 2010).

The total nickel production by electrowinning from chloride solutions is approximately 140 000 tonnes per year.

26.2.1. Chemistry of Electrowinning in Chloride Solutions

The cathode half-reaction is the deposition of nickel:

$$\underset{\substack{\text{in high-purity aqueous}\\ \text{electrolyte at}\\ \text{cathode surfcaces}}}{Ni^{2+}(aq)} + \underset{\substack{\text{electrons supplied to}\\ \text{cathodes through}\\ \text{external circuit}}}{2e^-} \xrightarrow{60^\circ C} \underset{\substack{\text{high-purity nickel}\\ \text{plated on cathode}\\ \text{surfaces}}}{Ni(s)} \qquad (26.1)$$

The standard reduction potential, E°, of the nickel deposition half-reaction is approximately –0.25 V.

The anode half-reaction is the evolution of chlorine gas:

$$\underset{\substack{\text{in aqueous}\\ \text{electrolyte at}\\ \text{anode surfaces}}}{2Cl^-(aq)} \xrightarrow{60^\circ C} \underset{\substack{\text{chlorine gas produced}\\ \text{at anode surfaces and}\\ \text{recycled to leach}}}{Cl_2(g)} + \underset{\substack{\text{electrons released at anodes}\\ \text{conducted around external}\\ \text{circuit by applied voltage}}}{2e^-} \qquad (26.2)$$

The standard reduction potential, E°, of the half-reaction for chlorine evolution is approximately –1.35 V.

The overall electrowinning reaction is the sum of the two half-reactions:

$$\underset{\text{in electrolyte}}{Ni^{2+}(aq) + 2Cl^-(aq)} \xrightarrow{60^\circ C} \underset{\substack{\text{electrodeposited}\\ \text{on cathode}}}{Ni(s)} + \underset{\text{evolved at anode}}{Cl_2(g)} \qquad (26.3)$$

The equilibrium potential for the electrowinning reaction is –1.6 V. The negative value indicates that the reaction is not spontaneous and has to be driven by an applied voltage that is greater than 1.6 V.

26.2.2. Applied Potential

At equilibrium, the rates of reaction are zero. In order to plate nickel at the cathode, the applied potential or cell voltage must be increased above the equilibrium potential.

The additional applied potential that is used to drive the reactions over and above the equilibrium potential is called the *overpotential*. In other words, the overpotential is the kinetic driving force. The overpotential drives the reactions, so that the electroplating and chlorine evolution reactions occur at a reasonable rate. In practice, the overpotential is about 0.6 V.

In addition to the equilibrium potential and the overpotential, there are a number of other resistances that increase the overall cell voltage. These are the *electrolyte resistance*, the *anode bag resistance* and the *contact resistance*.

The electrolyte resistance refers to the electrical resistance that the electrolyte offers during the transfer of the current by charged ions in solution.

The anode-bag resistance refers to the resistance that the anode bag offers due to the transfer of ionic charge being impeded by the pore structure of the bag.

The contact resistance refers to the resistance cause by the electrical contact between the nickel support straps and the hanger bars and between the hanger bars and the busbar.

In practice, the electrolyte, anode bag and contact resistances result in an increase in the cell voltage of approximately 0.6, 0.2 and 0.1 V, respectively.

As a result, the overall cell voltage is about 3.0 V, which is equal to the sum of the individual contributions to the voltage.

26.2.3. Electrodes

The anode is made up of titanium wire or mesh that is coated with ruthenium oxide with other metals. This ruthenium-metal oxide coating is essentially chemically inert (but electrically conductive) in the highly corrosive environment found in chloride electrowinning cells. This type of anode is often referred to as a '*dimensionally stable anode*'.[1] The anodes are 0.8 m wide and 1 m in height.

The anodes emit chlorine gas, which is safely collected for reuse in leaching. The system for collecting the chlorine is shown in Figure 26.1. Anode and cathode supports are shown in Figure 26.2.

Each anode is placed in a bag made of permeable polyester cloth. The ionic solution is able to move through the cloth, but, because of the size and surface

---

1. The name comes from the chlor-alkali industry, where these types of electrodes were first introduced. Prior to dimensionally stable anodes, carbon anodes were used in chlor-alkali cells. These anodes deteriorated with chlorine production. Titanium anodes with a coating of ruthenium oxide do not degrade, and were called 'dimensionally stable'. DSA is registered trademark of Industrie de Nora (www.denora.com).

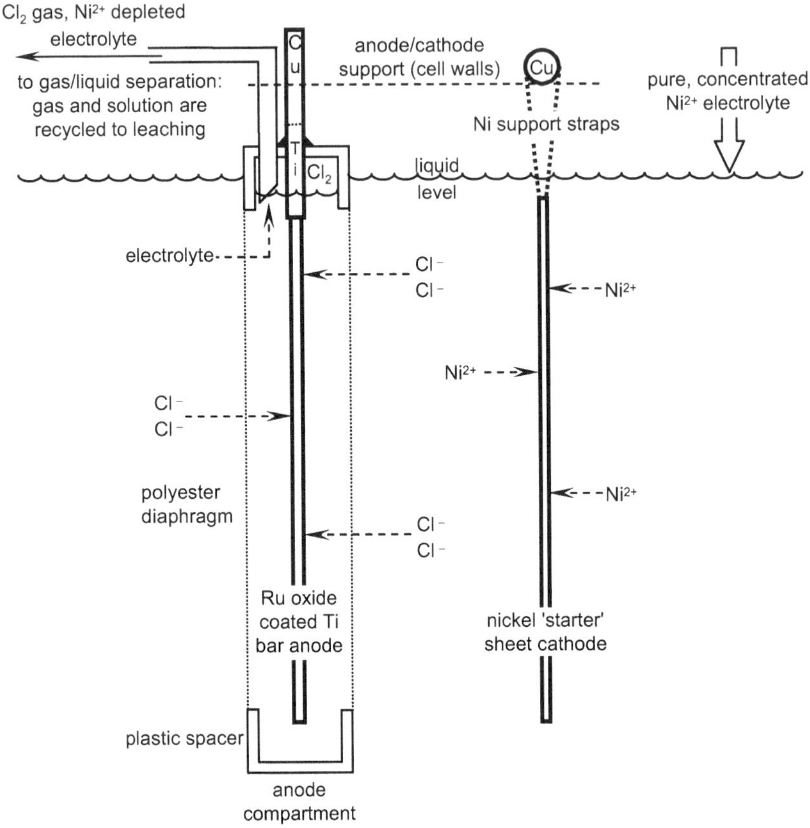

FIGURE 26.1 A schematic diagram of two electrodes in an aqueous nickel chloride electrowinning cell (Niihama, 2007). The cathode is shown on the right. The anode and its surrounding compartment are shown on the left. Solution in the anode compartment is referred to as the *anolyte*, while solution in the cathode compartment is referred to as the *catholyte*. Direct electric current flows between the anode and the cathode. A vacuum draws chlorine and nickel-depleted solution out of the anode compartment. Note that the liquid level in the anode compartment is lower than elsewhere in the cell. This causes continuous flow of solution past the cathodes and into the anode compartments. An electrowinning cell contains one more anode than cathode. Most nickel electrowinning cells have 45–55 cathodes per cell. Each cathode is typically 1 m deep ad 0.8 m wide.

tension of the gas bubbles, the gas remains on the anodic side of the bag. A solid 'lid', made of polypropylene, prevents the chlorine gas from escaping to the atmosphere.

Chlorine and chlorine-saturated solution (anolyte) are continuously extracted from each anode compartment by vacuum through an individual polymer tube. The mixture of gas and solution is then drawn into a manifold and on to liquid/gas separators and vacuum liquid dechlorinators. The gas is dried, compressed and sent to leaching.

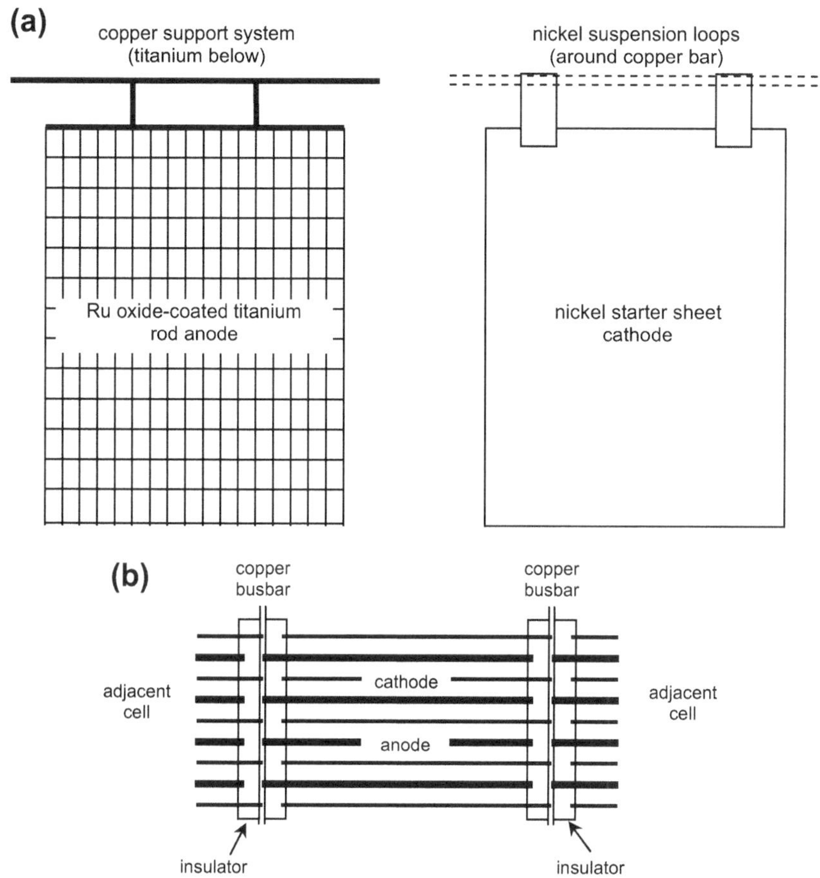

FIGURE 26.2 (a) Anode and cathode for electrowinning nickel from aqueous chloride solution. They are about 1 m long and 0.8 m wide. During electrowinning, the cathode grows from about 1 mm thick to 10 mm thick. The anode wires are about 4 mm diameter. (b) Sketch of electrowinning cell circuitry showing that current flow between the copper bars is copper bar–anode–electrolyte–cathode–copper bar. Current flow through the electrolyte is by ion transport.

The cathodes are generally 'starter' sheets, which are sheets of nickel, approximately 1 mm thick, that are hung on round copper bars using starter sheet loops (shown in Figure 26.1). The size of the cathode is the same as that of the anode, that is, 0.8 m wide and 1 m deep.

In order to make the starter sheets, nickel is plated on titanium cathode blanks in separate cells for between 1 and 2 days. Plastic edge and bottom strips are fitted on the blanks to permit easy stripping of the starter sheets from the blanks.

The cathode blanks are harvested and the deposit stripped from the blank. The stripping is mostly done by automated machinery (Stensholt *et al.*, 2001). The nickel plate is trimmed and nickel support straps are fitted to the deposit to form the starter sheet. Automated machinery is used for these activities.

26.2.4. Cells

The anodes and cathodes are placed in an electrolyte-filled rectangular tank, referred to as a cell. The cells are typically made of either (i) acid-resistant polymer concrete; or, (ii) concrete with fiberglass-reinforced polymer inserts. Newer cells are mostly made of polymer concrete.

Each cell contains 46 anodes and 45 cathodes. The anodes and cathodes are spaced evenly along the cell to equalize current among all anodes and cathodes. This ensures equal plating on all cathodes. The anodes in the cell are in contact with the same busbar, so that they are at the same electrical potential. In a similar manner, the cathodes are in contact with another busbar and at the same potential, which is about 3 V lower than the anode potential. This means that the anodes and cathodes within a single cell are electrically connected in parallel.

Direct current for electrowinning is provided from the power grid via thyristoformer rectifiers. The current density is 240 A/m^2 of cathode. The current to a cell is typically 23 kA.

The cells are electrically connected in series. During maintenance and repair, individual cells are electrically isolated using a shorting frame or shorting cables.

26.2.5. Production Cycle

Metal is plated on the starter sheet for about 7 days during which time the 1 mm thick starter sheets become 10 mm thick, weighing about 70 kg. At the end of the plating cycle, the cathodes are lifted out of the cell and washed. The cathodes are either packed or mechanically cut into 2.5, 5 or 10 cm square pieces for delivery to clients.

The cathodes that are harvested are then replaced with new starter sheets. This activity is scheduled so that about 15% of the cathodes are harvested and replaced each day, which leaves the rest of the plant operating under controlled, efficient, steady-state conditions.

26.2.6. Nickel Purity

The product of nickel chloride electrowinning is typically greater than 99.98% Ni, sometimes 99.99%. This purity is maintained by efficient electrolyte purification, by thorough washing of cathodes in clean water at 80°C immediately after they are lifted from the cells and by maintenance of a clean environment throughout the refinery.

26.2.7. Nickel Crowns

The Nikkelverk electrowinning plant produces nickel '*crowns*' as well as full cathode plates. Crowns are used by the nickel-plating industry. They are produced by masking titanium blanks with a polymer – except for unmasked circles where nickel electrodeposition can occur.

FIGURE 26.3 Nickel crowns produced by masking titanium mother blank cathodes with polymer then electrowinning. *Photograph courtesy of Xstrata Nikkelverk.*

The result is nickel crowns, shown in Figure 26.3, that are proportional to the size of the non-masked circles. The industrial products are between 1 and 2 cm diameter and about 1 cm thick.

26.3. ELECTROWINNING NICKEL FROM SULFATE SOLUTIONS

High-purity nickel is produced from sulfate solutions by electrowinning at the following operations: (i) Norilsk Nickel Finland Oy, Harjavalta, Finland (Latva-Kokko, 2006); and, (ii) Anglo American Platinum, Rustenburg, South Africa (Kruyswijk, 2009). Details of industrial operations are given in Table 26.2.

Nickel electrowinning from sulfate solutions was also used at Hartley Platinum in Zimbabwe and at the laterite plants at Cawse and Bulong in

Western Australia (Donegan, 2006). All of these tankhouses are no longer operational.

The total production of nickel by the electrowinning of sulfate solution is about 60 000 tonnes per year.

26.3.1. Electrowinning Chemistry

The cathode half-reaction is the deposition of nickel:

$$\underset{\substack{\text{in high-purity}\\ \text{aqueous solution at}\\ \text{cathode surfaces}}}{Ni^{2+}(aq)} + \underset{\substack{\text{electrons supplied}\\ \text{to cathodes}}}{2e^-} \xrightarrow{60°C} \underset{\substack{\text{high-purity}\\ \text{nickel plated at}\\ \text{cathode surfaces}}}{Ni(s)} \tag{26.4}$$

The standard reduction potential, $E°$, of the nickel deposition half-reaction is approximately –0.25 V.

The anode half-reaction is the evolution of oxygen:

$$H_2O(\ell) \xrightarrow{60°C} \underset{\substack{\text{evolved at}\\ \text{anode surfaces}}}{0.5O_2} + \underset{\substack{\text{in acidic}\\ \text{sulphate solution}}}{2H^+(aq)} + \underset{\substack{\text{electrons released}\\ \text{at anodes}}}{2e^-} \tag{26.5}$$

The standard reduction potential, $E°$, of the half-reaction for chlorine evolution is approximately –1.23 V.

The overall electrowinning reaction is the sum of the two half-reactions:

$$\underset{\text{feed solution}}{Ni^{2+}(aq) + H_2O(\ell)} \xrightarrow{60°C} \underset{\substack{\text{nickel metal}\\ \text{plated at cathode}}}{Ni(s)} + \underset{\substack{\text{oxygen evolved}\\ \text{at anode}}}{0.5O_2(g)} + \underset{\substack{H_2SO_4\text{-enriched}\\ \text{solution}}}{2H^+(aq)} \tag{26.6}$$

The equilibrium potential for the electrowinning reaction is –1.48 V. In practice, the cell potential is about 4.0 V, due to the electrolyte, cathode bag, contact resistances and the overpotential. The overpotential in sulfate solutions is significantly higher than that in chloride solutions because the kinetics of oxygen evolution on lead is less favorable than chlorine evolution on ruthenium oxide-coated titanium.

26.3.2. Products

The products from electrowinning are high-purity nickel (>99.9% Ni), oxygen gas and a solution that is depleted in nickel and enriched in sulfuric acid.

The solution in the feed to the tankhouse is often referred to as the *advance electrolyte* and that leaving the tankhouse as the *spent electrolyte*.

The nickel is carefully washed and packed. Some of the nickel is mechanically cut into approximately 3 cm squares for use in the electroplating industry. The oxygen is released to the environment. The spent solution is recycled to leaching (Kruyswijk, 2009).

26.3.3. Competing Reactions

The standard reduction potential for the evolution of hydrogen is more positive than that for the deposition of nickel. This means that it is thermodynamically more favorable to evolve hydrogen at the cathode than to plate nickel.

The evolution of hydrogen is undesirable because it wastes current and energy. The evolved hydrogen can also occlude into the nickel deposit, causing the deposit to blister, crack, tear or warp.

The half-reaction for the evolution of hydrogen, which competes at the cathode with the half-reaction for the plating of nickel, is given as follows:

$$\underset{\substack{\text{in } H^+\text{-enriched solution}\\ \text{at cathode surfaces}}}{2H^+(aq)} + \underset{\substack{\text{electrons supplied to cathodes}\\ \text{through external circuit}}}{2e^-} \xrightarrow{60^\circ C} \underset{\substack{\text{evolved at}\\ \text{cathode surfaces}}}{H_2(g)} \qquad (26.7)$$

Thus, the amount of hydrogen that is evolved must be minimized. This is achieved by ensuring that the concentration of acid is low, which lowers the rate of the reaction given in Equation (26.7). However, acid is generated at the anode, which means that the solutions in the vicinity of the anode and the cathode must be separated into anolyte and catholyte, and the acid from the anolyte must not be allowed to transfer to the catholyte. In practice, this is achieved by using a separator, like an anode or a cathode bag.

Since anodes bags were discussed in chloride electrowinning, cathode bags will be discussed in this section. Most sulfate electrowinning plants use cathode bags. At least one, Port Colburne, Canada, uses anode bags for cobalt electrowinning. The new plant at Long Harbor, Canada, plans to use anode bags.

The cathode is placed in a polypropylene frame and a bag made of permeable cloth is pulled over the frame. The frame and bag form a cathode compartment that separates the solution within the bag from the solution outside of it. This is shown in Figure 26.4. The bag provides a physical barrier, not a chemical barrier, between the anode and cathode compartments. Solution can move through the bag.[2]

The pH of the feed to the tankhouse is adjusted to a value of about 3.5 and is fed to each cell in parallel. The feed to each cell is further divided and introduced separately to each cathode compartment (Jayasekera & Kyle, 1999). The

2. Cells with separate anode and cathode compartments where the separator is cloth are called *'diaphragm'* cells. In other applications, for example, in the chlor-alkali industry, ion-exchange membranes are used and the cells are referred to as *'membrane'* cells.

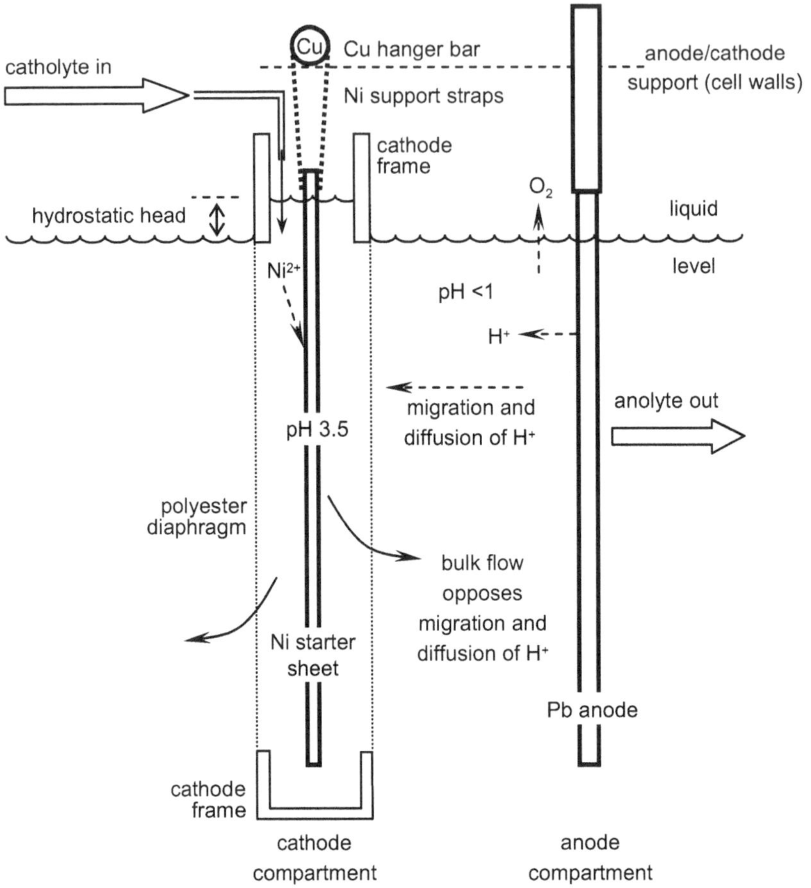

FIGURE 26.4 A schematic diagram of two electrodes in a nickel sulfate electrowinning cell. A diaphragm separates the cathode from the anode. Solution is fed to each cathode compartment at a rate that maintains a hydrostatic head of about 2 cm in the cathode compartment. The flow of solution from the cathode compartment opposes the migration of acid (H^+ ions) into the cathode compartment. In this manner, the pH in the cathode compartment is maintained at a value of about 3.5. The concentration of sulfuric acid in the anode compartment is about 45 g/L. In practice, a polyester bag is pulled over the cathode frame to form the cathode compartment. The polyester bag is referred to as a *cathode bag*.

level within the cathode compartment is maintained at a slightly higher level, about 2 cm higher, than that in the anode compartment. This creates sufficient head that the solution flows from the cathode compartment to the anode compartment. The flow through the cloth opposes the migration of H^+ from the anode compartment, where the concentration of acid is high, to the cathode compartment, where the concentration of acid is low.

Boric acid is also added to the solution as a buffer, so as to control pH. In addition, sodium lauryl sulfate may be added as a surfactant to help release hydrogen bubbles from the cathode surface, thereby improving the quality of the deposit.

26.3.4. Anodes

Lead alloy electrodes are used in nickel sulfate tankhouses. The typical composition of an anode is 99% Pb, 0.6% Sn and 0.05% Sr. A chemically inert layer of lead oxide forms on the surface of the anode during electrolysis. Tin and strontium are alloyed with lead to provide corrosion resistance and mechanical strength. Chemical lead may also be used as an anode.

26.3.5. Maximizing Cathode Purity

Cathode purity depends on many factors. Some of the factors that tend to maximize purity are:

(a) pure, strong, particle-free and organic-free electrolyte;
(b) straight, vertical anodes and cathodes;
(c) even, steady current distribution among the electrodes;
(d) avoidance of anode-cathode short circuiting;
(e) taut cathode bags;
(e) use of addition agents that gives dense, smooth nickel deposits, for example, sodium lauryl sulfate and boric acid;
(f) immediate and thorough washing of the cathode product with clean, 70°C–80°C water; and,
(g) pristine, clean operating conditions throughout the refinery.

Problems that are encountered during nickel electrowinning are burning straps,blinding of the cathode bag and nickel growth in the cathode bag. Burning straps are caused by poor electrical contact between the hanger bar and the nickel straps. The nickel straps heats up sufficiently for the straps to melt. The cathode then falls into the cathode compartment. Blinding of the cathode bag can occur if the feed solution contains solids, especially fine solids like filter aid, or if the pH is not controlled and nickel hydroxide precipitates in the pores of the bag. Nickel hydroxide precipitation in the pores of the bag is referred to as ‘green bags’.

26.4. NEW DEVELOPMENTS IN NICKEL ELECTROWINNING

Anglo American Platinum commissioned a new nickel tankhouse in 2011 using state-of-the-art equipment and technology. Permanent titanium cathodes, mechanical stripping, anode skirts and cell hoods are used to reduce worker exposure to aerosols and solutions that contain nickel sulfate and acid. The development and testing of the technology has been performed in-house (Bryson, Graham, Bogosi, & Erasmus, 2008).

26.4.1. Titanium Cathodes

Rather than plating a starter sheet on a cathode blank, nickel is plated directly on a titanium cathode. The titanium cathode is fitted with edge strips to facilitate stripping. Nickel that has been plated on titanium is stressed and the deposit may fall off the cathode on harvesting or cooling. Three holes, with a diameter of about 3 mm, are drilled through the titanium. During deposition, these holes are filled with nickel deposit and hold the two sides of the plated nickel together. As a result, the nickel plate does not fall if it detaches from the cathode. These supports snap open easily in the chiselling unit of the stripping machine.

Plating nickel onto a titanium blank simplifies the operating sequence in the tankhouse by removing a number of steps in the deposition cycle. The deposition cycle of the traditional nickel tankhouse is as follows:

(a) plate starter sheet;
(b) harvest starter sheet;
(c) wash, strip, trim and attach straps to starter sheet;
(d) prepare surface of starter sheet;
(e) plate in a production cell;
(f) harvest from production cells; and,
(g) wash.

The use of the titanium blanks reduces this cycle to:

(a) production plating;
(b) harvest from the production cells;
(c) wash; and,
(d) strip.

As a result of the reduction in operational steps and the use of automated mechanical stripping, worker exposure to solutions and aerosols containing nickel sulfate is reduced.

26.4.2. Mist Abatement

Two aerosol and mist abatement technologies are used at Anglo American Platinum: (i) anode skirts; and, (ii) cell hoods (Bryson, Graham, Bogosi, & Erasmus, 2008). The tankhouse is shown in Figure 26.5.

The tankhouse at Cawse used anode bags, similar to the operation of the chloride electrowinning. However, tests by Anglo American Platinum revealed that the concentration of acid in the catholyte was too high. Instead, anode skirts have been implemented at Anglo American Platinum in addition to cathode bags.

The anode skirt reduces the formation of mist and aerosols from the bursting of oxygen bubbles at the anode. The skirts are clipped onto the lead anode. A titanium strip is placed on the anode at the point where the skirt attaches. This prevents the corrosion of the anode at the point of

FIGURE 26.5 The nickel tankhouse at Anglo American Platinum, showing the mist abatement hoods and the crane for automatic harvesting of the nickel cathodes. *Source: Photograph by J. J. Taute, courtesy of Anglo American Platinum*

attachment. The skirt is not closed at the bottom, hence the term '*skirt*' and not '*bag*'.

The cell hoods are similar to those used in the copper industry. The hoods cover the whole cell and are vented to a scrubber. Flushing water is used to prevent an accumulation of deposits of nickel sulfate in the hood.

Anglo American Platinum reported that these technologies reduce the ambient concentration of nickel from about 0.5 mg/m^3 to less than 0.05 mg/m^3 (Bryson, Graham, Bogosi, & Erasmus, 2008). As a result of the use of titanium blanks, automated harvesting, automated stripping and these mist–abatement technologies, total occupational exposure to acid and nickel aerosols is significantly reduced. In the light of increasingly stringent legislation on occupational exposure limits arising from concerns about nickel dermatitis and nickel-induced carcinogenesis, this is an industry-leading operation.

26.5. OTHER ELECTROLYTIC NICKEL PROCESSES

High-purity nickel is also made by electrorefining from the following materials:

(a) impure nickel metal anodes; and,
(b) metallized nickel matte anodes.

These processes are discussed in Appendix F.

26.6. APPRAISAL

There are two ways to make nickel metal from high-purity electrolytes: electrowinning and hydrogen reduction.

Electrowinning is applicable to both chloride and sulfate solutions, while hydrogen reduction is only applicable to ammoniated sulfate solutions.

Electrowinning makes purer nickel than hydrogen reduction (described in Chapter 27). Electrowinning is, however, labor and energy intensive. It will continue to be used for chloride and sulfate electrolytes where very high-purity nickel is required.

Occupational exposure to dissolved nickel is likely to drive the choice of process in the future. Dissolved nickel is classified as a Class 1 carcinogen and causes dermatitis. Electrowinning of nickel from sulfate solutions has fewer problems of materials of construction than chloride solutions, but it is easier to contain aerosols in a chloride tankhouse. Anglo American Platinum's nickel tankhouse is a world leader in reducing occupational exposure to nickel.

26.7. SUMMARY

Electrowinning produces high-purity nickel metal by electroplating from purified leaching solutions. Chloride and sulfate electrolytes are used. It is mostly used to recover nickel from the solutions produced by leaching sulfide mattes and sulfide intermediate precipitates.

Chloride and sulfate electrowinning are both well-established. It seems, however, that future projects will be carefully assessed in terms of the occupational exposure to dissolved nickel salts.

REFERENCES

Bryson, L. J., Graham, N., Bogosi, E., & Erasmus, D. L. (2008). Nickel electrowinning tank house developments at Anglo Platinum's base metal refinery. In A. Taylor (Ed.), *ALTA conference 2008 Ni/Co pressure leaching and hydrometallurgy forum and autoclave design and operation symposium*. ALTA Metallurgical Services.

Collins, M. J., Buban, K. R., Holloway, et al. (2009). Ambatovy laterite ore preparation plant and high pressure acid leach pilot plant operation. In J. J. Budac, R. Fraser, I. Mihaylov, V. G. Papangelakis & D. J. Robinson (Eds.), *Hydrometallurgy of nickel and cobalt 2009* (pp. 499–510). CIM.

Donegan, S. (2006). Direct solvent extraction of nickel at Bulong operations. *Minerals Engineering, 19*, 1234–1245.

Dotterud, O. M., Peek, E. M. L., Stenstad, O., & Ramsdal, P. O. (2009). Iron control and tailings disposal in the Xstrata chlorine leach process. In J. J. Budac, R. Fraser, I. Mihaylov, V. G. Papangelakis & D. J. Robinson (Eds.), *Hydrometallurgy of nickel and cobalt 2009* (pp. 321–333). CIM.

Eramet. (2010). Le Havre-Sandouville Refinery.

Jayasekera, S., & Kyle, J. (1999). Electrowinning of nickel from sulphate electrolyte – A review. In A. Taylor (Ed.), *ALTA conference 1999 Ni/Co pressure leaching and hydrometallurgy forum and autoclave design and operation symposium*. ALTA Metallurgical Services.

Kobayashi, H., & Imamura, M. (2009). The study of mixed sulfide reaction in the chlorine leach system. In J. J. Budac, R. Fraser, I. Mihaylov, V. G. Papangelakis & D. J. Robinson (Eds.), *Hydrometallurgy of nickel and cobalt 2009* (pp. 27–37). CIM.

Kruyswijk, L. (2009). Anglo platinum base metal refinery. Paper presented at the SAIMM Southern African Hydrometallurgy Conference, SAIMM.

Latva-Kokko, M. J. (2006). Iron removal as part of the nickel matte leaching process. In J. E. Dutrizac & P. A. Riveros (Eds.), *Iron control technologies* (pp. 391–401). CIM.

Niihama. (2007). *Outline of MCLE (mixed sulfide chlorine leach electrowinning) process.* Visitors' guide. Sumitomo Metal Mining Co. Ltd.

Stensholt, E. O., Dotterud, O. M., Henriksen, E. E., et al. (2001). Development and plant practice of the Falconbridge chlorine leach process. *CIM Bulletin, 94*, 101–104.

Stensholt, E. O., Zachariasen, H., & Lund, J. H. (1986). Falconbridge chlorine leach process. *Trans, Instit. Min. Met. C, 5*, C10–C16.

Stensholt, E. O., Zachariansen, H., Lund, J. H., & Thornhill, P. G. (1988). Recent improvements in the Falconbridge Nikkelverk nickel refinery. In G. P. Tyroler & C. A. Landolt (Eds.), *Extractive metallurgy of nickel and cobalt* (pp. 403–413). TMS.

Stevens, D., Bishop, G., Singhal, A., et al. (2009). Operation of the pressure oxidative leach process for Voisey's Bay nickel concentrate at Vale Inco's hydromet demonstration plant. In J. J. Budac, R. Fraser, I. Mihaylov, V. G. Papangelakis & D. J. Robinson (Eds.), *Hydrometallurgy of nickel and cobalt 2009* (pp. 3–16). CIM.

SUGGESTED READING

Robinson, T., Weatherseed, M., Tuppa, E., & Heyting, E. (2002). Developments in nickel electrowinning cellhouse design. In A. Taylor (Ed.), *ALTA conference 2002*. ALTA Metallurgical Services.

Chapter 27

Hydrogen Reduction of Nickel from Ammoniacal Sulfate Solutions

An alternative process to electrowinning for the recovery of nickel from purified nickel solutions is hydrogen reduction. Approximately 240 000 tonnes of nickel are produced per year by hydrogen reduction.

The objectives of this chapter are:

(a) to describe the chemistry of hydrogen reduction of nickel from aqueous solutions;
(b) to discuss how this chemistry is applied safely and efficiently; and,
(c) to compare hydrogen reduction and electrowinning.

Hydrogen reduction is practiced at the following operations: (i) BHP Billiton, Kwinana, Australia; (ii) Minara Resources, Murrin Murrin, Leonora, Australia; (iii) Sherritt, Fort Saskatchewan, Canada; (iv) Ambatovy Joint Venture, Toamasina, Madagascar; (v) Norilsk Nickel, Harjavalta, Finland; and, (vi) Impala Platinum, Springs, South Africa.

Hydrogen reduction is carried out by injecting hydrogen into aqueous ammoniacal nickel sulfate solutions in stirred high-pressure autoclaves. A schematic flowsheet of the operation is shown in Figure 27.1.

Details of two industrial operations are given in Table 27.1.

27.1. PROCESS CHEMISTRY

The reduction of nickel by hydrogen may be represented by the reaction:

$$\underset{\substack{\text{in purified}\\\text{ammoniacal}\\\text{aqueous solution}}}{Ni(NH_3)_2^{2+}(aq)} + \underset{\substack{\text{30 bar hydrogen}\\\text{gas injected}\\\text{into solution}}}{H_2(g)} \xrightarrow{\substack{\sim 180^\circ\text{C, with}\\\text{nickel seeds}}} \underset{\substack{\text{nickel metal}\\\text{deposited on}\\\text{nickel seed particles}}}{Ni(s)} + \underset{\substack{\text{in aqueous}\\\text{sulfate solution}}}{2NH_4^+(aq)} \qquad (27.1)$$

Extractive Metallurgy of Nickel, Cobalt and Platinum-Group Metals. DOI: 10.1016/B978-0-08-096809-4.10027-9

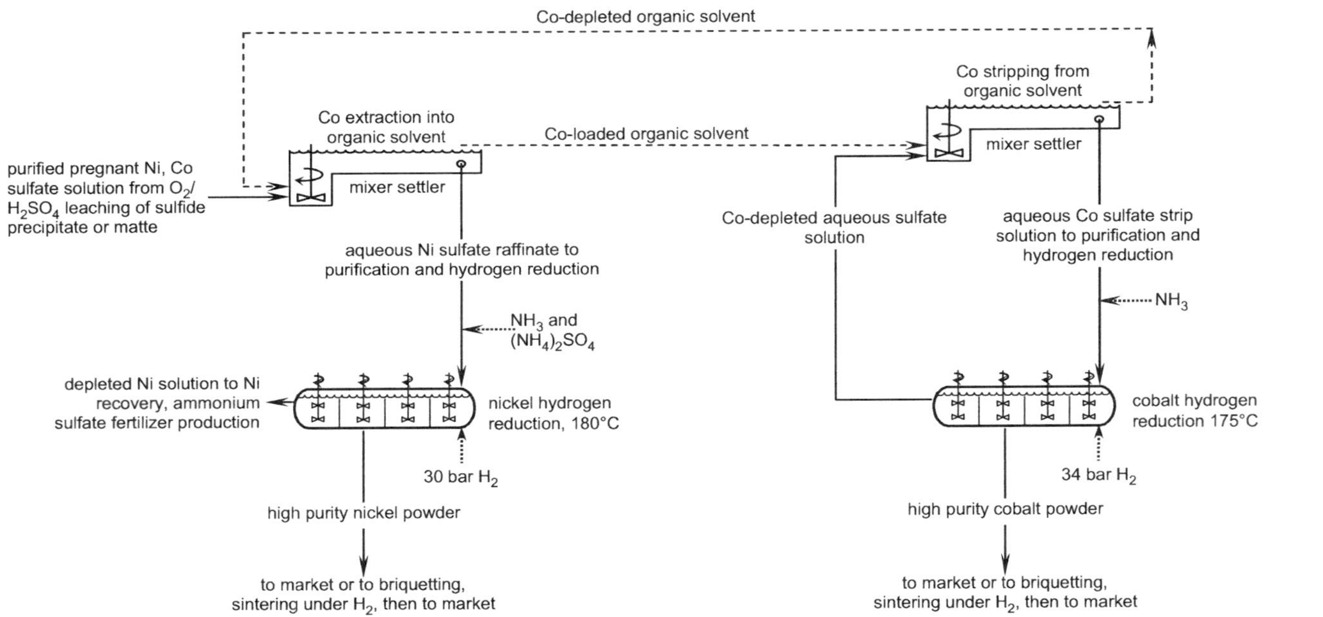

FIGURE 27.1 Schematic nickel and cobalt hydrogen reduction flowsheet based on the Murrin Murrin and Ambatovy operations. The input is pregnant leach solution from leaching of sulfide precipitates and/or mattes in sulfuric acid using oxygen. The products are (i) high-purity nickel and cobalt powders; and, (ii) ammonium sulfate fertilizer. Note the addition of ammonia to the nickel and cobalt sulfate solutions. The organic solvent is CYANEX 272, or LIX 272 extractant, in aliphatic diluent, discussed in Chapter 25.

TABLE 27.1 Operational Details of Nickel Hydrogen Reduction

Location	Ft. Saskatchewan, Canada	Springs, South Africa
Nickel production, tonnes/year	32 000	13 000
Starting solution		
Description	Purified pregnant solution from air/NH_3 leaching of Moa Bay sulfide precipitate	Purified pregnant solution from O_2/H_2SO_4 leaching of matte
Pre-treatment	Complete sulfation by air oxidation at 250°C	NH_3 addition
Composition, g/L		
Ni	70–75	55–65
Co	3	
Cu	<0.001	
NH_3	35–40	
$(NH_4)_2SO_4$	320–340	330–350
NH_3/Ni molar ratio	~1.9	1.8–1.95
Reductant	hydrogen	hydrogen
Source	Waste H_2-bearing gases	Petroleum refinery
Method of introduction into liquid	Injecting below agitator impeller	Injecting below agitator impeller
Autoclaves (number)	7	
Working volume, m^3	7–22, length/diameter ratio ≈5	
Construction material	Stainless steel	SAF 2205 stainless steel
Operating temperature, °C	185	180
Heating method	Steam coils	Steam coils
Operating pressure	30–35 bar gauge	28 bar gauge

(Continued)

TABLE 27.1 Operational Details of Nickel Hydrogen Reduction—cont'd

Location	Ft. Saskatchewan, Canada	Springs, South Africa
Operating method	Make 0.01 mm Ni(s) seed then reduce 60–80 'batches' of feed solution, remove 20–60 tonne high-purity nickel 'lot' then start again	Reduction of batches of feed solution, remove nickel, start again
Total batch time	~1 h, 25% for reduction	10 min for reduction
Post-reduction treatment	Leach Ni from autoclave walls	
Exit solution (composition, g/L)		
Ni	2–3	3
Co	2–3	0.5
Zn	2–3	
$(NH_4)_2SO_4$	500	550
Destination	Zn sold as ZnS; NiS and CoS precipitated, recycled to leach; $(NH_4)_2SO_4$ crystallization	To cobalt H_2 reduction then ammonium sulfate crystallization
Nickel product (composition, %)		
Ni	99.8	
Co	0.1	
Cu	0.004	
Fe	0.004	
S	0.03 (0.01 after H_2 sintering)	
C	0.01 (0.005 after H_2 sintering	
Particle size	0.1–0.2 mm	0.01–0.2 mm
Destination after drying	Sold or briquetted and sold, briquettes often H_2 sintered at 950°C	Briquetted, sintered under H_2, sold

Source: *Cordingley and Krentz, 2005; Impala, 2009.*

The oxidation half-reaction for this reaction is given as follows:

$$H_2(g) \rightarrow 2H^+(aq) + 2e^- \quad (27.2)$$

The reduction potential for this half-reaction, given by the Nernst equation, is linearly dependent on the pH of the solution. As the reaction proceeds, acid (H^+ ions) is produced, which increases the reduction potential for the hydrogen half-reaction, lowering the driving force for the overall reaction. Industrially, ammonia is used to maintain the pH (Collins *et al.*, 2009; Hayward, 2008; Impala, 2009):

$$\underset{\text{formed by Equation (27.2)}}{2H^+(aq)} + \underset{\text{in leach solution or added just before } H_2 \text{ reduction}}{2NH_3(g)} \rightarrow \underset{\text{contains } H^+ \text{ at low thermodynamic activity}}{2NH_4{}^+(aq)} \quad (27.3)$$

The amount of ammonia that is added is controlled based on the molar ratio of ammonia to nickel in solution, and a value of about 1.9 is optimum. Electrochemical studies have indicated that it is the $Ni(NH_3)_2^{2+}$ species that is reduced at the nickel surface, which concurs with the experience of the industrial operations that the molar ratio of ammonia to nickel must be about two.

At this value of the molar ratio, 98% of the nickel in solution is typically reduced to powder and hence removed from each batch of solution. The nickel that remains in solution is subsequently recovered by precipitation with hydrogen sulfide. The nickel sulfide precipitate is recycled to pressure leaching (Cordingley & Krentz, 2005).

An alternative method of removing of the nickel that remains in solution is the precipitation of a mixed double salt consisting of $Ni(NH_4)_2(SO_4)_2.6H_2O$ and $Co(NH_4)_2(SO_4)_2.6H_2O$ (Spandiel, 2008). The double salt is redissolved so that the concentration of nickel is about 85 g/L. Ammonia and manganese dioxide are then added to oxidize the cobalt, so that cobaltic pentammine, $Co(NH_3)_5^{3+}$, forms. On the addition of acid, only $(NH_4)_2Ni(SO_4)_2.6H_2O$ precipitates, leaving the cobaltic pentammine in solution. This cobalt is reduced with cobalt powder and then recovered from solution by hydrogen reduction (see Table 29.2 for more details on hydrogen reduction of cobalt).

27.2. INDUSTRIAL APPLICATIONS

The reduction of nickel by hydrogen was originally developed for the solutions produced by the leaching of nickel sulfide concentrates in ammonia solutions using air. Fort Saskatchewan, Canada, and Kwinana, Western Australia, still use the process, but now the solutions are produced by the leaching of sulfide mattes using air and ammonia (Palmer, Malone, & Loth, 2005) and intermediate sulfide precipitates (Egedahl & Collins, 2009).

New hydrogen reduction plants have also been built for the sulfate solutions produced from the leaching of sulfide mattes and intermediate precipitates using oxygen and sulfuric acid (Collins *et al.*, 2009; Hayward, 2008; Impala, 2009; Makinen, Fagerlund, Anjala, & Rosenback, 2005).

The leach solutions from these new plants are ammoniated before hydrogen reduction. An important advantage of hydrogen reduction is that the sulfur in the feed to these plants is removed as a saleable by-product, that is, as ammonium sulfate fertilizer.

27.3. INDUSTRIAL PRODUCTION OF NICKEL POWDER

The operation of the autoclave for hydrogen reduction occurs in the following steps:

(a) the preparation of the nickel 'seed' powder;
(b) the reduction of solution batches;
(c) the finishing of a 50 tonne 'lot' of nickel powder (~60 solution batches); and,
(d) the preparation of the autoclave for a new cycle.

Each of these steps is discussed further.

27.3.1. Nickel Seed Preparation

The initial seed powder, with a particle size of about 0.01 mm, is made in the autoclave by the reduction of nickel by hydrogen from a dilute solution of nickel ammonium sulfate. Ferrous sulfate, aluminum sulfate and an organic smoothing agent are added (Cordingley & Krentz, 2005). These seed particles provide a large surface area for the heterogeneous particle growth during hydrogen reduction and hence rapid production of nickel.

Once the seed has been produced, the solution is removed from the autoclave with the new 'seed' particles left in place. The first batch of production solution is then pumped into the autoclave.

27.3.2. Reduction of Solution Batches

A batch of hot, purified ammoniacal sulfate solution with about 70 g/L Ni is pumped into the hydrogen reduction autoclave with nickel seed particles already present in the autoclave. Hydrogen gas, at a pressure of about 30 bar, is sparged into the autoclave. The contents are stirred to suspend the powder and to ensure good gas–liquid mass transfer. The temperature is maintained at about 180°C with steam heating coils.

These conditions are maintained until the concentration of nickel in solution decreases to about 2 or 3 g/L. The supply of hydrogen is stopped and the hydrogen pressure is lowered to about 2 bar. The stirring is stopped so that the newly grown powder settles to the bottom of the autoclave. The solution, depleted in nickel, is discharged from the autoclave through a flash-cooling pressure-reduction tank.

This sequence of steps, that is the filling with fresh solution, the reduction of nickel and the discharging of the depleted solution, is referred to as a 'densification'. This sequence or densification is repeated 60–80 times to produce a 20–50 tonne 'lot' of nickel powder. Reduction of each batch of solution, that is, each densification, takes about an hour. This means that a lot of nickel is produced in 60–80 hours.

27.3.3. Finishing the Cycle

After reducing the last of the prescribed number of solution batches (about 60), most of the depleted solution is discharged through the flash tank, the autoclave is flushed with nitrogen, and a slurry containing the nickel powder is drained out through the bottom of the autoclave at ambient pressure.

The nickel powder, which now has a particle size of about 0.1–0.2 mm, is then filtered in a rotating pan filter, washed with clean water to remove ammonium sulfate solution and dried in a rotary drier (Cordingley & Krentz, 2005; Impala, 2009).

Some of the nickel is sold directly as nickel powder. Most of the nickel powder, however, is briquetted with polyacrylic acid binder. The product briquettes are sold as is or, more commonly, sintered in hydrogen at 950°C. Sintering under these conditions increases the strength and lowers the sulfur and carbon contents of the briquettes.

At the Sherritt refinery at Fort Saskatchewan, Canada, the content of the nickel powder is 99.8% Ni, 0.1% Co, 0.004% Cu, 0.004% Fe, 0.03% S and 0.01% C.

Subsequent hydrogen sintering of the briquetted nickel powder removes sulfur and carbon to 0.01% S and less than 0.005% C (Egedahl & Collins, 2009).

27.3.4. Preparing the Autoclave for the Next Cycle

Virtually all the nickel entering the reduction autoclave leave as metal powder. However, a small amount deposits on the autoclave surfaces. Fortunately, this deposit is easily removed by leaching with a solution of ammonia and ammonium sulfate under compressed air. Complete removal of the deposited nickel requires about 6–7 hours. Nickel leaching of the autoclave internals is done after every nickel production cycle (60–80 densifications or solution batches).

27.4. APPRAISAL

The nickel powder from hydrogen reduction is less pure than the nickel from electrowinning. However, it is sufficiently pure for most nickel applications.

Hydrogen reduction requires little labor, plant space or power. It is relatively inexpensive, especially if the hydrogen is obtained from waste gas.

It seems to be the process of choice for many new nickel projects.

27.5. SUMMARY

About 240 000 tonnes of nickel metal per year are produced by hydrogen reduction from aqueous solutions of nickel and ammonium sulfate. The feed solutions are produced by the leaching of sulfide mattes and precipitates.

Industrial hydrogen reduction entails reacting 30 bar H_2 gas with 180°C solution in the presence of 0.01 mm metallic nickel 'seed'. The hydrogen reacts with ammoniacal nickel sulfate in the solution to produce (i) nickel metal that plates on the nickel seeds; and, (ii) a solution of ammonium sulfate, which is purified and made into by-product ammonium sulfate fertilizer.

The principal product of the process is nickel powder with a particle size of between 0.1 and 0.2 mm. The powder is washed, dried and usually briquetted before being sold to customers.

REFERENCES

Collins, M. J., Buban, K. R., Holloway, P. C., et al. (2009). Ambatovy laterite ore preparation plant and high pressure acid leach pilot plant operation. In J. J. Budac, R. Fraser & I. Mihaylov, et al. (Eds.), *Hydrometallurgy of nickel and cobalt 2009* (pp. 499–510). CIM.

Cordingley, P. D., & Krentz, R. (2005). Corefco refinery – review of operations. In J. Donald & R. Schonewille (Eds.), *Nickel and cobalt 2005 challenges in extraction and production* (pp. 407–425). CIM.

Egedahl, R. D., & Collins, M. J. (2009). Vital status of Sherritt nickel refinery workers (1954–2003). In J. J. Budac, R. Fraser & I. Mihaylov, et al. (Eds.), *Hydrometallurgy of nickel and cobalt 2009* (pp. 689–699). CIM.

Hayward, K. (2008). Murrin Murrin leads the way. *Sulfuric Acid Today, 14*, 7–10.

Impala. (2009). Impala Platinum Limited: BMR [Base Metals Refinery] operation.

Makinen, T., Fagerlund, K., Anjala, Y., & Rosenback, L. (2005). Outokumpu's technologies for efficient and environmentally sound nickel production. In J. Donald & R. Schonewille (Eds.), *Nickel and cobalt 2005 challenges in extraction and production* (pp. 71–89). CIM.

Palmer, J., Malone, J., & Loth, D. (2005). WMCR – Kalgoorlie nickel smelter operations overview 1972–2005. In J. Donald & R. Schonewille (Eds.), *Nickel and cobalt 2005 challenges in extraction and production* (pp. 441–455). CIM.

Spandiel, T. (2008). Impala platinum base metal refinery. Southern African Hydrometallurgy Conference, SAIMM.

SUGGESTED READING

Saarinen, T., Lindfors, L. -E., & Fugleberg, S. (1998). A review of the precipitation of nickel from salt solutions by hydrogen reduction. *Hydrometallurgy, 47*, 309–324.

Part II

Extractive Metallurgy of Cobalt

Chapter 28

Cobalt – Occurrence, Production, Use and Price

About half the global output of primary cobalt is produced as a by-product of nickel extraction. As a result, cobalt has been discussed in previous chapters on the extraction of nickel. The other half of the global output of cobalt is from copper–cobalt mines, located mainly in Zambia and the Democratic Republic of Congo.

The purpose of this chapter is to discuss the occurrence, extraction methods, uses and price of cobalt.

28.1. OCCURRENCE AND EXTRACTION

28.1.1. Cobalt in Sulfide Ores

Approximately a quarter of the primary production of cobalt is from sulfide ores. Co is always present in nickel sulfide ores, mostly in pentlandite, $(Ni,Fe,Co)_9S_8$, which is by far the most common nickel sulfide mineral. On average, pentlandite contains about $1.1 \pm 0.3\%$ Co. The nickel content is about 36%, which means that the mass ratio of cobalt to nickel in pentlandite is typically 0.03.

The ore is processed by flotation to produce a pentlandite concentrate, which is further processed by sulfide smelting. The recovery of cobalt to the converter matte during sulfide smelting is between 50 and 75%; the remainder is mostly lost in slag. The matte is then refined where the recovery of cobalt is 95% or higher. This is discussed in more detail in Chapters 24–27.

28.1.2. Cobalt in Laterite Ores

Approximately one quarter of the primary production of cobalt is from laterite ores. Cobalt is always present in nickel laterite ores, because it is always present in the unlaterized precursor igneous rock (Freyssinet, Butt, Morris, and Piantone, 2005).

Most laterite ores are smelted to produce ferronickel. Cobalt cannot be recovered separately during ferronickel production, and for this reason, laterite ores account for a smaller fraction of cobalt production than the amount of cobalt contained in the ore.

Extractive Metallurgy of Nickel, Cobalt and Platinum-Group Metals. DOI: 10.1016/B978-0-08-096809-4.10028-0

Cobalt is, however, recovered from the limonite layer of laterite ores, which is leached in sulfuric acid. The cobalt in these ores is presented in goethite, (Fe,Ni,Co)OOH. The goethite is leached, and the cobalt is recovered by sulfide precipitation, re-dissolution, cobalt solvent extraction and then electrowinning or hydrogen reduction.

28.1.3. Copper-Cobalt Ores from the Democratic Republic of the Congo and Zambia

The ores of the Central Africa Copperbelt in the Democratic Republic of Congo and Zambia contain large amounts of cobalt. About half of the global production of cobalt originates from this region, although it is often refined elsewhere.

The deposits were originally sulfidic. The principal cobalt mineral was carrollite, Co_2CuS_4, which occurs with chalcocite, Cu_2S, and digenite, Cu_9S_5.

Rain, air and vegetative acids have extensively weathered some of these deposits. After weathering, the deposits contain mainly heterogenite, CoOOH, and sphaerocobaltite, $CoCO_3$, in association with malachite, $CuCO_3 \cdot Cu(OH)_2$, and chrysocolla, $CuO \cdot SiO_2 \cdot H_2O$, hosted in dolomitic rock. Material mined industrially from these weathered ores typically contains 0.3% Co and 3% Cu.

These weathered ores are readily dissolved by reductive leaching in sulfuric acid using sulfur dioxide to yield a solution of copper and cobalt sulfate. Solvent extraction can be used to separate the cobalt and copper in the leach solution. The separate solutions are purified, and high-purity copper and cobalt are electrowon from the separate solution.

Sulfide ores occur beneath these weathered ores. Mining and extraction of cobalt from these ores are usually more expensive than extraction from the weathered ores. As a result, exploitation of these deeper sulfide ores is likely to slower than the weathered deposits.

The sulfide ores are readily treated for cobalt and copper production by standard mineral processing, pyrometallurgical and hydrometallurgical techniques. Cobalt metal produced electrolytically in the Democratic Republic of Congo is shown in Figure 28.1.

28.1.4. Other Sources

A small amount of cobalt, between 1000 and 2000 tonnes/year, is produced from the arsenic ores mined near Bou Azzer, Morocco (Favreau, 2007; Managem, 2005). The major mineral in this ore is skutterudite, $(Co,Fe,Ni)As_x$, where x is between 2 and 3.

28.2. RECYCLING OF COBALT

In addition to primary cobalt production, considerable cobalt metal and chemicals are made from recycled end-of-use-scrap. It is estimated that, on

FIGURE 28.1 Cobalt metal, produced electrolytically in the Democratic Republic of Congo. Photograph courtesy of Katanga Mines Limited.

average, a cobalt chemical or alloy contains 80% primary cobalt and 20% recycled cobalt (Shedd, 2010).

28.3. USES OF COBALT

Cobalt minerals have been used as pigments for thousands of years. This was the only use of cobalt until the 20th Century. Cobalt metal was first isolated in 1730 (Habashi, 2009) and first used in alloy form in 1907.

The main uses of cobalt are listed in Table 28.1. The biggest uses are in battery chemicals and high-temperature superalloys.

28.4. GLOBAL MINE PRODUCTION

The global mine production of cobalt since 1995 is shown Figure 28.2. Production has grown from approximately 20 000 tonnes/year in 1995 to about 65 000 tonnes/year in 2010. This increase in production is driven mainly by increased cobalt use in developing countries, particularly China.

Mine production by country is shown in Table 28.2. The Democratic Republic of Congo is by far the biggest miner and China is the biggest refiner of cobalt.

28.5. PRICE

The cobalt price since 1982 is shown in Figure 28.3. The volatility in the price is notable. However, the average price adjusted for inflation has not increased. This is likely due to the increasing availability of cheaply mined cobalt ore from the Democratic Republic of Congo.

TABLE 28.1 Estimated Uses of Cobalt in 2008

Use (example)	Fraction of total use, %
Batteries (rechargeable)	27
Superalloys (jet engines)	19
Cutting tools (crushed tungsten carbide in a cobalt metal matrix)	14
Pigments (glass, enamels, plastics, ceramics, artist colors, fabrics)	10
Catalysts (petroleum desulfurization, polyethylene terephthalate manufacture)	9
Magnets, all types	7
Tire adhesives, drying agents, soaps (rubber-steel adhesion with Co boroacylate)	6
Wear-resistant alloys and coatings (CoCrW alloys, CoCr prosthetic implants)	4
Miscellaneous	4

Source: Cobalt Development Institute (2009) Batteries and alloys are the dominant use. World cobalt consumption in 2008 (all forms) was about 70 000 tonnes.

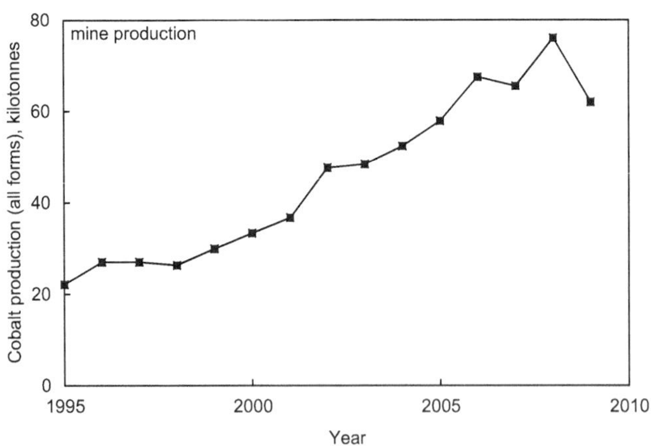

FIGURE 28.2 World production of primary cobalt since 1995. Ore destined for ferronickel smelting (which does not recover cobalt) is excluded. The data are from United States Geological Survey cobalt reports from 1996 to 2010 (Shedd, 2010).

TABLE 28.2 Estimated 2008 Cobalt Production (±20%) by Country

	Primary cobalt production, kilotonnes	
Country	Mined[a]	Refined
Australia	6	4
Belgium	0	3
Brazil	1	1
Canada	8	6
China	2	15–25
Congo, Democratic Republic of	20–30	1
Cuba	4	0
Finland	1	9
France	0	0.3
India	0	1
Japan	0	1
Morocco	2	2
New Caledonia	1	0
Norway	0	4
Philippines	1	0
Russia	6	3
South Africa	0.3	0.3
Uganda	1	1
Zambia	8	4
Total	65	60

[a]*Excludes cobalt in nickel laterite ores from which cobalt is not extracted.*
Source: Cobalt Development Institute, 2009; Shedd, Khatri, Roberts, and Wallace, 2009; Yager, 2009.

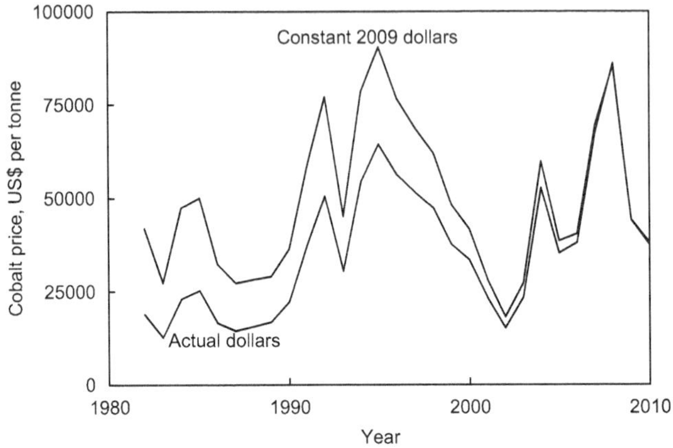

FIGURE 28.3 Price of cobalt since 1982 (Sahr, 2009; Shedd, 2010). The prices are for United States spot cathode as reported in Metals Week (1982–1992) and Platts Metals Week (1993–2009). *London Metals Exchange* has instituted (February, 2010) a cobalt contract which specifies cobalt with a minimum of 99.3% Co, containing impurities specified by producers of each accepted brand. The 2010 price in the graph is the London Metal Exchange July 22, 2010 cash buyer price.

28.6. SUMMARY

Cobalt has a myriad of uses, such as chemicals for rechargeable batteries and high-temperature superalloys. About half of the cobalt is used in chemical form and half in alloy form.

About 60 000 tonnes of cobalt are mined per year. An additional 10 000 or 15 000 tonnes are produced by recycling end-of-use cobalt, mainly in recycled alloys.

About half of the mined cobalt originates from copper-cobalt ores mined in the Copperbelt of the Democratic Republic of Congo and Zambia. The remainder comes equally from sulfide ores containing pentlandite, $(Ni,Fe,Co)_9S_8$, and from lateritic ores containing goethite, (Fe,Ni,Co)OOH.

REFERENCES

Cobalt Development Institute (2009). *Cobalt supply and demand 2008*. Cobalt Development Institute.

Favreau, G. (2007). Famous mineral localities: Bou Azzer, Morocco. *The Mineralogical Record, 38*, 345–407.

Freyssinet, Ph., Butt, C. R. M., Morris, R. C., & Piantone, P. (2005). Ore-forming processes related to lateritic weathering. In J. W. Hedenquist, J. F. H. Thompson, R. J. Goldfarb & J. P. Richards (Eds.), *Economic geology 100th anniversary volume* (pp. 681–722). Society of Economic Geologists Inc.

Habashi, F. (2009). History of cobalt. In J. J. Budac R. Fraser & I. Mihaylov, et al. (Eds.), *Hydrometallurgy of nickel and cobalt 2009* (pp. 157–165). CIM.

Managem. (2005). Managem's global expansion begins in west Africa. *Mining Review Africa, 2*, 10–15.

Sahr, R. C. (2009). Consumer price index (CPI) conversion factors 1774 to estimated 2009 to convert to dollars of 2009. Oregon State University http://oregonstate.edu/cla/polisci/faculty-research/sahr/sahr. Accessed 28.05.11.

Shedd, K. B. (2010). *Cobalt (Mineral Commodity Survey)*. United States Geological Survey.

Yager, T. R. (2009). Congo (Kinshasa). In *2007 United States Mineral Yearbook*. United States Geological Survey.

SUGGESTED READING

Cailteux, J. L. H., Kampunzu, A. B., Lerouge, C., Kaputo, A. K., & Milesi, J. P. (2005). Genesis of sediment-hosted stratiform copper-cobalt deposits, central African Copper belt. *Journal of African Earth Sciences, 42*, 134–158.

Habashi, F. (2009). History of cobalt. In J. J. Budac, R. Fraser, & I. Mihaylov, et al. (Eds.), *Hydrometallurgy of nickel and cobalt 2009* (pp. 157–165). CIM.

Shedd, K. B., Khatri, R. C., Roberts, L., & Wallace, G. J. (2009). Cobalt. In *2007 United States Mineral Yearbook*. United States Geological Survey.

Chapter 29

Extraction of Cobalt from Nickel Laterite and Sulfide Ores

About half of the global production of primary cobalt is extracted as a by-product from nickel laterite and nickel sulfide ores. This chapter describes the processes by which this cobalt is made and recoveries of cobalt by these processes.

29.1. COBALT EXTRACTION FROM NICKEL LATERITE ORE

Most nickel laterite ore is smelted to ferronickel from which cobalt is not recovered. Some nickel laterite ore is smelted to sulfide matte. Cobalt is recovered from a small quantity of this matte (Eramet, 2010) but most of the matte is oxidation and reduction roasted to melting-grade nickel without extracting cobalt.

Cobalt is, however, produced as a by-product from the leaching of the limonite layer of the laterite ore, where the cobalt is present in goethite, (Fe,Ni,Co)OOH, at high temperatures in sulfuric acid. The leaching reaction is:

$$\underset{\text{in laterite}}{Co(OH)_2(s)} + \underset{\text{98\% sulfuric acid}}{H_2SO_4(\ell)} \xrightarrow{250^\circ C} \underbrace{Co^{2+}(aq) + SO_4^{2-}(aq)}_{\text{leach solution}} + 2H_2O(\ell) \quad (29.1)$$

After counter-current decantation and solution purification, the pregnant solution contains about 0.4 g/L Co^{2+} and 6 g/L Ni^{2+}.

About 95% of the cobalt in the laterite feed is recovered to this purified pregnant solution (Chalkley & Toirac, 1997). The remainder is lost in leaching residues and precipitated iron hydroxides.

Most of the cobalt and nickel in the pregnant solution is first precipitated as a mixed sulfide that is then refined to metallic nickel and cobalt. The rationale for the choice of this process route is discussed in Chapter 10.

The precipitation of cobalt by hydrogen sulfide is given by the reaction:

$$\underbrace{Co^{2+}(aq) + SO_4^{2-}(aq)}_{\text{in purified leach solution}} + \underset{\text{bubbled into leach solution}}{H_2S(g)} \xrightarrow{80-120^\circ C} \underset{\text{in sulfide precipitate}}{CoS(s)} + \underbrace{2H^+(aq) + SO_4^{2-}(aq)}_{\text{sulfuric acid in exit solution}} \quad (29.2)$$

Extractive Metallurgy of Nickel, Cobalt and Platinum-Group Metals. DOI: 10.1016/B978-0-08-096809-4.10029-2

The mixed sulfide precipitate that results typically contains 5% Co and 55% Ni. The recovery of cobalt from solution to precipitate is 98–99% and the remainder is lost in discard solutions (Kofluk & Freeman, 2006).

29.2. REFINING OF COBALT

The process steps for the production of high-purity cobalt from mixed sulfide precipitates are:

(a) leaching using chlorine in hydrochloric acid, or using air in ammonia solutions, or using oxygen in sulfuric acid;
(b) solvent extraction for the separation of cobalt from nickel;
(c) solution purification (before and after solvent extraction); and,
(d) electrowinning or hydrogen reduction for the recovery of cobalt metal.

Steps (a)–(c) were described in Chapters 24 and 25. Cobalt electrowinning and hydrogen reduction are described here.

29.2.1. Electrowinning of Cobalt from Chloride Solutions

The operating conditions for the electrowinning of cobalt from aqueous chloride solutions are given in Table 29.1. These conditions are similar to those for the electrowinning of nickel from chloride solutions. The process produces cobalt with a purity greater than 99.95% Co.

The cathodic reaction for the deposition of cobalt is given as follows:

$$\underset{\substack{\text{in high-purity aqueous}\\\text{electrolyte at}\\\text{cathode faces}}}{Co^{2+}(aq)} + \underset{\substack{\text{electrons supplied to}\\\text{cathodes through}\\\text{external circuit}}}{2e^-} \xrightarrow{60^\circ C} \underset{\substack{\text{high-purity cobalt}\\\text{plated on}\\\text{cathode faces}}}{Co(s)} \qquad (29.3)$$

The reduction potential for this half-reaction is $E^\circ \approx -0.28$ V *vs* standard hydrogen electrode.

A problem encountered during the electrowinning of cobalt in chloride solutions is the formation of CoOOH on the ruthenium-metal oxide-coated titanium anode. The formation of this product appears to corrode the oxide coating and shorten anode service life.

Electrowinning from sulfate solutions is similar to that of electrowinning of nickel, and is discussed in Chapter 26. The major difference between nickel and cobalt electrowinning is that cobalt can be plated from sulfate solutions in cells with or without a separator between anode and cathode compartments.

29.2.2. Hydrogen Reduction of Cobalt

Operating conditions for the hydrogen reduction of cobalt from ammoniacal sulfate solutions are given in Table 29.2 (Cordingley & Krentz, 2005). These conditions are similar to the conditions for hydrogen reduction of nickel.

TABLE 29.1 Industrial Electrowinning of Cobalt from Chloride Electrolyte

Location	Niihama, Japan	Kristiansand, Norway
Cathode production, tonnes per year	2000	4000
Chloride electrolyte		
Composition, g/L		
Co	50	50
Ni	10–50	0.01
Cl		70
Cu	0.0001	
Fe	0.0001	
Pb	<0.0001–0.0003	
Zn	<0.0001	
Temperature, °C	60 (in cells)	60
pH		1
Flowrate in and out of each cell, m^3/min		0.06
Anodes		
Material	Ru oxide-coated Ti wire	Ru oxide-coated Ti wire
Depth × width × thickness, m	1 × 0.8 × 0.002	1 × 0.8 × 0.002
Anode compartments (Figure 22.1)		
Construction materials		
Frame	Polypropylene	Plastic
Width, m	0.03	0.03
Membrane	Polyester	Polyester
Cathodes		
Material	Cobalt 'starter' sheets	Cobalt 'starter' sheets
Depth × width × thickness, m	1 × 0.8 × 0.001	1 × 0.7 × 0.001

(Continued)

TABLE 29.1 Industrial Electrowinning of Cobalt from Chloride Electrolyte—cont'd

Location	Niihama, Japan	Kristiansand, Norway
Center to center distance apart, m	0.11	0.14
Number per cell	53 anodes, 52 cathodes	49 anodes, 48 cathodes
Plating time, days	7	7
Cells		
Number	8	
Materials	Concrete with fiber-reinforced polymer inserts	Reinforced monolithic concrete with fiber-reinforced polymer inserts
Inside length × width × height, m	6 × 1 × 1.1	7 × 1 × 1.6
Electrical		
Cell voltage, V		3.7
Cell current, A		15 000
Cathode current density, A/m^2		220
Current efficiency, %		90
Cathode deposit (composition%)		
Co	99.98	>99.95
C		<0.002
Cu	0.0001–0.0005	<0.0005
Fe	<0.0001–0.0008	<0.001
Ni	0.005–0.02	<0.03
Pb	0.0001–0.0005	<0.0002
Zn	<0.0001	<0.0002

The feed is high-purity cobalt chloride solution. The product is high-purity sheets of electrowon cobalt, greater than 99.95% Co.
Source: Akre *et al.*, 2005; Niihama, 2007.

TABLE 29.2 Industrial Cobalt Hydrogen Reduction Data

Location	Fort Saskatchewan, Canada	Springs, South Africa
Cobalt production, tonnes/year	4000	300
Starting solution		
Description	Purified Co-rich solution from air/NH_3 leaching of Moa Bay sulfide precipitate	Purified Co-rich solution
Pre-treatment	Hexamine precipitation	Pentamine precipitation
Composition, g/L		
Ni	<0.03	
Co	75–80	
Cu	Estimated <0.001	
NH_3	50	
$(NH_4)_2SO_4$	550	
NH_3/Ni molar ratio	2.4	
Reductant		
Source	Waste H_2-bearing gases	Petroleum refinery
Method of introduction into liquid	Bubbling below surface	Bubbling below surface
Autoclaves (number)	2	
Working volume, m^3	7	
Construction material	Stainless steel	Stainless steel
Operating temperature, °C	175	
Heating method	Steam coils	Steam coils
Operating pressure	35	
Operating method	Make 0.01 mm Co seeds then reduce 60–80 batches of feed solution, remove cobalt lot, start again	Make 0.01 mm seeds then reduce 60–80 batches of feed solution, remove cobalt lot, start again

(*Continued*)

TABLE 29.2 Industrial Cobalt Hydrogen Reduction Data—cont'd

Location	Fort Saskatchewan, Canada	Springs, South Africa
Total batch time	~45 min, 25% for reduction	
Post-reduction treatment	Dissolve Co from autoclave walls	
Exit solution (composition, g/L)		
Ni		
Co		
Zn		
$(NH_4)_2SO_4$		
Destination		
Cobalt product (composition, %)		
Ni	<0.01	
Co	99.9	
Cu	0.005	
Fe	0.002	
S	0.03[a]	
C	0.07[a]	
Particle diameter	0.2 mm	0.2 mm
Destination after drying	Sold or briquetted and sold, briquettes often H_2 sintered at 950°C	Sold or briquetted and sold, briquettes often H_2 sintered at 950°C

[a]*<0.002% S and <0.004% C after powder sintering under hydrogen.*
Source: Cordingley and Krentz, 2005.

The hydrogen reduction of cobalt takes place by the following reaction:

$$\underset{\text{in leach solution}}{Co(NH_3)_2^{2+}(aq)} + \underset{\substack{\text{hydrogen gas}\\ \text{35 bar bubbled}\\ \text{into solution}}}{H_2(g)} \xrightarrow{175^\circ C} \underset{\substack{\text{cobalt metal}\\ \text{deposited on}\\ \text{cobalt seed}}}{Co(s)} + \underset{\substack{\text{sent to ammonium}\\ \text{sulfate crystallizer}}}{2NH_4^+(aq)} \quad (29.4)$$

The operational differences between the hydrogen reduction of cobalt and nickel are as follows:

(a) cobalt seed is made by precipitation from cobaltous diamine solution using sodium sulfide particles to provide nucleation sites and sodium cyanide as a catalyst;
(b) cobalt reduction is slightly faster than nickel reduction;
(c) the cobalt powder has a slightly larger particle size and the particles are more irregular due to particle agglomeration; and,
(d) inadvertently plated cobalt inside the autoclave is removed by circulating cobaltic hexamine solution through the autoclave after each lot of cobalt powder has been removed from the autoclave.

In other ways, the chemical behaviors of nickel and cobalt are quite similar. The recovery of cobalt from the sulfide precipitate to metal is about 99%. The cobalt that is not recovered is lost in iron residues (Kofluk & Freeman, 2006).

29.2.3. Overall Recovery of Cobalt

Since the recovery of cobalt from laterite to the leach solution is approximately 95%, the recovery from leach solution to sulfide precipitate is about 99%, and the recovery from sulfide precipitate to metal is about 99%, the overall recovery of cobalt to metal is approximately 93%.

29.3. EXTRACTION OF COBALT FROM NICKEL SULFIDE ORES

Cobalt occurs in all nickel sulfide ores, mostly in pentlandite, $(Fe,Ni,Co)_9S_8$. The concentration of cobalt in the ore varies from 0.01% Co in sulfide ores mined for platinum-group metals to 0.15% Co in high-grade nickel-copper sulfide ores, such as that mined at Voisey's Bay (Wells *et al.*, 2007).

The cobalt in these sulfide ores is almost always recovered by the following process steps:

(a) producing a flotation concentrate;
(b) smelting and converting the concentrate to cobalt-nickel-copper-sulfur matte that has a low iron content;
(c) leaching the matte using chlorine in hydrochloric acid, or using air in ammonia solutions, or using oxygen in sulfuric acid; and,
(d) producing cobalt metal by electrowinning or hydrogen reduction from purified leach solutions.

These processes are described in Chapters 11 through 23 for nickel; they are reexamined here from the point of view of the recovery of cobalt.

29.3.1. Behavior of Cobalt in Flotation Concentration

Fortunately, efforts made to increase the recovery and grade of nickel in flotation concentrate are rewarded by identical increases in recovery grade of cobalt. The direct correspondence between the recoveries of nickel and cobalt is shown in Figure 29.1. Typical industrial recoveries to the flotation concentrate of cobalt are 90 ± 5%.

29.3.2. Behavior of Cobalt in Smelting

Cobalt lies between iron and nickel on the Periodic Table and its smelting behavior also falls between that of iron and nickel. For example, cobalt tends to oxidize more than nickel but less than iron.

The economic result of this observation for smelting is that the recovery of cobalt to matte is always less than the recovery of nickel to matte.

29.3.3. Behavior of Cobalt during Converting

The optimal sulfide product from the smelter for refining to high-purity metals is a nickel-cobalt-copper-iron sulfide matte that is low in iron (0.5–4%). The metals in this type of matte can be efficiently recovered by leaching and other extraction processes.

The low-iron matte is almost always made by oxidizing the iron from the furnace matte with air (or air enriched with oxygen) in Peirce-Smith converters. Since the objective of converting is to oxidize iron from molten sulfide matte and since cobalt behaves somewhat like iron (thermodynamically), it is

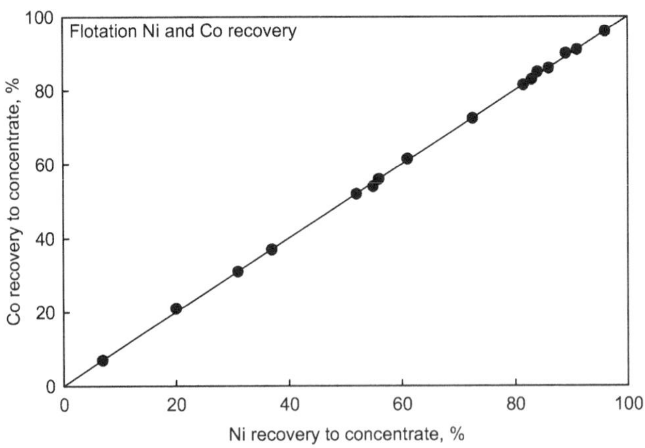

FIGURE 29.1 Experimental flotation recoveries of nickel and cobalt to concentrate (Kelebek & Tukel, 1999). They are nearly identical, suggesting that cobalt occurs in the same minerals as nickel, mostly in pentlandite.

inevitable that some cobalt is oxidized and, like the iron, reports in the converter slag. Converter slags in nickel smelters typically contain 0.5–1.5% Co. The longer the converter blow, the higher the cobalt in the slag, because most of the iron in the feed matte has been already been oxidized (Matousek, 1993).

29.3.4. Recovery of Cobalt from the Converter Slag

Cobalt is mostly recovered from converter slag in an electric furnace, either by (i) sending the slag to a separate slag-settling furnace (slag-cleaning furnace); or by, (ii) recycling the slag to the electric smelting furnace.

Cobalt is recovered in the molten nickel-copper-cobalt sulfide matte, which is solidified and then leached.

29.3.5. Overall Smelter Recoveries of Cobalt

Overall recoveries of cobalt in nickel sulfide smelters are given in Table 29.3. They vary from 30% to 80%. This is significantly less than that of nickel (97%) and copper (95%). However, it does demonstrate that, with care and attention, recoveries of cobalt of up to 80% can be attained.

29.3.6. Behavior of Cobalt in Matte Leaching

The recovery of cobalt in the refining of sulfide precipitate (by leaching, solution purification and electrowinning/hydrogen reduction) is about 99%. The loss of cobalt is mainly to leach residue where the cobalt remains undissolved as the ferrite, $CoFe_2O_4$. The loss of cobalt increases with increasing iron in the feed to the sulfide precipitation plant (Kofluk & Freeman, 2006).

The recovery of cobalt in the hydrometallurgical refining of converter matte is similar to that of sulfide precipitates.

29.3.7. Overall Recovery of Cobalt from Sulfide Ores

The industrial recoveries of cobalt from sulfide ores are $90 \pm 5\%$ in the concentrator, $50 \pm 20\%$ in the smelter and $95 \pm 2\%$ in the matte-to-metal refinery, which results in an overall recovery of about 40%.

The smelter is the worst area in terms of the recovery of cobalt. This indicates that companies processing sulfide ores would benefit most by improving the recovery of cobalt in the smelter.

29.4. SUMMARY

About half the global production of primary cobalt is from nickel laterite and sulfide ores. Most of this cobalt is produced by the leaching of laterites and the

TABLE 29.3 Cobalt Recoveries in Smelting and Converting. Matte Destinations for Cobalt Recovery are also Outlined

Smelter	Cobalt recovery from concentrate to converter matte, %	Converter matte destination
Flash furnace smelting		
Harjavalta, Finland	70	Oxygen-sulfuric acid leach. Cobalt is removed from pregnant solution by solvent extraction. Cobalt solution is sold to OMG, Kokkola, Finland
Jinchuan, China	54	Various processes leading to electrowon cobalt
Kalgoorlie, Australia		Air-ammonia leach in Kwinana, Australia, where Co is recovered in a 30% Co, 24% Ni, 32% S (remainder Fe) sulfide precipitate, which is toll refined in Europe
Nadezda (flash), Russia	69	Nickel refining, cobalt recovery
Selebi-Phikwe, Botswana	26	Chlorine leach in Kristiansand, Norway. Cobalt is removed from pregnant solution by solvent extraction. Cobalt metal is electrowon
Sudbury, Canada	47	Matte separation, oxidation, carbonyl refining in Sudbury, Canada, and Clydach, Wales. Cobalt residue is leached. Cobalt is precipitated as Co-Ni carbonate, which is sent to Port Colborne, Canada, for electrowinning of cobalt (Sabau & Bech, 2007)
Electric furnace smelting		
Doniambo, New Caledonia		Chlorine leach in Le Havre-Sandouville, France. Cobalt is removed from pregnant solution by solvent extraction. Cobalt is produced as cobalt chloride solution (Eramet, 2010)

TABLE 29.3 Cobalt Recoveries in Smelting and Converting. Matte Destinations for Cobalt Recovery are also Outlined—cont'd

Smelter	Cobalt recovery from concentrate to converter matte, %	Converter matte destination
Falconbridge, Canada	75 (estimated)	Chlorine leach in Kristiansand, Norway. Cobalt is removed from pregnant solution by solvent extraction. Cobalt metal is electrowon
Norilsk, Russia nickel plant	66	Nickel refining, cobalt recovery
Pechenga, Russia	75	Nickel refining, cobalt recovery
Thompson, Canada	51	$Co(OH)_2$ precipitation from matte anode refining electrolyte. Purified precipitate sent to Co electrowinning in Port Colborne, Canada (Jebbink, Stefan, Neff, & Tomlinson, 2006)

Source: Warner *et al.*, 2007.

smelting of sulfide concentrates. Both these processes produce an intermediate sulfide that is refined by hydrometallurgical techniques.

Cobalt is not recovered during laterite to ferronickel smelting; instead, it remains in the ferronickel product.

REFERENCES

Akre, T., Haarberg, G. M., Thonstad, J., et al. (2005). Electrowinning of cobalt from chloride solution – a pilot plant study. In J. Donald & R. Schonewille (Eds.), *Nickel and cobalt 2005 challenges in extraction and production* (pp. 359–374). CIM.

Chalkley, M. E., & Toirac, I. L. (1997). The acid pressure leach process for nickel and cobalt laterite. In W. C. Cooper & I. Mihaylov (Eds.), *Nickel/cobalt 97, Vol. I. Hydrometallurgy and refining of nickel and cobalt* (pp. 341–353). CIM.

Cordingley, P. D., & Krentz, R. (2005). Corefco refinery – review of operations. In J. Donald & R. Schonewille (Eds.), *Nickel and cobalt 2005 challenges in extraction and production* (pp. 407–425). CIM.

Eramet (2010). Le Havre-Sandouville Refinery. www.eramet.fr/us/PRODUTION_GALLERY_CONTENT/DOCUMENTS/Nos_metiers/Nickel/UK/HavreSandouville_Eng7.pdf. Accessed 28.05.11.

Jebbink, P., Stefan, R., Neff, D., & Tomlinson, M. (2006). Expanding the cobalt recovery circuit at the Thompson nickel refinery. *JOM Journal of the Minerals, Metals and Materials Society, 58*, 37–42.

Kelebek, S., & Tukel, C. (1999). The effect of sodium metabisulfite and triethylenetetramine system on pentlandite–pyrrhotite separation. *International Journal of Mineral Processing, 57*, 135–152.

Kofluk, R. P., & Freeman, G. K. W. (2006). Iron control in the Moa Bay laterite operation. In J. E. Dutrizac & P. A. Riveros (Eds.), *Iron control technologies* (pp. 573–589). CIM.

Matousek, J. W. (1993). The behavior of cobalt in pyrometallurgical processes. Fundamental Aspects. In R. G. Reddy & R. N. Weizenbach (Eds.), *Proceedings of the Paul E. Queneau International Symposium, Extractive Metallurgy of Copper, Nickel and Cobalt, Vol. I* (pp. 129–142). TMS.

Niihama. (2007). Outline of MCLE [Mixed Sulfide Chlorine Leach Electrowinning] Process. Company handout May 29, 2007. Sumitomo metal mining Co., Ltd., Niihama Nickel Refinery, Shikoku, Japan.

Sabau, M., & Bech, K. (2007). Status and improvement plans in Inco's electrowinning tankhouse. In G. E. Houlachi, J. D. Edwards & T. G. Robinson (Eds.), *Copper 07-Cobre 07 Proceedings of the Sixth International Conference, Vol. V. Electrorefining and Electrowinning* (pp. 439–450). CIM.

Warner, A. E. M., Diaz, C. M., Dalvi, A. D., et al. (2007). JOM world nonferrous smelter survey part IV: nickel sulfide. *JOM Journal of the Minerals, Metals and Materials Society, 59*, 58–72.

Wells, P., Langlois, P., Barrett, J., Holmes, J., Xu, M., & Labonte, G. (2007). Flowsheet development, commissioning and start-up of the Voisey's Bay Mill. In J. Folinsbee (Ed.), *Proceedings of the 39th Annual Meeting of the Canadian Mineral Processors* (pp. 3–18). Ottawa, Canada: Canadian Mineral Processors, a division of CIM.

SUGGESTED READING

Hossain, M. R., Alam, S., & Abdi, M. A. (2009). Removal of impurities from cobalt electrolyte solution. In J. J. Budac, R. Fraser & I. Mihaylov, et al. (Eds.), *Hydrometallurgy of nickel and cobalt 2009* (pp. 623–632). CIM.

Louis, P. E. (2009). Cobalt electrowinning. In J. J. Budac, R. Fraser & I. Mihaylov, et al. (Eds.), *Hydrometallurgy of nickel and cobalt 2009* (pp. 553–569). CIM.

Morimitsu, M., & Uno, K. (2009). A novel electrode for cobalt electrowinning to suppress CoOOH deposition. In J. J. Budac, R. Fraser & I. Mihaylov, et al. (Eds.), *Hydrometallurgy of nickel and cobalt 2009* (pp. 571–580). CIM.

Chapter 30

Production of Cobalt from the Copper–Cobalt Ores of the Central African Copperbelt

About half of the global mine production of cobalt comes from the Copperbelt region of the Democratic Republic of Congo and Zambia (Shedd, 2010).

The objectives of this chapter are the following:

(a) to describe copper and cobalt ore deposits in the Copperbelt;
(b) to indicate how high-purity cobalt and copper are made from these deposits; and,
(c) to suggest how the deposits will be exploited as mining moves down into the sulfide layer.

The locations of the main producers of cobalt from this region are shown in Figure 30.1. The production data for these operations is given in Table 30.1.

30.1. TYPICAL ORE DEPOSIT

The Copperbelt in Zambia and the Democratic Republic of the Congo contains over one third of the mineral reserves of cobalt and one tenth of the mineral reserves of copper in the world.

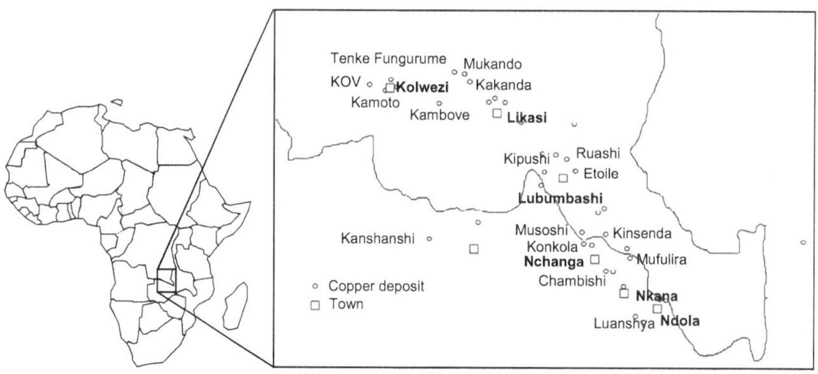

FIGURE 30.1 The location of the main deposits of the central African Copperbelt.

Extractive Metallurgy of Nickel, Cobalt and Platinum-Group Metals. DOI: 10.1016/B978-0-08-096809-4.10030-9

TABLE 30.1 The Main Producers of Cobalt from the Central African Copperbelt

Owner, site	Cobalt production[a], tonnes/year	Copper production[a], tonnes/year
DRC		
Metorex, Ruashi	4000	30 000
Freeport McMoRan, Tenke Fungurume	9000	120 000
Kakanga Mining, KOV, Luilu	3400	55 000
Chemaf, Etoile	1200	11 000
ENRC, Luita	Under construction	20 000
FQM, Lonshi, Kingamyambo Musonoi Tailings (KMT)	Currently under dispute	
Zambia		
Konkola Copper Mines, Mufulira and Nkana	2000	134 000
Metorex, Sable	400	10 000
ENRC, Chambishi	6000	25 000

[a]*Approximate production.*

It is a 500-million-year-old mountain chain, which formed when two large pieces of continental crust, the Kalahari Craton and the Congo Craton, collided. The collision also resulted in tectonic movement, which caused significant folding. In some regions, the deposits are vertical or even inverted.

This collision is thought to have remobilized base metals that were already present in the marine sediments that had accumulated in the basin between the two cratons. These brines then concentrated the base metals, either along stratigraphic boundaries, fractures and faults or within traps, such as the nose of a fold.

The mineralization within the nose of a weathered fold, which now forms a mountain top, is illustrated in Figure 30.2. There are two zones: (i) weathered oxide deposits near the surface; and, (ii) an unweathered sulfide deposits below these oxide deposits. The weathered zone generally extends to a depth between 70 m and 150 m, but varies considerably between deposits. With increasing depth, the weathered zone gives way to a transitional zone, consisting of mixed oxides and sulfides, which, in turn, gives way to the sulfide ore at depths greater than 250 m.

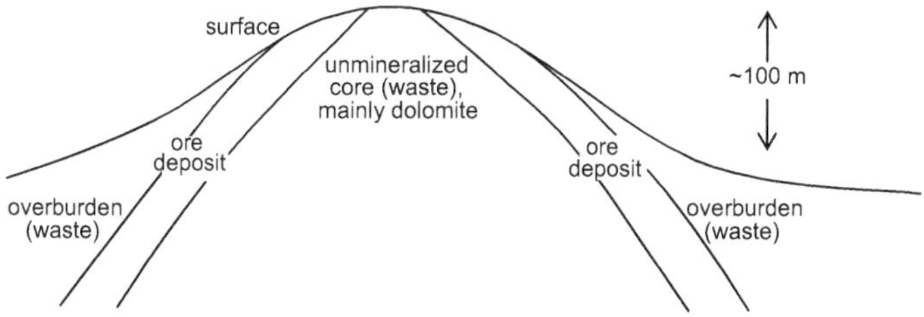

FIGURE 30.2 A schematic diagram of cobalt–copper ore deposit in the Democratic Republic of Congo. Mining starts at the top of the mound and moves downwards, so that the waste-to-ore ratio increases with time (Nilsson *et al.*, 2009). Ore grade and mineralization vary throughout the deposit. Representative concentrations of cobalt and copper are 0.4% and 4%, respectively. The ore deposit is about 40 m thick.

The main copper and cobalt minerals in the oxide zone are chrysocolla, $CuO \cdot SiO_2 \cdot 2H_2O$, malachite, $CuCO_3 \cdot Cu(OH)_2$, and heterogenite, CoOOH. In the sulfide zone, the main minerals are chalcocite, Cu_2S, digenite, $Cu_{1.8}S$, and carrollite, Co_2CuS_4.

Ore grades commonly vary between 4% and 6% Cu and about 0.4% Co, with the ratio of copper to cobalt of the order of 8:1.

The main gangue minerals are siliceous dolomite, $MgCO_3 \cdot CaCO_3$, and quartz.

30.2. MINING

The weathered part of the Figure 30.2 deposit is easily mined. Surface scrapers, as shown in Figure 30.3, are used at Tenke Fungurume. Bulldozers and diggers are used at other operations. Even though the mining of these ores is relatively straightforward, the amount of waste material that must be removed is large, that is the strip ratio is high.

Sulfide minerals are generally (i) deeper in the deposit; and, (ii) more expensive to mine and to treat for metal recovery. Most mines plan to exploit the sulfides after the oxide cap is exhausted.

30.3. EXTRACTION OF COBALT AND COPPER FROM WEATHERED ORE

Cobalt and copper metals or chemicals are produced by the following hydrometallurgical steps:

(a) reductive leaching using sulfur dioxide gas or sodium metabisulfite as a reductant;
(b) solid/liquid separation and clarification of the solution;
(c) separation of copper from cobalt by solvent extraction;
(d) purification of the aqueous solution;
(e) precipitation of cobalt hydroxide or electrowinning of cobalt; and,
(f) copper electrowinning.

FIGURE 30.3 Surface mining machine, with newly mined ore discharging on the right hand side of the photograph.

30.3.1. Leaching

As mined ore is typically less than 0.25 m in size. It is ground to a particle size of 80% passing 100 μm in a semi-autogenous grinding mill then sent on to leaching using sulfuric acid and sulfur dioxide under atmospheric conditions. A schematic flowsheet of this part of the process is shown in Figure 30.4.

Cobalt is in the trivalent form in heterogenite. Because trivalent cobalt is unstable in solution, heterogenite does not dissolve without reducing the cobalt to the divalent state. This reduction is achieved by bubbling sulfur dioxide with minimal air into the sulfuric acid solution during leaching (Miller, 2009). The concentration of the sulfur dioxide is approximately 10%. The leaching of heterogenite is given by the following reaction:

$$\underset{\text{heterogenite}}{2\text{CoOOH}(s)} + \underset{\substack{\text{sulfuric acid}\\ \text{in water}}}{\text{H}_2\text{SO}_4(\ell)} + \underset{\substack{\text{in 10\% SO}_2\\ \text{gas from}\\ \text{sulfur burning}}}{\text{SO}_2(g)} \xrightarrow{30^\circ\text{C}} \underset{\text{aqueous leach solution}}{2\text{CoSO}_4(aq) + 2\text{H}_2\text{O}(\ell)} \quad (30.1)$$

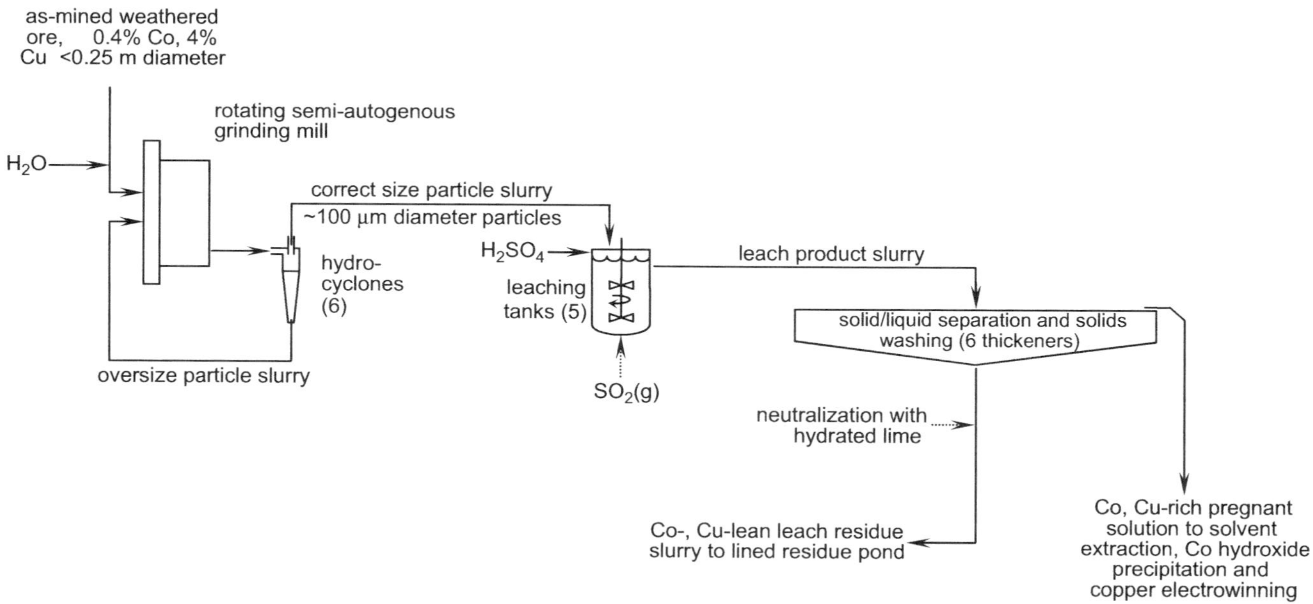

FIGURE 30.4 Flowsheet for grinding and leaching of the weathered cobalt–copper ore. Note the reductive leaching using sulfuric acid and sulfur dioxide.

Copper in the ore is dominated by malachite. The reaction for the leaching of malachite is given as follows:

$$\underset{\text{malachite}}{CuCO_3\cdot Cu(OH)_2(s)} + \underset{\substack{\text{sulfuric acid}\\ \text{in water}}}{2H_2SO_4(\ell)} \overset{30^{\circ}C}{\rightarrow} \underset{\text{aqueous leach solution}}{2CuSO_4(aq) + 3H_2O(\ell)} + \underset{\text{vent}}{CO_2(g)} \tag{30.2}$$

The resulting solution is then separated from unleached solids by counter-current decantation, washing then filtration. The solids are discarded and the solution is transferred to solvent extraction to separate Cu^{2+} from Co^{2+}.

30.3.2. Separating Copper from Cobalt by Solvent Extraction

Copper and cobalt are separated from each other by extracting copper into an organic extractant while leaving cobalt in the aqueous raffinate. A schematic diagram of the process is shown in Figure 30.5. The extraction reaction is:

$$\underset{\text{organic extractant}}{2RH(\ell)} + \underset{\substack{\text{aqueous leach}\\ \text{solution}}}{Cu^{2+}(aq)} \overset{30^{\circ}C}{\rightarrow} \underset{\substack{\text{Cu-loaded}\\ \text{organic extractant}}}{R_2Cu(\ell)} + \underset{\text{sulfuric acid}}{2H^+(aq)} \tag{30.3}$$

The stripping reaction is the reverse of this reaction.

The extractants most commonly employed are oximes, such as LIX 984N, which is a mixture in equal parts of aldoxime and ketoxime (Cognis, 2010), or AGORGA M5640, which is an ester-modified aldoxime (Cytec, 2010). The concentration of the extractant is between 20% and 35%. The diluent is usually ShellSol 2325, a petroleum distillate with a high-flash point (Shell, 2009).

A flowsheet for the process is shown in Figure 30.5. It entails the following operations:

(a) extracting copper from clarified aqueous leach solution into organic extractant;
(b) separating the copper-loaded organic phase from the aqueous phase;
(c) pumping the copper-rich organic phase to stripping and hence to copper electrowinning; and,
(d) pumping the cobalt-rich raffinate to impurity removal and cobalt hydroxide precipitation.

30.3.3. Purification of the Cobalt-Rich Raffinate

The major impurities in the cobalt-rich raffinate are iron, aluminum, manganese and copper. All of these impurities must be removed in order to produce high-purity cobalt hydroxide (Fisher & Treadgold, 2009).

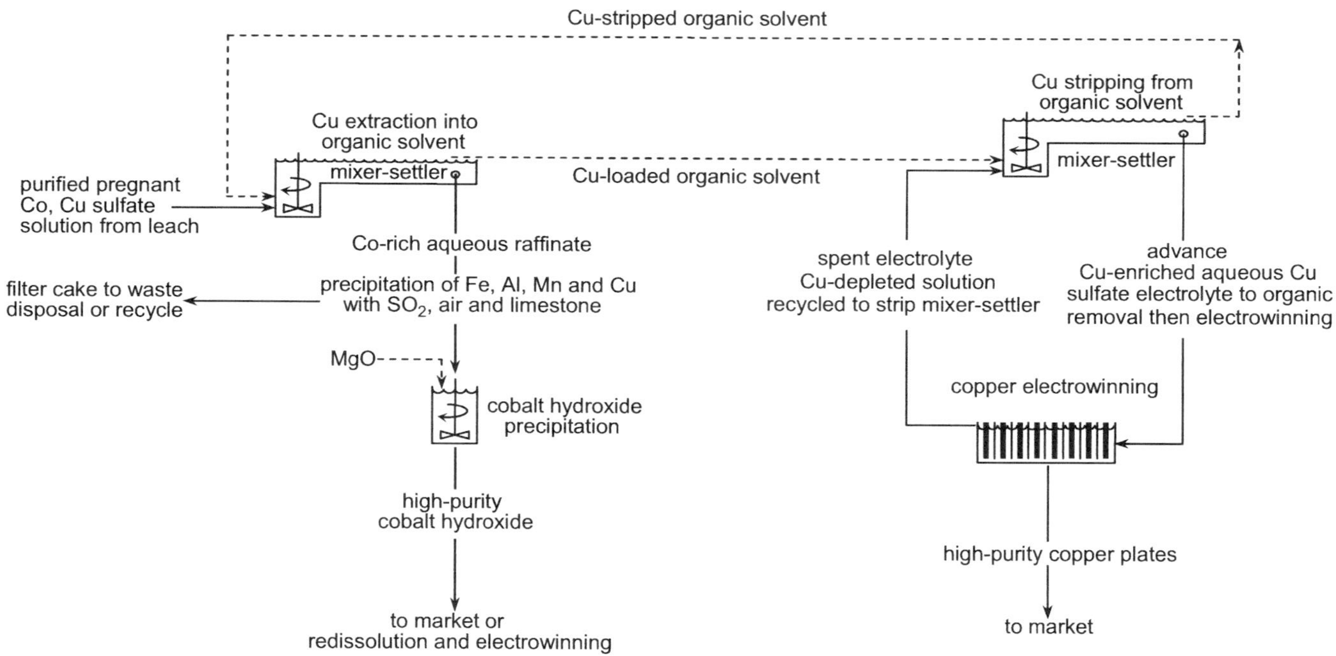

FIGURE 30.5 Production of high-purity cobalt and copper from the purified pregnant solution containing copper and cobalt sulfate, based on the Tenke Fungurume plant, which produces about 100 000 tonnes of copper per year and 8000 tonnes of cobalt per year. A typical plant has three to four extraction mixer-settlers and two strip mixer-settlers for the copper extraction circuit. A portion of the raffinate is recycled to cobalt leaching (not shown). Note the precipitation of cobalt hydroxide (lower left). Cobalt hydroxide is a suitable starting point for production of cobalt chemicals and/or redissolution and cobalt electrowinning.

Iron, aluminum and manganese are removed in three stages, with increasing values of the pH at each stage. In the first stage, air and sulfur dioxide are sparged into agitated tanks at a pH of 3.2 to oxidize the iron, and particularly, the managanese. Limestone is used to maintain the pH. The ratio of air to sulfur dioxide is critical because if too much sulfur dioxide is added, the conditions become reductive and none of these impurities precipitate.

In the second stage, the pH is raised to a value of 4.7 using lime. Aluminum and copper are precipitated in this stage.

In the third stage, the pH is raised to 6.0 using lime. At this pH, copper precipitates as the basic sulfate, $CuSO_4 \cdot 2Cu(OH)_2$.

30.3.4. Precipitation of Cobalt Hydroxide

Cobalt is precipitated from the solution as high-purity cobalt hydroxide, $Co(OH)_2$, at values of the pH between 7.8 and 8.5 using MgO. Cobalt hydroxide product is shown in Figure 30.6. The precipitate is recovered by settling and filtration then dried, bagged, and sent to market for cobalt chemical production or redissolved and electrowon as cobalt metal.

FIGURE 30.6 Cobalt hydroxide produced from weathered copper–cobalt ore from Democratic Republic of Congo. It is sent to market for chemical production or redissolution and cobalt electrowinning.

The advantage of precipitating cobalt hydroxide before electrowinning is that it is readily redissolved with a small amount of aqueous solution of sulfuric acid to make a concentrated cobalt electrolyte.

30.3.5. Costs

The capital cost for a new weathered ore project that mines and delivers 0.5% Co, 4% Cu ore to a hydrometallurgical plant to produce about 100 000 tonnes/year of high-purity electrowon copper cathodes and about 8000 tonnes/year of cobalt in high-purity cobalt hydroxide is about US$1.8 billion (Nilsson, Simpson, and McKenzie, 2009). This capital cost is comprised of the following components:

	US$ billion
Plant and infrastructure	00.9
Taxes, import duties and other	00.07
Capital spares and first fills	00.02
Site preparation	00.3
Owners indirect costs and contingencies	*00.5*
Total	**1.8**

Operating costs for the project are about US$ 85/tonne of ore milled, made up of the following components (Nilsson *et al.*, 2009):

	US$/t of milled ore
Mining	13
Ore-to-product processing	42
Overhead, accommodations, community services	16
Raw material and product transportation	*14*
Total	**85**

These costs are for guidance only. They will vary among sites and ore grades.

30.4. EXPLOITING THE COBALT–COPPER SULFIDE ORE LAYER

Cobalt–copper sulfide ores are being exploited in several Democratic Republic of Congo and Zambia mines (Katanga, 2010). Cobalt and copper are extracted in the following steps:

(a) flotation;
(b) sulfation roasting;

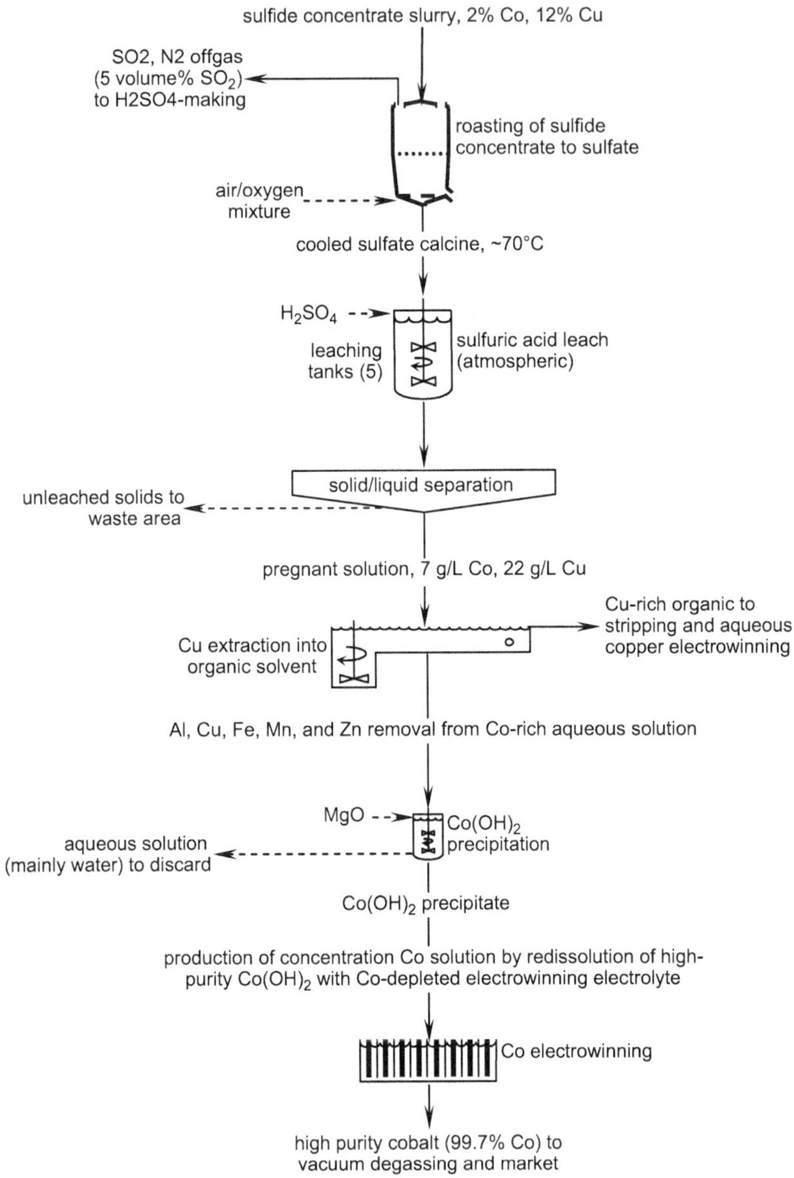

FIGURE 30.7 Production of high-purity cobalt from cobalt–copper sulfide concentrate. Note the sulfate roasting, the precipitation of cobalt hydroxide and the redissolution of the cobalt hydroxide followed by cobalt electrowinning. Precipitation with MgO and redissolution produces cobalt-rich, high-purity electrolyte. *(Adham *et al.*, 2010).

TABLE 30.2 Extraction of Cobalt and Copper from Sulfide and Mixed Sulfide-oxide Ores

Carrollite Co_2CuS_4 – Chalcocite Cu_2S ore	
Feed	Co_2CuS_4–Cu_2S ore from underground mining: 0.4% Co, 4% Cu in dolomite gangue
Process description	Flotation of bulk concentrate by froth flotation with xanthate collector
Equipment	Autogenous grinding mills, particle size control cyclones, 8.5 m^3 mechanical flotation cells, middlings regrind, cleaner flotation
Product	45% Cu, 4% Co concentrate. Recoveries from ore to concentrate: Cu 85%; Co 60%
Mixed sulfide-oxide ore	
Feed	Carrollite, Co_2CuS_4; malachite $CuCO_3 \cdot Cu(OH)_2$; sphaerocobaltite, $CoCO_3$; kolwezite, $Co_2Cu[(OH)_2 \cdot CO_3]$ in dolomite gangue
Description	Flotation of sulphide concentrate as described above. Flotation of separate oxide concentrate as described below
'Oxide' flotation	Flotation of non-sulfide minerals is based on coating the non-sulfide minerals with a superficial sulfide surface (using NaHS), then flotation with xanthate collector
Products	Sulfide concentrate: 40% Cu, 4% Co (80% Cu recovery, 50% Co recovery). Oxide concentrate: 20% Cu, 1.3% Co (60% Cu recovery, 40% Co recovery).

This table describes concentrate production. The data are approximate.

(c) leaching (sometimes along with additional oxide concentrate or ore);
(d) solvent extraction separation of copper from cobalt;
(e) copper electrowinning;
(f) purification of the cobalt-rich raffinate;
(g) precipitation of cobalt hydroxide;
(h) redissolution of cobalt hydroxide; and,
(i) cobalt electrowinning.

A flowsheet for a typical process is given in Figure 30.7 and operating details are given in Tables 30.2–30.6.

TABLE 30.3 Roasting of Cobalt–Copper Sulfide Concentrates to Soluble Sulfate Calcine

Feed to roaster	Sulfide concentrates ~2% Co, 12% Cu, 21% S with dolomite gangue (not the concentrates in Table 30.2)
Feed rate, tonnes/day	600 max; 480 average
Form of feed	Concentrate/water slurry, specific gravity = 1.9
Roaster temperature	695–705°C, chosen to give soluble sulfate calcine
Controlled by	O_2 content of input gas, H_2O in feed slurry
Input gas	30 volume% O_2, 70 volume% N_2 (oxygen-enriched air)
Roaster description	Fluidized bed
Height × diameter	
Construction	Refractory-lined steel
Tuyeres	~0.1 m apart
Products	
Calcine	Sulfate calcine (90% removed from roaster offgas: 10% as bed overflow), quenched from 700°C to 75°C and sent to cobalt–copper leaching plant (90% of sulfides are oxidized to soluble sulfates rather than oxides)
Offgas	3–5 volume % SO_2 (remainder N_2) cooled to 40°C, dedusted then sent to 250 tonnes/day H_2SO_4 single-contact acid plant. Acid is used for leaching and sold.

Industrial details are based on a site visit to Nkana in 2008. Note the roasting temperature of 695–705°C. It is designed to give soluble sulfate calcine rather that insoluble oxide calcine. Annual Co production is about 2000 tonnes.

TABLE 30.4 Details of Industrial Leaching of Copper–Cobalt Sulfate Roaster Calcine, Figure 30.7*

Type of leaching	Open agitation leach in pachuca air-stirred leach tanks
Feed	Roaster calcine, ~2% Co, 12% Cu, Table 30.2
Leachant	Copper solvent extraction raffinate, ~30 g/L H_2SO_4 plus fresh sulfuric acid (made from roaster gas)
Control	pH maintained at <1.5 by adding fresh sulfuric acid, as needed
Product	Sulfate leach solution 7 g/L Co, 22 g/L Cu, <50 ppm solids (after settling, filtration and clarification)
Destination	Copper solvent extraction
Residue	Undissolved calcine, sent to washing and waste disposal

Cooled roaster calcine (~75°C) is leached in aqueous sulfuric acid solution to give a leach sulfate solution of ~22 g/L Cu and 7 g/L Co. The pregnant solution goes to solvent extraction to separate Cu^{2+} from Co^{2+} and eventually to copper electrowinning and cobalt electrowinning.

**Based on a site visit to Nkana in 2008*

TABLE 30.5 Preparation of Advance Electrolyte for Cobalt Electrowinning*

Step 1: Precipitation of high-purity cobalt hydroxide	
Starting solution	Solvent extraction raffinate (~7 g/L Co) from solvent extraction (about 30% of raffinate flow, remainder back to calcine leach)
Impurity removal	
Copper	Removal of copper using secondary solvent extraction with LIX 984N
Iron and residual copper removal	Raise pH to 3.5 with limestone and quicklime
Zinc removal	Cycle 30% of solution through Zn solvent extraction plant (using D2EHPA[a])
Final iron removal	Raise pH to 6.7
Precipitation of high-purity cobalt hydroxide	Raise pH to 8.8 with quicklime to precipitate $Co(OH)_2$
Step 2: Cobalt electrolyte preparation	
Starting materials	(1) High-purity $Co(OH)_2$ prepared as described above; and, (2) Recycled spent electrolyte from cobalt electrowinning
Procedure	Dissolve $Co(OH)_2$ in spent electrolyte, add sulfuric acid to control pH between 6.2 and 6.5 (to prevent $Zn(OH)_2$ from dissolving)
Undissolved solids removal	Clarifiers and pre-coat filters
Soluble sulfides and entrained organic removal	Filtered solution is passed through carbon columns
Nickel removal	Cycle 20% of solution through nickel-removal ion-exchange column
Product	Cobalt sulfate electrolyte >20 g/L Co

The starting solution is solvent extraction raffinate, containing approximately 7 g/L Co. Advance electrolyte is prepared from the raffinate by purification, precipitation of cobalt hydroxide then redissolution of the cobalt hydroxide in recycle spent electrolyte from cobalt sulfate electrowinning. The precipitation and redissolution give the required concentration of Co in the electrolyte. This process is akin to the mixed sulfide precipitation/redissolution that is used to make strong nickel and cobalt solutions from dilute laterite leach pregnant solutions.

[a]*di-(2-ethylhexyl)phosphoric acid.*

**Based on a Visit to Site Nkana in 2008*

TABLE 30.6 Industrial Cobalt Electrowinning Details*

Tankhouse	
Annual cobalt production	~2000 tonnes
Electrolyte	Cobalt sulfate, ~20 g/L Co^{2+}
Cells	92
Size: length × width × depth	
Construction material	Polymer concrete
Anodes/cathodes/cell	13/12
Anodes	
Dimensions	~1 m × 1 m
Material	94% Pb, 6% Sb alloy
Cathodes	
Dimensions	~1 m × 1 m
Material	316L stainless steel
Electrical	
Applied voltage	4–4.5 V
Cathode current density	300 A/m^2 of cathode surface
Operation	
Plating cycle	4–5 days
Cobalt from cathode stripping	Manual
Cobalt plate thickness	~0.003 m
Mist suppression	0.01 m diameter polymer balls
Product cobalt destination	Cathodes are crushed to 0.02 m × 0.04 m flakes in roll crusher, hydrogen gas is removed by degassing under vacuum at 800–840°C in ~1 m diameter, 7 m high electric furnaces
Final steps	Polishing, assaying, grading, then shipping in drums ~230 kg cobalt each
Product grade	99.65–99.7% Co

High-purity cobalt metal is electrowon from cobalt sulfate electrolyte. The input is high-purity cobalt sulfate electrolyte ~20 g/L Co^{2+}. The product is electrowon cobalt metal, 99.65–99.7% Co. Cobalt recovery from concentrate to metal is ~62%.

**Based on a Site Visit to the Nkana in 2008*

30.5. SUMMARY

About half of the global mine production of cobalt comes from the layered sedimentary cobalt–copper ore deposits of central Africa. The deposits usually consist of a weathered surface cap over unweathered sulfide ore below. The host rock is usually siliceous dolomite.

As-mined oxide ores typically contain 0.3–0.5% Co and 3–5% Cu.

The 'oxide' cap is usually mined first. The copper is recovered as high-purity cathodes by leaching, solvent extraction separation from cobalt, solution purification and electrowinning. The cobalt is recovered from solvent extraction raffinate by solution purification followed by precipitation of $Co(OH)_2$.

The cobalt hydroxide precipitate is sold for chemical production or cobalt electrowinning.

REFERENCES

Adham, K., Buchholz, T., Kokourine, A., et al. (2010). Design of copper-cobalt sulphating roasters for Katanga Mining Limited in D.R Congo. In J. Harre (Ed.), *Copper 2010, Pyrometallurgy 1, Vol. 2* (pp. 587–600). GDMB.

Cognis. (2010). LIX 984N. Technical brochure. www.products.cognis.com. Accessed 29.05.11.

Cytec. (2010). ACORGA M5640. Technical brochure. www.acorga.us/specialty_chemicals/acorga.html. Accessed 29.05.11.

Fisher, K. G., & Treadgold, L. G. (2009). Design considerations for the cobalt recovery. Circuit of the KOL (KOV) copper/cobalt refinery, DRC. In *ALTA Nickel-Cobalt Conference (ALTA 2009)*. ALTA Metallurgical Services.

Katanga. (2010). Our operations. www.katangamining.com/kat/operations/. Accessed 29.05.11.

Miller, G. (May 25–30, 2009). (2009). Design of copper-cobalt hydrometallurgical circuits. Paper presented to the ALTA Nickel-Cobalt Conference (ALTA 2009). *Perth Australia*.

Nilsson, J., Simpson, R. G., & McKenzie, W. (2009). *Technical report for the Tenke Fungurume Project, Katanga Province, Democratic Republic of Congo*. Toronto, Canada: Prepared for Lundin Mining Corporation.

Shedd, K. (2010). Cobalt, United States Geological Survey (Mineral Commodity Survey).

Shell. (2009). ShellSol 2325. Technical brochure.

SUGGESTED READING

Cole, P. M., & Feather, A. M (2008). Processing of African copper belt copper-cobalt ores: flowsheet alternatives and options. In B. Moyer (Ed.), *International Solvent Extraction Conference ISEC 2008*. CIM.

Miller, G. (2009). Design of copper-cobalt hydrometallurgical circuits. *ALTA Nickel-Cobalt Conference (ALTA 2009)*. ALTA Metallurgical Services.

Part III

Extractive Metallurgy of the Platinum-Group Metals

Chapter 31

Platinum-Group Metals, Production, Use and Extraction Costs

This chapter describes the sources, production, properties, uses and prices of platinum-group metals. The platinum-group metals are platinum (Pt); palladium (Pd); rhodium (Rh); ruthenium (Ru); iridium (Ir); and osmium (Os).

Platinum-group metals have many useful properties such as:

(a) their ability to catalyze chemical reactions;
(b) their ability to resist corrosion;
(c) their visual appeal;
(d) the ease with which they can be worked; and,
(e) their high conductivities, densities and melting points.

Osmium is, however, rarely used (Shott, 2003). It found use in early production of incandescent light bulbs, but this use was replaced by tungsten (the lighting company OSRAM derives its name from *os*mium and wolf*ram*, another name for tungsten).

31.1. USES OF THE PLATINUM-GROUP ELEMENTS

The uses of platinum, palladium, rhodium, ruthenium and iridium are shown in Tables 31.1–31.5.

Two features stand out from these tables:

(a) the consumptions of palladium and platinum are similar, whereas those for rhodium, ruthenium and iridium are considerably lower; and,
(b) the biggest single use of platinum-group metals is as metallic catalysts for minimizing car and truck exhaust emissions.

A catalytic converter is shown in Figure 31.1.

The role of platinum in vehicle emissions is to catalyze (i) the oxidation of carbon monoxide and hydrocarbons to carbon dioxide; and, (ii) the reduction of NO_x to nitrogen. Representative reactions that are catalyzed by platinum-group

Extractive Metallurgy of Nickel, Cobalt and Platinum-Group Metals. DOI: 10.1016/B978-0-08-096809-4.10031-0

TABLE 31.1 Palladium Uses in 2009*

Use	Tonnes[a]
Car and truck emission-control catalysts	126.0
Electronics	39.5
In conductive Pd-Ag paste for multi-layer ceramic capacitors	
Plated on connectors, circuit boards and copper lead frames	
In resistor chips and hybrid integrated circuits	
Jewelry	25.3
Investment (bars and coins)	19.4
Dental	19.1
Chemical industry	10.1
Catalyst for producing terephthalic acid (feedstock for polyester and polyethylene terephthalate manufacture)	
Catalyst for production of vinyl acetate monomer for paints, adhesives, *etc.*	
Corrosion-resistant gauzes for nitric acid manufacture	
Other	
(Stationary power source emission-control catalysts, oxygen sensors)	2.2
Total	241.6

Car and truck emission-control catalysts dominate but there are many other uses. Typical palladium purity is 99.95% Pd.
[a]*For production in troy ounces, multiply tonnes* $\times 3.215 \times 10^4$.
**Jollie, 2010*

elements in vehicle catalysts are the following (Jollie, 2007; Jollie, 2009; Kendall, 2004):

$$\underset{\text{unburnt hydrocarbon fuel}}{8HC(s)} + \underset{\text{in unused input engine air}}{10O_2(g)} \xrightarrow{\text{platinum-group metal catalyst}} \underset{\text{inert tailpipe exit gases}}{4H_2O(g) + 8CO_2(g)} \quad (31.1)$$

$$\underset{\text{partially combusted fuel}}{2CO(g)} + \underset{\text{in unused input engine air}}{O_2(g)} \xrightarrow{\text{platinum-group metal catalyst}} \underset{\text{inert tailpipe exit gas}}{2CO_2(g)} \quad (31.2)$$

TABLE 31.2 Platinum Uses in 2010*

Use	Tonnes[a]
Jewelry	93.6
Car and truck emission-control catalysts	69.4
Investment bars and coins	20.5
Chemical industry	9.2
Catalyst for silicone manufacture	
Corrosion-resistant gauzes for nitric acid manufacture	
Medical	7.8
Dental, anticancer drugs, medical implant components, angioplasty	
Pacemakers, defibrillators, tips of heart electrodes	
Petroleum refining	6.4
Reforming and isomerization catalyst	
Electrical	5.9
Hard disks, platinum thermocouple wire, fuel cell catalysts	
Glass, liquid crystal displays and flat screens	0.3
Other:	5.9
Stationary pollution control catalysts and gas safety sensors	
Platinum aluminide coatings on turbine blades	
Wire for positioning turbine blade casting cores	
Platinum-tipped spark plugs, oxygen sensors	
Total	219.0

By far the largest uses are automotive catalysts and jewelry. Typical platinum purity is 99.95% Pt.
[a]*For production in troy ounces, multiply tonnes* $\times 3.215 \times 10^4$.
**Jollie, 2010*

TABLE 31.3 Rhodium Uses in 2009*

Use	Tonnes[a]
Car and truck exhaust emission-control catalysts	19.3
Especially for catalyzing $NO_x \rightarrow N_2$ reduction	
Chemical industry catalysts	1.7
For producing oxo-alcohols and acetic acid	
Glass industry	0.6
In furnaces for making liquid-crystal display glass and fiberglass	
Electrical	0.1
Other	0.7
Total	22.4

Car and truck emission-control catalysts dominate. Typical rhodium purity is 99.9% Rh.
[a]*For production in troy ounces, multiply tonnes × 3.215 × 10^4.*
**Jollie, 2010*

TABLE 31.4 Ruthenium Uses in 2009*

Use	Tonnes[a]
Electronics	10.5
In perpendicular magnetic recording disks	
In plasma display panels (enhances image quality when included in inner surface conductive paste)	
In conductive pastes	
In resistor components	
Chemical and electrochemical industries	5.7
In catalysts, especially for producing acetic acid	
Other (including electrochemical anode coatings)	1.7
Total	17.9

Electronic uses dominate. Typical ruthenium purity is 99.9% Ru.
[a]*For production in troy ounces, multiply tonnes × 3.215 × 10^4.*
**Jollie, 2010*

TABLE 31.5 Iridium Uses in 2005*

Use	Tonnes[a]
Chemical and electrochemical industries	1.7
Catalyst for producing acetic acid	
Electrochemical anode coatings	
Electronics	0.2
Crucibles for growing lithium-based scintillator crystals and yttrium aluminum garnet (YAG) laser crystals	
Other	
(Including car and truck engine spark plugs)	0.9
Total	2.8

Usage is small but important. Typical iridium purity is 99.9% Ir.
[a] *For production in troy ounces, multiply tonnes × 3.215 × 10⁴.*
**Jollie, 2010*

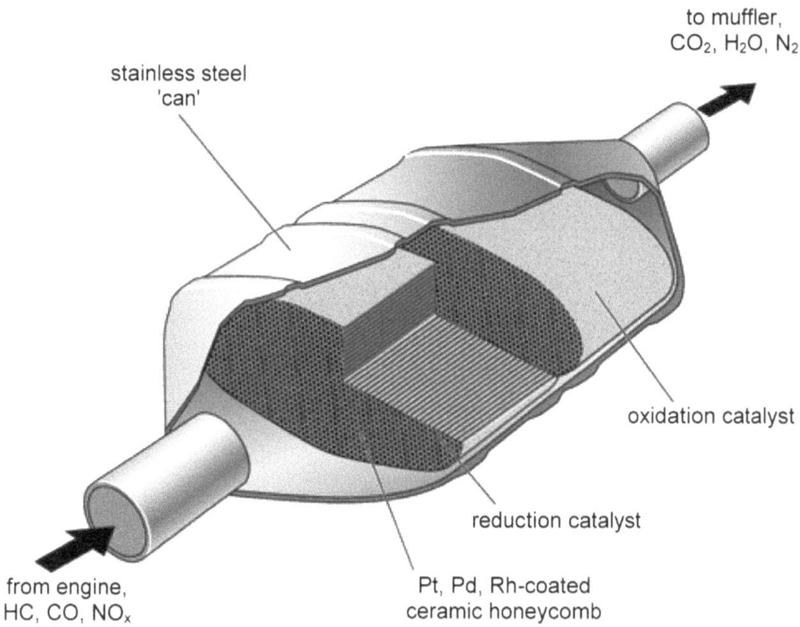

FIGURE 31.1 Sketch of automobile catalytic converter, which catalyzes the reduction of NO_x and hydrocarbons and the oxidation of carbon monoxide. It operates at ~200°C during start-up and 800°C during cruising. Converters are typically oval, ~0.35 m long and ~0.25 m across the oval (sometimes round).

$$\underset{\substack{\text{accidentally oxidized}\\\text{input engine } N_2\\\text{(from input air)}}}{2NO_x(g)} + \underset{\substack{\text{in unused input}\\\text{engine air}}}{(2-x)O_2(g)} \xrightarrow{\substack{\text{platinum-group}\\\text{metal catalyst}}} \underset{\substack{\text{fully oxidized nitrogen}\\\text{for Reaction (31.4)}}}{2NO_2(g)} \qquad (31.3)$$

$$\underset{\substack{\text{from}\\\text{Reaction}\\\text{(31.3)}}}{10NO_2(g)} + \underset{\substack{\text{unburnt}\\\text{hydrocarbon}\\\text{fuel}}}{8HC(s)} \xrightarrow{\substack{\text{platinum-group}\\\text{metal catalyst}}} \underset{\text{inert tailpipe exit gases}}{4H_2O(g) + 8CO_2(g) + 5N_2(g)} \qquad (31.4)$$

These reactions take place at 200°C at the start-up of the engine and increase to 800°C during cruising (Catalytic, 2010).

31.1.1. Specific Uses

Palladium and platinum are the most used of the platinum-group metals (Figures 31.2 and 31.3). This is because:

(a) they are present at higher concentrations in platinum-group ores than the other platinum-group elements;

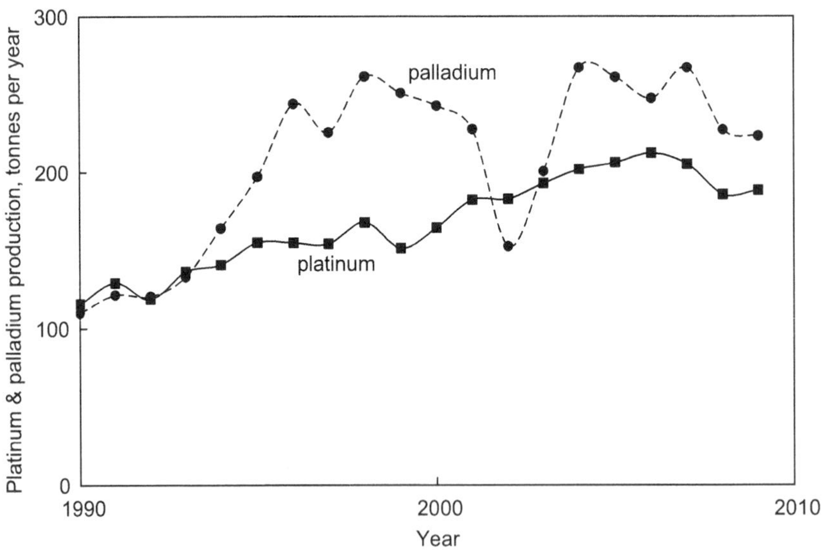

FIGURE 31.2 Annual primary production of palladium and platinum (Jollie, 2010). Growth has been steady except for the wild swings in palladium production from about 1995 on. These swings may be due to rapid growth in demand for palladium catalyst and subsequent releases of palladium from the Russian stockpile.

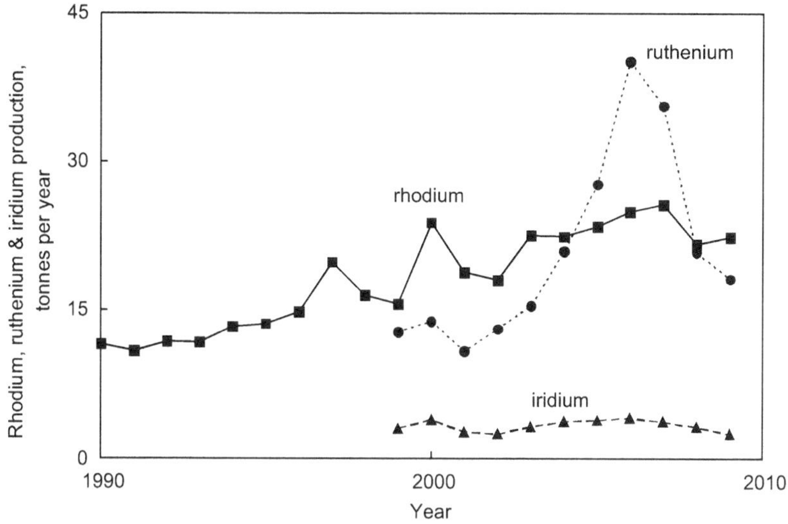

FIGURE 31.3 Annual primary production of rhodium, ruthenium and iridium (Jollie, 2010). Ruthenium and iridium production comfortably meets demand (Jollie, 2009).

(b) they are powerful catalysts, especially useful as car and truck emission reduction catalysts (Jollie, 2007; Jollie, 2009; Kendall, 2004);
(c) they are soft and easily worked into useful shapes, for example jewelry; and,
(d) they are readily hardened by alloying with other metals, such as copper, cobalt, iridium, tungsten and other platinum-group metals (Kendall, 2002).

Rhodium is also a key ingredient in car and truck emission reduction catalysts (Jollie, 2007; Kendall, 2004). However, it is always more dilute in ore than platinum, which restricts its supply and increases its price, which in turn limits its use.

Ruthenium is used extensively in vertical magnetic recording hard disks. It is also the third most common platinum-group element, which ensures a sufficient supply for all its uses.

Iridium has few uses and occurs at low concentration in ores. Its production and consumption are very small.

The primary production of platinum-group metals is small when compared to the primary production of nickel (1.6 million tonnes/year, 2008) and cobalt (70 000 tonnes/year, 2008). However, the value of the production of platinum-group metals is comparable to that of nickel.

31.2. MINING OF PLATINUM-GROUP ELEMENTS

South Africa and Russia dominate the mining of platinum-group elements, as shown in Table 31.6. South Africa mines the most platinum and rhodium. Russia

TABLE 31.6 Primary Production of Platinum-Group Metals by Region*

	2009 production, tonnes[a]				
Region	Pt	Pd	Rh	Ru	Ir
South Africa	141	074	21		
Russia	024	084	02		
North America	008	024	000.5		
Zimbabwe	008	006	000.6		
Other	004	005	000.1		
Total	185	193	24	18[b]	3[b]

Platinum and rhodium are dominated by South Africa. Global productions are shown in Figures 31.2 and 31.3. A further 20%–30% is made from recycle end-of-use scrap (Chapter 38).
[a]*For production in troy ounces, multiply tonnes × 3.215 × 10^4.*
[b]*Demand.*
**Jollie, 2010*

mines the most palladium. The ores from these countries also contain ruthenium and iridium, which are mined together with platinum, palladium and rhodium.

31.2.1. Ore Grade

The concentrations of platinum-group elements in proven South African and Russian deposits are shown in Table 31.7. All of these concentrations are less than 10 g/tonne (0.001%) in ore.

Platinum-group elements also occur:

(a) at higher concentrations in small deposits, for example, at Stillwater, USA, where the concentration is about 20 g/tonne total PGE, mostly palladium (Norilsk Nickel, 2008); and,
(b) at lower concentrations, about 1 g/tonne, in nickel-copper sulfide deposits, from which they are recovered as by-products (Kerr, 2002).

31.2.2. Mineralogy

Platinum-group elements occur in ore:

(a) as small (10–50 μm) platinum-group mineral grains, for example, braggite [(Pt,Pd)S] isoferroplatinum [Pt_3Fe] laurite [$(Ru,Ir,Os)S_2$] near mineral grains of nickel and copper sulfides (Cabri, 2002); and,
(b) as atoms in pentlandite [$(Fe,Ni,Co)_9S_8$].

TABLE 31.7 Concentrations of Platinum-Group Elements in South African and Russian *in situ* Ores*

Element	Concentration, or grade, of '*in situ*' ore, g/tonne
Pd	2–7
Pt	2–4
Rh	0.2–0.5
Ru	0.3–0.7
Ir	0.1–0.2
Os	0.04–0.1
Total	7–10

0.0001% ≡ 1 g/tonne of ore. 'As-mined' concentrations of ore are as much as 40% lower than *in situ* concentrations of ore, because the ore is diluted with rock during underground mining.
**Anglo American Platinum, 2009; Impala Platinum, 2008; Jones, 2005; Norilsk Nickel, 2008*

Efficient platinum-group element extraction requires, therefore, fine grinding of ore and efficient flotation of platinum-group element minerals and base-metal sulfides. This is discussed further in Chapter 32.

31.2.3. Mine Locations

The locations of mines where the ore is extracted primarily for platinum-group metals are shown in Figure 31.4.

The South African mines are located in the Bushveld region of South Africa. There are about 20 active mines in the region that mine about 250 tonnes of platinum-group elements per year (Jollie, 2010).

Most of the mines are underground, although several are open pit. They also produce significant amounts of nickel, copper, cobalt, silver and gold as by-products.

The main Russian mining region is around Norilsk and Talnakh, in northern Siberia. It has six underground mines and one open pit mine (Kendall, 2004). These mines produce about 120 tonnes of platinum-group elements per year along with large quantities of nickel, copper, cobalt, silver and gold.

Russia also has several placer platinum metal deposits, in Kondyor and Koryak, in Siberia and in the central Urals. These deposits account for about one quarter of the Russian production.

Small amounts of platinum-group metals are also mined in North America and Zimbabwe (Table 31.6).

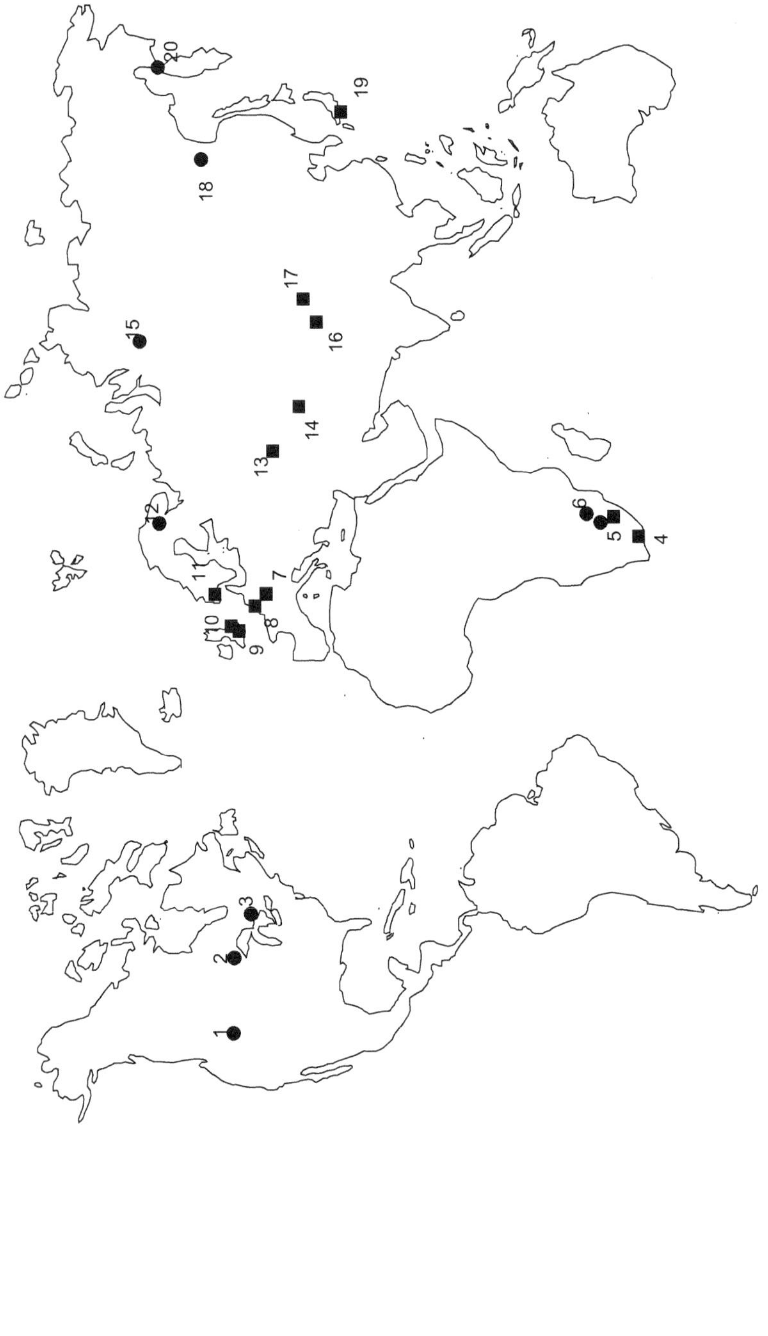

FIGURE 31.4 Platinum-group metal mines, concentrators, primary smelters and primary refineries. ● mines, concentrators, smelters, Cu-Ni refineries. ■ Primary platinum-group metal refineries. The mines and refineries are identified in Table 31.8.

TABLE 31.8 Locations of Primary (from Ore) Mines, Concentrators, Smelters and Refineries in the Platinum-Group Metal Industry

	Location	Operations	Approximate platinum-group metal production, tonnes/year
1	Stillwater, USA	M(2), C, S, B	016
2	Lac des Iles, Canada	M, C	007
3	Sudbury Basin, Canada	M(~10), C(2), S(2), B	020
4	Port Elizabeth, South Africa	R	
5	Bushveld and Springs, South Africa	M(~20), C(~20), S(6), B(4), R (3)	250
6	Selous, Zimbabwe	M, C, S	05
7	Hanau, Germany	S, B, R	
8	Hoboken, Belgium	S, B, R	
9	Acton, England	R	010
10	Royston, England	R	
11	Kristiansand, Norway	B, R	010
12	Kola Peninsula, Russia	M, C, S, B	010
13	Prioksk, Russia	R	
14	Yekaturinburg, Russia	R	
15	Norilsk/Talnakh, Russia	M(7), C(2), S(3), B	120
16	Novosibirsk, Russia	R	
17	Krasnoyarsk, Russia	R	
18	Kondyor, Russia	Alluvial platinum mine	004
19	Niihama, Japan	R	
20	Koryak, Russia	Alluvial platinum mine	003

M = mine; C = concentrator; S = smelter, B = base metal (nickel, copper, cobalt) refinery; R = platinum-group metal refinery. The bracketed numbers indicate number, for example, M(2) = 2 mines. Their locations are shown on the map given in Figure 31.4. South Africa and Russia dominate production.

31.3. EXTRACTION OF PLATINUM-GROUP METALS

Extraction of platinum-group metals consists of the following steps:

(a) mining ore rich in platinum-group metals while leaving rock lean in platinum-group metals behind;
(b) isolating the platinum-group elements in the ore into a flotation concentrate consisting of nickel-copper-iron sulfides that is rich in platinum-group elements;
(c) smelting and converting this concentrate to a nickel-copper sulfide matte that is richer than the concentrate in platinum-group metals;
(d) separating the platinum-group elements in the converter matte from the base metals, either by magnetic concentration or by leaching, to produce a very rich platinum-group metal concentrate containing about 60% platinum-group elements; and,
(e) refining this concentrate to individual platinum-group metals with purities in excess of 99.9%.

Concentrating and smelting/converting are done in or near the mining region. Refining is done in the region or in distant refineries (see Figure 31.4).

31.4. PRICES

The recent platinum-group metal prices are shown in Figures 31.5 and 31.6. The following features of these figures are worth noting:

(a) the platinum price has increased steadily;

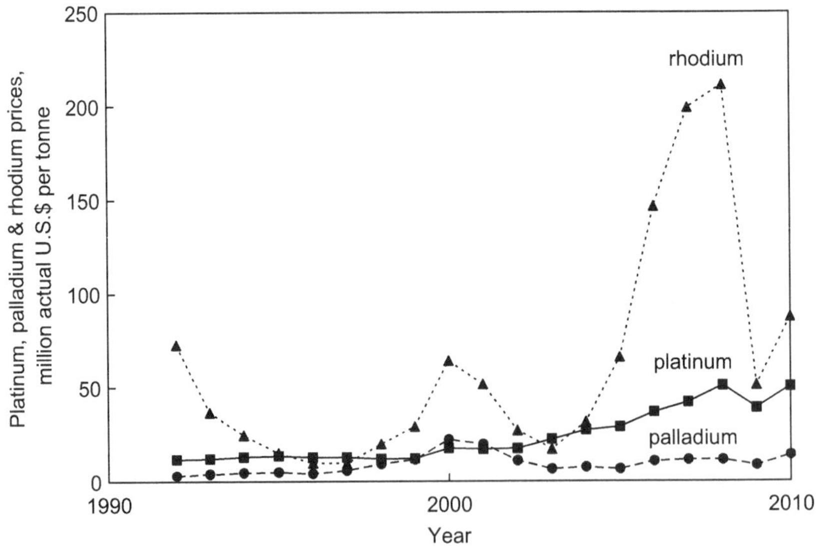

FIGURE 31.5 Palladium, platinum and rhodium prices, 1992–2010 (Johnson Matthey, 2010).

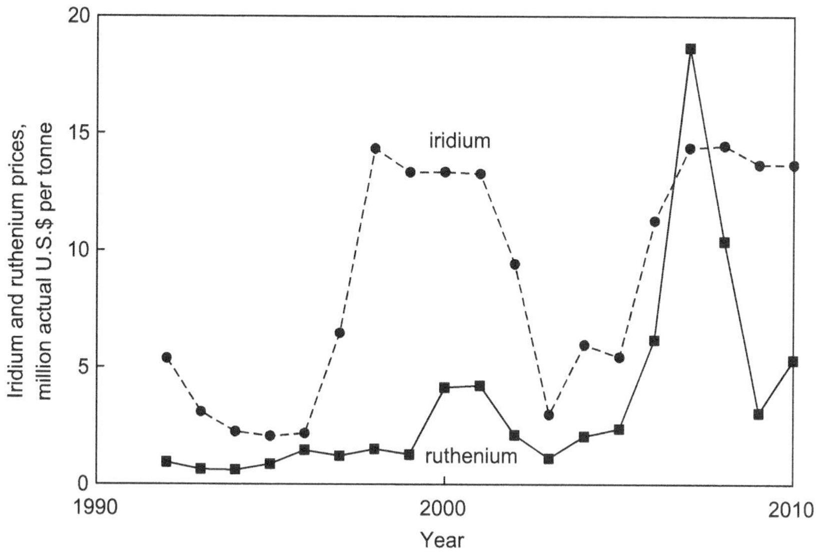

FIGURE 31.6 Ruthenium and iridium prices, 1992–2010 (Johnson Matthey, 2010).

(b) the palladium price peaked in 2000 and since remained steady; and,
(c) the rhodium, ruthenium and iridium prices have been volatile over the years due to imbalances in supply and demand.

31.5. COSTS OF EXTRACTION OF PLATINUM-GROUP METALS

31.5.1. Operating Costs

Annual reports of South African platinum mining companies indicate that operating costs for making platinum-group metals are of the order of US$ 20 million/tonne of refined platinum metals (Impala Platinum, 2009). These operating costs are divided between plant areas. This itemization is given in Table 31.9. The high cost of the front-end (mining, concentration and smelting) is in contrast with the low cost of hydrometallurgical refining.

These costs may be further divided as shown in Table 31.10.

31.5.2. Analysis

The above costs are expressed per tonne of total precious metals (that is, the sum of Pt, Pd, Rh, Ru, Ir and Au) (Impala Platinum, 2009). A representative make up of this tonne of product metals is typically made up of 0.52 tonnes of platinum, 0.25 tonnes of palladium, 0.07 tonnes of rhodium, 0.11 tonnes of ruthenium, 0.03 tonnes of iridium and 0.02 tonnes of gold. This depends, of

TABLE 31.9 Breakdown of Operating Costs Per Area

Area	Contribution to cost, %
Mining	70
Concentration	10
Smelting/converting	10
Leaching (platinum concentrate)	5
Refining to metal	5

TABLE 31.10 Components of Cash Operating Costs for Making Metallic Platinum-Group Metals from South African Ores

Item	Mining/ concentration %	Smelting/ converting %	Leaching/refining %
Labor	41	21	32
Stores	26	22	27
Utilities	06	26	09
Contracting	14	01	
Sundry	13	25	15
Toll smelting/refining		05	17

course, on the composition of the original ore. The combined sales value of 1 tonne of these metals at the end 2010 was US$ 40 million, compared with an operating cost US$ 20 million/tonne of product platinum-group metals.

31.6. SUMMARY

Platinum-group metals are found mainly in South Africa and Russia, and are mostly also produced in those countries.

About 500 tonnes of platinum-group metals are produced per year. Their biggest use is as automotive emission-control catalysts. They catalyze reduction of NO_x to nitrogen and oxidation of carbon monoxide and hydrocarbons.

Platinum-group metals are also used in jewelry and many other applications.

As-mined platinum-group ores typically contain 4–5 g/tonne of ore platinum-group elements. Platinum-group metals are, therefore, costly to produce and expensive to buy. In early 2011, their prices range from US$ 6 million/tonne of ruthenium to US$75 million/tonne of rhodium.

REFERENCES

Anglo Platinum. (2009). 2008 Anglo Platinum Annual Report.

Cabri, L. J. (2002). The platinum-group minerals. In L. J. Cabri (Ed.), *The geology, geochemistry, mineralogy and mineral beneficiation of platinum-group elements* (pp. 13–129). CIM, Special Volume 54.

Catalytic. (2010). Catalytic converter.

Impala Platinum. (2008). Fact sheet.

Impala Platinum. (2009a). Impala Platinum fact sheet.

Impala Platinum. (2009b). Review of operations – Impala Platinum Annual Report 2008. pp. 52–55.

Johnson Matthey. (2010). Current and historical [platinum group metal] prices.

Jollie, D. (2007). Heavy duty diesel: a growing source of PGM demand. In *Platinum 2007*. Johnson Matthey.

Jollie, D. (2009). Palladium use in diesel oxidation catalysts. In *Platinum 2009*. Johnson Matthey.

Jollie, D. (2010). *Platinum 2010*. Johnson Matthey.

Kendall, T. (2002). Platinum jewellery alloys. In *Platinum 2002*. Johnson Matthey.

Kendall, T. (2004). 30 years in the development of autocatalysts and PGM mining in Russia. In *Platinum 2004*. Johnson Matthey.

Kerr, A. (2002). An overview of recent developments in flotation technology and plant practice for nickel ores. In A. L. Mular, D. N. Halbe & D. J. Barratt (Eds.), *Mineral processing, plant design, practice and control proceedings, Vol. 1* (pp. 1142–1158). SME.

Norilsk Nickel. (2008). Mineral reserves and resources statement.

Shott, I. (2003). *Osmium, chemical and engineering news*. Available from. http://pubs.acs.org/cen/80th/osmium.html.

SUGGESTED READING

Cabri, L. J. (2002). *The geology, geochemistry, mineralogy and mineral beneficiation of platinum-group elements. CIM Special Volume 54*. CIM.

Jones, R. T. (2005). An overview of Southern African PGM smelting. In J. Donald & R. Schonewille (Eds.), *Nickel and cobalt 2005 challenges in extraction and production* (pp. 147–178). CIM.

Kendall, T., & Jollie, D. (2000–2010). *Platinum 2000 to Platinum 2010*. Johnson Matthey.

London Platinum and Palladium Market. (2011). History and market practices.

Chapter 32

Overview of the Extraction of Platinum-Group Metals

An overview of the process for the recovery of platinum-group metals (PGMs) is shown in Figure 32.1. The ore that is mined contains between 3 g/tonne and 10 g/tonne (0.0003–0.001%)[1] PGMs. The PGMs occur either as sulfides or are associated with the sulfides of copper and nickel. As a result, the froth flotation of the sulfide component of the ore yields a concentrate containing between 35 g/tonne and 150 g/tonne PGMs. This means that flotation provides a 20 times upgrade from the ore.

The flotation concentrate is smelted and converted, which removes silica, iron and sulfur, to produce a sulfur-deficient matte. The PGM content of the matte is between 1000 g/tonne and 2000 g/tonne, which means that smelting provides a 10–20 times upgrade from the flotation concentrate. Typical matte composition is as follows:

45–49% Ni
28–30% Cu
1.5–3% Fe
0.5% Co
15–23% S
1000–2000 g/tonne PGM

Mineralogically, the matte consists of heazlewoodite, Ni_3S_2, chalcocite, Cu_2S, troilite FeS, and bornite, Cu_5FeS_4.

The next major step is the removal of base metals and the sulfur associated with them from the PGM-containing material. There are two process routes that are used to achieve this:

(a) slow cooling followed by magnetic concentration; and,
(b) whole-matte leaching.

Only Anglo American Platinum's operations in Rustenburg, South Africa, use the slow-cooling process. In this process, converter matte is allowed to cool

1. 0.0001% = 1 g/tonne of ore.

Extractive Metallurgy of Nickel, Cobalt and Platinum-Group Metals. DOI: 10.1016/B978-0-08-096809-4.10032-2

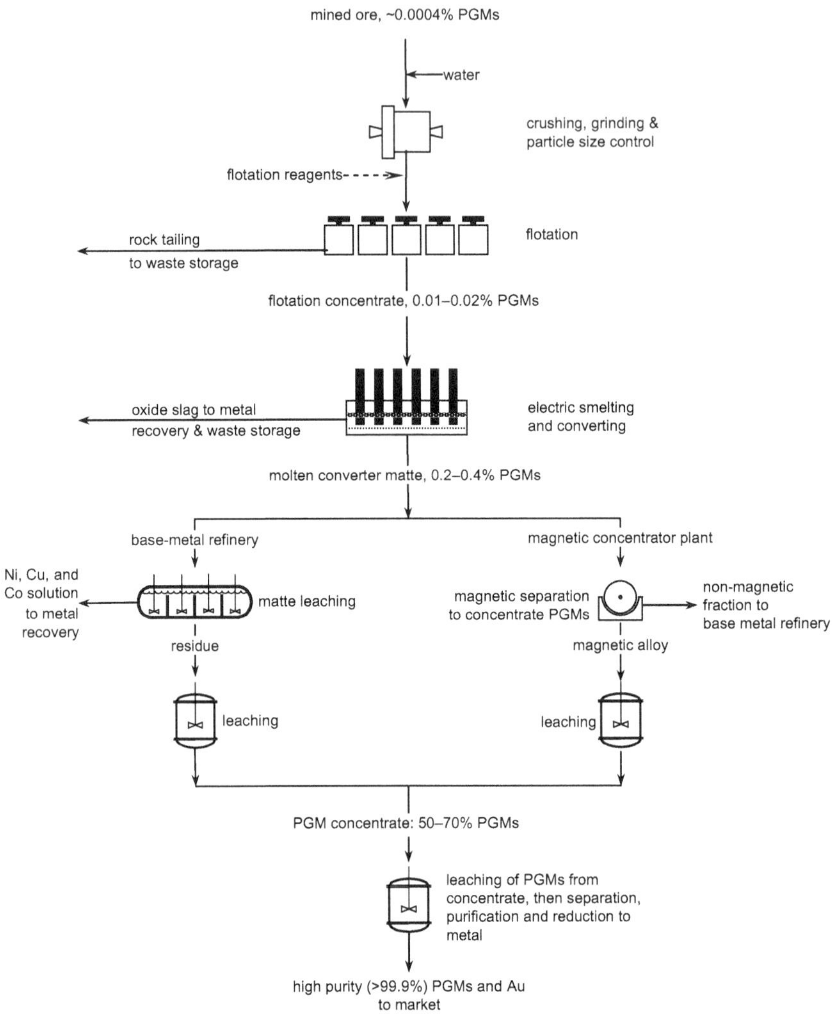

FIGURE 32.1 Generalized flowsheet for producing pure PGMs and gold from ore in South Africa. The major steps are flotation concentrate production, matte production, platinum concentrate production and high-purity metal production.

over several days, so that relatively large crystals of heazlewoodite, Ni_3S_2, chalcocite, Cu_2S, and metal alloy develop. During slow cooling, the PGMs concentrate in the alloy phase, which is separated from the base metal sulfides by magnetic separation.

This magnetic fraction is then treated in leaching steps to remove copper, nickel and iron, so that the platinum content is upgraded to between 50 and 60% PGMs. The overall upgrading from converter matte to platinum concentrate is between 100 and 400 times, by far the largest upgrade of all the steps in the extractive metallurgy of platinum.

The platinum concentrate is then transferred to the precious metals refinery for the separation of the individual precious metals.

In the whole-matte leaching process, nickel and copper are leached from the converter matte, leaving the PGMs in the insoluble residue. The PGM content of this leach residue is about 30%. In most of the operations, the residue is treated to remove residual copper, nickel, iron, selenium and tellurium prior to it being transferred to the precious metals refinery so that it contains between 50% and 70% PGMs.

The final step in the production of PGMs is the separation of the metals into individual metals, and their refining to high-purity products for sale.

There is no standard flowsheet for the refining of these metals. The most important separation step in the refinery is that between platinum and palladium because these two metals are the most abundant in the platinum concentrate. As a result, the types of refining processes can be classified according to this separation.

The older technology is based on precipitation of ammonium chloride salts of platinum and palladium. The next generation of technologies, developed in the 1970s and 1980s, is based on solvent extraction. The most recent refinery uses ion exchange or molecular recognition technology for this crucial separation.

The preparation of flotation concentrates is described in Chapter 33 and the extraction of PGMs from Russian ores is described in Chapter 34. Smelting and converting of sulfide concentrates containing PGMs are described in Chapter 35. The production of a platinum metal concentrate by separation of the base metals from the converter matte is described in Chapter 36. Finally, the refining of PGMs is described in Chapter 37.

Chapter 33

Production of Flotation Concentrates Containing Platinum-Group Metals

The production of nickel sulfide concentrates is described in Chapters 14 through 16. This chapter describes the production of concentrates containing the platinum-group metals PGMs. The objectives of this chapter are:

(a) to describe platinum-group ores and minerals; and,
(b) to show how they are made into flotation concentrates enriched in platinum-group metals.

South Africa is the largest primary producer of PGMs, as shown in Table 31.6. This chapter focuses on the concentration technology used in South Africa. Russia is the second largest primary producer of PGMs; the technology used to produce concentrates containing PGMs in Russia is described in Chapter 34.

Concentration is required because platinum ores contain only 3–4 g/tonne of ore.[1] This concentration is too low in PGMs for direct smelting and refining. Heating, melting and leaching of the huge amounts of waste rock in ore would require far too much energy and equipment.

33.1. ORES AND CONCENTRATES

As delivered to a concentrator, South African ores typically contain 3–4 g of platinum-group elements per tonne of ore. After flotation, their equivalent concentrates typically contain 100–200 g/tonne PGMs (Cramer, 2008).

This upgrading is obtained by isolating most of the minerals that are rich in PGMs in a small amount of flotation concentrate while rejecting most of the rock that is barren in PGMs to waste.

1. 0.0001% ≡ 1 g/tonne of ore.

Extractive Metallurgy of Nickel, Cobalt and Platinum-Group Metals. DOI: 10.1016/B978-0-08-096809-4.10033-4

33.2. ORES CONTAINING PLATINUM-GROUP METALS

PGMs always occur along with iron, nickel and copper sulfide minerals. The sulfide minerals and the PGMs are isolated together in concentrates, mostly by froth flotation. This is because the PGMs occur:

(a) as mineral grains of PGMs in or near mineral grains of iron, nickel and copper sulfides; and,
(b) in solid solution in the sulfide minerals containing iron, nickel and copper – particularly in pentlandite (Godel, Barnes, & Maier, 2007).

As a result, the efficient isolation of PGMs into a concentrate requires maximum recovery of both the minerals containing PGMs and the sulfide minerals containing iron, nickel and copper.

33.3. SOUTH AFRICAN ORES CONTAINING PLATINUM-GROUP METALS

There are two main ore types in the South African deposits. They are the Merensky Reef, which is feldspathic pyroxenite ore, and the Upper Group 2 Reef, which is chromitite ore. The Upper Group 2 layer is usually referred to as the UG2 ore. A section of the UG2 layer is shown in Figure 33.1.

Currently, the Merensky and UG2 ores each account for almost half of mined PGM production from South Africa. About 5% of production comes

FIGURE 33.1 Section of the platiniferous UG2 chromitite, $(Fe,Mg)Cr_2O_4$, layer (Bushveld Field Photography, 2010). The layer is about 1 m thick and extends thousands of square kilometers around. It is bounded top and bottom mainly by pyroxenite $(Mg,Fe)_2Si_2O_6$ (Barnes & Maier, 2002). The concentrations of platinum-group metals of the layer are shown. They average about 6 g/tonne of ore. Mining dilutes the ore to 3 g/tonne or 4 g/tonne of ore platinum-group elements with barren rock because it must capture the platinum-group elements at the top and bottom edges of the ore layer.

from Platreef ores. The rate of production from UG2 ore is increasing faster than that of the Merensky ore. (Cramer, 2008; Jones, 2005; & Schouwstra, Kinloch, and Lee 2000).

The platinum-bearing ores mostly occur in roughly horizontal layers that are between 0.5 m and 2 m thick. They are spread over a 10 000 km^2 circular area in northeastern South Africa. The layers are part of a 7–9 km thick saucer-shaped stack of intrusive igneous layers whose circumferences lie near the surface of the earth, from where they are mined. The Merensky reef is 20–400 m above the UG2 layer (Schouwstra, Kinloch & Lee, 2000).

33.3.1. Chemical Composition of the Main Ore Types

The chemical compositions of the main ore types are given in Table 33.1. Concentrations vary significantly throughout the reefs, but the average is approximately 6 g/tonne of ore. The as-mined concentrations of ore are 10–40% lower than *in situ* concentrations due to dilution with barren rock during mining (amount depending on mining method).

TABLE 33.1 Generalized *in situ* Concentrations of PGMs, Nickel, Cobalt, Copper, Sulfur and Iron in Merensky Reef, UG2 and Platreef Ores*

Element	Merensky Reef (pyroxenite ore), g/tonne of ore	UG2 (chromitite ore), g/tonne of ore	Platreef (pyroxenite ore), g/tonne of ore
Pt	3.3	2.5	1.3
Pd	1.4	2.0	1.4
Rh	0.2	0.5	0.09
Ru	0.4	0.7	0.12
Ir	0.1	0.1	0.02
Au	0.2	0.02	0.1
Ni	1300	700	3600
Co	30	20	
Cu	800	200	1800
S	4000	1000	
Fe	5000	1200	

*Jones, 2005

The UG2 ores contain significantly less nickel and copper than the Merensky ores. The host rock in the Merensky reef is pyroxenite, whereas the host rock in UG2 reef is chromitite.

33.3.2. Mineralogy of Platinum-Group Metals

The most common minerals containing platinum-group elements in South African ores are (Schouwstra Kinloch & Lee, 2000):

Sulfides	Braggite, (Pt,Pd)S, cooperite, PtS, laurite, Ru,[Os,Ir]S_2, and unnamed Pt, Rh, Cu, S minerals
Pt-Fe metallics	Isoferroplatinum, Pt_3Fe, tetraferroplatinum, PtFe
Arsenides	Sperrylite, $PtAs_2$
Tellurides	Moncheite, $PtTe_2$

All of these minerals, together with iron, nickel and copper sulfides, can be concentrated by froth flotation using copper sulfate as an activator and xanthate or dithiophosphate as collectors (see Chapter 15).

33.3.3. Iron, Nickel and Copper Sulfide Minerals

Ores from the Merensky Reef contain about 1% sulfide minerals, principally pyrrhotite, Fe_8S_9, pentlandite, $(Ni,Fe,Co)_9S_8$, and chalcopyrite, $CuFeS_2$.

Ores from the UG2 layer also contain these minerals, but at a lower concentration, approximately 0.4%. Iron, nickel, copper, cobalt and sulfur occur mainly in pyrrhotite, Fe_8S_9, pentlandite, $(Fe,Ni,Co)_9S_8$, and chalcopyrite, $CuFeS_2$.

The minerals containing PGMs are associated with the sulfide minerals. As a result, they report to the flotation concentrate along with the nickel and copper sulfides. The flowsheets for the extraction of these metals are therefore similar to those used to extract nickel.

33.4. PRODUCTION OF FLOTATION CONCENTRATE FROM THE MERENSKY REEF

The ores from the Merensky are treated by conventional processing to produce high-grade flotation concentrates. The following processing steps are used:

(a) the ore is ground to liberate the sulfide mineral grains, containing platinum group metals, iron, nickel and copper from the host rock; and,
(b) the liberated mineral grains are separated, mostly by froth flotation, into a concentrate that is rich in platinum-group metals.

No attempt is made to separate platinum-group minerals from iron, nickel and copper sulfide minerals during flotation because these always contain PGMs, either in solid solution or as unliberated mineral grains.

A schematic flowsheet for concentrating ores for the recovery of PGMs is shown in Figure 33.2. The concentration of PGMs in the feed material is

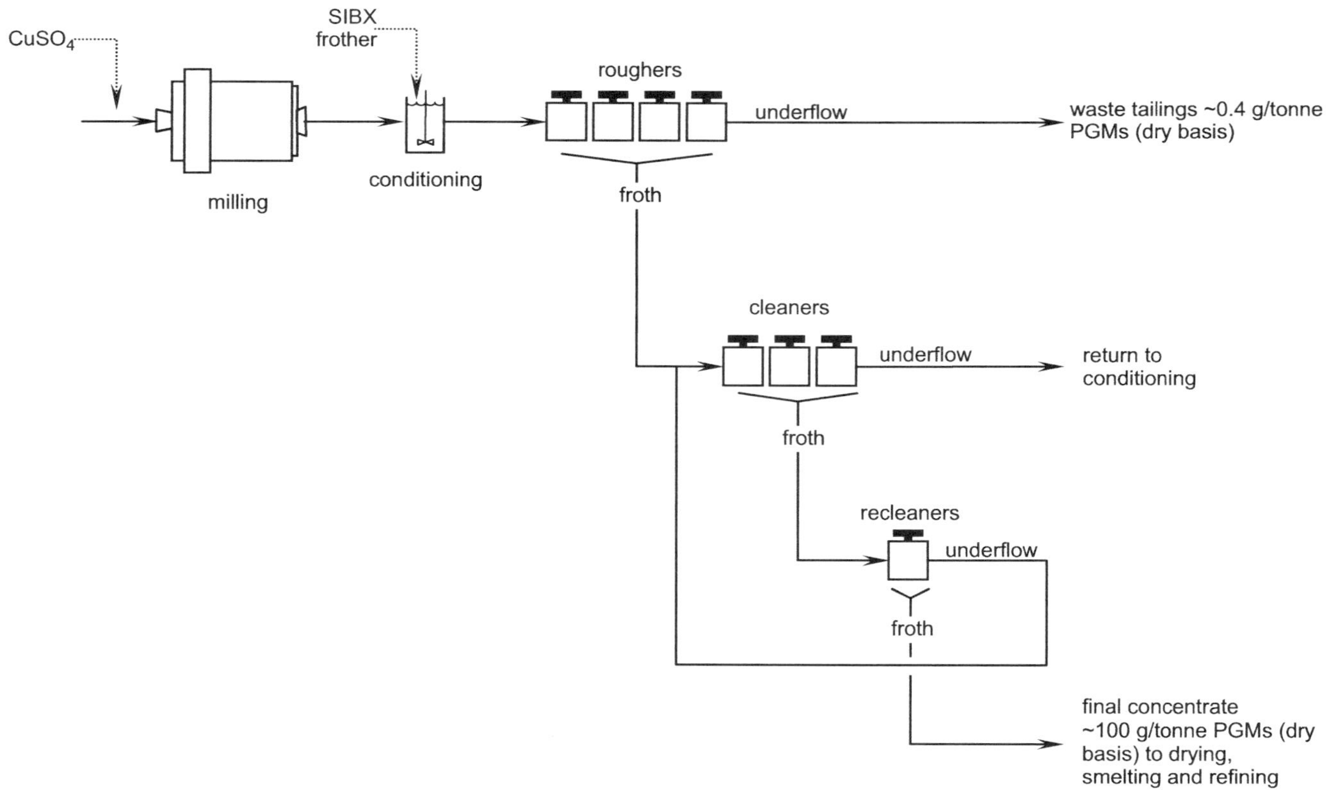

FIGURE 33.2 A schematic diagram of the flowsheet using conventional flotation for making high-grade PGM concentrate (100–200 g/tonne PGM) from South African ore containing 4 g/tonne PGMs from the Merensky reef (Deepaul & Bryson, 2004; Rule & Anyimadu, 2007; Rule *et al.*, 2008a). Reagents and their dosages are described in Table 33.2. Grinding and flotation are also discussed in Chapters 13–15.

about 4 g/tonne of ore and the flotation concentrate contains between 100 g/tonne and 200 g/tonne of concentrate.

The process entails the following steps:

(a) crushing and grinding;
(b) primary or rougher flotation of sulfide minerals; and,
(c) cleaning and recleaning of the rougher concentrate.

The flotation underflow is discarded. The concentrate is dewatered, and sent to the smelter for further treatment. The target recovery of PGMs to the concentrate is 90% (Rule & Anyimadu, 2007), although actual recovery is usually somewhat lower, at about 85% (Rule, Knopjes, & Atkinson, 2008a). Further details on the concentration of conventional nickel-sulfide ores are given in Chapters 15 and 16.

33.5. UG2 ORES AND THE PROBLEM OF CHROMITE

As mentioned earlier in this chapter, the UG2 ore occurs as a layer of chromitite. Since the conditions in smelting are sensitive to the level of chromite in the feed, the objectives of the flotation of UG2 ore are to recover the PGMs hosted in the sulfide minerals and to reject the chromite, so that there is less than 3% Cr_2O_3 in the concentrate.

The chromite content in the UG2 ores varies from 20% to 60%. Although chromite is hard, the ore matrix is friable, and chromite is easily liberated. Chromite is hard, and has a high Bond Work Index, that is, it requires a large amount of specific energy to grind the chromite smaller.

The main mechanism is by which chromite reports to the concentrate is entrainment (Knights and Bryson, 2009). Chromite is entrained by the rising bubbles in the flotation cell, and is carried by the froth into the concentrate launders. Because chromite is dense, only particles smaller than about 50 μm are entrained. This means that grinding the feed too fine must be prevented.

Coupled with this understanding is the observation that most of the PGM-bearing minerals are liberated by a relatively coarse grind, 35% passing 75 μm. However, there is a fraction of the PGM-bearing minerals that are locked in the silicate phases and require finer grinding.

These observations, the recognition of faster and slow-floating fractions, and the pilot plant test work that followed led to the development of the mill-float-mill-float circuit, commonly referred to as the '*MF2 circuit*', shown in Figure 33.3. The principle embodied in the MF2 circuit is to float the PGM-bearing minerals that are liberated by the coarse grind into a high-grade concentrate. The tailings from the rougher circuit are reground and then processed again through a similar rougher-cleaner-recleaner circuit.

An important advantage of this circuit is that roughing and scavenging are both done immediately after fresh mineral surfaces have been created by grinding, giving rapid, efficient flotation.

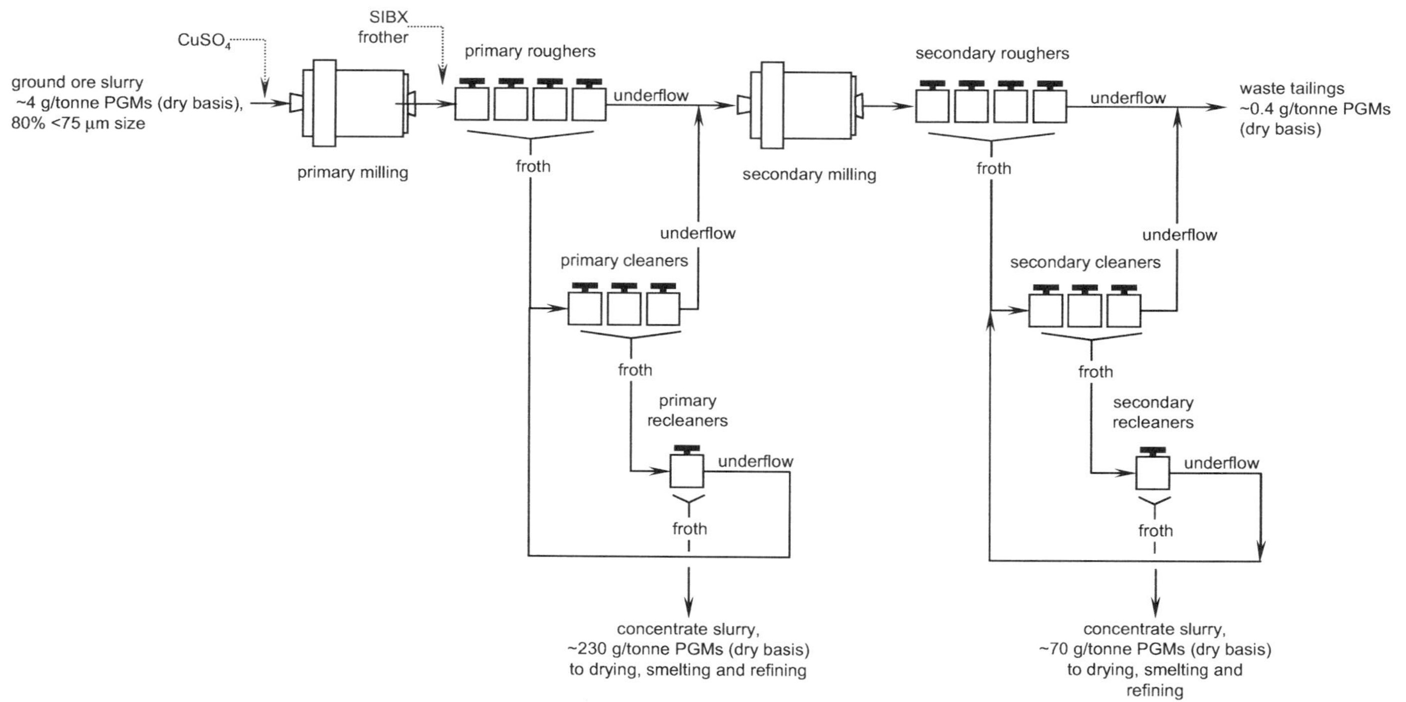

FIGURE 33.3 Flowsheet using the MF2 concept for making high-grade PGM concentrate (100–200 g/tonne PGMs) from South African ore containing 4 g/tonne PGMs (Deepaul & Bryson, 2004; Rule & Anyimadu, 2007; Rule *et al.*, 2008a). The two concentrates are mixed and then sent to a smelter. New developments in this circuit are (i) inert grinding in the cleaning circuits; and, (ii) column cells for recleaning. The secondary ball mill might be profitably replaced with an Isamill, which gives the small particle sizes needed for the efficient recovery of PGMs in the scavenger circlet.

The amount of chromite in the concentrate is minimized by avoiding excessive grinding, minimizing water recovery and recycles, carefully controlling pulp density, and preferentially recovering silicate gangue rather than chromite. Over-grinding can be prevented by operating the mills in open circuit, or by using a screen instead of a hydrocyclone to classify oversize particles from the mill (Knights and Bryson, 2009).

The addition of frother must be controlled to avoid an overly stable froth, which will encourage entrainment of chromite. If depressant is added to suppress flotation of talc, the amount of depressant added must avoid complete depression of the silicate gangue, because this will result in an increase in the chromite grade.

33.6. FLOTATION REAGENTS AND CONDITIONS

Flotation is done mostly in mechanical cells from a slurry of ground ore and water. The slurry concentration is 60% solids and has a natural pH between 8 and 9 (Woodcock, Sparrow, & Bruckard, 2007).

The reagents that are used, their purposes and addition points are outlined in Table 33.2.

33.7. IMPROVING RECOVERY OF PLATINUM-GROUP ELEMENTS TO CONCENTRATE

Mining of platinum-group element ore is costly. It is important, therefore, that the recovery of these elements into concentrate should be as high as possible. Maximum recovery is obtained by paying close attention to the following aspects of flotation:

(a) grinding the ore fine enough to maximize liberation of the platinum-group metals and base metal minerals as much as possible from the host rock;
(b) adding collector as soon as possible (perhaps in the grinding mills) to ensure efficient flotation of slow-floating pentlandite, $(Ni,Fe)_8S_9$, and its associated platinum minerals (Weise, Becker, Bradshaw, & Harris, 2007);
(c) providing a long enough slurry residence time in the flotation cells (~1 h, Rule & Anyimadu, 2007) for all the sulfide and platinum minerals to float;
(d) determining and maintaining the optimum combination of flotation reagents; and,
(e) maintaining a constant froth layer thickness in each cell, by controlling (i) frother input rate; and, (ii) slurry height in each cell by adjusting the flow rate of the underflow.

Continuous on-stream chemical analysis of ore–water pulp streams (particularly feed, concentrate and tailings) by x-ray dispersive analysis (Blue Cube Systems, 2010; Outotec, 2010) is also used to determine optimum flotation conditions while ore characteristics and input rates vary.

TABLE 33.2 Reagents Used in the Concentrator Circuit*

Reagent	Purpose	Chemical (all added in solution)	Dosage, g/tonne of dry ore
Collectors	Create hydrophobic conditions on PGM mineral surfaces so that they attach to rising air bubbles[a]	Xanthates: sodium isobutyl xanthate, sodium isopropyl xanthate; dithiophosphates: ethyl, butyl, isobutyl	200
Activator	Creates a copper sulfide surface coating on minerals to give strong mineral-collector attachment hence efficient flotation	$CuSO_4 \cdot 5H_2O$	50
Frother	Creates strong but short-lived bubble-mineral particle froth at the top of flotation cells, preventing the particles from falling back until the froth overflows	Polyglycol ethers, crecylic acid, triethoxybutane	20
Depressant	Minimizes flotation of talc and chromite	Guar, carboxy methyl cellulose, starch, dextrose	100

The reagents are added before each flotation step, sometimes in the grinding mills.
[a]*Chapter 14 explains the flotation of sulfide minerals. Flotation of platinum alloy minerals (Pt_3Fe, PtFe) may be due to the attraction these metals may have to the sulfur end of xanthate molecules, that is, they may want to form PtS or FeS.*
**Steyn, Knopjes, Goodall, & Harris, 2005; Weise et al., 2007; Woodcock et al., 2007*

An important tool for maximizing the recovery of PGMs to concentrate is quantitative scanning electron microscopy. This technique is essentially the automation of scanning electron microscopy. It provides, for any assemblage of ground ore, a map of the particles in the ore. In particular, it indicates the makeup of the particles in the sample, showing liberated, partially liberated and unliberated particles. A typical map of the particles is shown in Figure 33.4.

The results from quantitative scanning electron microscopy indicate whether further grinding will give improved recovery of PGMs. This technique is used extensively by the South African platinum industry (Rule & Anyimadu, 2007).

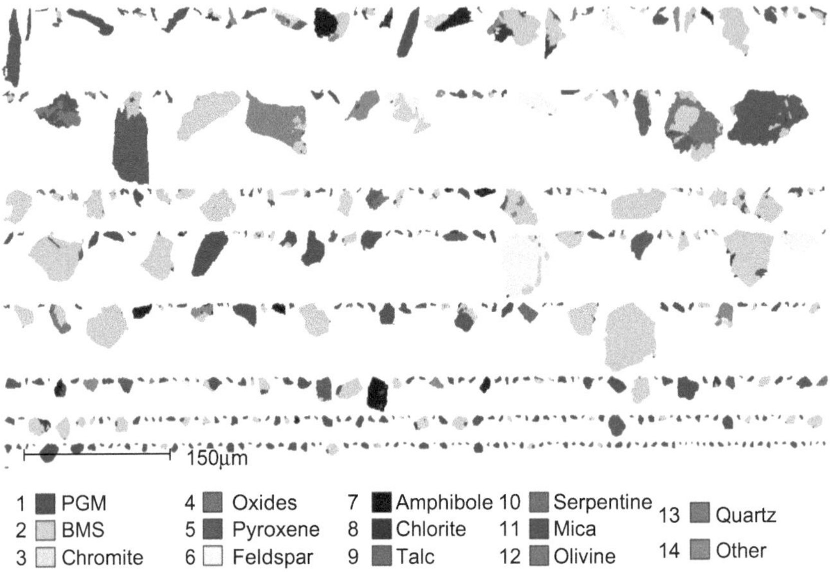

FIGURE 33.4 Quantitative scanning electronic microscope image of PGM flotation concentrate. The objective of flotation is maximizing the flotation of minerals containing PGMs and base metal sulfide. The latter is important because base metal sulfide minerals often contain dissolved and entrapped PGMs. Minerals containing PGMs are blue. Minerals containing base metal sulfides are green. Note the very small (but occasionally large) PGM particles and the larger base metal sulfide (pentlandite, chalcopyrite, pyrrhotite) particles. This concentrate might benefit from further grinding to liberate grains of base metal sulfide from gangue (for example, feldspar [yellow]) grains. *Courtesy of Anglo American Research.*

33.8. GRAVITY SEPARATION

Platinum-containing minerals are much denser than the host rock and the iron, nickel and copper sulfide minerals. For example, the specific densities of isoferroplatinum, Pt_3Fe, tetraferroplatinum, PtFe, and cooperite, PtS, are 18, 16 and 10, respectively. In contrast, the specific densities of pentlandite, $(Ni,Fe)_9S_8$, chromite, $(Fe,Mg)Cr_2O_4$, pyroxene, $(Mg,Fe)_2Si_2O_6$, and quartz, SiO_2, are 5, 5, 4 and 3, respectively.

This difference is occasionally exploited to produce a gravity concentrate containing about 20% PGMs, which is sent directly to the precious metals refinery (bypassing flotation, smelting and base-metal refining). The remainder of the ore is then treated by flotation as described in Figure 33.2.

An example of this is the production of a gravity concentrate containing 20% platinum obtained by cycloning the primary crusher fines. The dense gravity concentrate is collected from the cyclone apex (see Figure 11.5). The principle is that dense particles move more quickly to the cyclone wall than less dense particles.

Russian concentrators also exploit density differences to make gravity concentrates. Knelson centrifuges are used in these operations, discussed in Chapter 34.

33.9. RECENT DEVELOPMENTS

Recent developments in South African concentrators have focussed on the implementation of high-pressure grinding rolls and stirred ceramic media fine grinding.

High-pressure grinding rolls have been installed in several locations in place of gyratory crushers (Rule Minnaar, & Sauermann, 2008b). This has reduced circulating ore loads (increasing concentrator productivity) and improved platinum-group mineral liberation (through grain-boundary fracturing).

Likewise, several ultrafine stirred ceramic media grinding mills have been installed in existing concentrators (Anyimadu, Rule, & Knopjes, 2007; Clermont, De Haas, & Hancotte, 2008). Their purpose is to liberate very fine platinum-group minerals. Use of inert media in these concentrators has been particularly rewarding, as have real-time mill management techniques (Clermont, De Haas & Hancotte, 2008).

For flotation, introduction of column flotation cells is proving to be useful for cleaning magnesium-bearing pyroxene and talc from the concentrates (Northam Platinum Limited, 2009).

33.10. SUMMARY

South African platinum ores contain platinum-group elements either as discrete mineral grains, for example braggite, (Pt,Pd)S, or in solid solution in pentlandite, $(Ni,Fe)_9S_8$.

As-mined ores, delivered to a concentrator, typically contain 3 g/tonne or 4 g/tonne PGMs.

The ores are made into concentrates by the following steps:

(a) crushing and grinding to particle size of about 80% passing 50 μm; and,
(b) bulk flotation of sulfide minerals from the host rock.

The concentrates are then smelted and converted to a molten sulfide matte, which is discussed in Chapter 35.

Recovery to concentrate is typically 85% at a grade of 100–200 g/tonne PGMs.

REFERENCES

Anyimadu, A. K., Rule, C. M., & Knopjes, L. (2007). The development of ultra-fine grinding at Anglo Platinum. *Journal Southern African Institute of Mining and Metallurgy, 107*, 15–22.

Barnes, S.-J., & Maier, W. D. (2002). Platinum-group element distributions in the Rustenburg layered suite of the Bushveld complex, South Africa. In L. J. Cabri (Ed.), *The geology, geochemistry, mineralogy and mineral beneficiation of platinum-group elements* (pp. 431–458). CIM.

Blue Cube Systems. (2010). In-line and on-line analyzers. Available from. www.bluecube.co.za.

Bushveld Field Photography. (2010). *Rustenburg layered suite*. South Africa: Department of Geological Sciences, University of Cape Town.

Cramer, L. A. (2008). What is your PGM concentrate worth? In M. Rogers (Ed.), *Platinum in transformation, Third International Platinum Conference* (pp. 387–394) SAIMM.

Clermont, B., De Haas, B., & Hancotte, O. (2008). Real time mill management tools stabilizing your milling process. In M. Rogers (Ed.), *Proceedings of Platinum in Transformation, Third International Platinum Conference* (pp. 13–20). SAIMM.

Deepaul, V., & Bryson, M. (2004). Mintek, a national resource of minerals processing expertise for platinum ores. In *International Platinum Conference 'Platinum Adding Value'* (pp. 9–14). SAIMM.

Godel, B., Barnes, S.-J., & Maier, W. D. (2007). Platinum-group elements in sulphide minerals, platinum-group minerals, and whole-rocks of the Merensky Reef (Bushveld Complex, South Africa): implications of the formation of the reef. *Journal of Petrology, 48*, 1569–1604.

Jones, R. T. (2005). An overview of Southern African PGM smelting. In J. Donald & R. Schonewille (Eds.), *Nickel and cobalt 2005, challenges in extraction and production* (pp. 147–178). CIM.

Knights, B. D. H., & Bryson, M. A. W. (2009). Current challenges in PGM flotation of South African ores. In C. O. Gomez, J. E. Nesset & S. R. Rao (Eds.), *Advances in minerals processing* (pp. 285–396). CIM.

Northam Platinum Limited. (2009). Northam annual report 2009. p. 11. www.northam.co.za. Accessed 29.05.11

Outotec. (2010). On-line elemental analyzers. Company brochure. www.outotec.com. Accessed 29.05.11.

Rule, C. M., & Anyimadu, A. K. (2007). Flotation cell technology and circuit design – an Anglo Platinum perspective. *Journal Southern African Institute of Mining and Metallurgy, 107*, 615–622.

Rule, C. M., Knopjes, L., & Atkinson, R. J. (2008a). Ultra fine grinding of intermediate flotation concentrates in the PGM industry at the Pt Mile operation on Anglo Platinum tails. In M. Rogers (Ed.), *Proceedings of platinum in transformation, Third International Platinum Conference* (pp. 37–44). SAIMM.

Rule, C. M., Minnaar, D. M., & Sauermann, G. M. (2008b). HPGR – revolution in platinum? In M. Rogers (Ed.), *Proceedings of platinum in transformation, Third International Platinum Conference* SAIMM. 21-8.

Schouwstra, R. P., Kinloch, E. D., & Lee, C. A. (2000). A short geological review of the Bushveld complex. *Platinum Metals Review, 44*(1), 33–39.

Steyn, J., Knopjes, B., Goodall, C., & Harris, M. (2005). An evaluation of the effect of multiple grinding and flotation stages on flotation performance of a platinum-bearing ore. In G. J. Jameson (Ed.), *Centenary of flotation symposium* (pp. 1037–1043). AusIMM.

Weise, J. G., Becker, M., Bradshaw, D. J., & Harris, P. J. (2007). Interpreting the role of reagents in the flotation of platinum-bearing Merensky ores. *Journal of the Southern African Institute of Mining and Metallurgy, 107*, 29–36.

Woodcock, J. T., Sparrow, G. J., & Bruckard, W. J. (2007). Flotation of precious metals and their minerals. In M. C. Fuerstenau, G. Jameson & R.-H. Yoon (Eds.), *Froth flotation, a century of innovation* (pp. 575–609). SME.

SUGGESTED READING

Cabri, L. J. (Ed.), (2002). *The geology, geochemistry, mineralogy and mineral beneficiation of platinum-group elements*. CIM. CIM Special Volume 54.

Rogers, M. (Ed.), (2008). *Platinum in transformation.Third International Platinum Conference. Symposium Series S52*. The Southern Africa Institute of Mining and Metallurgy.

Woodcock, J. T., Sparrow, G. J., Bruckard, W. J., Johnson, N. W., & Dunne, R. (2007). Plant practice: sulfide minerals and precious metals: section IV: flotation of platinum-group metals. In M. C. Fuerstenau, G. Jameson & R.-H. Yoon (Eds.), *Froth flotation, a century of innovation* (pp. 832–843). SME.

Chapter 34

Extraction of Platinum-Group Metals from Russian Ores

Russia is the largest primary producer of palladium (~45% of world production) and the second largest producer of platinum (~15%), rhodium (~10%), ruthenium and iridium.

Almost all the production comes from mines in the Norilsk-Talnakh region of north Siberia (see Figure 31.4).

A schematic flowsheet of the process for the extraction of these metals is shown in Figure 34.1. The process steps are:

(a) mining;
(b) concentrate production;
(c) concentrate smelting;
(d) refining to produce copper and nickel metal anode slimes that are rich in platinum-group metals (PGMs);
(e) enrichment of the anode slimes by roasting and leaching; and,
(f) refining of the enriched residues to high-purity platinum, palladium, rhodium, ruthenium and iridium.

This chapter describes Russian ores that contain PGMs and the methods used to concentrate the PGMs from these ores. The refining of the concentrate will be described in Chapter 37.

34.1. NICKEL COPPER AND PLATINUM ORES FROM NORILSK-TALNAKH

Norilsk-Talnakh ores are rich in nickel, copper and PGMs. The metal grades of the major ore types in the Norilsk-Talnakh region are given in Table 34.1. Metal production from these ores is limited by the remoteness and harsh Arctic climate of the region.

34.1.1. Differences between the Russian and South African ores

The concentrations of platinum, palladium and rhodium for the Russian and South African ores are given in Table 34.2. Ores from the Norilsk-Talnakh

Extractive Metallurgy of Nickel, Cobalt and Platinum-Group Metals. DOI: 10.1016/B978-0-08-096809-4.10034-6

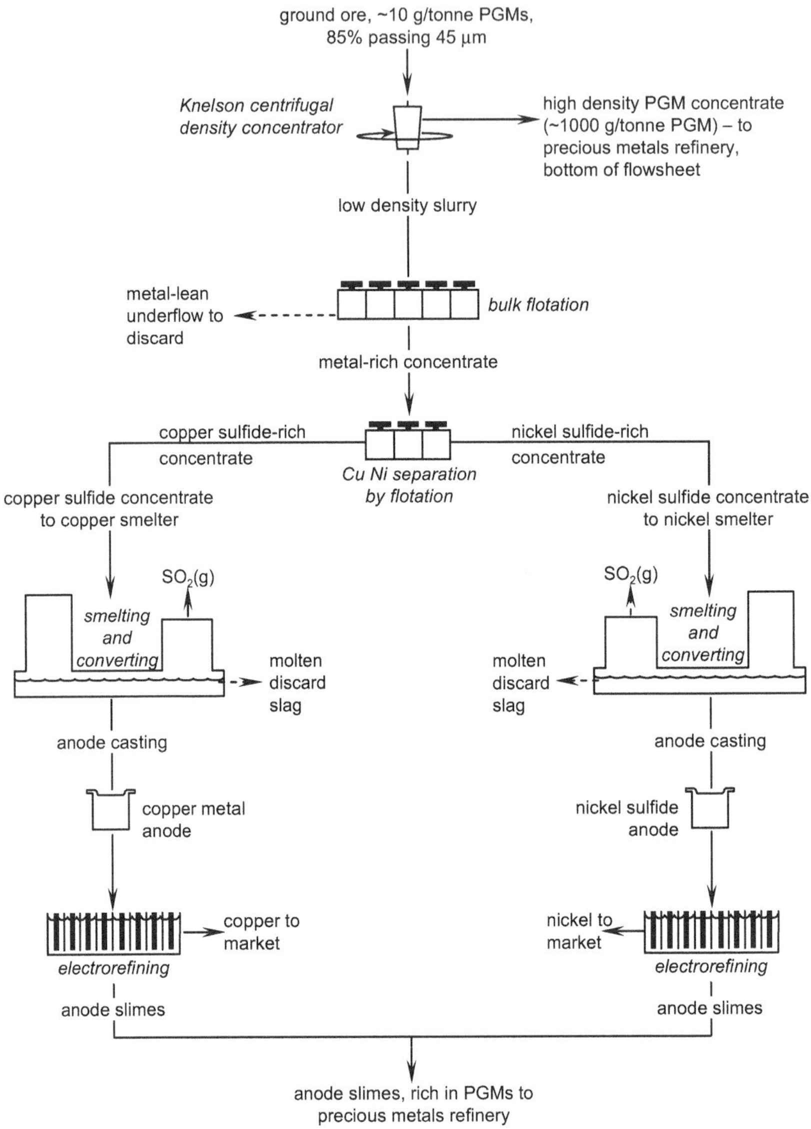

FIGURE 34.1 Schematic flowsheet for producing PGMs from Norilsk-Talnakh ores. Production of high-density concentrate at the top of the flowsheet is notable. The density concentrate joins the electrorefining slimes, thereby bypassing flotation, smelting, converting and electrorefining.

region contain up to 50 times more nickel, copper and cobalt than South African ores. The Norilsk-Talnakh ores contain about 7% nickel, copper and cobalt, whereas the South African ores only contain about 0.15% of these metals. The Russian ores are also more concentrated in PGMs. The Russia ores

TABLE 34.1 The *in situ* Concentrations of the Major Ore Types in the Norilsk-Talnakh Region[*,a]

Ore type (Kozyrev *et al.*, 2002)	Proven ore reserves, million tonnes	Cu, %	Ni, %	Co, %	Pt, g/tonne	Pd, g/tonne	(Rh + Ru + Ir + Os), g/tonne	Au, g/tonne
Massive rich ore (>70 volume% sulfides) some smelted directly without concentration	49	4.0	2.9	0.09	1.5	7.0	0.5	0.2
Disseminated (dispersed) cuprous ore (10–15 volume% sulfides)	16	4.6	1.1	0.04	2.6	1.1	0.2	0.7
Stringer disseminated ores	31	0.9	0.5	0.02	1.5	4.0	0.2	0.3
Weighted average		3.2	1.7	0.06	1.7	7.0	0.3	0.3

Average concentration of rhodium in the ore is about 0.15 g/tonne, based on Russian production data for 2008 (Jollie, 2009). The high nickel and copper concentrations, compared to 0.1–0.2% (nickel plus copper) in South African ores, are notable. The ores contain isoferroplatinum, $(Pt)_3Fe$, cooperite, PtS, sperrylite, $PtAs_2$, and many other minerals (Kozyrev *et al.*, 2002).

[a] *As-mined ore concentrations are lower due to dilution with rock during mining.*

* *Norilsk, 2010.*

TABLE 34.2 Concentrations of Platinum, Palladium and Rhodium in the Russian and South African Ores

	Country	
Element	Russia, g/tonne of ore	South Africa, g/tonne of ore
Pt	2	3
Pd	7	2
Rh	0.2	0.5

contain about 10 g/tonne of ore versus 6 g/tonne for the South African ores.[1] (All concentrations are for ore *in situ*.)

Another important difference between the ores is the relative concentrations of platinum and palladium. In the Russian ores, palladium is more concentrated, whereas platinum, which usually commands a significantly higher price, is more concentrated in the South African ores.

34.1.2. Platinum-Group Elements and Minerals

Most of the PGMs (especially platinum) occur as grains of minerals that are either enclosed in, intergrown with, attached to or in close proximity to mineral grains of nickel and copper sulfides (Kozyrev *et al.*, 2002). All of the PGMs, except platinum, also occur in solid solution in pentlandite (Komarova, Kozyrev, Simonov, & Lulko, 2002).

34.2. RUSSIAN PRODUCTION

The flowsheet for the production of materials suitable as a feed to a precious metals refinery is shown in Figure 34.1.

The following steps are used:

(a) separation of a gravity concentrate using Knelson concentrators;
(b) bulk flotation;
(c) separation of nickel and copper by flotation;
(d) smelting and converting of both the nickel and the copper concentrates;
(e) electrorefining of both the copper and nickel (copper is cast as metal anodes, while nickel is cast as matte anodes); and,

1. 0.0001% = 1 gram/tonne of ore.

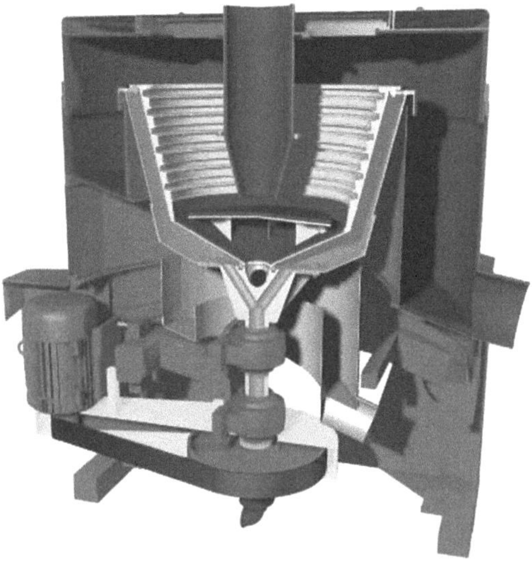

FIGURE 34.2 Knelson centrifugal separator for producing a gravity concentrate containing ~1000 g/tonne PGMs. Ore slurry ~10 g/tonne PGMs enters the top. The slurry is rapidly rotated by the rotating bowl (Knelson, 2004). Centrifugal forces cause high-density particles to reach the bowl wall and move out through perforations. The less dense material exits through the bottom of the separator and on to froth flotation. The high-density PGM concentrate is sent directly to the precious metals refinery. It is treated for the recovery of PGMs along with anode slimes from the nickel and copper electrorefineries.

(f) collecting the anode slimes from both of these electrorefining activities as feed to the precious metals refinery.

34.2.1. Production of Gravity Concentrates

Most of the Norilsk-Talnakh ore is passed through centrifugal density separators before proceeding to froth flotation. The objectives of the density separation are:

(a) to improve the recovery of PGMs to concentrate; and,
(b) to bypass the nickel and copper production steps directly to refining.

The introduction of gravity concentrators reduces the amount of PGMs held up in processing and improves the overall economics of the process. In addition, this partially uncouples the production of nickel and copper from the production of PGMs.

Six, 1.2-m diameter, Knelson centrifugal density separators (Figure 34.2) recover about 70% of the platinum, about 25% of the gold and 15% of the palladium to the gravity concentrate (Blagodatin, Distler, Zakharov, & Sluzhenikin, 2000). The platinum in the gravity concentrate is mainly as isoferroplatinum, Pt_3Fe, and tetraferroplatinum, PtFe. The gold in the gravity concentrate is mainly as metallic gold. The grade of the gravity concentrate is about 1000 g/tonne (PGMs and gold) (Blagodatin *et al.*, 2000).

This gravity concentrate is combined with the nickel and copper electrorefinery residues and sent to a precious metals refinery for refining to pure metal.

34.2.2. Bulk Nickel–Copper Flotation

A bulk flotation concentrate containing nickel, copper and PGMs is made by grinding ore to a particle size of about 50–75 μm and then floating the sulfides with butyl dithiophosphate collector and pine oil frother at the natural pH of ore, that is, at about 9. The flotation is mostly done in mechanical flotation cells, as shown in Figure 15.3. The recovery of PGMs to the concentrate is about 90% (Kozyrev *et al.*, 2002).

34.2.3. Nickel–Copper Separation

Copper sulfide minerals are floated from nickel sulfide minerals by regrinding then:

(a) depressing the nickel sulfides with CaO and steam (Kerr, 2002; Kozyrev *et al.*, 2002); and,
(b) floating the copper sulfides with potassium butyl xanthate collector and T80 frother.

34.2.4. Anode Slimes

The flotation concentrates are processed by conventional pyrometallurgical techniques for nickel and copper concentrates (smelting, converting, fire-refining and casting) for production of nickel matte and copper metal anodes. (See Chapters 17–19 for more details on smelting, converting, fire-refining and casting and Appendix F for the electro refining of matte anodes.)

These concentrates all contain PGMs, which means that the anodes contain these elements. The PGMs do not dissolve during copper and nickel electro-refining. Instead, they fall to the bottom of the cell, where they form part of the anode slimes. The anode slimes are collected from the refining cells.

The anode slimes are further enriched by oxidation roasting, followed by the leaching of nickel, copper and cobalt. The undissolved solids containing the PGMs and gold are collected as the final concentrate and sent to the Kraznoyarsk precious metals refinery for platinum, palladium, rhodium, ruthenium and iridium production (see Figure 31.4 for a map).

34.3. RECENT DEVELOPMENTS

Recent developments in Norilsk-Talnakh concentrators have been:

(a) the increased use of Knelson density separators;
(b) the use of nitrogen instead of air in pentlandite and pyrrhotite flotation cells, significantly improving metals recoveries (Woodcock, Sparrow, & Brickyard, 2007); and,
(c) the oxidative pressure leaching of pyrrhotite concentrates to efficiently recover the contained PGMs at high concentration in the residue.

34.4. SUMMARY

The mines near Norilsk, Siberia, contain large quantities of nickel and copper sulfide ores that also contain PGMs. The PGMs are recovered from these ores by:

(a) crushing and grinding to a particle size of less than 50–75 μm;
(b) isolation of minerals containing PGMs in high-density concentrate by centrifugal density separations;
(c) froth flotation of nickel, copper and iron sulfides to a concentrate that is contains PGMs;
(d) smelting and electrorefining of these concentrates to produce nickel, copper and cobalt metals and anode slimes that are rich in PGMs; and,
(e) refining these anodes slimes to individual high-purity PGMs.

The rate at which these metals are produced is limited only by the remoteness of the mines and the harsh Arctic climate.

REFERENCES

Blagodatin, Y. V., Distler, V. V., Zakharov, B. A., & Sluzhenikin, S. F. (2000). Disseminated ores of the Norilsk ore district as a potential for increasing output of platinum metals at the Norilsk Mining Company. *Obogashcheniye Rud*. Special issue.

Jollie, D. (2009). Production statistics. In *Platinum 2009*. Johnson Matthey plc.

Kerr, A. (2002). An overview of recent developments in flotation technology and plant practice for nickel ores. In A. L. Mular, D. N. Halbe & D. J. Barratt (Eds.), *Mineral processing, plant design, practice and control proceedings, Vol. 1* (pp. 1142–1158). SME.

Knelson. (2004). Knelson Gravity Solution's Platinum Approach. http://knelsongravity.xplorex.com/page186.htm. Accessed 02.06.11.

Komarova, M. Z., Kozyrev, S. M., Simonov, O. N., & Lulko, V. A. (2002). The PGE mineralization of disseminated sulphide ores of the Noril'sk-Taimyr Region. In L. J. Cabri (Ed.), *The geology, geochemistry, mineralogy and mineral beneficiation of platinum-group elements* (pp. 547–567). CIM.

Kozyrev, S. M., Komarova, M. Z., Emelina, L. N., et al. (2002). The mineralogy and behaviour of PGE during processing of the Noril'sk-Talnakh PGE-Cu-Ni ores. In L. J. Cabri (Ed.), *The geology, geochemistry, mineralogy and mineral beneficiation of platinum-group elements* (pp. 757–791). CIM.

Norilsk. (2010). Polar Division. http://www.norilsk.ru/en/our_products/polar_divisions/. Accessed 02.06.11.

Woodcock, J. E. T., Sparrow, G. J., & Brickyard, E. G. (2007). Flotation of precious metals and their minerals. In M. C. Fuerstenau, G. Jameson & R.-H. Yoon (Eds.), *Froth flotation, a century of innovation* (pp. 575–609). SME.

SUGGESTED READING

Kendall, T. (2004). PGM mining in Russia. *Platinum, 2004*, 16–22.

Chapter 35

Smelting and Converting of Sulfide Concentrates Containing Platinum-Group Metals

This chapter describes smelting and converting of flotation concentrates containing nickel, copper and iron sulfides that are rich in platinum-group metals (PGMs) [150 g/tonne PGMs]. The product of smelting and converting is a converter matte that is low in iron (1%–3%) and further enriched in platinum-group metals (about 2000 g/tonne). The converter matte is furnace processed by separating the PGMs from the base metals in a base metals refinery (Chapter 36) and purifying the individual PGM in a precious metals refinery (Chapter 37).

35.1. MAJOR PROCESS STEPS

The main steps in the smelting and converting process are as follows:

(a) drying of the concentrate;
(b) smelting of the concentrate to a furnace matte that is enriched in platinum-group elements; and,
(c) converting of the furnace matte to produce a final converter matte.

The flowsheets for this process are shown in Figures 32.1 and 35.1. The compositions of the concentrate, the electric furnace matte, the converter matte and the slag are given in Tables 35.1a–35.1e.

35.2. CONCENTRATE DRYING

The concentrations of typical concentrates containing platinum-group metals are given in Table 35.1a. The concentrates are delivered to the smelter either by truck, or by slurry pipeline or by tank train (Hundermark, de Villiers, & Ndlovu, 2006; Mabiza, 2006; Sima & Legoabe, 2006). The concentrates that

Extractive Metallurgy of Nickel, Cobalt and Platinum-Group Metals. DOI: 10.1016/B978-0-08-096809-4.10035-8

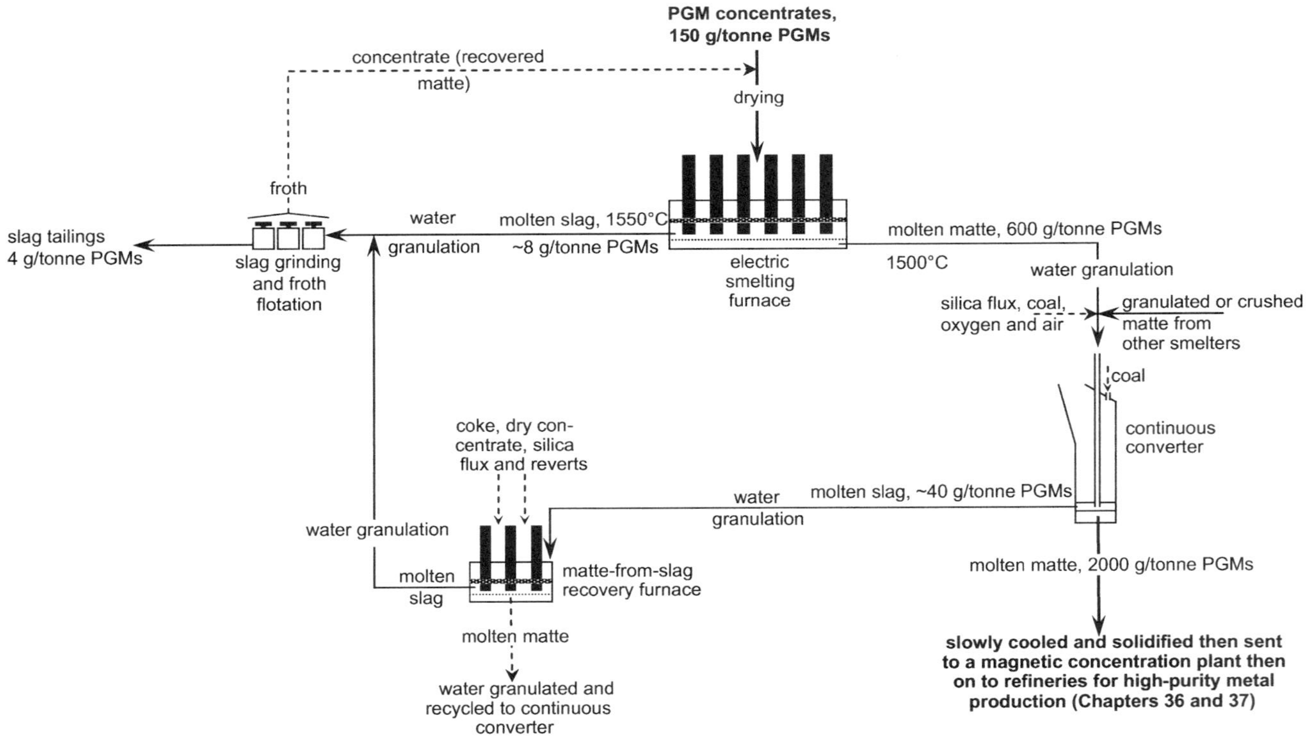

FIGURE 35.1 Electric concentrate smelting and Ausmelt converting at Anglo American Platinum (Jacobs, 2006). The feed is sulfide concentrate containing PGMs. The product is converter matte greatly enriched in PGMs. Sulfur dioxide gas is evolved by all three furnaces (data not shown). Gases from the smelting furnace and the converter are sent to gas cleaning and sulfuric acid manufacture (Chapter 20). Sulfur dioxide from the slag-cleaning furnace is captured by scrubbing with caustic solution. Note the three granulation steps.

TABLE 35.1a Compositions of Concentrates Produced From Merensky and UG2 Ores*

Smelter receiving concentrate	PGMs, g/tonne	Concentration in concentrate, %									
		Ni	Cu	Co	S	SiO_2	FeO	MgO	Al_2O_3	CaO	Cr_2O_3
Anglo American Platinum Waterval (Merensky)	143	3.6	2.1	0.08	9	34	20	15	3.2	4.7	0.8
Anglo American Platinum Union (Merensky)	142	2.2	1.1	0.04	5	38	15	20	3.8	2.5	2.6
Impala (Merensky)	138	2.1	1.3	0.06	6	42	18	18	4.1	2.9	1.1
Lonmin (Merensky)	130	3.0	2.0	0.08	9	41	23	18	1.8	2.8	0.4
Lonmin (UG2 blend)	340	2.1	1.2	0.06	4	47	15	21	3.6	2.7	2.8
Northam (Merensky)	132	2.5	1.3	0.05	5	47	17	18	2.6	3.0	0.9

In 2010, Merensky Reef and UG2 concentrates typically contained 100–200 g/tonne PGMs, Platreef concentrates somewhat less, 75–150 g/tonne. Higher grades, up to 400 g/tonne are occasionally produced as a consequence of minimizing chromite concentration in concentrate. Note: 1 g/tonne = 0.0001%.

**Jones, 1999*

TABLE 35.1b Compositions of Electric Furnace Mattes Produced by Smelting the Concentrates Given in Table 35.1a

Smelter	PGMs, g/tonne	Concentration in electric furnace matte, %				
		Ni	Cu	Co	Fe	S
Anglo American Platinum Waterval	600	17	9	0.5	41	27
Anglo American Platinum Union	800	12	7	0.3	37	25
Impala Rustenburg	1100	20	16	0.4	34	28
Lonmin: (Merensky)	1000	17	10	0.5	37	28
Lonmin: UG2 blend	2500	17	10	0.5	35	28
Northam	700	16	8	0.4	41	27

The enrichment is approximately 5–8 times, obtained mostly by removing gangue rock into molten slag.

TABLE 35.1c Compositions of Electric Furnace Slags from Smelting the Concentrates Given in Table 35.1a

	Concentration in electric furnace slag, %									
Smelter	Ni	Cu	Co	S	Al_2O_3	CaO	Cr_2O_3	FeO	MgO	SiO_2
Anglo American Platinum Waterval	0.19	0.11	0.05	0.5	3	6	1	31	15	46
Anglo American Platinum Union	0.16	0.10		0.3	3	6	3	20	13	41
Impala Platinum Rustenburg	0.11	0.11	0.03	0.3	6	8	1	21	18	47
Lonmin: Merensky	0.15	0.09	0.05		2	10	1	28	19	44
Lonmin: UG2	0.11	0.13	0.02		4	13	2	9	22	47
Northam	0.2	0.10	0.03		2	10	1	21	20	44

It is estimated that about 5% of the PGMs in concentrate feed reports to this slag. About half of it is subsequently recovered by slag cleaning (see Figure 35.1).

TABLE 35.1d Composition of Converter Mattes from Oxidizing the Electric Smelting Furnace Mattes Given in Table 35.1b

		Concentration in converter matte, %				
Smelter	**PGMs, g/tonne**	**Ni**	**Cu**	**Co**	**Fe**	**S**
Anglo American Platinum Waterval	2100	47	26	0.5	2.9	21
Impala Platinum Rustenburg	3400	47	31	0.4	0.5	21
Lonmin	6000	48	29	0.6	1.4	20
Northam	2600	51	27	0.5	1.0	19

PGM enrichment (electric furnace matte to converter matte) is 3–4 times, obtained mostly by oxidizing iron and sulfur. Note: 1 g/tonne = 0.0001%.

TABLE 35.1e Composition of Converter Slags from Oxidizing the Electric Smelting Furnace Mattes Given in Table 35.1b

	Concentration in converter slag, %									
Smelter	**Ni**	**Cu**	**Co**	**S**	**Al_2O_3**	**CaO**	**Cr_2O_3**	**FeO**	**MgO**	**SiO_2**
Anglo American Platinum Waterval	2.3	1.2	0.5	2.4	0.7	0.4	0.4	63	1.1	27
Impala Platinum Rustenburg	1.9	1.1	0.4	1.0	1.8	0.3	1.4	64	0.7	27
Lonmin	1.4	0.9	0.4	1.7	0.7	0.5	1.4	65	0.8	28
Northam	2.2	1.4	0.4		1.3	0.7	0.4	64	0.8	27

Converter slags also contain significant quantities of PGMs, most of which is recovered by slag cleaning (refer to Figure 35.1).

arrive by truck contain 10–18% moisture. They are dried in fuel-fired flash or spray driers (Coetzee, 2006; Jacobs, 2006). The concentrates that arrive as slurries are thickened and filtered then flash or spray dried.

The dried product contains less than 0.5% of water. This low water content is imperative because water can cause hydrogen explosions in electric furnaces.

35.3. SMELTING THE CONCENTRATES

The dried concentrates are smelted in electric furnaces because the sulfide content in the concentrate is too low (sometimes less than 15%) for autothermal (flash) smelting. In other words, the amount of heat produced by the reactions is insufficient to heat and melt the feed.

In addition, the slags from the smelting of these concentrates have high melting points (up to about 1600°C) due to the presence of magnesium and chromium in the pyroxenite and chromitite gangue (Nelson *et al.*, 2005).[1] As a result, the temperatures that are required are most easily obtained in electric furnaces.

35.3.1. Objectives of Smelting

The principal objective of smelting the concentrate is to produce molten matte that is considerably richer in platinum-group metals than the feed concentrate. This objective is achieved by rejecting the gangue rock in the concentrate to molten slag by (i) melting the concentrate feed in the furnace with electrical energy; and, (ii) allowing the melted concentrate to separate by gravity into two immiscible molten layers. These two immiscible layers are (i) a dense sulfide matte layer with a density of about 5 tonne/m^3 (Jones, 2005), which is enriched in PGMs; and, (ii) a less-dense silicate slag layer, with a density of about 2.7 tonne/m^3 (Jones, 2005), which is low in PGMs.

South African concentrates usually contain enough silica to ensure that the two phases are immiscible.

35.3.2. Partitioning of the Platinum-Group Elements into the Matte

All of the PGMs have a much greater affinity for the sulfide matte phase rather than the oxide slag phase in the electric furnace.

1. Chromitite is an igneous rock that contains mostly the mineral chromite (75–90%). The remainder is plagioclase.

This assertion is underscored by the observation that the PGMs occur in nature as sulfides (such as cooperite, PtS), alloys (such as isoferroplatinum, Pt_3Fe) and arsenides (such as sperrylite, $PtAs_2$). Oxide minerals of the PGMs are easily found in nature.

35.3.3. Collection of Platinum-Group Metals by Descending Sulfide Droplets

The mineral assemblies containing the platinum-group metals are small, sometimes less than 10 μm in size. Droplets of this size would take a long time to settle through the slag into the matte (Nelson *et al.*, 2005). In practice, they coalesce with the sulfide droplets as they both descend towards the matte layer.

In order for the sulfides to efficiently collect the platinum-group metals, the feed to the smelting furnace must contain sufficient sulfides. Concentrates from the Merensky Reef contain 40% or 50% sulfides, and, as a result, they are efficient collectors. Concentrates from the UG2 ores may, however, contain only about 10% sulfides, which may be too low for efficient collection of the platinum-group elements. This can be remedied either by blending UG2 and Merensky concentrates, or by including high-grade nickel or copper concentrates, which contain about 90% sulfides, into the charge.

35.3.4. Chromium Problems

Concentrates made from UG2 ores contain considerable chromite, $FeCr_2O_4$. The melting point of the slag is raised because of the chromite content. In addition, the chromium from the chromite may cause precipitation of the $(Fe,Mg)Cr_2O_4$ spinel, which has a melting point of about 1600°C, in cooler parts of the furnace (Nelson *et al.*, 2005). This precipitated spinel may, in turn, prevent efficient settling of matte (Jones, 2005).

Problems associated with chromite are minimized by (i) rejecting chromite to tailings during flotation; and, (ii) minimizing the internal recycle in the smelter complex of high-chromium materials, such as converter slag, to the smelting furnace.

These procedures permit steady, efficient smelting with between 2 and 3% of chromite in the feed concentrate.

35.3.5. Industrial Smelting

Industrial smelting of concentrates containing platinum-group metals is mostly done in rectangular electric furnaces with six large carbon electrodes. The operating details of some furnaces are given in Tables 35.2a and 35.2b. Circular and smaller rectangular furnaces with three carbon electrodes are also used (Warner *et al.*, 2007). In all cases, the energy for smelting is provided by

TABLE 35.2a Dimensions and Production Details of Smelting Furnaces for PGM Flotation Concentrates*

Smelter	Waterval (Anglo American Platinum)	Mortimer (Anglo American Platinum)	Polokwane (Anglo American Platinum)	Rustenberg (Impala Platinum)
Smelting furnaces	2 Rectangular	1 Rectangular	1 Rectangular	4 Rectangular, 2 operating
Length × width, m	26 × 8 (inside)	25 × 7	29 × 10	26 × 8
Electrode diameter, m	1.1 (self-baking)	1.25 (self-baking)	1.6 (self-baking)	1.14 (self-baking)
Electrode consumption, kg/tonne of feed	2		3	2
Power rating, MW	34	20	68	35 and 38
kWh/tonne of feed	700	820–850	750–850	721
Wall cooling	Water-cooled copper fingers and plates	Forced air-cooled copper waffle coolers	Water-cooled copper waffle coolers	Water-cooled copper plate coolers
Throughput tonnes/h dry concentrate	71 (total for both furnaces)	20	87 (capacity)	92 (total for both furnaces)
Concentrate composition				
g/tonne PGMs	70	70	70	130
% Ni; % Cu; % Co	3.6; 2.1; 0.08	2.2; 1.1; 0.04	2.1; 1.2; 0.04	1.7; 1.1; 0.05
% Fe; % S; % MgO	16; 9; 15	12; 5; 20	12; 5; 18	12; 5; 18
Flux	Limestone	Limestone 3% of charge	Limestone (sometimes)	None

(Continued)

Table 35.2a Dimensions and Production Details of Smelting Furnaces for PGM Flotation Concentrates*—cont'd

Smelter	Waterval (Anglo American Platinum)	Mortimer (Anglo American Platinum)	Polokwane (Anglo American Platinum)	Rustenberg (Impala Platinum)
Temperatures, °C				
Slag tapping	1550	1650	1600	1460
Matte tapping		1550	1550	1300
Tonnes matte; slag produced per tonne of concentrate feed	0.22; ~1	0.15; 0.7	0.15; 0.9	0.12; 0.9
Matte composition				
g/tonne PGMs	600	600	600	1000
% Ni; % Cu; % Co	17; 9; 0.5	12; 7; 0.3	14; 8; 0.3	14; 9; 0.3
% Fe; % S	41; 27	37; 25	40; 30; Cr 2%	45; 30
Destination	Continuous converter	Continuous converter	Continuous converter	Peirce-Smith converter
Slag composition				
% SiO_2; % Fe; % MgO	46; 24; 15	41; 16; 13	52; 8; 23; Cr_2O_3 2%	47; 9; 21
Destination	Granulation, grinding, flotation	Granulation, grinding, flotation	Discard	Granulation, grinding, flotation
Offgas volume% SO_2	0.5–1.3	0.5–1	0.5–1.5	0.9
Destination	H_2SO_4 plant			Electrostatic precipitator H_2SO_4 plant

All PGM flotation concentrates are dried. The smelting is continuous. Product destinations and other details are shown in Figures 1.6 and 35.1.
Jones, 2004; Warner et al., 2007

TABLE 35.2b Dimensions and Production Details of Smelting Furnaces for PGM Flotation Concentrate*

Smelter	Lonmin Platinum	Northam Platinum	Zimplats, Zimbabwe	Stillwater, Montana
Smelting furnaces	1 Circular, 3 electrodes	1 Rectangular (6 electrodes)	1 Circular (3 electrodes)	1 Rectangular (3 electrodes)
Length × width, m	11 m diameter	26 × 9	12 m diameter steel shell	9 × 5
Electrode diameter, m	1.4	1	1.2 (self-baking)	0.3 (pre-baked)
Electrode consumption, kg/tonne of feed	2.6	2.6	3.1	3.5
Power rating, MW	28	15	12.5	5
kWh/tonne of feed	850	1000	900	900
Wall cooling	Copper waffle coolers, water-cooled shell	None	Copper plate coolers	Water-cooled copper plates
Throughput, tonne/h dry concentrate	30	10	10	1
Concentrate composition				
g/tonne Pt group elements	300	130	80	90
% Ni; % Cu; % Co	2.5; 1.5; 0.13	2.5; 1.3; 0.05	2.0; 1.5; 0.07	5.3; 3.2; 0.01
% Fe; % S; % MgO	17; 6; 17	13; 5; 18	13; 6; 24	15; 13; 12
Flux			Limestone	Limestone, catalyst scrap

(*Continued*)

Table 35.2b Dimensions and Production Details of Smelting Furnaces for PGM Flotation Concentrate*—cont'd

Smelter	Lonmin Platinum	Northam Platinum	Zimplats, Zimbabwe	Stillwater, Montana
Temperatures, °C				
Slag tapping	1650	1500	1600	1500
Matte tapping	1550	1400	1400	1200
Tonnes matte; slag produced per tonne of concentrate feed	0.14; ~0.9	0.18; ~0.9	0.12; 1.1	0.14–0.25; ~0.9
Matte composition				
g/tonne PGMs	2000	700	600	400
% Ni; % Cu; % Co	15; 9; 0.5	16; 8; 0.4	15; 10; 0.7	17; 11; 0.02
% Fe; % S	43; 28	41; 27	43; 28	43; 27
Destination	Peirce-Smith converter	Peirce-Smith converter	Peirce-Smith converter	Top-blown rotary converter
Slag composition, mass%				
%SiO_2; %Fe; %MgO	45; 22; 20	44; 16; 20	54; 14; 22	45; 10; 14
Destination			Granulated then discarded	Slow cooling, grinding, flotation
Offgas volume% SO_2			0.1	4
Destination	Precipitator and H_2SO_4 plant			Baghouse then SO_2 scrubber

The smelting is continuous. Product destinations and other details are shown in Figures 1.6 and 35.1.
Jones, 2004; Warner et al., 2007

passing electric current between carbon electrodes through molten slag. Resistance of the slag to current flow heats the slag, which, in turn, heats and melts the concentrate, which is charged from the roof.

Smelting is continuous. The slag is always covered with concentrate. Concentrate is charged as needed to maintain this condition.

Slag flows out of the furnace almost continuously through a water-cooled copper taphole. It is granulated in water and either sent to a flotation plant for the recovery of platinum-group metals, as shown in Figure 35.1, or discarded.

Matte is tapped intermittently from a water-cooled taphole as needed to maintain a specified level of matte in the furnace. The furnace produces much less matte than slag, that is, the production ratio of matte to slag is typically 0.2. As a result, the matte outflow is intermittent.

35.3.6. Steady-State Operation

The start-up, steady-state and shutdown procedures of electric furnaces used for smelting nickel–copper matte were described in Chapter 17. The smelting of matte containing platinum-group metals is similar except that its charge is dried sulfide concentrate rather than hot oxidized calcine.

Smelting of concentrates containing platinum-group elements requires more energy than smelting of hot nickel–copper calcine. Smelting platinum-containing concentrates requires about 800 kWh/tonne of concentrate, whereas smelting calcine requires about 500 kWh/tonne of calcine (Warner *et al.*, 2007). The reasons for this difference are:

(a) the feed concentrate to a PGM furnace is at ambient temperature while the nickel–copper calcine is hot, about 400°C; and,
(b) the operating temperature of smelting furnaces is about 1550°C, which is about 250°C hotter than nickel–copper smelting furnaces, which are operated at about 1300°C. The higher operating temperature is required because of the high melting point of the MgO and Cr_2O_3 slags.

The molten matte is sent to converting, either as molten matte or as solid matte that has been granulated using water.

35.4. CONVERTING THE FURNACE MATTE

Converting the furnace matte entails oxidizing the iron and sulfur in the matte using air or an air–oxygen mixture. The objective is to produce a molten matte that is enriched in platinum-group metals and is relatively low in iron and sulfur.

This converter matte is a suitable feed for the refining of the metals. Peirce-Smith, top-blown rotary and Ausmelt continuous converters are used. Operating data for converters used in the converting of furnace mattes rich in platinum-group metals are given in Table 35.3.

TABLE 35.3 Dimensions and Production Details of PGM Converters*

Smelter	Waterval (Anglo American Platinum)	Rustenburg (Impala Platinum)	Stillwater Montana
Converter type	Continuous (Anglo American Platinum Converting Process)	Peirce-Smith	Top-blown rotary converter
Number of converters	2 (1 operating)	6 (3 or 4 operating)	2 (1 operating)
Dimensions	4.5 m inside diameter, 4 m hearth height, ~10 m total height	2 Converters 3.6 m diameter × 7.3 m long	0.8 m diameter × 1.5 m deep inside
		4 Converters 3 m diameter× 4.5 m long	
Vertical lance			
Number	1 per converter		1 per converter
Diameter outside, m	0.45		~0.1
Blowing rate, air + O_2	25 000 Nm^3/h (up to 40% O_2 in blast)		2600 Nm^3/h, 93% O_2
Tuyeres			
Number per converter		32 in each large converter	
		26 in each small converter	
Diameter, outside m		0.05	
Blowing rate		11 000–22 000 Nm^3/h, no O_2 enrichment	
Feed	Solid granulated electric furnace matte	Molten electric furnace matte 1200°C	Solid granulated electric furnace matte
Feed rate	14 000, tonnes per month (2006)	~8000	~50
Composition			
g/tonne PGMs	600	1100	
%Ni	17	14	17

Table 35.3 Dimensions and Production Details of PGM Converters*—cont'd

Smelter	Waterval (Anglo American Platinum)	Rustenburg (Impala Platinum)	Stillwater Montana
% Cu	9	9	11
% Co	0.5	000.3	
% Fe	41	45	43
% S	27	30	20
Product matte	1300°		
Production rate, tonnes per month	5000	~2500	~20
Composition			
g/tonne PGMs	2000	3000	
% Ni	47	47	42
% Cu	26	31	33
% Co	0.5	0.3	
% Fe	3.5	0.5	~2
% S	22	21	20
Destination	Slow freezing, matte separation, refining	Water granulated then sent to refineries	Water granulated then sent to refineries
Product slag			
Production rate, tonnes per month	~12 000	~7000	50
Composition			
% Fe, % SiO_2	46, 26	50, 27	50% Fe, 20% CaO
Destination	Water granulation then electric furnace settling recovery of matte	Water granulation then mineral processing recovery of matte	Water granulation then recycled to smelting furnace

All treat electric furnace matte. Product destinations and other details are shown in Figures 32.1 and 35.1.

**Jones, 2004; Warner et al., 2007*

35.4.1. Converter Feeds and Products

The feed to the converter is a furnace matte that typically contains 40% Fe, 27% S and 0.1% platinum-group metals, while the product from the converter, referred to as the converter matte, typically contains 0.6%–3% of Fe, 20% S and 0.25% platinum-group metals.

The low iron content of this converter matte is advantageous, since it minimizes the precipitation of iron residues, such as jarosite, during hydrometallurgical refining. This, in turn, minimizes the loss of platinum to waste residues.

The converter matte is also sulfur deficient, that is, it contains less sulfur than is stoichiometrically needed to form Ni_3S_2 and Cu_2S. A sulfur-deficient converter matte has the following benefits:

(a) it encourages precipitation and isolation of an alloy phase rich in platinum-group metals during slow cooling of the matte; and,
(b) it minimizes the sulfur input to the hydrometallurgical refinery, which minimizes the amount of acid that must be neutralized in the base metals refinery.

35.4.2. Converting Methods

The traditional matte-converting equipment is Peirce-Smith converters, which were described in Chapter 19. Both Lonmin Platinum and Impala Platinum use Peirce-Smith converters.

Top-blown rotary converters are used at the Stillwater smelter, USA. They are described in Appendix G.

A continuous top-blown converter (Ausmelt) is used to convert all the furnace matte from three smelters at Anglo American Platinum. This type of furnace is described in the following section.

35.4.3. Ausmelt Continuous Converting

Ausmelt converting entails continuously blowing dry furnace matte, oxygen and air, silica flux and coal into the hot molten bath of a vertical furnace (Coveney, Baldock, Hughes, & Reuter, 2009). The furnace design is shown in Figure 35.2. The matte granules and pieces melt in the molten bath. Most of the iron and sulfur are oxidized by oxygen in feed air and oxygen.

Energy required for heating is provided by the oxidation of iron and sulfur and by the burning of the coal in the feed (Jacobs, 2006).

35.4.4. Advantages and Disadvantages of Continuous Converting

The principal advantage of continuous converting at Anglo American Platinum is the production of a continuous stream of an offgas that is rich in

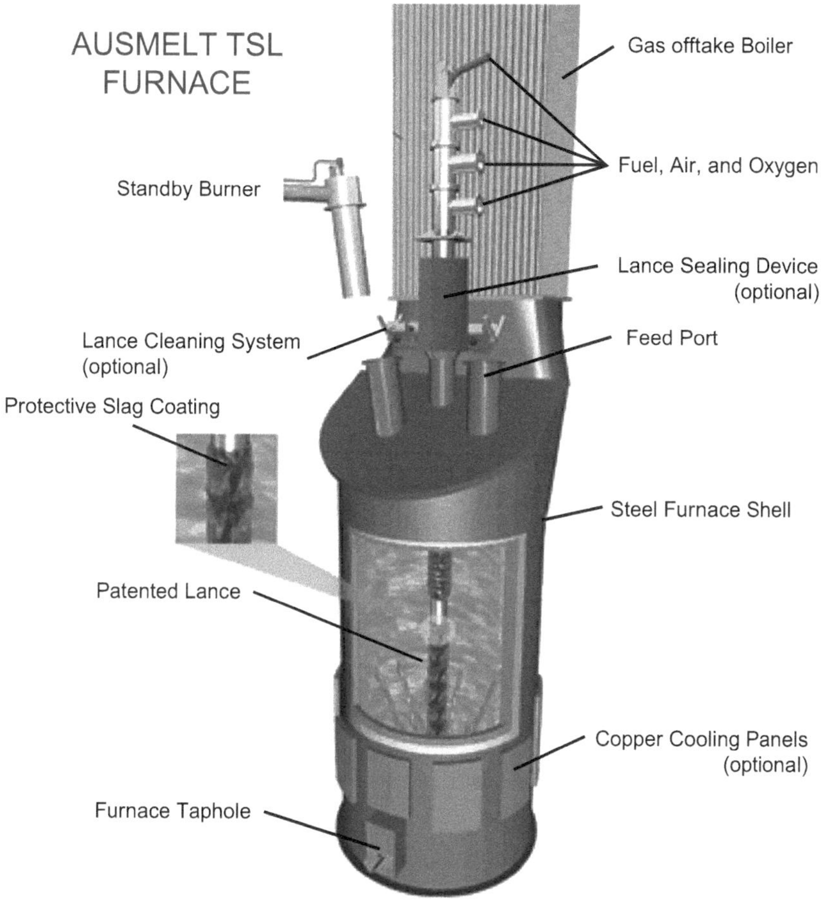

FIGURE 35.2 Sketch of the continuous converter at Anglo American Platinum (Ausmelt, 2007). Note the central lance that consists of (i) a central pipe through which solidified electric furnace matte, flux and coal are blown into the bath; (ii) an annulus through which oxygen and air are blown; and, (iii) an outer tube. Coal is also added through the feed ports on the roof (shown). See Table 35.3 for details.

sulfur dioxide, which is ideal for sulfur capture in a sulfuric acid plant (Chapter 20).

The principal disadvantage of the process is the rapid wear of the refractory lining of the furnace, which necessitates a second standby converter.

The slag also contains considerable quantities of payable metals, that is PGMs, nickel, copper and cobalt. As a result, the slag needs to be processed in a separate slag-cleaning furnace, as shown in Figure 35.1.

35.4.5. Vertical Lance

The heart of the Ausmelt furnace is the lance, which consists of the following features:

(a) a central pipe through which solid furnace matte, silica flux and coal are blown into the furnace; and,
(b) an annulus between this pipe and an outer tube through which air/oxygen gas is blown into the furnace.

The outer tube extends into the hot matte–slag emulsion of the converter. The inner pipe ends about 1 m above the emulsion. Coal is also added through a port on the roof.

A curved blade gas 'swirler' is attached around the outside of the central pipe near the bottom end, extending outwards towards the outside tube. These blades force descending cool feed gas against the outer tube, extracting heat from the tube and causing slag to freeze on the outer tube where it enters the molten bath. This solid slag layer protects the lance and is a key design feature of the converter. Without it, the lance would be rapidly consumed.

35.4.6. Operation and Control

Steady-state converting entails the following (Jacobs, 2006):

(a) supplying the solid and gas feeds at a constant rate;
(b) continuously removing offgas and tapping slag; and,
(c) periodically tapping matte, through a separate taphole, to maintain a specified level in the furnace.

The iron content in the product matte is controlled by adjusting the feed rate of oxygen (in both air and oxygen) while maintaining the feed rate of the furnace matte.

The product temperature is mostly controlled by the feed rate of coal.

35.5. RECENT DEVELOPMENTS IN SMELTING AND CONVERTING IN THE PLATINUM INDUSTRY

Virtually all PGMs are produced via matte smelting and converting of sulfide concentrates. A departure from this is the *ConRoast* process, which is still in development. The process entails the following steps:

(a) the oxidization of all the sulfur in the concentrate in a fluidized-bed roaster, with excellent capture of sulfur dioxide;
(b) the reduction smelting of the resulting calcine into a molten iron alloy that contains approximately 80% of Fe, 8% Ni, 3% Cu, 1% Co and 600 g/tonne PGMs (McCullough, Geldenhuys, & Jones, 2008); and,

(c) the converting (oxidation) of iron from this molten alloy to make feed for hydrometallurgical production of PGMs, nickel, copper and cobalt.

Planning for a commercial plant is underway (Jones, 2009).

35.6. SUMMARY

Platinum-group element concentrates are made into a matte that is relatively low in iron and sulfur, and enriched in platinum-group elements by smelting and converting. Smelting removes most gangue minerals as molten slag. Converting removes most iron and sulfur by oxidation with air and oxygen.

In combination, these two steps upgrade the feed concentrate containing 150 g/tonne platinum-group elements into a converter matte containing 2000 g/tonne PGMs. The converter matte is suitable for further processing to high-purity PGMs, nickel, copper and cobalt (discussed in Chapters 36 and 37).

REFERENCES

Ausmelt. (2007). Generation 3 lances and Ausmelt furnace [Company brochure].

Coetzee, V. (2006). Common-sense improvements to electric smelting and Impala Platinum. In R. T. Jones (Ed.), *Pyro 2006 – Southern African pyrometallurgy 2006 international conference* (pp. 43–62). SAIMM.

Coveney, J. A., Baldock, B. R., Hughes, S., & Reuter, M. A. (2009). Ausmelt technology for nickel and PGM processing. In J. Liu, J. Peacey & M. Barati, et al. (Eds.), *Pyrometallurgy of nickel and cobalt 2009* (pp. 169–180). CIM.

Hundermark, R., de Villiers, B., & Ndlovu, J. (2006). Process description and short history of Polokwane Smelter. In R. T. Jones (Ed.), *Pyro 2006 – Southern African pyrometallurgy 2006 international conference* (pp. 35–41). SAIMM.

Jacobs, M. (2006). Process description and abbreviated history of Anglo Platinum's Waterval Smelter. In R. T. Jones (Ed.), *Pyro 2006 – Southern African pyrometallurgy 2006 international conference* (pp. 17–34). SAIMM.

Jones, R. T. (1999). Platinum smelting in South Africa. *South African Journal of Science, 95*, 525–534.

Jones, R. T. (2004). JOM world nonferrous smelter survey, Part II: platinum group metals. *JOM, 56*, 59–63.

Jones, R. T. (2005). An overview of Southern African PGM smelting. In J. Donald & R. Schonewille (Eds.), *Nickel and cobalt 2005, challenges in extraction and production* (pp. 147–178). CIM.

Jones, R. T. (2009). Towards commercialization of Mintek's Conroast Process for platinum smelting. In J. Liu, J. Peacey & M. Barati, et al. (Eds.), *Pyrometallurgy of nickel and cobalt 2009, proceedings of the international symposium* (pp. 159–168). CIM.

Mabiza, L. (2006). An overview of PGM smelting in Zimbabwe – Zimplats operations. In R. T. Jones (Ed.), *Pyro 2006 – Southern African pyrometallurgy 2006 international conference* (pp. 63–75). SAIMM.

McCullough, S. D., Geldenhuys, I. J., & Jones, R. T. (2008). Pyrometallurgical iron removal from a PGM-containing alloy. In M. Rogers (Ed.), *Proceedings of platinum in transformation, third*

international platinum conference (pp. 169–176). The Southern Africa Institute of Mining and Metallurgy.
Nelson, L. R., Stober, F., Ndlovu, J., et al. (2005). Role of technical innovation on production delivery at the Polokwane smelter. In J. Donald & R. Schonewille (Eds.), *Nickel and cobalt 2005, challenges in extraction and production* (pp. 91–116). CIM.
Sima, C., & Legoabe, M. (2006). Mortimer smelter: Operations description. In R. T. Jones (Ed.), *Pyro 2006 – Southern African pyrometallurgy 2006 international conference* (pp. 29–34). SAIMM.
Warner, A. E. M., Diaz, C. M., Dalvi, A. D., et al. (2007). JOM world nonferrous smelter survey, part IV: Nickel: sulfide. *JOM, 59*, 58–72.

SUGGESTED READING

Cole, S., & Ferron, C. J. (2002). A review of the beneficiation and extractive metallurgy of the platinum-group elements, highlighting recent process innovations. In L. J. Cabri (Ed.), *The geology, geochemistry, mineralogy and mineral beneficiation of platinum-group elements* (pp. 811–844). CIM.
Jones, R. T. (Ed.), (2006). *Pyro 2006 – Southern African pyrometallurgy 2006 international conference*. SAIMM.
Kendall, T. (2004). PGM mining in Russia. In *Platinum 2004* (pp. 16–22). Johnson Matthey.
Merkle, R. K. W., & McKenzie, A. D. (2002). The mining and beneficiation of South African PGE ores – An overview. In L. J. Cabri (Ed.), *The geology, geochemistry, mineralogy and mineral beneficiation of platinum-group elements* (pp. 793–809). CIM.
Rogers, M. (Ed.), (2008). *Proceedings of platinum in transformation, third international platinum conference*. SAIMM.

Chapter 36

Separation of the Platinum-Group Metals from Base Metal Sulfides, and the Refining of Nickel, Copper and Cobalt

Converter matte consists mainly of nickel and copper alloys and sulfides. The platinum-group metals (PGMs) in the matte are separated from the base metals in the base metals refinery. The processing routes are described in this chapter. In the next chapter, the refining of the PGMs to produce the pure metals is described.

Operating details for some base metals refineries are given in Table 36.1.

36.1. OVERVIEW OF THE REFINING OF THE PLATINUM-GROUP METALS

The concentration of the sulfide minerals by froth flotation upgrades the PGM content by a factor of about 20 times. The smelting of the concentrate to a converter matte further upgrades the PGM content by a factor of 20 times. The next step, the base metals refinery, results in an upgrade factor of about 400 times, by far the largest in the production of PGMs from ores.

There are two process routes that are used to achieve this upgrade in PGM content. These process routes are:

(a) slow cooling followed by magnetic separation; and,
(b) whole-matte leaching.

36.1.1. Slow-Cooled Matte and Magnetic Separation

Of the major platinum producers, only Anglo American Platinum's refinery in Rustenburg, South Africa, uses the slow-cooling process. This is the largest producer of PGMs. The details of the slow-cooling process were presented in Chapter 21.

Extractive Metallurgy of Nickel, Cobalt and Platinum-Group Metals. DOI: 10.1016/B978-0-08-096809-4.10036-X

TABLE 36.1 Operating Details for Selected Base Metals Refineries

Location	Stillwater, USA/ Northam, South Africa/ Lonmin, South Africa	Impala Platinum, South Africa	Anglo American Platinum, South Africa
Year of first production	1986/ 1985/ 1971	1969	1954, upgraded in 1966, 1981, and 2011
Converter matte throughput, tonnes/year	1250/ 4000/ 10 000	30 000	45 000
Nickel leaching			
Temperature, °C	85	150	NAL: 85–90 NOX: 150*
Pressure, bar	Atmospheric	4.5	NAL: ambient NOX: 6
Nickel extraction, %	75	85	80–85
Solution composition leaving			
Ni^{2+}, g/L	80–100	100	100
Cu^{2+}, g/L	0–2	5	NAL: 12 NOX: 20–30
H_2SO_4, g/L	0–2	15	NAL: 10–12 NOX 20
pH	3	1.5	1
Iron removal step			
Product	Ferric hydroxide	Ammonium jarosite	Sodium jarosite
Temperature, °C	85	140	140
Cobalt removal step			
Precipitation method	None	$Ni(NH_4)_2(SO_4)_2 \cdot 6H_2O$	$Co(OH)_3$
Product	None	Cobalt powder or briquette	$CoSO4 \cdot 6H_2O$
Product process		Hydrogen reduction	Crystallization

TABLE 36.1 Operating Details for Selected Base Metals Refineries—cont'd

Location	Stillwater, USA/ Northam, South Africa/ Lonmin, South Africa	Impala Platinum, South Africa	Anglo American Platinum, South Africa
Nickel product			
Production rate, tonnes/year	640/ 1800/ 3500	16 000	33 000 (capacity)
Process route	Nickel crystallizer	Hydrogen reduction	Electrowinning
Product type	$NiSO_4 \cdot 6H_2O$	Nickel powder or briquette	Nickel cathode
Copper leaching stage			
Temperature, °C	140	130–140	140
Pressure, bar	6	6	6
Copper extraction, %	90	>95	>95
Solution composition leaving			
Ni^{2+}, g/L	30–35	20–30	20–30
Cu^{2+}, g/L	60–65	90	90–110
H_2SO_4, g/L	15–20	15	10–15
Selenium and tellurium removal			
Solution composition leaving			
Se, mg/L	<5	<10	<10
Te, mg/L	<10	<10	<10
Copper electrowinning			
Production rate, tonnes/year	400/ 1000/ 2500	9000	18 000

(Continued)

TABLE 36.1 Operating Details for Selected Base Metals Refineries—cont'd

Location	Stillwater, USA/ Northam, South Africa/ Lonmin, South Africa	Impala Platinum, South Africa	Anglo American Platinum, South Africa
Third-stage leach			
Temperature, °C	140	140	–
Pressure, bar	6	6	–
PGM upgrade			
Composition of PGM concentrate			
Ni, %	1–2	1–2	
Cu, %	2–3	3–4	
S, %	1	<1	
PGMs, %	60–65	50–60	50–60
SiO_2, %	1–3	10	

**NAL, nickel atmospheric leaching; NOX, non-oxidative leaching.*

In this process, converter matte is allowed to cool over several days so that relatively large crystals of heazlewoodite, Ni_3S_2, chalcocite, Cu_2S, and metal alloy develop. Since the converter matte is deficient in sulfur, about 10% of the converter matte is in the form of an alloy, which is magnetic.

During slow cooling, the PGMs concentrate in the alloy phase. This means that they can be separated from the base metal sulfides by magnetic separation. This magnetic fraction is then treated in leaching steps to remove copper, nickel and iron so that the platinum content is upgraded. It is then transferred to the precious metals refinery for the separation of the individual precious metals.

The non-magnetic fraction of the material from the magnetic concentration plant is transferred to the base metals refinery, where the copper and nickel are leached into solution. The solutions are purified. From the purified solutions, saleable copper, nickel and cobalt products are produced.

The advantage of the slow-cooling process is that the production of the PGMs is less dependent on the production of the base metals. In other words, the non-magnetic fraction could be stockpiled, or refined elsewhere,

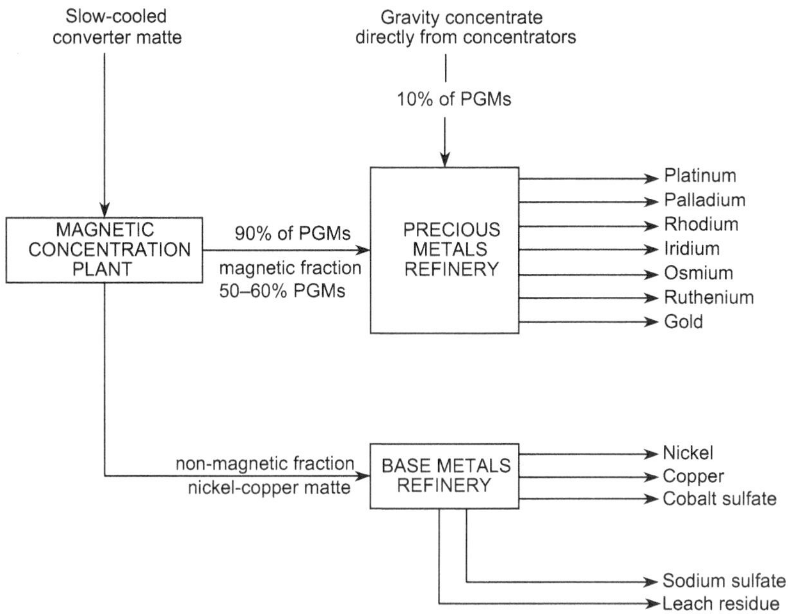

FIGURE 36.1 The magnetic concentration process used by Anglo American Platinum for separating the PGMs from the base metals. Note that about 10% of the PGM feed to the precious metals refinery comes directly from the concentrators.

or even sold if the base metals refinery became a bottleneck to platinum production.

The flowsheet for the separation of the metallic alloy from the base metal sulfides is shown in Figure 36.1.

36.1.2. Whole-matte Leaching

Most of the other primary producers, that is, Impala Platinum, Lonmin Platinum, Northam Platinum and Stillwater, use the whole-matte leaching process. In this process, nickel and copper are leached from the converter matte, leaving the PGMs in the insoluble residue. The PGM content of this leach residue is about 60%. In most of the operations, the residue is treated to remove residual copper, nickel, iron, selenium and tellurium prior to it being transferred to a precious metals refinery.

The whole-matte leaching process is illustrated in Figure 36.2.

36.2. OBJECTIVES OF THIS CHAPTER

This chapter describes the role of the base metal refineries in the process of producing PGMs. The feed material is the converter matte. The products

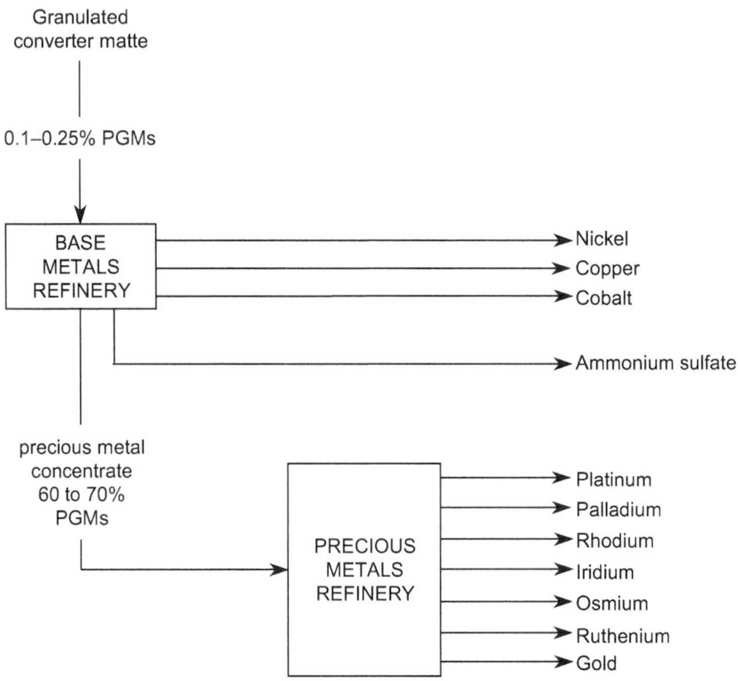

FIGURE 36.2 The overall process for the refining of PGMs from the smelter matte by the whole-matte leaching process.

are saleable nickel, copper and cobalt and a platinum concentrate suitable for the refining into the individual PGMs. In addition, sulfur has to be removed.

The objectives for this chapter are:

(a) to describe the manner in which the PGMs are separated from the nickel, copper and cobalt sulfides;
(b) to describe the manner in which salable nickel, copper and cobalt products are refined; and,
(c) to describe the manner in which sulfur is removed.

The focus is on the processes that are used by the industrial producers of primary PGMs.

There is both similarity and variety in the flowsheets of the different producers. However, there are two major distinguishing factors. Firstly, there is a distinction between the whole-matte leaching and the magnetic concentration, as discussed earlier. Secondly, there is a distinction between the different manner in which the processes produce nickel and remove sulfur. There are three different ways that are currently employed for the production of nickel and the removal of sulfur. These process routes are:

(a) both nickel and sulfur are removed as nickel sulfate;
(b) nickel is produced by hydrogen reduction and sulfur is removed as ammonium sulfate; and,
(c) nickel is produced by electrowinning and sulfur is removed as sodium sulfate.

The final forms of the nickel and sulfur determine the choice of the process for the winning of the nickel. If nickel sulfate is the choice for the removal of sulfur, then this is also the nickel product. If hydrogen reduction is the preferred method for the production of nickel, then the sulfur will be removed as ammonium sulfate. If nickel electrowinning is the choice for the production of nickel, then sulfur will be removed as sodium sulfate.

Each of these processes is discussed. The process used by all the smaller producers, in which nickel is produced as nickel sulfate crystals, is discussed first. After this, the process used by Impala Platinum, in which nickel is produced as briquetted nickel powder by hydrogen reduction, is presented. Finally, the process used at Anglo American Platinum, in which nickel is produced as nickel cathodes by electrowinning, is discussed.

36.3. BASE METAL REFINERIES WHERE NICKEL IS PRODUCED AS NICKEL SULFATE

There are three base metals refineries in which the nickel is produced as nickel sulfate:

Lonmin, Marikana, South Africa;
Northam, South Africa, and
Stillwater, Montana, USA.

The base metals refineries of Lonmin, Northam and Stillwater are very similar. A schematic flowsheet of the process is shown in Figure 36.3.

In the description of the process, the flow of solids, which contain the PGMs, is described first. After this, the solution streams are described.

Granulated matte is received from the converter, milled typically to a particle size of 80% less than 75 μm and transferred to the first leaching stage. The matte consists of heazlewoodite, Ni_3S_2, chalcocite, Cu_2S, troilite, FeS, bornite, Cu_5FeS_4, and nickel–copper–PGM alloy.

36.3.1. First Leaching Stage

In the first leaching stage, milled matte is contacted with solution from copper electrowinning (referred to as *copper spent*) at 85°C. The equipment consists of five tanks in series. Oxygen gas is injected into the first three tanks. The

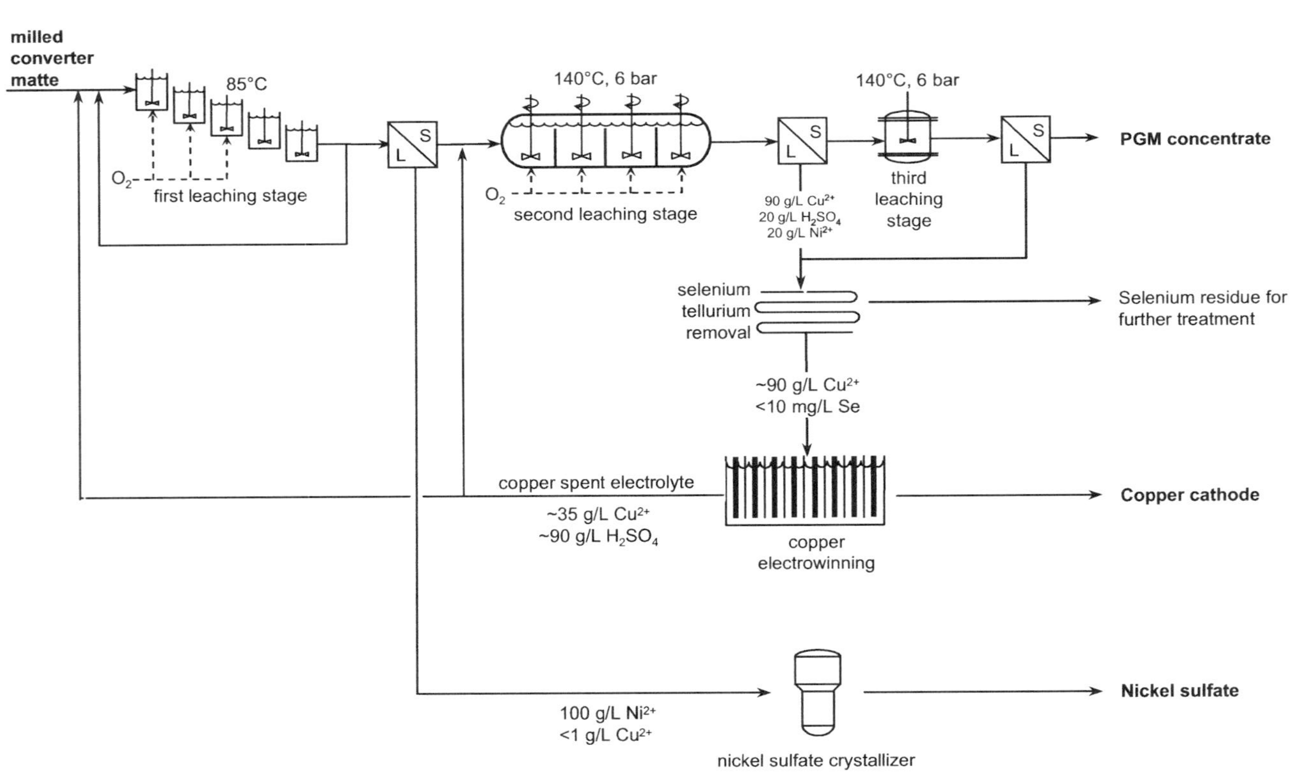

FIGURE 36.3 A schematic flowsheet of the process used by Lonmin, Northam and Stillwater in which nickel is produced as nickel sulfate.

objective of the first stage is to leach about 75% of the nickel, reduce the concentration of copper in solution is less than 0.5 g/L and reduce the concentration of sulfuric acid in solution is less than 1 g/L.

Nickel is leached by metathesis[1] with copper ions and by dissolution with oxygen. These two reactions are as follows:

$$\underset{\text{feed}}{Ni_3S_2(s)} + \underset{\text{in copper spent}}{2CuSO_4(aq)} \xrightarrow{85^\circ C} \underset{\substack{\text{nickel sulfate}\\ \text{solution to crystallizer}}}{2NiSO_4(aq)} + \underset{\substack{\text{undissolved solids to}\\ \text{second leaching stage}}}{Cu_2S(s) + NiS(s)} \quad (36.1)$$

$$\underset{\text{feed}}{Ni_3S_2(s)} + \underset{\substack{\text{acid in copper}\\ \text{spent}}}{H_2SO_4(aq)} + \underset{\substack{\text{gas supplied}\\ \text{to reactors}}}{0.5O_2(g)} \xrightarrow{85^\circ C} \underset{\substack{\text{nickel sulfate}\\ \text{solution to crystallizer}}}{NiSO_4(aq)} + \underset{\substack{\text{undissolved solids to}\\ \text{second leaching stage}}}{2NiS(s)} + H_2O(\ell) \quad (36.2)$$

All the metals present as alloys in the milled matte are leached in a manner similar to the following reaction:

$$\underset{\text{feed}}{Ni(s)} + \underset{\substack{\text{acid in copper}\\ \text{spent}}}{H_2SO_4(aq)} + \underset{\substack{\text{gas supplied}\\ \text{to reactors}}}{0.5O_2(g)} \xrightarrow{85^\circ C} \underset{\substack{\text{nickel sulfate}\\ \text{solution to}\\ \text{crystallizer}}}{NiSO_4(aq)} + H_2O(\ell) \quad (36.3)$$

Some copper might also be leached by the following reaction:

$$\underset{\text{feed}}{Cu_2S(s)} + \underset{\substack{\text{sulfuric acid in}\\ \text{copper spent}}}{H_2SO_4(aq)} + \underset{\substack{\text{gas supplied}\\ \text{to reactors}}}{0.5O_2(g)} \xrightarrow{85^\circ C} \underset{\substack{\text{copper sulfate to}\\ \text{be precipitated by}}}{CuSO_4(aq)} + \underset{\substack{\text{undissolved solids}\\ \text{to second leaching}}}{CuS(s)} + H_2O(\ell) \quad (36.4)$$

The operation of the first leaching stage requires a balance between achieving a high leaching efficiency for nickel and removing all, or almost all, the copper from solution by metathesis-type reactions, such as Equation 36.1.

The solid residue from the first stage typically contains 13–18% Ni and 48–55% Cu. The solution contains about 100 g/L Ni and less than 1 g/L Cu and 1 g/L sulfuric acid.

1. The term metathesis is used to denote these reactions. However, the reactions may not be true metathesis reactions; rather they may be a combination of dissolution and precipitation. Little research has been done on the mechanisms of these reactions. See Chapter 23 for other examples of similar reactions.

The slurry from the first stage leach is separated in a thickener. The overflow stream is filtered and transferred to the nickel sulfate crystallizer. Part of the underflow slurry is recycled back to the first tank and the rest of it is transferred to the second leaching stage.

36.3.2. Second Leaching Stage

Part of the underflow slurry from the first-stage thickener is mixed with solution from copper electrowinning in the autoclave feed tank. The combined slurry is pumped into an autoclave that is pressurized with oxygen at 6 bar and 140°C. The objective of the second leaching stage is to dissolve as much of the nickel and copper as possible.

The nickel in the residue from the first leaching stage is mainly in the form of NiS, which dissolves in the autoclave as follows:

$$\underset{\text{from first leaching stage}}{NiS(s)} + \underset{\text{gas feed to autoclave}}{2O_2(g)} \xrightarrow{140^\circ C} \underset{\text{to selenium removal}}{NiSO_4(aq)} \tag{36.5}$$

Copper in the residue from the first leaching stage is mainly in the form of Cu_2S, which dissolves in the autoclave by the following reactions:

$$\underset{\text{feed}}{Cu_2S(s)} + \underset{\text{acid in copper spent}}{H_2SO_4(aq)} + \underset{\text{gas supplied to autoclave}}{0.5O_2(g)} \xrightarrow{140^\circ C} \underset{\text{to selenium removal}}{CuSO_4(aq)} + \underset{\text{to reaction Equation (36.7)}}{CuS(s)} + H_2O(\ell) \tag{36.6}$$

$$\underset{\text{from reaction Equation (36.6)}}{CuS(s)} + \underset{\text{gas feed to autoclave}}{2O_2(g)} \xrightarrow{140^\circ C} \underset{\text{to selenium removal}}{CuSO_4(aq)} \tag{36.7}$$

The leaching efficiency for copper is greater than 90% and the mass loss, which is the proportion of mass removed from the feed solids, exceeds 95%.

The slurry from the autoclave is filtered. The filter cake contains about 10 to 30% PGMs and is transferred to the third stage leach for further processing.

36.3.3. Third Leaching Stage

The filter cake from the second stage is repulped with copper spent, water and sulfuric acid and leached again under conditions that are similar to the second stage. The purpose of this leaching stage is to remove as much of the undissolved copper and nickel as possible, so that the PGM content of the residue is as high as possible. The operation of the third leaching stage involves a balance between maximizing the extraction of copper and nickel and minimizing the dissolution of rhodium and ruthenium.

At Stillwater and Northam, the residue from the third leaching stage is the final PGM-containing product from the base metals refinery. At Lonmin, this residue is transferred to the high-security area for further upgrading before being transferred to the precious metals refinery.

The filtrate is transferred to selenium and tellurium removal. Alternatively, it can also be pumped back to the second leaching stage.

36.3.4. Selenium and Tellurium Removal

Selenium, particularly selenium present as Se^{4+}, co-deposits with copper at the cathode during copper electrowinning. Selenium has a detrimental effect on the quality of the copper cathodes.

The filtrates from the second and the third stage leaching stages are therefore contacted with sulfur dioxide, sulfurous acid or sodium sulfite to precipitate selenium and tellurium from solution. The equipment used is a pipe reactor, which has a short residence time, followed by a tank with a relatively large residence time. The objective of this operation is to reduce the concentration of Se^{4+} in solution to less than 1 mg/L and the concentration of Se^{6+} in solution to less than 10 mg/L.

The effect of the sulfurous acid is to reduce the copper to the cuprous state, Cu^+. This reaction, which occurs rapidly in the pipe reactor, is given as follows:

$$\underset{\text{in solution from second stage}}{2CuSO_4(aq)} + \underset{\text{introduced in pipe reactor}}{H_2SO_3(aq)} + H_2O(\ell) \xrightarrow{90^\circ C} \underset{\text{to reaction Equation (36.9)}}{Cu_2SO_4(aq)} + 2H_2SO_4(aq) \tag{36.8}$$

The cuprous ions, Cu^+, react with Se^{4+} to produce copper selenide by the following reaction:

$$\underset{\text{from reaction Equation (36.8)}}{4Cu_2SO_4(aq)} + \underset{\text{selenium in solution}}{H_2SeO_3(aq)} + 2H_2SO_4(aq) \xrightarrow{90^\circ C} \underset{\text{precipitate for disposal}}{Cu_2Se(s)} + \underset{\text{solution to copper electrowinning}}{6CuSO_4(aq)} + 3H_2O(\ell) \tag{36.9}$$

Some of the selenium may be present as Se^{6+}, which react with the Cu^+ ions by the following reaction:

$$\underset{\text{from reaction Equation (36.8)}}{5Cu_2SO_4(aq)} + \underset{\text{selenium in solution}}{H_2SeO_4(aq)} + \underset{\text{acid in solution}}{3H_2SO_4(aq)} \xrightarrow{90^\circ C} \underset{\text{precipitate for disposal}}{Cu_2Se(s)} + \underset{\text{solution to copper electrowinning}}{8CuSO_4(aq)} + 4H_2O(\ell) \tag{36.10}$$

The precipitation of selenium from solution by these reactions is relatively slow. The tanks that are used for the reaction, therefore, have a long residence time, typically about 24 h, and are commonly referred to as aging tanks.

The slurry from the aging tanks is filtered. The residue is further treated to recover any precipitated PGMs.

Air is blown through the solution prior to it being transferred to copper electrowinning. This strips sulfur dioxide from the solution and oxidizes excess Cu^{+} to Cu^{2+}.

36.3.5. Copper Electrowinning

The solution feed to the copper tankhouse contains between 60 and 75 g/L Cu and about 15 g/L H_2SO_4. The objective of copper electrowinning is to produce copper metal that is of a high quality.

Copper is plated onto stainless steel cathodes. The half-reaction for copper plating is given as follows:

$$\underset{\text{copper in feed}}{Cu^{2+}(aq)} + \underset{\text{electrical current}}{2e^{-}} \rightarrow \underset{\text{copper cathode}}{Cu(s)} \quad (36.11)$$

The half-reaction at the anode is the following:

$$\underset{\text{solution}}{2H_2O} \rightarrow \underset{\text{acid generated}}{4H^{+}(aq)} + \underset{\substack{\text{oxygen vented}\\ \text{to atmosphere}}}{O_2(g)} + \underset{\text{electrical current}}{4e^{-}} \quad (36.12)$$

As a result of these half-reactions, the concentration of copper in solution drops between 25 and 35 g/L and the concentration of acid increases between 80 and 100 g/L.

The solution from copper electrowinning is commonly called copper spent and is recycled to the three leaching stages.

The copper cathodes from the electrowinning tankhouse are the copper product from the process.

36.3.6. Nickel Crystallization

The last remaining stage in the process used by Lonmin, Northam and Still-water is the crystallization of nickel sulfate.

The solution from the first leaching stage contains essentially nickel sulfate. Impurities that may be present are copper, iron, rhodium and ruthenium. The objective of this operation is to remove all the nickel and sulfur from the circuit as nickel sulfate crystals, $NiSO_4 \cdot 6H_2O$.

The crystallizer circuit is a two-stage operation. The first stage consists of an evaporator and the second stage is the crystallizer. In both stages, water is evaporated by heating the solution and by drawing a vacuum. The circuit is optimized so that water vapor is recompressed and used as the heating medium in the evaporation.

Nickel sulfate crystals are separated from the mother liquor in a cyclone and a centrifuge, then dried and bagged for shipment. A bleed from the circuit is returned to the first stage leach. This bleed removes acid and other impurities that may accumulate in the crystallizer.

36.3.7. Circuit Modifications and Improvements

The base metals refinery described above was developed by Sherritt (Kerfoot, Kofluk, & Weir, 1986); however, the individual operations have modified that design to better suit their needs. In this section, four of these modifications are discussed. The first three are implemented at Lonmin and the fourth at Stillwater. These improvements are as follows:

(a) PGM upgrading;
(b) cooling on second stage leaching by flash recycle;
(c) continuous third-stage leaching; and,
(d) iron removal prior to nickel crystallization.

PGM Upgrading

The residue from the third leaching stage is the PGM concentrate. It typically contains about 30% PGMs. The more the residue is upgraded, the lower the burden on the PGM refinery. The upgrading might include one or more of the following steps:

(a) a further pressure-leaching step under similar conditions to the second-stage leach to remove copper;
(b) a leaching step using a reducing agent, such as formic acid, to remove magnetite and other iron oxides; and,
(c) a pressure-leaching step using oxygen and caustic to remove selenium, tellurium and arsenic.

These additional steps produce a product from the base metals refinery, whose content of PGMs is greater than 60%. The concentrate produced at Lonmin has the highest concentration in the industry, between 65% and 70% PGMs. Such a high PGM content reduces the purification requirements in the precious metals refinery.

Cooling of the Second Leaching Stage by Flash Recycle

A limiting factor in the operational capacity of the second leaching stage is the removal of the heat generated by exothermic reactions (Crundwell, 2010; Dunn, 2009; Steenekamp & Dunn, 1999). One method for doing this without dilution or other losses in capacity is to recycle part of the flow from the first compartment back to the feed tank. Since the autoclave is operated at 6 bar

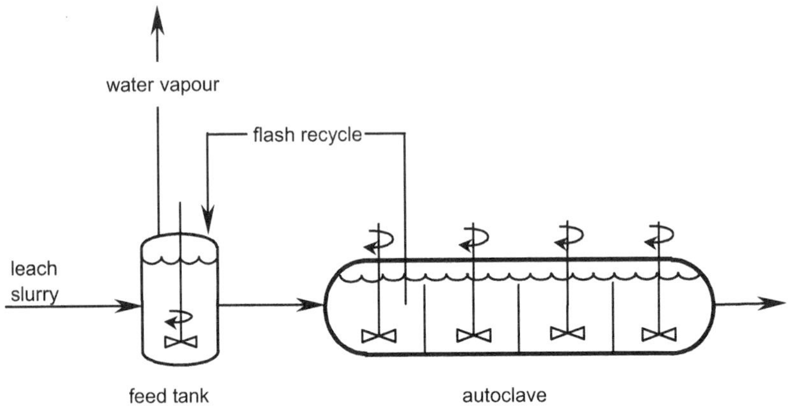

FIGURE 36.4 A schematic diagram of the flash recycle system for removal of heat from the second-stage autoclave.

and 140°C and the feed tank is at atmospheric pressure, the recycled portion will flash, or boil, generating steam. The amount of steam generated is essentially the amount of cooling that is achieved. This innovation was first implemented at Impala Platinum. A schematic diagram of the flash recycle system is shown in Figure 36.4.

Continuous Third Leaching Stage

Productivity of the installed equipment can be increased by continuous operation rather than batch operation. The third leaching stage can be run as a continuous operation in much the same manner as the first and second stages. Lonmin use the final compartment of the second stage autoclave as a third stage. Slurry is withdrawn from the second last compartment, centrifuged and the residue repulped and pumped into the last compartment of the autoclave. As a result, both second and third stages of the leach occur in the same autoclave. A schematic diagram of the continuous third-stage leach at Lonmin is shown in Figure 36.5.

Iron Removal Prior to Crystallization

Ammonia is added to the solution stream from the first stage to precipitate iron as a hydroxide. The slurry is filtered to remove the iron cake and the solution proceeds to nickel crystallization.

36.3.8. Appraisal

The advantage of the process used at Lonmin, Northam and Stillwater is its simplicity. There are only three stages of leaching, one of which is at

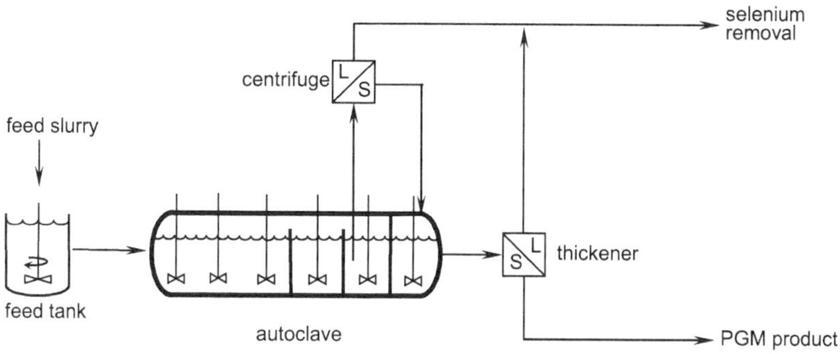

FIGURE 36.5 A schematic diagram of the continuous third-stage leach at Lonmin Platinum's base metals refinery. Slurry is removed from the compartment before the last compartment, centrifuged, repulped in acid and fed into the last compartment.

atmospheric conditions. No attempt is made to recover cobalt. Nickel is produced in the simplest manner possible. All of the sulfur is removed with the nickel in a single operation, rather than in the two steps required if nickel is produced as metal.

The disadvantage is the limited market for nickel sulfate, especially if it is relatively impure. Expansion of production beyond this limit will require the installation of either hydrogen reduction or electrowinning and an associated crystallizer for the removal of either ammonium sulfate or sodium sulfate.

In the next section, the flowsheet in which nickel is produced by hydrogen reduction and sulfur is removed as ammonium sulfate is discussed.

36.4. BASE METAL REFINERIES WHERE NICKEL IS PRODUCED AS POWDER BY HYDROGEN REDUCTION

There is one base metals refinery in the PGM industry in which nickel is produced by hydrogen reduction and sulfur is removed as ammonium sulfate. This refinery is Impala Platinum, Springs, South Africa.

The leaching circuit (first and second stage leaching) and the copper circuit (selenium removal and copper electrowinning) are similar in concept to the circuit described in Section 36.3. However, it is in the nickel circuit where major differences exist.

A schematic flowsheet of the process is shown in Figure 36.6.

36.4.1. First Leaching Stage

Milled matte is mixed with solution from copper electrowinning (copper spent) in the autoclave feed tank. Oxygen is also added to the feed tank, so some

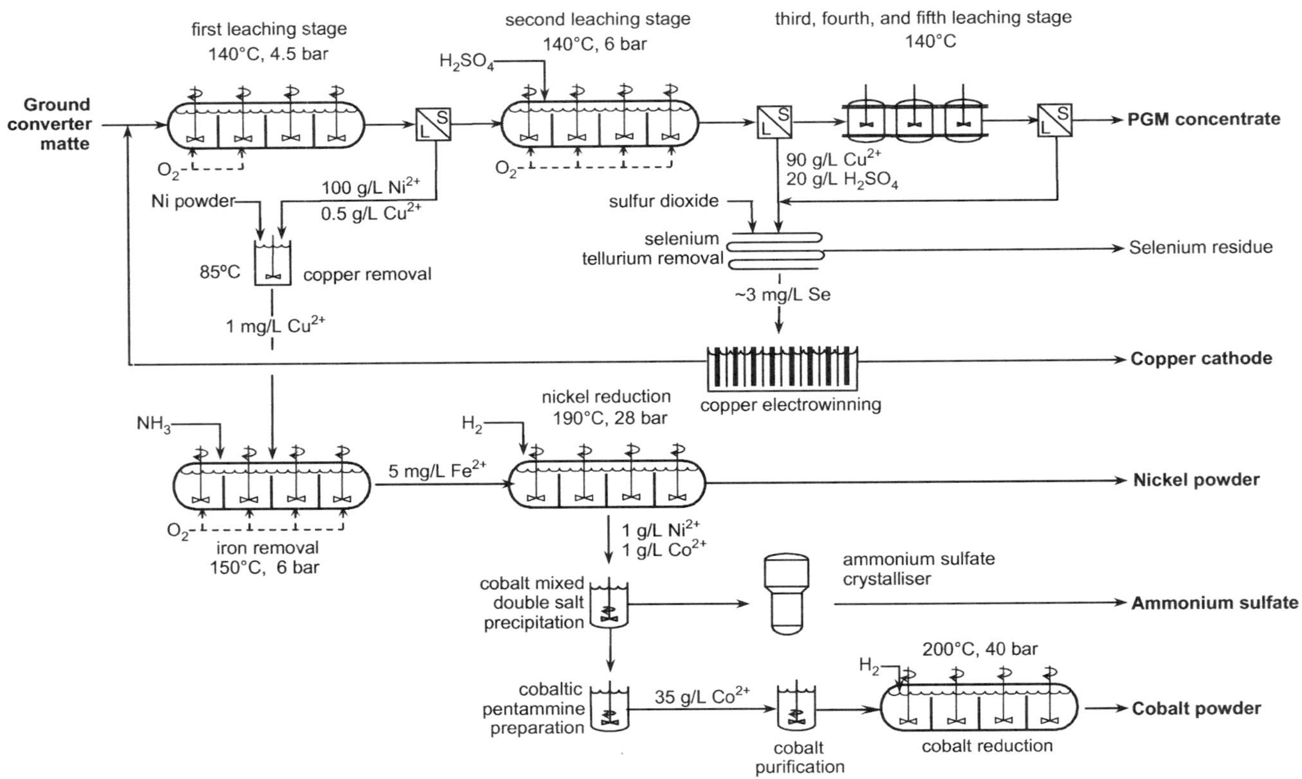

FIGURE 36.6 A schematic flowsheet of the Impala Platinum base metals refinery in which nickel is produced by hydrogen reduction.

FIGURE 36.7 Two of the autoclaves used for leaching at Impala Platinum. *Photograph by M. Fox, used with permission of Impala Platinum.*

leaching occurs here. The slurry is pumped from the feed tank into the first leaching stage autoclave.

The autoclave is divided into an oxidizing section, where oxygen is added, and a non-oxidizing section, where no oxygen is added. The autoclave is operated at about 130°C and 9.5 bar.

The objective of the first stage is to leach as much nickel as possible. At the same time, the concentration of copper in solution must be less than 0.5 g/L and the concentration of acid about 15 g/L. The extraction of nickel is typically about 85% in the first leaching stage.

In addition to reactions given in Equations (36.1) and (36.2), nickel is leached by the following metathesis reaction with copper ions by dissolution with oxygen:

$$\underset{\text{feed}}{NiS(s)} + \underset{\text{in copper spent}}{CuSO_4(aq)} \xrightarrow{130^\circ C} \underset{\text{to nickel circuit}}{NiSO_4(aq)} + \underset{\text{undissolved solids to second leaching stage}}{CuS(s)} \quad (36.13)$$

$$\underset{\text{feed}}{NiS(s)} + \underset{\text{gas supplied to reactors}}{2O_2(g)} \xrightarrow{130^\circ C} \underset{\text{solution to nickel circuit}}{NiSO_4(aq)} \quad (36.14)$$

The slurry from the autoclave is thickened. The thickener underflow is pumped to the second stage leach, while the overflow is pumped to the nickel purification circuit.

The autoclaves used for leaching are shown in Figures 36.7 and 36.8.

FIGURE 36.8 Closer view of one of the leaching autoclaves. *Photograph by M. Fox, used with permission of Impala Platinum.*

36.4.2. Second Leaching Stage

The slurry from the first leaching stage is mixed with process water and pumped into the density adjustment tank. This ensures that the correct ratio of solids and liquid is fed to the autoclave. The slurry is pumped to the autoclave feed tank.

Acid is added to the autoclave feed tank and this slurry is pumped into the autoclave.

The autoclave operates at about 140°C and 9.5 bar. The objectives for the second stage are to leach as much copper and nickel as possible. The reactions are similar to those discussed in Section 36.3.2.

The slurry is thickened. The underflow is pumped to the third-stage leach, while the overflow is filtered and pumped to selenium tellurium removal.

36.4.3. Third, Fourth and Fifth Leaching Stages

The objectives of these third, fourth and fifth leaching stages are to remove iron, copper and amphoteric elements, respectively. The amphoteric elements are selenium, tellurium, arsenic and sulfur. All of these leaching stages are operated in batch mode.

Iron is removed by leaching with formic acid, copper is removed by leaching in a solution containing sulfuric acid using oxygen and the selenium, tellurium, arsenic and sulfur are removed from the PGM residue by leaching in a solution containing sodium hydroxide using oxygen.

Interestingly, the oxidative conditions used in the fifth stage, which uses sodium hydroxide, leave about 50% of the PGMs as oxides or hydroxides. The oxides and hydroxides of the PGMs are easily dissolved in hydrochloric acid in the precious metals refinery (Asamoah-Bekoe, 1998).

The residue, containing about 60% PGMs, is transferred to the precious metals refinery. The solutions from each of these stages are returned to the second leaching stage.

36.4.4. Selenium and Tellurium Removal

The purification of the solution from the second leaching stage is essentially the same as that described in Section 36.3.4. The solution, after filtration, is pumped to copper electrowinning.

36.4.5. Copper Electrowinning

The electrowinning of copper from solution is that same as that described in Section 36.3.5. The copper cathodes from the electrowinning tankhouse are the copper product from the process.

Solution from the electrowinning of copper is pumped to the first leaching stage.

In the next section, the description returns to the solution from the first leaching stage, which undergoes a number of purification steps prior to hydrogen reduction.

36.4.6. Solution Purification

The solution from the first leaching stage contains 100 g/L Ni^{2+}, 5 g/L Cu^{2+}, 2 g/L Fe^{2+} and some other impurities, such as arsenic and lead.

Copper Removal

The first of these two purification steps is copper removal. The objective of copper removal is to reduce the concentration of copper in solution to less than 1 mg/L. The copper is removed from the solution by cementation with nickel powder. The slurry is filtered. The cemented copper is returned to the first leaching stage and the filtrate is pumped to iron removal.

Iron Removal

In the iron removal stage, ammonia is added to the solution, the solution is heated and pumped into an autoclave that operates at 150°C and 7 bar. Oxygen is added to the autoclave. Iron precipitates as ammonium jarosite, $NH_4Fe_3(SO_4)_2(OH)_6$, lead as plumbojarosite, $Pb_{0.5}Fe_3(SO_4)_2(OH)_6$ and arsenic as scorodite, $FeAsO_4 \cdot 2H_2O$.

The slurry is filtered. The filter cake is sent for disposal, and the filtrate is pumped to nickel reduction.

36.4.7. Nickel Recovery Using Hydrogen Reduction

Ammonia and water are added to the solution to adjust the concentration of nickel to 60 g/L, and the concentration of ammonium sulfate to 300 g/L. The molar ratio of nickel to free ammonia is about 1.9.

The nickel is reduced using hydrogen gas in an autoclave that is operated at 190°C and 28 bar. The objectives of the operation are to reduce the nickel concentration to about 1.5 g/L and to produce nickel powder of greater than 99.9% purity. Almost all of the impurity content in the nickel powder is cobalt.

The details of the production of nickel powder by hydrogen reduction have been discussed in Chapter 27.

The slurry is filtered and the solids are briquetted. The nickel briquettes are nickel product from the refinery. The filtrate is pumped to the cobalt purification.

36.4.8. Cobalt Purification

The solution from nickel reduction contains about 1.5–2 g/L of Ni and about 0.5 g/L of Co. Before the cobalt in this solution can be reduced with hydrogen, the concentrations must be increased and the nickel must be removed.

Mixed Double-Salt Precipitation

The nickel and cobalt are precipitated from solution by the addition of ammonia as a mixed salt composed of $Ni(NH_4)_2(SO_4)_2 \cdot 6H_2O$ and $Co(NH_4)_2(SO_4)_2 \cdot 6H_2O$. Most of the nickel and all the cobalt are removed from the solution.

The slurry is filtered. The solids are transferred to the cobalt pentammine purification circuit. The filtrate is pumped to the ammonium sulfate crystallizers.

Cobalt Separation

The mixed double salt is dissolved in water, so that the concentration of nickel in solution is 85 g/L. The concentration of cobalt in solution is about 20 g/L.

Ammonia is added to the solution so that cobaltic pentammine, $Co(NH_3)_5^{3+}$, forms. This complex is sufficiently stable that when the solution is acidified and cooled to 30°C, only $Ni(NH_4)_2(SO_4)_2.6H_2O$ precipitates; no $Co(NH_4)_2(SO_4)_2 \cdot 6H_2O$ precipitates. The solution after filtration contains about 15 g/L of Co.

The filter cake, which consists of $Ni(NH_4)_2(SO_4)_2 \cdot 6H_2O$, is dissolved in water and ammonia and returned to nickel reduction.

Manganese and Copper Removal

Any manganese in solution is oxidized to MnO_2 by reaction with the cobaltic pentammine.

Copper is cemented from solution by the addition of cobalt powder. The cobaltic pentammine complex is also decomposed by the addition of cobalt powder. The solution is pumped to cobalt reduction.

At the end of these steps, the solution has been sufficiently concentrated and purified for the reduction of the cobalt from solution with hydrogen.

36.4.9. Cobalt Reduction

The cobalt in solution is reduced to cobalt metal under pressure using hydrogen. The operating conditions are 200°C and 40 bar. The operation of cobalt reduction is similar to that of hydrogen reduction of nickel, which was discussed in Chapter 27.

The cobalt powder is briquetted. These briquettes are the cobalt product from the refinery.

The final product from the refinery is ammonium sulfate, which is discussed in the next section.

36.4.10. Ammonium Sulfate Crystallizer

The solution from the cobalt purification section where the mixed double salt is precipitated contains about 450 g/L of ammonium sulfate. This solution is the feed to the ammonium sulfate crystallizers. The objective of the crystallizers is to produce ammonium sulfate crystals, $(NH_4)_2SO_4$, that can be sold to the fertilizer industry.

36.5. BASE METALS REFINERY WHERE NICKEL IS PRODUCED AS NICKEL CATHODE

There is one base metals refinery in which nickel is produced as nickel cathode. This operation is Anglo American Platinum, Rustenburg, South Africa.

The process used by Anglo American Platinum is also the only process that uses slow cooling of the matte followed by magnetic separation. The magnetic concentrate is transferred to the precious metals refinery and the non-magnetic fraction to the base metals refinery.

In this section, the magnetic concentrator plant is described first and then the base metals refinery is discussed.

36.5.1. Magnetic Concentrator Plant

The objective of the magnetic concentrator plant is to produce a PGM concentrate with a PGM content of about 50% with as little loss of PGMs as possible to the base metals refinery. A schematic diagram of the process is shown in Figure 36.9.

The slow-cooled matte is crushed and milled to 75% less than 70 μm. There are three stages of milling and magnetic concentration to ensure the highest recovery with the lowest contamination of the magnetic fraction by base metals. The mass of the magnetic concentrate is between 13% and 17% of the mass of the converter matte.

The non-magnetic fraction, referred as the nickel-copper matte by Anglo American Platinum, is the feed to the base metals refinery.

The magnetic concentrate is leached in three different solutions to remove nickel, copper and iron.

The objective for the primary leaching stage is to remove copper. The reactions are performed at 160°C and 9 bar with sulfuric acid and oxygen.

The objective of the secondary leach is to dissolve iron oxides, such as the spinel trevorite, $NiFe_2O_4$. The reactor is operated at atmospheric pressure under reducing conditions using formalin and sulfuric acid, which make formic acid.

36.5.2. Overview of the Base Metals Refinery

The feed to the base metals refinery is the nickel-copper matte, which has the following composition: 48% Ni, 29% Cu, 24% S and 3% Fe.

The leaching solution from the magnetic concentrator plant is also fed to the base metals refinery. This solution, referred to as the *pressure vessel liquor*, typically has the following composition: 100 g/L Ni^{2+}, 30 g/L Cu^{2+}.

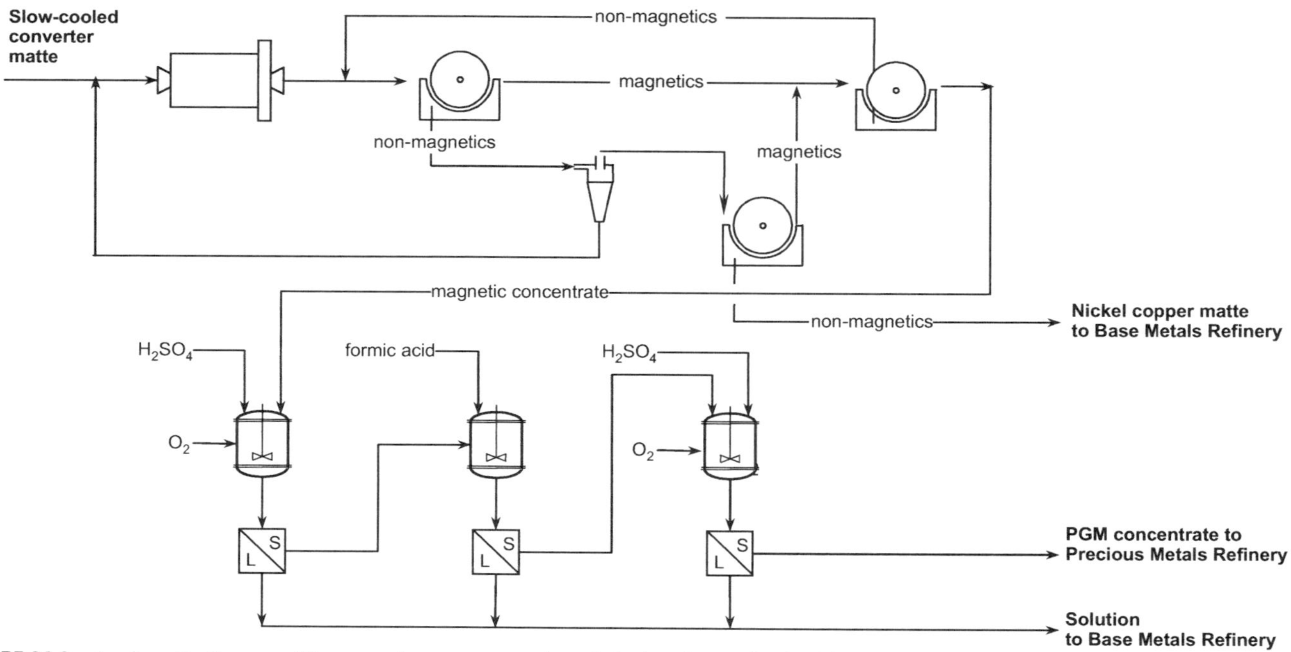

FIGURE 36.9 A schematic diagram of the magnetic concentrator plant. Only the primary circuit of the magnetic separation plant is shown. There are secondary and tertiary circuits. The objective of the tertiary leaching stage is to dissolve any remaining copper, nickel or iron. This leaching stage operates under conditions similar to those of the primary leach.

Because the alloy phase has been removed, the matte feed to the base metals refinery contains significantly more sulfur than the feed to the whole-matte processes. This means that the process does not require as much sulfuric acid as in the previously describe processes (Sections 36.3 and 36.4).

An overview of the process used at Anglo American Platinum's Rustenburg Base metal Refinery is shown in Figure 36.10. Leaching of the solids through the process is describe first, followed by description of the copper and nickel circuits.

36.5.3. Copper Removal from Nickel Solution

Nickel-copper matte is contacted with solution from iron removal. The solution contains about 70 g/L Ni^{2+}, 12 g/L Cu^{2+} at a pH of about 2. The objective of this section of the plant is to precipitate the copper, so that the concentration of copper in solution is less than 0.5 g/L.

The reactions that occur are the hydrolysis of copper to form basic copper sulfate [$CuSO_4 \cdot 2Cu(OH)_2$], which produces some acid, and the consumption of the acid by oxidation of the fresh matte. The hydrolysis reaction is as follows:

$$\underset{\substack{\text{in solution}\\\text{from leaching}}}{3CuSO_4(aq)} + 4H_2O(\ell) \xrightarrow{85^\circ C} \underset{\substack{\text{returned to nickel}\\\text{atmospheric leaching}}}{CuSO_4 \cdot 2Cu(OH)_2(s)} + \underset{\text{hydrolysis acid}}{2H_2SO_4(aq)} \tag{36.15}$$

The acid produced during hydrolysis of copper is consumed by the partial leaching of heazlewoodite, Ni_3S_2, which is given as follows:

$$\underset{\text{in feed matte}}{Ni_3S_2(s)} + \underset{\text{from hydrolysis}}{H_2SO_4(aq)} + \underset{\substack{\text{gas supplied}\\\text{to reactors}}}{0.5O_2(g)} \xrightarrow{85^\circ C} \underset{\substack{\text{purified solution}\\\text{to nickel circuit}}}{NiSO_4(aq)} + \underset{\substack{\text{undissolved solids to}\\\text{nickel atmospheric leaching}}}{2NiS(s)} + H_2O(\ell) \tag{36.16}$$

The hydrolysis reaction is in equilibrium, which means that the concentration of copper is dependent only on the temperature and the pH. However, if the pH increases above 6, nickel precipitates from solution as $NiSO_4 \cdot 2Ni(OH)_2$. As a result, careful control of pH is required. The reactions occur in four continuous stirred tanks, each of which has a residence time of 1 h. The slurry leaving the tanks is thickened. The solution is pumped to lead removal, while the underflow solids are pumped to nickel atmospheric leaching.

36.5.4. Nickel Atmospheric Leaching

The slurry from copper removal is mixed with copper spent, which contains about 30 g/L of Cu^{2+} and 90 g/L of H_2SO_4, and solution from the non-oxidizing leaching, which contains about 20 g/L of Cu^{2+} and 35 g/L of H_2SO_4. The objectives of this section are to reduce the concentration of

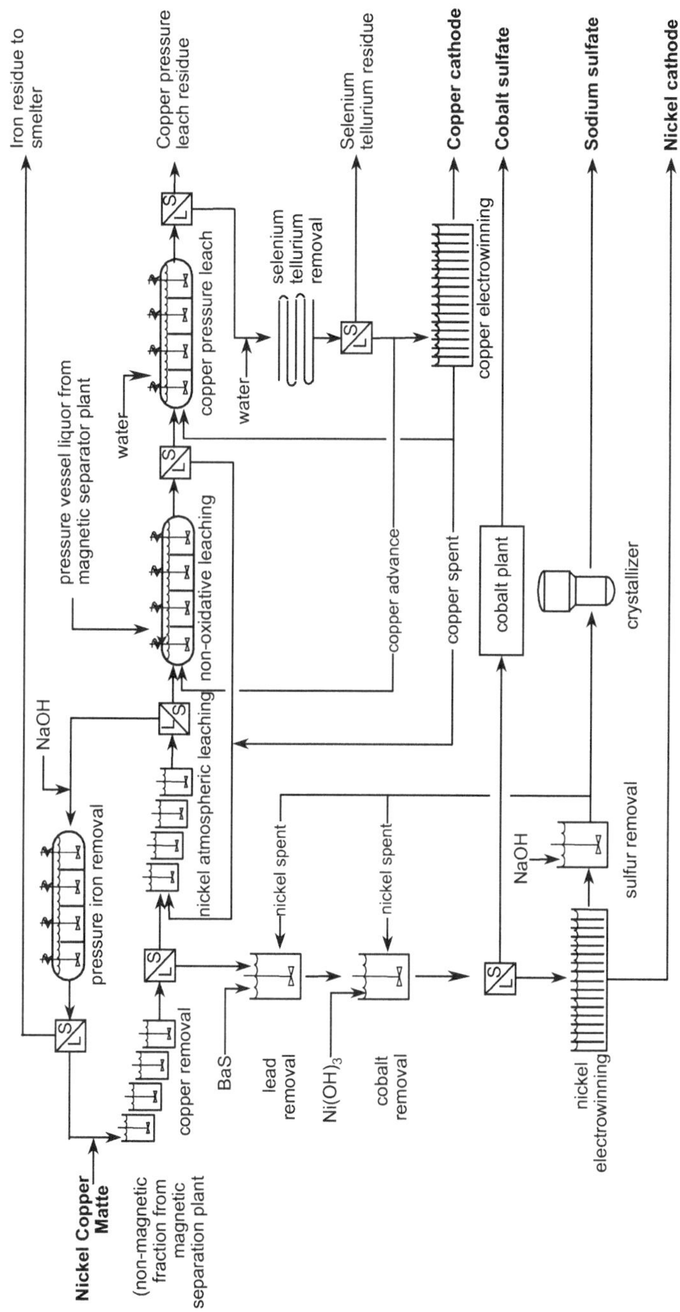

FIGURE 36.10 A schematic diagram of the process used at Rustenburg Base Metal Refiners, the base metals refinery operation of Anglo American Platinum. The nickel is produced as nickel cathode.

copper in solution to about 12 g/L, and to dissolve all the remaining heazlewoodite.

The leaching takes place in four continuous stirred tanks, each with a residence time of about 1 h, between 85°C and 90°C. Air is added to the last two tanks (not shown in Figure 36.10).

The reaction in the first two tanks is the metathesis leaching of heazlewoodite:

$$\underset{\substack{\text{in feed from}\\\text{copper removal}}}{Ni_3S_2(s)} + \underset{\substack{\text{in solution from}\\\text{non-oxidation leaching}}}{2CuSO_4(aq)} \xrightarrow{85^\circ C} \underset{\text{to copper removal}}{2NiSO_4(aq)} + \underset{\substack{\text{undissolved solids to}\\\text{non-oxidative leaching}}}{Cu_2S(s) + NiS(s)} \quad (36.17)$$

The reaction in the last two tanks is the dissolution of heazlewoodite with oxygen:

$$\underset{\substack{\text{in feed from}\\\text{copper removal}}}{Ni_3S_2(s)} + \underset{\substack{\text{in solution from}\\\text{non-oxidation leaching}}}{H_2SO_4(aq)} + \underset{\substack{\text{gas supplied}\\\text{to last two tanks}}}{0.5O_2(g)} \xrightarrow{85^\circ C} \underset{\text{to copper removal}}{NiSO_4(aq)} + \underset{\substack{\text{undissolved solids to}\\\text{non-oxidative leaching}}}{2NiS(s)} + H_2O(\ell) \quad (36.18)$$

In addition, all of the basic copper sulfate that is precipitated in copper removal redissolves in nickel atmospheric leaching.

In order to maximize the leaching efficiency of nickel in this stage, it is necessary to promote the first of these two reactions. This is the reason why air is added only in the last two tanks.[2]

The slurry is thickened. The solution is pumped to iron removal and the underflow solids to the non-oxidative leaching stage.

36.5.5. Iron removal

The objective of this section is to remove the iron from solution so that its concentration is less than 10 mg/L.

The solution from nickel atmospheric leaching is neutralized with sodium hydroxide to a pH of 4. Sulfur dioxide is added to reduce all the iron to Fe^{2+}, so that iron does not precipitate in the feed tank or scale the heat exchangers. The solution is heated to 150°C and pumped into a four-compartment autoclave. Oxygen and sodium hydroxide are added to each compartment.

---

2. Nickel atmospheric leaching should be compared with the first leaching stage of the processes used by other base metals refineries, where oxygen is added to the initial tanks or autoclave compartments and not to the tanks or compartments at the back-end of the leaching stage.

Iron precipitates as sodium jarosite, $NaFe_3(SO_4)_2(OH)_6$ (s).

The slurry is filtered. The filtrate is pumped to copper removal and the filter cake is transferred to the smelter.

36.5.6. Nickel Non-Oxidative Leaching

The slurry from the underflow of the thickener at nickel atmospheric leaching is mixed with solution from the magnetic concentrator plant, referred to as pressure vessel liquor, and with solution obtained from the feed to copper electrowinning, referred to as copper advance. The objective of this operation is to dissolve as much of the nickel as possible.

The equipment used is a four-compartment autoclave. The operating conditions are 150°C and 6 bar.

The reactions that occur are the metathesis of millerite (NiS):

$$\underset{\substack{\text{in feed from nickel}\\\text{atmospheric leaching}}}{NiS(s)} + \underset{\substack{\text{in solution from}\\\text{copper advance}}}{CuSO_4(aq)} \xrightarrow{150^\circ C} \underset{\substack{\text{in solution to nickel}\\\text{atmospheric leaching}}}{NiSO_4(aq)} + \underset{\substack{\text{undissolved solids to}\\\text{copper pressure leaching}}}{CuS(s)} \quad (36.19)$$

$$\underset{\substack{\text{in feed from nickel}\\\text{atmospheric leaching}}}{6NiS(s)} + \underset{\substack{\text{in solution from}\\\text{copper advance}}}{9CuSO_4(aq)} + 4H_2O(\ell) \xrightarrow{150^\circ C} \underset{\substack{\text{in solution to nickel}\\\text{atmospheric leaching}}}{6NiSO_4(aq)} + \underset{\substack{\text{undissolved solids to}\\\text{copper pressure leaching}}}{5Cu_{1.8}S(s)} + \underset{\substack{\text{in solution to nickel}\\\text{atmospheric leaching}}}{4H_2SO_4(aq)} \quad (36.20)$$

The last reaction produces acid, so the concentration of acid rises across non-oxidative leaching. The solution leaving non-oxidative leaching contains about 20 g/L of Cu^{2+} and 30 g/L of H_2SO_4.

The slurry is thickened and filtered. The filter cake is transferred to copper pressure leaching and the solution is pumped to nickel atmospheric leaching.

36.5.7. Copper Pressure Leaching

The solids from non-oxidative leaching are mixed with solution from copper electrowinning (copper spent) and water. A solution containing ferrous sulfate is also added.

This slurry is pumped into a four-compartment autoclave, which is operated at a temperature of 140°C and a pressure of 6 bar. Oxygen is sparged into each of the compartments through sparge rings positioned below the agitator. The agitators are Smith turbines, designed for efficient gas mass transfer. Heat is removed efficiently using a flash recycle, which was described in Section 36.3.7

The reactions are similar to those in the second leaching stage of the other processes. Typical reactions were given in Section 36.2.3. The leaching efficiency is high, above 90%, and the overall mass loss is about 98%.

The solution leaving the copper pressure leaching autoclave contains about 120 g/L Cu^{2+}, 40 g/L Ni^{2+} and 15 g/L H_2SO_4. The slurry is diluted with water to about 95 g/L Cu^{2+}.

The slurry is filtered. The solids are collected for further processing elsewhere to recover the value of residual PGMs that may be present. The solution is pumped to selenium and tellurium removal.

36.5.8. Selenium and Tellurium Removal

The purification of the solution from the copper pressure leaching is essentially the same as that described in Section 36.3.3. The equipment used is a pipe reactor followed by two large aging tanks with a total residence time of 33 h.

The slurry is filtered. The solids in the filter cake are combined with the residue from copper pressure leaching and are sent for further processing elsewhere.

Part of the solution is pumped to copper electrowinning. The other part, referred to as copper advance, is pumped to the non-oxidative leaching stage.

36.5.9. Copper Electrowinning

The electrowinning of copper from solution is the same as that described in Section 36.3.4.

The copper cathodes from the electrowinning tankhouse are the copper product from the process.

Solution from the electrowinning of copper is pumped to copper pressure leaching and to nickel atmospheric leaching.

In the next section, the description returns to the solution leaving copper removal, which undergoes lead removal and cobalt separation prior to nickel electrowinning.

36.5.10. Lead Removal

Lead dissolves from the anodes in nickel electrowinning and, if not removed, it will contaminate the nickel product. It is removed as lead sulfate by the addition of barium hydroxide.

The lead sulfate is precipitated and removed by filtration. The filter cake is recycled to the Waterval smelter. The filtrate is pumped to cobalt removal.

36.5.11. Cobalt Removal

Cobalt is separated from the nickel solution by precipitation with nickelic hydroxide. The nickelic hydroxide is produced electrolytically from an alkaline solution containing nickel hydroxide. The reaction of nickelic hydroxide with the cobalt in solution is given as follows:

$$\underset{\text{nickelic hydroxide from electrolysis}}{Ni(OH)_3(s)} + \underset{\text{cobalt in solution}}{CoSO_4(aq)} \rightarrow \underset{\text{nickel to electrowinning}}{NiSO_4(aq)} + \underset{\text{precipitated cobalt to cobalt crystallization}}{Co(OH)_3(s)} \quad (36.21)$$

The slurry is filtered. The filter cake is transferred to the cobalt sulfate plant and the solution is pumped to nickel electrowinning.

36.5.12. Nickel Electrowinning

The practice of nickel electrowinning was described in detail in Chapter 26.

The solution from cobalt removal is filtered to remove fine material. This solution contains about 90 g/L of Ni^{2+} and 140 g/L of Na_2SO_4.

The value of the pH of the solution is adjusted to 3.5 by the addition of sodium hydroxide. Boric acid is also added to buffer the solution. This solution is pumped to the cell house where it is distributed to each of the cathode compartments of the cells.

The anodes and cathodes are separated by a cloth or bag, which separates the cell into anode and cathode compartments. The solution has a slightly higher level on the cathode side, so the flow is from the cathode to the anode. This flow opposes the concentration gradient of H^+, and therefore impedes the movement of H^+ from the anode compartment to the cathode compartment. As a result, the pH of the cathode compartment can be maintained at a much higher value than the anode compartment.

At a pH value of about 3.5 in the cathode compartment, the current efficiency of the deposition of nickel is greater than 95%. The nickel is plated on titanium cathodes, which have permanent edge strips. The cathode that is produced has a purity of greater than 99.9%.

The practice of nickel electrowinning at Anglo American Platinum is focused on reducing worker exposure to nickel sulfate aerosols and solutions. The cells have mist abatement hoods and anode skirts to reduce the acid mist below the environmental limits. The cathodes are harvested by automatic crane, and stripped mechanically.

Anolyte from the tankhouse, referred to as nickel spent, is pumped to nickel atmospheric leaching or to sulfur removal.

36.5.13. Sulfur Removal

Anolyte from the nickel tankhouse contains about 60 g/L of Ni^{2+}, 45 g/L H_2SO_4 and 140 g/L Na_2SO_4.

The stream is split into two streams. One stream is neutralized using sodium hydroxide so that nickel hydroxide precipitates, removing all the nickel from solution. The reaction is given as follows:

$$\underset{\text{nickel in solution}}{NiSO_4(aq)} + \underset{\text{hydroxide added}}{2NaOH(aq)} \rightarrow \underset{\substack{\text{nickel hydroxide}\\ \text{precipitate}}}{Ni(OH)_2(s)} + \underset{\substack{\text{sodium sulfate}\\ \text{to crystallizer}}}{Na_2SO_4(aq)} \quad (36.22)$$

The slurry is filtered. The filtrate, which contains only Na_2SO_4, is pumped to the sodium sulfate crystallizer.

The nickel hydroxide is then mixed with the second stream from nickel used to neutralize the acid. This reaction is given as follows:

$$\underset{\text{acid in nickel spent}}{H_2SO_4(aq)} + \underset{\substack{\text{nickel hydroxide}\\ \text{from precipitation}}}{Ni(OH)_2(s)} \rightarrow \underset{\substack{\text{to lead or}\\ \text{copper removal}}}{NiSO_4(aq)} + 2H_2O(\ell) \quad (36.23)$$

This solution is returned to copper removal or to lead removal.

The next section is the sodium sulfate crystallizer.

36.5.14. Sodium Sulfate Crystallizer

The feed to the sodium sulfate contains about 300 g/L Na_2SO_4. The objective of the crystallizer is to produce Na_2SO_4 crystals that can be sold, mainly to detergent manufacturers.

36.5.15. Cobalt Plant

The feed to the cobalt plant is cobaltic hydroxide, which contains a significant amount of nickel (Section 36.5.11). The objective of the cobalt plant is to purify this cobalt and produce pure crystals of cobalt sulfate hexahydrate, $CoSO_4 \cdot 6H_2O$.

The cobaltic hydroxide contains about 40% nickel and several other impurities. The cake is dissolved in sulfuric acid with formalin as a reducing agent. Iron is precipitated as a ferric hydroxide, and barium sulfide is added to precipitate the copper from solution as copper sulfide.

The cobalt is then removed from solution by solvent extraction using di(2-ethylhexyl) phosphoric acid, commonly referred to as D2EHPA. The raffinate, containing nickel, is returned to the leaching stages. The extractant is present in the sodium form, rather than as an acid. There are seven extraction stages, six scrubbing stages and three stripping stages. Cobalt is stripped with sulfuric acid. The stripped organic is regenerated to the sodium form using sodium hydroxide.

The strip solution forms the feed to the cobalt crystallizer, where the cobalt is crystallized as cobalt sulfate hexahydrate, $CoSO_4 \cdot 6H_2O$.

36.6. APPRAISAL

The advantage of the magnetic separation plant is that the production PGM-bearing material is not as dependent on the production of nickel and copper. However, this separation is not completely efficient.

The advantage of the whole-matte process is that it is simple and straightforward. It eliminates the need for slow cooling and magnetic separation. Instead, the converter matte can be granulated, milled and fed directly to the base metals refinery.

36.7. SUMMARY

The primary purpose of the base metals refinery is to produce a PGM concentrate that contains about 60% PGMs. This concentrate is the feed to the PGM refinery.

There are two different routes to meet this primary objective. They are as follows: (i) the slow cooling of matte, followed by magnetic separation; and, (ii) the whole-matte leaching process.

The magnetic concentrate process has the advantage that it separates the main PGM production from the refining of base metals.

The secondary objective is to produce nickel, copper and cobalt products for sale. The copper and nickel are separated from one another by metathesis reactions in the first leaching stage, where nickel dissolves and copper precipitates.

The solution from the first stage is purified to remove, lead, arsenic, iron and cobalt, among others. The nickel in solution after purification is crystallized, redissolved and reduced using hydrogen to nickel powder or electrowon.

The residue from the first stage contains mainly copper, which is dissolved in the second leaching stage. The copper solution is purified and copper is recovered by electrowinning.

REFERENCES

Asamoah-Bekoe, Y. (1998). *Investigation of the leaching of the platinum-group metal concentrate in hydrochloric acid by chlorine (MSc thesis).* Johannesburg: University of the Witwatersrand.

Crundwell, F. K. (2010). *Optimizing the performance of exothermic autoclaves. ALTA 2010 Nickel and Cobalt Pressure Leaching and Hydrometallurgy Forum.* ALTA Metallurgical Services.

Dunn, G. M. (2009). Increasing the capacity of existing and new exothermic autoclave circuits. In *Hydrometallurgy Conference 2009.* SAIMM. pp. XX–XX.

Kerfoot, D. G. E., Kofluk, R. P., & Weir, D. R. (Feb 18, 1986). *Recovery of platinum group metals from nickel-copper-iron matte.* U.S. Patent No. 4571262.

Steenekamp, N., & Dunn, G. M. (1999). *Operations of and improvements to the Lonrho Platinum Base Metal Refinery, EPD Congress 1999.* TMS. (pp. 356–390).

SUGGESTED READING

Bryson, L. J., Hofirek, Z., Collins, M. J., Stiksma, J., & Berezowsky, R. M. (2008). New matte leaching developments at Anglo Platinum's base metal refinery. In C. A. Young, P. R. Taylor, C. G. Anderson & Y. Choi (Eds.), *Hydrometallurgy 2008, proceedings of the sixth international symposium* (pp. 570–579). SME.

Faris, M. D., Moloney, M. J., & Pauw, O. G. (1992). Computer simulation of the Sherritt nickel-copper matte acid leach process. *Hydrometallurgy, 29*, 261–273.

Fugleberg, S., Hultholm, S. E., Rosenback, L., & Holohan, T. (1995). Development of the Hartley Platinum leaching process. *Hydrometallurgy, 39*, 1–10.

Hofirek, Z., & Kerfoot, D. G. E. (1992). The chemistry of the nickel-copper matte leach and its application to process control and optimization. *Hydrometallurgy, 29*, 357–387.

Hofirek, Z., & Noval, P. J. (1995). Pressure leach capacity expansion using oxygen-enriched air at RBMR (Pty) Ltd. *Hydrometallurgy, 39*, 91–116.

Lamya, R. M., & Lorenzen, L. (2006). Atmospheric acid leaching of nickel-copper matte from Impala Platinum Refineries. *Journal of the South African Institute of Mining and Metallurgy, 106*, 385–395.

Rademan, J. A. M., Lorenzen, L., & van Deventer, J. S. J. (1999). The leaching characteristics of Ni–Cu matte in the acid-oxygen pressure leach process at Impala Platinum. *Hydrometallurgy, 52*, 231–252.

van Schalkwyk, R.F., Eksteen, J.J., Petersen, J., Thyse, E.L., & Akdogan, G. (in press). An experimental evaluation of the leaching kinetics of PGM-containing Ni–Cu–Fe–S Pierce-Smith converter matte, under atmospheric conditions. Minerals Engineering.

Chapter 37

Refining of the Platinum-Group Metals

The feed to a precious metals refineries is a concentrate containing between 50% and 70% precious metals, which include platinum, palladium, rhodium, iridium, ruthenium, osmium, gold and silver. In this chapter, the technologies and techniques used to separate the concentrate into individual metals of high purity, typically greater than 99.9%, are described.

37.1. OBJECTIVES OF THIS CHAPTER

The objectives of this chapter are

(a) to describe the different technologies for separating the platinum-group metals (PGMs) by the different producers; and,
(b) to describe the manner in which the individual metals are purified and prepared for sale.

37.2. CONCENTRATE COMPOSITION

An overview of the process for the extraction of PGMs from ore was presented in Chapter 32. A major difference exists between the processes that use whole-matte leaching and the processes that use slow-cooling and magnetic separation. However, both routes produce a concentrate that contains between 50% and 70% PGMs. The composition of a typical feed to a refinery is shown in Table 37.1.

37.3. SEPARATION TECHNIQUES USED IN THE REFINING OF THE PLATINUM-GROUP METALS

The refining of PGMs generally occurs in three separation stages:

(a) primary separation;
(b) secondary purification; and,
(c) reduction to metal.

Primary separation refers to the first time that a particular PGM is separated from impurities and the remaining PGMs. The primary separation seldom produces metal of purity suitable for sale. As a result, the product of the

Extractive Metallurgy of Nickel, Cobalt and Platinum-Group Metals. DOI: 10.1016/B978-0-08-096809-4.10037-1

TABLE 37.1 Composition of the Feed to a South African Precious Metals Refinery*

Element	Concentration, %
Pt	24.60
Pd	13.30
Au	1.12
Rh	4.12
Ru	5.03
Ir	1.49
Ag	2.20
Ni	1.20
Cu	3.00
Fe	2.60
Pb	1.90
SiO_2	12.9
As	0.43
Se	1.20
Te	2.40
Ba	0.20
Na	1.70
Os	~0.20
O, H (as oxides and hydroxides)	~20.00

**Asamoah-Bekoe, 1998*

primary separation is purified in secondary processing. Often, the product of the primary separation and secondary purification is a metal salt, such as $(NH_4)_2PtCl_6$. The metal salts need to be reduced to metal for sale. Consequently, reduction is the final step in refining.

A number of techniques are used, or have been used in the past, to separate and purify the individual PGMs, such:

(a) dissolution;
(b) crystallization and precipitation;
(c) hydrolysis;

(d) distillation;
(e) organic precipitation;
(f) solvent extraction;
(g) ion exchange; and,
(h) metal reduction.

In this section, these general separation techniques are described and then each of the different processes used industrially is described in more detail. Essentially all the processes are performed on a batch basis.

37.3.1. Dissolution

The PGMs display different reactivities toward oxidizing acids, such as aqua regia (which is a mixture of nitric acid and hydrochloric acid in equal parts).

Dissolution in aqua regia or in hydrochloric acid using chlorine gas can be used as a coarse separation between primary (platinum, palladium and gold) and secondary precious metals (rhodium, iridium, ruthenium, osmium and silver).[1] The concentrate is roasted prior to dissolution so that iridium, rhodium and ruthenium are oxidized, making them insoluble in aqua regia. If this is not done, significant dissolution of these metals may occur in this step.

On dissolving either in aqua regia or in hydrochloric acid using chlorine gas, the PGMs form stable complexes with chloride ions in solution. For example, platinum forms the $PtCl_6^{2-}$ chloro complex.

Modern refineries do not use dissolution as a separation technique. Instead, they prefer to dissolve all the PGMs in hydrochloric acid using chlorine gas.

Glass-lined vessels or titanium vessels with heating and cooling jackets are used. A titanium vessel is shown in Figure 37.1.

37.3.2. Crystallization and Precipitation

The differences in solubilities of the chloro complexes of the PGMs are frequently exploited in both the primary separation and in the purification stages of refining. For example, ammonium chloride (NH_4Cl) can be used to precipitate ammonium hexachloroplatinate [$(NH_4)_2PtCl_6$] from a solution that contains the other PGMs:

$$PtCl_6^{2+}(aq) + 2NH_4^{+}(aq) \rightarrow (NH_4)_2PtCl_6(s) \qquad (37.1)$$

Ammonium hexachloroplatinate is commonly referred to as '*yellow salt*'.

The solubility of the metals is strongly dependent on the particular chloro complex and the valence state of the metal. As a result, the redox pontential of the solution, the pH and the concentration of chloride ions is adjusted to ensure that

1. The secondary PGMs are often referred to either as the OPMs, which is an abbreviation for "other precious metals", or the IMs, which is an abbreviation for "insoluble metals".

FIGURE 37.1 A Grade 2 titanium vessel used for the dissolution of a PGM concentrate. The vessel has a volume of about 2.2 m^3. The vessel is agitated and has a heating/cooling jacket. Chlorine is intermittently passed through the contents. *Photograph courtesy of Impala Platinum.*

the ions in solution are in their correct form so that the correct metal preciptates. This adjustment of the solution conditions is commonly refered to as 'Conditioning'.

Hydrolysis of the chloro complexes of the PGMs is frequently a reason for poor separation efficiency during crystallization. For example, $PtCl_6^{2+}$ may hydrolyze to form $Pt(OH)_4^{2-}$, resulting in incomplete precipitation of ammonium hexachloroplatinate from solution.

Glass-lined vessels are frequently used in crystallization and precipitation of PGM salts.

37.3.3. Hydrolysis

Base metals, principally nickel, copper and iron, are frequently precipitated from solution by increasing the pH so that the hydroxides of these base metals are formed.

The hydrolysis of rhodium and iridium is frequently used to remove these metals from solution. $Rh(OH)_4$ and $Ir(OH)_4$ are precipitated by increasing the pH to between 5 and 6 after ruthenium distillation.

37.3.4. Distillation

Ruthenium and osmium in solution can be oxidized to ruthenium tetroxide (RuO_4) and osmium tetroxide (OsO_4) using a strong oxidizing agent, such as sodium chlorate or sodium bromate at 80°C–90°C. The vapor pressures of ruthenium and osmium tetroxide are high, even at room temperature.

Air is drawn through the solution, removing the ruthenium and osmium from solution. The air containing the ruthenium and osmium is then drawn through a scrubbing column or a smaller vessel ('trap') where these metals are reduced by reaction with hydrochloric acid.

This is a potentially dangerous process due to the formation of chlorine oxides, which are explosive. The equipment is designed to protect employees in case of accidental explosion, and great care is taken to monitor and control the process.

Ruthenium dioxide, a decomposition product of ruthenium tetroxide, results in the formation of 'Ru blacks', a glassy deposit of amorphous ruthenium dioxide that can choke and damage the glass exchangers and the reflux condensers.

Both ruthenium and osmium are routinely produced by distillation. The equipment used is a glass-lined vessel, as shown in Figure 37.2.

37.3.5. Organic Precipitation

Organic reagents can be used to separate metals by selective precipitation. Palladium can be precipitated with dimethylglyoxime. This reaction is used to remove palladium in platinum purification.

Iridium and rhodium may be precipitated with diethylene triamine. These precipitation reactions can be used to separate rhodium and iridium from each other.

37.3.6. Solvent Extraction

Gold, palladium, platinum and iridium can be separated at high efficiency using solvent extraction. Columns, mixer-settlers and centrifuge-type contactors are used.

37.3.7. Ion Exchange and Molecular-Recognition Technology

Molecular recognition technology is used to separate palladium efficiently from the bulk of the other PGMs. Ion exchange is used to remove gold and base metals. These separations are performed in columns, as shown in Figure. 37.3.

FIGURE 37.2 The ruthenium distillation section at Lonmin's Western Platinum refinery. The distillation is carried out in a glass-lined vessel. The top of this vessel is shown is in the bottom center of the picture. The glass condenser can be seen in the top right. *Photograph courtesy of Lonmin plc.*

37.3.8. Reduction to Metal

In most cases, the PGMs are produced as a salt, such as $(NH_4)_2PtCl_6$, which needs to be reduced to metal for sale. Two methods are used: (i) wet reduction, in which the dissolved salt is reduced in solution by a reductant such as hydrazine; and, (ii) ignition, in which the salt is heated above the its thermal decomposition temperature.

Ignition of the ammonium chloride salts has been used from the earliest times in the study of PGMs and their chemistry, and it is still used universally in refining of the PGMs.

The decomposition temperature of most of the ammonium chloride salts is in the region of 200°C–500°C. In practice, temperatures of about 1000°C are used. For example, the decomposition of ammonium hexachloroplatinate, $(NH_4)_2PtCl_6$, on ignition in air is given by the following reaction:

$$\underset{\text{platinum slat}}{3(NH_4)_2PtCl_6(s)} \xrightarrow{1000^\circ C} \underset{\text{platinum metal}}{3Pt(s)} + \underset{\text{decomposition gases, collected and scrubbed}}{2NH_4Cl(g) + 16HCl(g) + 2N_2(g)} \quad (37.2)$$

Under the oxidizing conditions, ruthenium salts will form volatile ruthenium tetroxide, RuO_4. The salts of ruthenium must therefore be reduced in reducing conditions and cooled with an absence of oxygen.

FIGURE 37.3 Ion-exchange columns for the separation of gold from solution. The columns are arranged in a lead–middle–trail configuration. Three columns are used in series – the first is the lead and the last is the trail column. Once the lead column is fully loaded, the middle column becomes the new lead, the trail column becomes the new middle, and the lead is eluted and becomes the new trail column. *Photograph courtesy of Impala Platinum.*

Industrially, the PGM salt is loaded into a tray. The trays are made of a refractory material, such as zirconia, that can withstand the temperature and not contaminate the metal product. The trays are placed in the ignition furnace and heated to the required temperature. The metal is reduced by reaction, and forms a sponge-like material in the tray. The gas products are drawn from the furnace by vacuum through gas scrubbers.

Once ignition is complete, the trays are withdrawn from the furnace. The metal sponge is crushed. The crushed metal is either packaged or melted and poured into ingots, depending on customer requirements.

An ignition furnace is shown in Figure. 37.4.

37.4. REFINING EFFICIENCY

Two measures are used to assess the efficiency of a precious metals refinery:

(a) recovery; and,
(b) first-pass yield.

The recovery is the fraction of metals entering the refinery that are sold as product. Generally, the recovery of all of the precious metals from the refinery

FIGURE 37.4 An ignition furnace at Impala Platinum. Note the off-take ducting on the left-hand side of the furnace for the extraction of gases.

exceeds 95%. For metals such as platinum, palladium and rhodium, recovery is as high as 99%.

The first-pass yield is the fraction of each metal that is produced in a single pass through the refinery. In other words, any intermediates or products that are recycled because they do not meet the required purity levels reduce the first-pass yield. In assessing different refining options, the first-pass yield is the most critical criterion. This is because the hold up of high-value products greatly increases the working capital of the refinery, and hence the overall capital requirements.

A summary of refining processes used is given in Table 37.2, including indicative values of the first-pass yields.

37.5. CLASSIFICATION OF REFINING PROCESSES

The different processing routes used to refine PGMs are classified by the technique used to separate platinum and palladium. As shown in Table 37.3, these two metals are the most abundant in the ore that is mined, and hence in the feed to the refinery. As a result, the efficient separation of platinum and palladium from one another is one of the most crucial factors in the successful operation of the refinery.

Three different processing philosophies for the refining of the PGMs are employed:

(a) the precipitation process, represented by Lonmin Platinum and Krastsvetmet;
(b) the solvent-extraction process, represented by Anglo American Platinum, Johnson Matthey, Heraeus and Vale; and,

TABLE 37.2 Methods of Extraction of the PGMs in Several Refineries

	Lonmin	Krastsvetmet	Vale (Acton)	Johnson Matthey	Anglo American Platinum	Impala Platinum
Capacity, kg	32 000		32 000	32 000	112 000	71 000
Capacity, oz	1 million		1 million	1 million	3.5 million	2.2 million
Year started	1975/1987	1943	1975		1989	1998
Production	Primary	Primary	Mainly toll refining	Mainly toll refining	Primary	Primary with large-scale toll refining
Technology	Precipitation	Precipitation	Solvent extraction	Solvent extraction	Solvent extraction	Ion exchange
Feed quality	65–75% PGMs		<15% PGMs	<15% PGMs	30–50% PGMs	60–65% PGMs
Dissolution	HCl/Cl_2 at 65°C and atm pressure	HCl/Cl_2 at 70°C, 3 bar pressure	HCl/Cl_2 at 65°C and atm pressure	HCl/Cl_2 at atm pressure	HCl/Cl_2 at 120°C, 4 bar pressure	HCl/Cl_2 at 85°C, 1 bar pressure
Order of extraction	Gold	Silver	Ruthenium	Gold	Gold	Gold
	Ruthenium	Gold	Gold	Palladium	Palladium	Palladium
	Platinum	Platinum	Palladium	Platinum	Platinum	Ruthenium
	Palladium	Palladium	Platinum	Ruthenium	Ruthenium	Platinum
	Rh and Ir separated after Ru distillation (pH 5 cake)	Rh and Ir separated after smelting, dissolution	Iridium Rhodium	Rh and Ir separated after Ru distillation (pH 5 cake)	Rh and Ir separated after Ru distillation (pH 5 cake)	Rh and Ir separated after Ru distillation (pH 5 cake)

(Continued)

TABLE 37.2 Methods of Extraction of the PGMs in Several Refineries—cont'd

	Lonmin	Krastsvetmet	Vale (Acton)	Johnson Matthey	Anglo American Platinum	Impala Platinum
Gold extraction	Crude sponge produced by hydrazine reduction. Secondary processing required	Crude precipitation with fresh concentrate. Secondary processing required	Solvent extraction with dibutyl carbitol, reduction with oxalic acid. No reprocessing	Solvent extraction with MIBK, reduction with oxalic acid. No reprocessing	Solvent extraction with MIBK, reduction with oxalic acid. No reprocessing	Ion exchange, reduction with hydroquinone. No reprocessing
Palladium extraction	Precipitation with ammonium acetate. Secondary refining required	Precipitation with ammonium chloride. Secondary reprocessing required	Solvent extraction with di-*n*-octyl sulfide, strippning in ammonium hydroxide or aqueous ammonia. Minimal reprocessing	Solvent extraction with β-hydroxyl oxime, stripping with ammonium hydroxide. Minimal reprocessing	Solvent extraction with β-hydroxyl oxime, stripping with ammonium hydroxide. Minimal reprocessing	Ion exchange with SüperLig 2, stripping with ammonium bisulfate and precipitation with ammonium hydroxide. Minimal reprocessing
Platinum extraction	Precipitation with ammonium chloride. Secondary reprocessing required	Precipitation with ammonium chloride. Secondary reprocessing required	Solvent extraction with TBP	Solvent extraction with a tertiary amine, stripped with HCl, and precipitated with ammonium chloride	Solvent extraction with a tertiary amine, tri-*n*-octylamine, stripped with HCl, and precipitated with ammonium chloride	Precipitation with ammonium chloride

Ruthenium extraction	Distillation with sodium chlorate and bromate	Solvent extraction with TBP, precipitation as RuO_2. Secondary reprocessing required	Distillation with sodium chlorate and bromate	Distillation with sodium chlorate and bromate	Distillation with sodium chlorate and bromate	Distillation with sodium chlorate and bromate
Iridium extraction	Precipitation with ammonium chloride	BaO_2 fusion, precipitation with ammonium chloride, further secondary purification	Precipitation with ammonium chloride	Solvent extraction with secondary ammine organic	Solvent extraction with secondary ammine organic, *n*-iso-tridecyl tri-decanamide	Ion exchange, precipitation with ammonium chloride
Rhodium extraction	Precipitation as Claus salt, $Rh(NH_3)_5Cl_3$, purification, precipitation with ammonium chloride	Precipitation as rhodium ammonium nitrite	Reduction with formic acid to form rhodium black	Ion-exchange, followed by precipitation with diethylene tri-ammine	Ion-exchange, followed by precipitation with diethylene tri-ammine	Precipitation with an organic ammine
First-pass yields[a]						
Pt	86	96	92		94	95
Pd	90	89	90		92	95
Au	92	–	98		84	98
Rh	77	–	80		82	82
Ru	86	–	92		90	82
Ir	70	–	80		–	66

(Continued)

TABLE 37.2 Methods of Extraction of the PGMs in Several Refineries—cont'd

	Lonmin	Krastsvetmet	Vale (Acton)	Johnson Matthey	Anglo American Platinum	Impala Platinum
Typical recoveries						
Pt	99		99	99.5	99.5	99.5
Pd	98		99	99.5	99.7	99.5
Au	99		98	99	99.5	94
Rh	93		97	98	99.5	98
Ru	98		95	90	99.5	99
Ir	90		95	90	99.5	94

TBP, tributyl phosphate; MIBK, methyl isobutyl ketone.
[a]*Best estimate based on industry interviews.*

TABLE 37.3 The Ratio of Each of the Precious Metals to the Amount of Platinum in a Typical Feed to a South African Precious Metals Refinery

Element	Abundance ratio with respect to platinum in feed to precious metals refinery
Pt	1.00
Pd	0.46
Ru	0.17
Rh	0.10
Ir	0.05
Os	0.01
Au	0.08

The ratio of platinum to palladium is about 2.2 (Cramer, 2008), whereas in Russian ores it is about 0.3.

(c) the ion-exchange process, represented by Impala Platinum.

The separation of the metals in the 'classical' precipitation process is based on the differences in the solubility of the ammonium chloride salts of the various metals, used repeatedly to produce a pure salt.

The second group of processes uses solvent extraction to effect a separation between platinum and palladium.

Finally, the third group of processes uses ion exchange as the principle method for the separation of palladium from the platinum stream.

The refining processes are described in the rest of this chapter. First, the precipitation process used by Lonmin and Krastsvetmet is described. After this, the solvent extraction processes used by Vale, Anglo American Platinum, Johnson Matthey and Heraeus are described. Finally, the ion-exchange process used by Impala Platinum is described.

37.6. LONMIN'S WESTERN PLATINUM REFINERY

Lonmin's Western Platinum Refinery was commissioned in 1974. The concentrate fed to the plant at that time contained 26% PGMs and it was processed to produce platinum, palladium and gold. Initially, the OPMs (rhodium, iridium and ruthenium) were toll refined elsewhere. The refinery was enhanced in 1987, so that these metals could also be refined. Currently, the feed to the refinery contains excess of 70% PGMs.

The flowsheet for the process at Lonmin's Western Platinum Refinery is shown in Figure 37.5. A summary of the operation is given in Table 37.2.

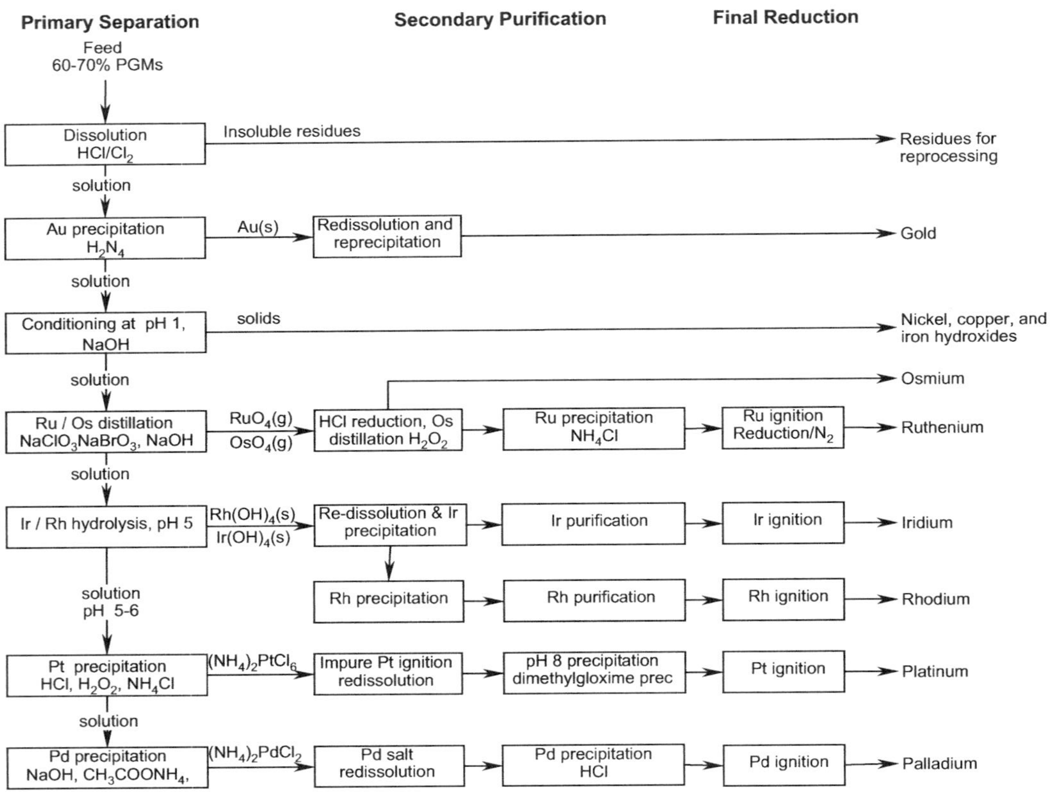

FIGURE 37.5 The refining of the PGMs by the precipitation process used at Lonmin's Western Platinum Refinery in Brakpan, South Africa. The process is divided into three steps: (i) primary separation; (ii) secondary purification; and, (iii) final reduction.

FIGURE 37.6 A view of the production facilities at Lonmin's Western Platinum Refinery. The glass-lined reactors are arranged on either side of the production hall of the primary separations plant. Generally, the filtration steps are carried out on the level below this. However, a variable chamber filter can be seen at the back of this hall. Note that the requirements for personal protection equipment have become more stringent since this photograph was taken. *Photograph courtesy of Lonmin plc.*

A photograph of the separation hall of the Lonmin refinery is shown in Figure 37.6. The dissolution and primary separation steps of the process are done in this hall.

37.6.1. Dissolution of the Concentrate in HCl/Cl_2

The concentrate that is received from the base metals refinery contains 65%–70% PGMs. The material is dissolved in a solution of 6 M hydrochloric acid at 65°C for about 6 h. Chlorine is continuously sparged through the reactor, which is a 2.2-m^3 glass-lined vessel. The slurry contents of the vessel are agitated with an agitator fitted with a pitched-blade turbine.

Once the dissolution is complete, air is blown through the solution to remove the dissolved chlorine. The slurry is filtered in a filter press, and the remaining solids are redissolved using similar conditions. The dissolution efficiency of the primary PGMs is in excess of 99%. The dissolution efficiency of the OPMs is in excess of 97%.

The next step in the process is the removal of the gold from solution.

37.6.2. Precipitation of Gold

The solution from chlorine leaching is treated to remove gold. Hydrazine, N_2H_4, a powerful reductant, and hydrochloric acid are added to the solution. This results in the formation of crude gold. The equipment used is a glass-lined vessel, and the reactants are agitated with a pitched blade turbine.

The slurry is filtered. Because the purity of the precipitated gold is below saleable specification, it is retreated by dissolution and precipitation in order to recover any co-precipitated PGMs. The solution is transferred to base metal precipitation.

37.6.3. Conditioning at pH 1

The acid in the solution from gold precipitation is neutralized to a pH of about 1 with sodium hydroxide at ambient temperature in a glass-lined vessel. Between 20% and 40% of the base metals in solution precipitate.

The slurry is filtered in a filter press. The filter cake contains mainly base metals and is reprocessed to recover any included PGMs. The filtrate is transferred to ruthenium distillation.

37.6.4. Distillation of Ruthenium

Ruthenium is removed from solution by reactive distillation. The equipment consists of a glass-lined vessel, a reflux condenser and two scrub columns, one for the extraction of ruthenium using dilute hydrochloric acid and the other for the scrubbing of chlorine using sodium hydroxide.

The initial charge per still is 12 kg of ruthenium. The solution is at a temperature of 60°C and a pH of approximately 1. Sodium chlorate, $NaClO_3$, is added to oxidize the ruthenium to ruthenium tetroxide, RuO_4, which is volatile.

Air is drawn through the solution by vacuum, and the volatile RuO_4 and OsO_4 are stripped from the aqueous phase into the gas phase. The gas stream moves through the reflux condenser, which condenses the water vapor in the gas stream, and then through the two scrubbing columns.

Later in the distillation, sodium bromate, $NaBrO_4$, is added to remove about 10% of the ruthenium that has not reacted sodium chlorate. Distillation continues until the concentration of ruthenium in solution is less than 250 mg/L.

In the first scrubbing column, hydrochloric acid reduces RuO_4 and OsO_4 to non-volatile aqueous state. As a result, the ruthenium and osmium are removed from the gas phase to the aqueous phase.

In the second scrubbing column, chlorine that is produced in the first column is scrubbed from the gas phase using sodium hydroxide.

Hydrogen peroxide is added to the solution collection tank after the first column to re-oxidize the osmium back to OsO_4, where it is reduced by the hydrochloric acid in the solution of the second column. As a result, ruthenium and osmium are separated.

Ruthenium from the scrub liquor is precipitated with ammonium chloride, and the salt is ignited in a muffle furnace at 1000°C and cooled under an atmosphere of nitrogen to produce pure ruthenium metal.

The next step in the process is the removal of iridium and rhodium from the solution remaining after distillation.

37.6.5. Removal of Iridium and Rhodium from Solution

The solution remaining in the distillation vessel is at a pH value of about 2. The pH of this solution is raised by the addition of sodium bicarbonate. At a pH of about 5, the rhodium and iridium in solution precipitate as $Rh(OH)_4$ and $Ir(OH)_4$, respectively. The slurry is filtered in a variable chamber filter. The filter cake is transferred to the OPM circuit for refining. The filtrate contains the platinum and palladium, and remaining impurities. The next step is the precipitation of platinum.

37.6.6. Precipitation of Platinum

After rhodium and iridium removal, the pH of the filtrate is about 5. Hydrochloric acid is added to this solution for acidification, and hydrogen peroxide is added to adjust the redox potential to between 900 and 1000 mV (*vs.* the saturated calomel electrode). Then, ammonium chloride is added, which results in the precipitation of ammonium hexachloroplatinate, $(NH_4)_2PtCl_6$. The slurry is then filtered in a glove-box filter.

The filter cake is ignited, which produces an impure platinum sponge that contains between 92% and 97% of platinum. This material is then transferred to the platinum purification section, which is described later in Section 37.6.9.

The filtrate contains mainly palladium. The next step in the process is the precipitation of this palladium.

37.6.7. Precipitation of Palladium

The pH of the solution is increased to a value of 0.9 by the addition of sodium hydroxide. The solution is then boiled to the point where the pH of the solution rises to a value of 3. Ammonium acetate is added, causing the pH to increase further. The solution is boiled for 30 min to increase the pH to 4.2 and then allowed to cool. As a result, the palladium precipitates as diamino-palladous dichloride, $(NH_3)_2PdCl_2$. Once the slurry cools below 30°C, it is filtered. The filter cake is transferred to the palladium purification section and the solution is transferred to barrens treatment.

37.6.8. Secondary Purification and Metal Reduction Steps

The precipitation and distillation steps described thus far are the primary separation steps. They have resulted in impure metals or metal salts, which need to be purified. The next steps in the process are the purification of the crude platinum sponge, the impure palladium salt and the pH 5 cake containing the rhodium and iridium. These purification steps are discussed in the same order.

37.6.9. Platinum Purification Section and Metal Production

The impure platinum sponge contains 92% and 97% platinum. Most of the impurity in the sponge is palladium. This platinum needs to be purified before it can be sold.

Purification of the impure platinum sponge takes place in the following steps:

(a) *Redissolution*: The impure platinum sponge is dissolved in hydrochloric acid with chlorine gas. This step is carried out at similar conditions to the initial dissolution step, including dechlorinating the solution with air once the dissolution reaction has ceased;

(b) *pH 8 precipitation*: The solution is boiled, and sodium bromate and sodium bicarbonate are added so that the pH rises to 3. Rhodium and iridium are precipitated as hydroxides. The pH is further increased to a value of 8 to precipitate the remaining base metals. The slurry is filtered, and the solution is transferred to dimethylglyoxime precipitation;

(c) *Dimethylglyoxime precipitation*: Dimethylglyoxime is added to precipitate the palladium from the solution. The concentration of palladium in solution after this treatment is less than 10 mg/L;

(d) *Ammonium chloride precipitation*: The solution from dimethylglyoxime treatment should contain only platinum. The pH is adjusted with hydrochloric acid and the redox potential is adjusted with hydrogen peroxide. Once this conditioning is complete, ammonium chloride is added, which precipitates the platinum as ammonium hexachloroplatinate, $(NH_4)_2PtCl_6$. The reaction continues until the concentration of platinum in solution is less than 300 mg/L. The slurry is filtered. The filtrate is treated with hydrazine to precipitate the remaining platinum. The salts are transferred to reduction;

(e) *Reduction*: The cake is dissolved with water, and the platinum metal is reduced from the solution by chemical reduction with hydrazine. The metal is calcined at 1000°C in air to produce platinum sponge of 99.99% purity.

37.6.10. Palladium Purification Section and Metal Production

The main impurity in the impure palladium salt is platinum. Platinum is removed from the impure palladium salt in the palladium purification section, which consists of the following steps:

(a) *Salt dissolution*: The impure palladium salt is dissolved in a solution of ammonium hydroxide with a pH of about 9. The redox potential of the solution is adjusted using hydrogen peroxide. Under these conditions, the palladium dissolves, whereas the platinum does not. The slurry is filtered, and the solution, containing only palladium, is transferred to precipitation;

(b) *Palladium precipitation*: The pH of the solution is adjusted to a value of about 0.6 using hydrochloric acid. Under these conditions, the palladium precipitates as diamino-palladous dichloride, $(NH_3)_2PdCl_2$. The reaction is complete when the solution contains less than 300 mg/L of palladium. The slurry is filtered, and the salt is transferred to ignition;

(c) *Salt ignition*: The diamino-palladous dichloride salt is ignited in air to produce palladium of between 99.98% and 99.99% purity.

37.6.11. Rhodium and Iridium Purification Section

The pH 5 cake (Section 37.6.5) is dissolved, and iridium is precipitated with ammonium chloride from the solution as crude ammonium hexachloroiridate, $(NH_4)_3IrCl_6$. Once the iridium has been removed from solution, rhodium is precipitated as the Claus salt, $(NH_3)_5RhCl_3$, dissolved in caustic soda, and precipitated again with ammonium chloride. These salts are purified by repeated precipitation to produce pure salts, which are then ignited to produce pure metals.

37.7. KRASTSVETMET'S REFINERY AT KRASNOYARSK

The Krasnoyarsk refinery receives a variety of feed materials that vary greatly in amount, metal content and grade. The material is sampled and the routing is determined. The two main avenues for processing are as follows: (i) material rich in platinum and palladium is routed to the platinum–palladium circuit; (ii) material rich in OPMs or of low grade, including all reworked material, is routed to the upgrade smelting circuit.

The flowsheet of the Krasnoyarsk refinery is given in Figure 37.7.

The main platinum and palladium circuit is described first, after which the smelting and rhodium, ruthenium and iridium circuit is discussed.

37.7.1. Dissolution

Material rich in platinum and palladium is dissolved in hydrochloric acid using chlorine gas as the oxidant in a batch operation. The vessel is about 1 m^3 in volume and is made of titanium. The slurry is stirred using a hollow-shaft impeller that draws gas from the headspace into the slurry. The temperature is 70°C and the pressure is 3 bar. The acid concentration is about 17.5% HCl. The redox potential is monitored during dissolution. Once the redox potential reaches 1100 mV (*vs.* Ag/AgCl), the reaction is deemed to be complete. The dissolution time is about 2 h.

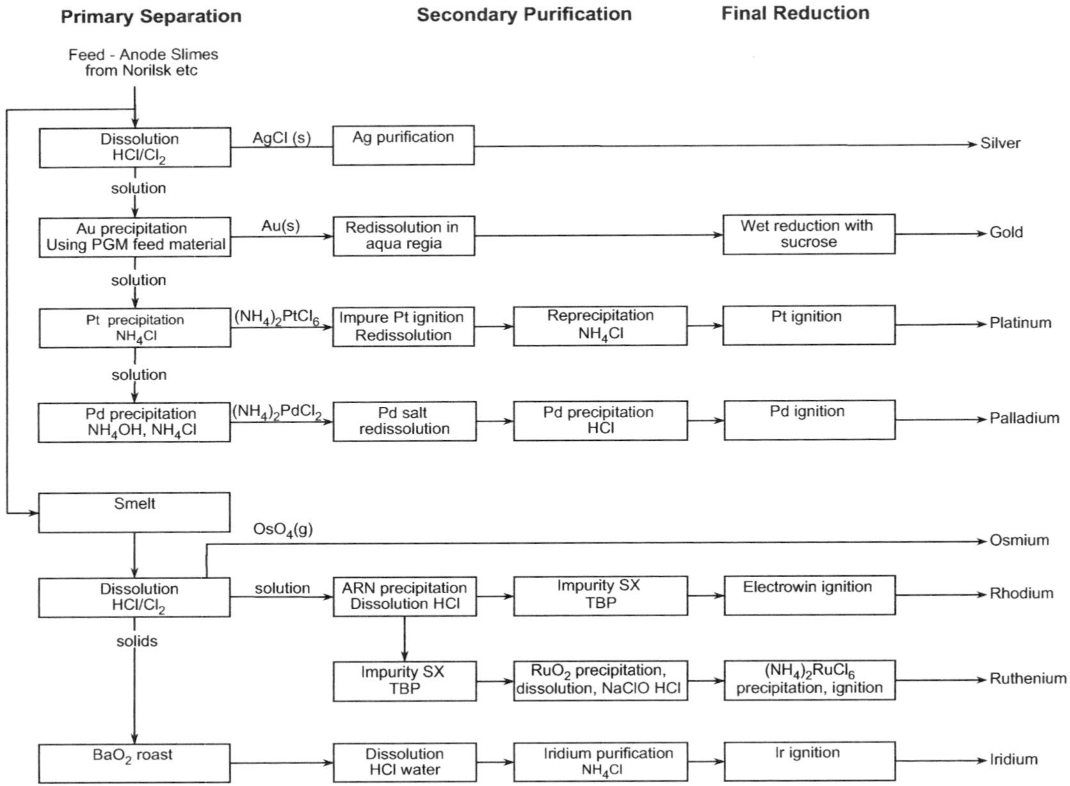

FIGURE 37.7 The overall flowsheet of Krastsvetmet's refinery in Krasnoyarsk, Russia. The anode slime feed is mainly a byproduct of nickel and copper electrorefining at Norilsk. ARN refers to ammonium rhodium nitrite, and TBP refers to tri-n-butyl phosphate.

The residue contains mainly silver in the form of silver chloride, which is purified and sold. The solution is transferred to gold precipitation.

37.7.2. Precipitation of Gold

The solution from the dissolution step is contacted with a small quantity of fresh concentrate. Most of the concentrate dissolves, causing gold to precipitate as a crude gold. The crude gold is filtered from solution. The solution proceeds to platinum precipitation, and the solids to the gold purification circuit.

The crude gold is redissolved in aqua regia. Gold sponge is precipitated from solution using sucrose. This gold sponge is melted and electrorefined to produce 99.99% pure gold.

37.7.3. Precipitation of Platinum

Ammonium chloride is added to the solution from the gold precipitation step. As a result, platinum precipitates as ammonium hexachloroplatinate, $(NH_4)_2PtCl_6$. The solution is filtered and transferred to palladium precipitation.

The filter cake, consisting of impure ammonium hexachloroplatinate, is ignited to form an impure platinum sponge, which is the feed material to the platinum purification section.

37.7.4. Precipitation of Palladium

Ammonium hydroxide is added to the solution from the platinum precipitation step to adjust the pH to between 4 and 5. After this precipitate has been filtered from solution, hydrochloric acid is added to adjust the pH to about 1. Palladium precipitates as diamino-palladous dichloride, $(NH_3)_2PdCl_2$. The solution is boiled for about 30 min. The solution is cooled, filtered and transferred to barrens recovery. The salt is transferred to the palladium purification circuit.

37.7.5. Platinum Purification Circuit

The impure platinum sponge that is produced from the initial precipitation needs to be purified. This is performed in the platinum purification circuit.

The impure platinum sponge is dissolved in hydrochloric acid using chlorine gas as the oxidant. The solution is filtered. The undissolved solids are sent to smelting, and the solution is boiled down, diluted and filtered. Ammonium chloride is then added to the solution to precipitate ammonium hexachloroplatinate. The solution is then filtered. The solution is transferred to barrens treatment, and the salt is ignited to form pure platinum sponge. The sponge is crushed, melted and cast as ingots.

37.7.6. Palladium Purification Circuit

The diamino-palladous dichloride salt from the palladium precipitation step is dissolved in ammonium hydroxide. Ferrous sulfate is added to the reactor and the contents react for a few hours. The contents are filtered thoroughly. Hydrochloric acid is added to the solution to precipitate pure diamino-palladous dichloride. The solution is filtered, and the salt is ignited to form pure palladium sponge.

This completes the description of the platinum and palladium circuits. The smelting rhodium, ruthenium and iridum circuits are discussed next.

37.7.7. Smelting

Feed materials to the refinery that are low in platinum and palladium, as well as the internal residues from the refinery, are smelted. This smelting process is key to the success of the rhodium, ruthenium and iridum circuits.

Smelting is performed in a rotating oil-fired furnace. Metallic alloys are formed from different feeds. Collectors such as iron, copper or lead are not needed. The smelting techniques have been developed to maximize the recovery of rhodium, ruthenium and iridum.

37.7.8. Alloy Dissolution

The alloy from the smelting operation is dissolved in hydrochloric acid using chlorine gas as the oxidant. The concentration of acid is stronger than is used to dissolve the platinum-rich materials. The efficiency of the dissolution is high, with the exception of iridium.

Osmium is also recovered in this dissolution step if the alloy that is dissolved is predominantly osmiridium, an alloy of osmium and iridium. The osmium forms osmium tetroxide, OsO_4, and volatizes. It is recovered from the gas by scrubbing the gas with a caustic solution, followed by the precipitation of the ammine salt, $OsO_2(NH_3)_4Cl_2$.

37.7.9. Rhodium Circuit

The solution from alloy dissolution is boiled down, and the pH is raised to a value of about 4.5. The base metals precipitate as hydroxides and are separated out using a centrifuge.

Ammonium nitrite is added to the centrifuged solution to precipitate the rhodium in the form of ammonium rhodium nitrite, $Rh(NH_4)_3(NO_2)_6$.

The slurry is filtered, and the solution is transferred to ruthenium removal. The salt is dissolved in water, which forms the feed to a solvent-extraction process. Tri-*n*-butyl phosphate (TBP) is used to extract impurities from the solution. Rhodium is electrowon from the raffinate as a fine rhodium powder. The rhodium powder is melted to form rhodium sponge.

37.7.10. Ruthenium Circuit

Solution from the precipitation of ammonium rhodium nitrite, forms the feed to the ruthenium circuit. Platinum, palladium and iridium in this solution are extracted by solvent extraction using tri-*n*-butyl phosphate. The strip solution is returned to the platinum circuit. Sodium hypchlorite, NaClO, is added to the raffinate, which results in the precipitation of ruthenium dioxide, RuO_2. The solution is filtered, and the ruthenium dioxide residue is dissolved in hydrochloric acid. Ruthenium is precipitated from solution ammonium ruthenium chloride by the addition of ammonium chloride. This salt is ignited under hydrogen to produce ruthenium metal.

37.7.11. Iridium Circuit

The residue from the alloy dissolution step is rich in iridium. This residue is fused with barium peroxide, BaO_2, and then dissolved in hydrochloric acid with chlorine gas. Sulfuric acid is added to precipitate barium as barium sulfate, which is filtered out and discarded.

Ammonium chloride is added to the solution from barium sulfate removal. This results in the precipitation of an impure salt of ammonium chloroiridate. The salt is redissolved in water, and sodium sulfite is added. The salt that is formed is redissolved and oxidized, and reprecipitated. Finally, this salt is ignited to form pure metal.

37.8. APPRAISAL OF THE PRECIPITATION PROCESSES

The advantages of the precipitation processes are as follows:

(a) the methods are established and well known;
(b) effluent volumes are relatively low compared with other process routes;
(c) products of the required purity can be produced; and,
(d) capital cost is relatively low.

The disadvantages of the process are as follows:

(a) The separations are not completely efficient, which means that there is a significant hold up of valuable material in the refinery as dissolution and precipitation steps are repeated to achieve the purity specifications. The overall hold-up in the precious metal refinery exceeds 1 month on average and 6 months for some individual metals. This is a significant contributor to the cost of operating a refinery;
(b) The separations are difficult and expensive to automate; and,
(c) The risk of exposure to platinum chemicals is high, which increases the risk of staff turnover due to platinum-salt sensitivity, an allergy to platinum chemicals (Niezborala & Garnier, 1996).

Because of the shortcomings of the 'classical' or precipitation separation process, a number of companies were started in the 1970s to develop a more efficient process for the refining of PGMs.

A major reason for developing the resulting solvent-extraction processes was to increase first-pass yields in the separation operation and thereby decrease the hold-up of material in the refinery.

Most of these efforts led to processes based on solvent extraction, such as:

Anglo American Platinum's refinery in Rustenburg, South Africa;
Johnson Matthey's refinery in Royston, UK;
Vale's refinery at Acton, UK; and,
Heraeus's refinery in Hannau, Germany.

Each of these processes is discussed in turn.

37.9. THE JOHNSON MATTHEY/ANGLO AMERICAN PLATINUM PROCESS

Anglo American Platinum, part of the Anglo American group, and Johnson Matthey work closely together on the development and optimization of their technology. The main features of their processes are similar and are presented together.

A schematic representation of the process used at Anglo American Platinum is shown in Figure 37.8. Many of the details of this process are closely guarded secrets of the two companies.

The Johnson Matthey/Anglo American Platinum process shown in Figure 37.8 is a major shift from the precipitation process for the refining of PGMs. The primary separation of the metals is performed by solvent extraction in four cases where the separations by precipitation are particularly difficult.

37.9.1. Concentrate Dissolution

The concentrate feed to the refinery contains between 50% and 60% PGMs. The concentrate, consisting mostly of leached magnetic concentrate from the magnetic concentrator plant (see Chapter 36), is dissolved in hydrochloric acid solution using chlorine gas. The temperature is 120°C and the pressure is 4 bar. This step, referred to as the 'high-intensity dissolve', ensures that all the PGMs and base metals are dissolved.

In this dissolution step, osmium is oxidized to OsO_4, which is volatile. The osmium is recovered from the offgas as potassium osmate, K_2OsO_4, using KOH.

In addition, there is a separate circuit for the leaching of the gravity concentrates that are obtained from the mill cyclones of the company's ore concentrators. Because of its grade, this material bypasses smelting and the

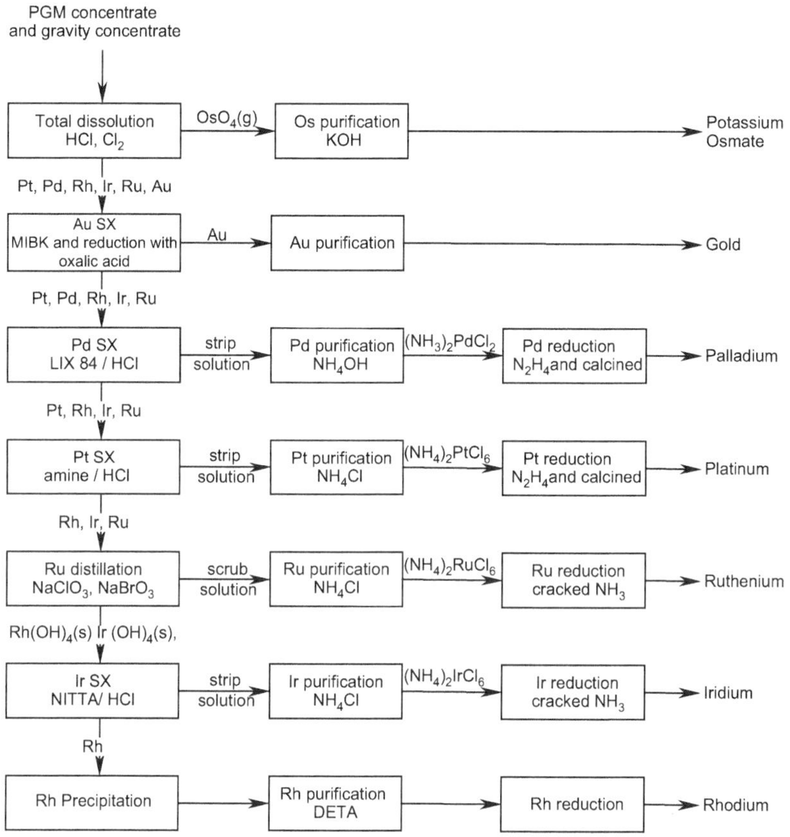

FIGURE 37.8 An overview of the processing of PGMs at Anglo American Platinum's refinery in Rustenburg, South Africa.

magnetic concentrator plant, and it is fed directly into the precious metals refinery. These gravity concentrates account for about 10% of the total feed of PGMs to the refinery.

The gravity concentrates are roasted prior to dissolution, which is carried out under similar conditions to the main refinery feed.

At the end of the dissolution period, the slurry is filtered. The residue is processed for silver, while the solution is transferred to the solvent extraction of gold.

37.9.2. Solvent Extraction of Gold

A solvent-extraction step is used to remove the gold from the filtrate from dissolution. This is achieved by solvent extraction at high acidity. The solvent used is methyl isobutylketone (MIBK), which has an advantage that most of the selenium, tellurium and antimony are co-extracted, thus removing them from the main production stream.

Although solvent extraction with MIBK is advantageous because it removes the selenium, tellurium and antimony from the stream carrying the main products, there are significant disadvantages to this process step. MIBK has a low flash point. As a result, the mixer-settlers are enclosed for fire protection. The material of construction is glass.

The loaded organic is scrubbed with hydrochloric acid to remove co-extracted impurities, such as iron, antimony, selenium and tellurium.

Gold is selectively recovered from the organic phase by direct reduction using oxalic acid. The direct reduction produces a gold powder, which is filtered, dried and sold.

The raffinate is transferred to palladium extraction.

37.9.3. Solvent Extraction of Palladium

Following solution conditioning, palladium is removed from the gold raffinate by solvent extraction with a ketoxime, β-hydroxyoxime, which is sold commercially as LIX 84A.

The kinetics of the extraction of palladium are slow. An organic amine compound is added as an accelerator, but this decreases the selectivity of the extraction. This decreased selectivity results in decreased purity of the palladium extracted. It has been reported that although the reaction is slow, it does go to completion, so that the raffinate contains less than 5 mg/L palladium.

Palladium is stripped from the loaded solvent using 6 M HCl. The strip solution is treated in a manner similar to that of the precipitation process: palladium is precipitated from solution with ammonium hydroxide as diamino-palladous dichloride, $(NH_3)_2PdCl_2$. This salt is dissolved and reduced to metal with hydrazine, N_2H_4, and then calcined to produce palladium metal with a purity in excess of 99.95%.

37.9.4. Solvent Extraction of Platinum

Following solution conditioning, platinum is removed from the palladium raffinate by solvent extraction with a secondary amine. The iridium in solution must be in its lower oxidation state (+3) to prevent it from co-extracting with the platinum.

The loaded organic is scrubbed with weak acid to remove base metals that may have been co-extracted. The platinum is then stripped from the loaded organic with 11 M HCl.

Ammonium chloride is then added to the strip solution to precipitate the platinum as ammonium hexachloroplatinate salt. The salt is redissolved, and the platinum is reduced to metal using hydrazine. The metal is melted and cast to produce pure platinum.

37.9.5. Ruthenium Distillation

The raffinate solution from the platinum solvent extraction is pre-treated to remove any entrained organic. This is because the organic may react with the strong oxidants used in ruthenium distillation, possibly resulting in an explosion.

The solution is neutralized, after which strong oxidants, sodium chlorate and sodium bromate, are added. These reagents oxidize ruthenium and any residual osmium to their tetroxides, RuO_4 and OsO_4, which are volatile. Air is drawn through the solution by vacuum and results in the stripping of the volatile tetroxides from the aqueous phase to the gas phase.

The gas stream passes through a packed column in a countercurrent direction to the flow of a solution of hydrochloric acid. The ruthenium is reduced from the 8+ state to the 4+ state and absorbed by the hydrochloric acid in this solution.

If this solution contains any small amounts of osmium, it may be treated with hydrogen peroxide, H_2O_2, to selectively volatilize osmium from the hydrochloric acid solution. This osmium is recovered in a similar manner using a hydrochloric acid trap and precipitated using potassium hydroxide as potassium osmate, K_2OSO_4. The potassium osmate is filtered from solution, leaving a solution of ruthenium in hydrochloric acid.

Ruthenium is then precipitated from the hydrochloric acid solution using ammonium chloride as ammonium hexachlororuthenate, $(NH_4)_2RuCl_6$.

The precipitate is calcined to form ruthenium dioxide, RuO_2, which is then reduced to ruthenium sponge using cracked ammonia. The final ruthenium sponge has a purity that is greater than 99.95%.

Once the ruthenium and osmium are removed from solution, the pH is increased to about 5.8 to precipitate the iridium and rhodium as $Ir(OH)_4$ and $Rh(OH)_4$, respectively, from the remaining base metals. The slurry is filtered. The filter cake, referred to as the 'pH 5 cake', is transferred to iridium extraction.

37.9.6. Extraction of Iridium

The 'pH 5 cake' is redissolved, and the solution is purified to remove trace amounts of base metals, such as copper and nickel.

The solution is treated with hydrogen peroxide so that the iridium is in the higher oxidation state in a concentrated solution of hydrochloric acid. Iridium is then removed from solution by solvent extraction using a novel amide solvent, n-iso-tridecyltri-decanamide. The iridium is stripped from the loaded organic using dilute hydrochloric acid.

The iridium is then precipitated from the strip solution using ammonium chloride. The salt is redissolved using a reductant, reprecipitated and calcined to form iridium dioxide, IrO_2. Iridium dioxide is reduced to iridium sponge using cracked ammonia. The purity of the iridium product is 99.95%.

37.9.7. Extraction of Rhodium

Finally, the rhodium is precipitated from the raffinate from iridium extraction using diethylene triammine. The rhodium then redissolved and is precipitated using ammonium chloride to form chloropentammine rhodium(III) dichloride, $[RhCl(NH_3)_5]Cl_2$.

The $[RhCl(NH_3)_5]Cl_2$ salt is calcined and reduced to give a rhodium sponge with a purity of greater than 99.98%. The rhodium is sold as metal.

37.10. THE ACTON REFINERY PROCESS

The precious metals refinery at Acton, England, is part of the Vale group of companies. It processes primary concentrates, anode slimes and secondary materials such as spent catalysts and electronic scrap (Barnes & Edwards, 1982).

The process at Vale's refinery in Acton in the UK is shown in Figure 37.9.

37.10.1. Concentrate Preparation

The materials containing PGMs are smelted with iron, which acts as a collector for the precious metals. The iron is leached from the smelter solids in hydrochloric acid. This material forms the feed to the refinery.

37.10.2. Concentrate Dissolution

The raw material is dissolved in hydrochloric acid using chlorine gas. The process takes place at 90°C–98°C and at ambient pressure.

The residues from the first dissolution step contain silver and lead. As a result, these residues are treated with nitric acid to dissolve the silver and lead. The residues remaining from the nitric acid treatment are fused with sodium hydroxide at a temperature of between 500°C and 600°C. This material is subject to a second dissolution step with hydrochloric acid using chlorine gas as the oxidant.

The solution from the first and second dissolution steps is neutralized with sodium hydroxide and conditioned with sodium bromate. The solution is then transferred to ruthenium distillation.

37.10.3. Ruthenium Distillation

Sodium bromate is added to oxidize the ruthenium and osmium to their respective tetroxides. The volatile tetroxides are removed by a gas sparge at a temperature of between 90°C and 95°C. The concentration of ruthenium remaining in solution is less than 10 mg/L.

The ruthenium and osmium are absorbed from the gas into a solution of dilute hydrochloric acid.

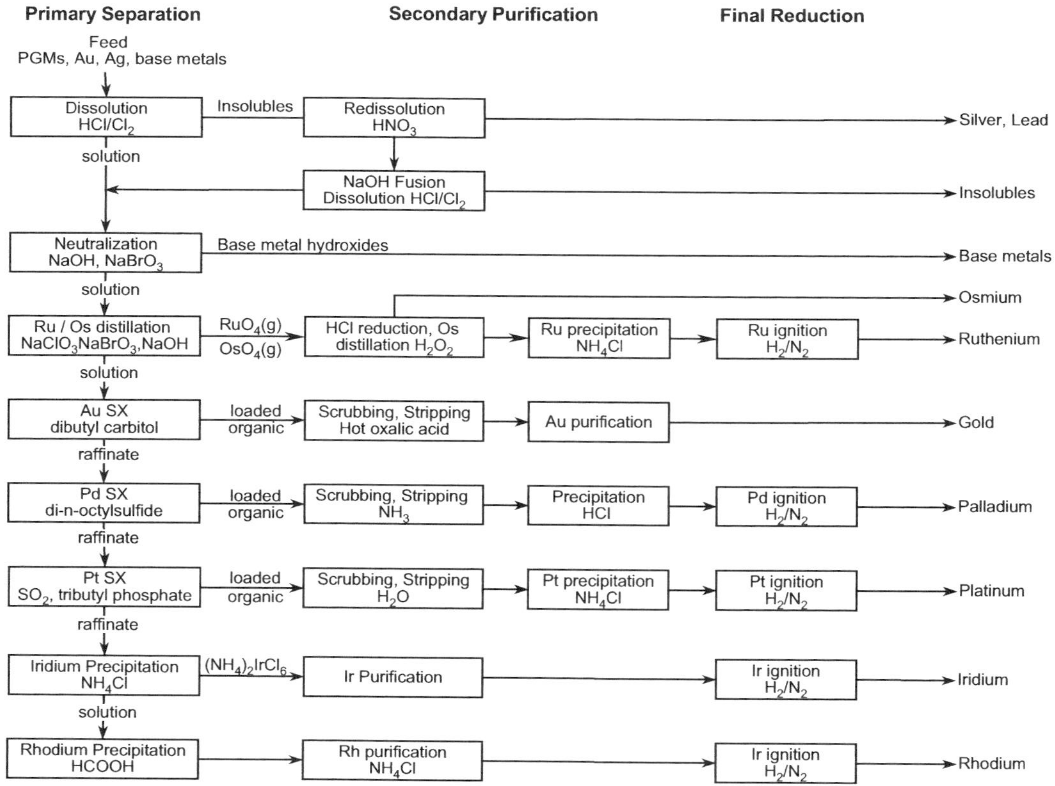

FIGURE 37.9 An overview of the processing of PGMs at Vale refinery in Acton, United Kingdom.

Ammonium chloride is added to the solution to precipitate ammonium ruthenium chloride, $(NH_4)_3RuCl_6$. The slurry is filtered. The salt is heated in air and reduced in an atmosphere of hydrogen and nitrogen to form ruthenium powder with a purity of 99.9% or greater.

37.10.4. Removal of Base Metals

Sodium hydroxide is added to the solution to precipitate base metal hydroxides, such as copper and nickel. The solution is filtered and the solids are sent for re-treatment. The filtrate is transfered to gold extraction.

37.10.5. Extraction of Gold

Hydrochloric acid is added to the solution to increase the concentration to between 3 and 4 M. This solution is the feed to the gold solvent-extraction circuit.

Gold is extracted using dibutyl carbitol in two stages. The loaded organic is scrabbed with a solution containing 1–2 M HCl in multiple stages to remove any base metals that may have been extracted with the gold.

The gold is stripped from the loaded organic with a solution of hot oxalic acid. The oxalic acid directly reduces the gold to gold powder. The gold powder is melted and granulated to form gold grains of 99.99% purity.

37.10.6. Extraction of Palladium

Palladium is extracted from the solution by solvent extraction with di-*n*-octyl sulfide. Even though the kinetics of extraction are slow, the reaction goes to completion so that the concentration of palladium in solution is less than 1 mg/L.

The loaded organic is scrubbed with hydrochloric acid. The palladium is then stripped from the loaded organic with ammonium hydroxide.

The palladium is precipitated from the strip solution by adding hydrochloric acid so that diamino-palladous chloride, $(NH_3)_2PdCl_2$, precipitates from solution.

The salt is heated to 1000°C and then slowly cooled under an atmosphere of nitrogen and hydrogen to produce palladium metal sponge of greater than 99.95% purity.

37.10.7. Extraction of Platinum

The raffinate from palladium extraction is conditioned using sulfur dioxide to reduce the Ir^{4+} to Ir^{3+}.

The platinum in solution is extracted by solvent extraction using tri-*n*-butyl phosphate. The extraction occurs in four stages. The raffinate contains less than 100 mg/L of platinum.

The loaded organic is scrubbed with 5–6 M HCl to remove impurites and then stripped with water. Platinum is recovered by adding ammonium chloride to precipitate ammonium hexachloroplatinate, $(NH_4)_2PtCl_6$.

The ammonium hexachloroplatinate is ignited in air to produce platinum metal sponge of greater than 99.95% purity.

37.10.8. Extraction of Rhodium and Iridium

The raffinate from platinum extraction is conditioned with sulfur dioxide so that the iridium is oxidized from Ir^{3+} to Ir^{4+}. Iridium is precipitated from solution as $(NH_4)_2IrCl_6$ using ammonium chloride. The slurry is filtered. The salt is heated under an atmosphere of nitrogen and hydrogen to produce iridium metal with a purity of greater than 99.9%.

Rhodium is then precipitated from solution using formic acid resulting in rhodium black. The rhodium black is heated and reduced to form rhodium powder of greater than 99.9% purity.

37.11. APPRAISAL OF THE SOLVENT-EXTRACTION PROCESSES

The advantages of the solvent-extraction process are as follows:

(a) the selectivity of the solvent-extraction steps is much higher than the precipitation methods;
(b) the first-pass yields are significantly higher than the precipitation methods; and,
(c) the handling of materials, and hence exposure by refinery workers to allergenic platinum chemicals, is greatly reduced.

The disadvantages of the process are as follows:

(a) the solutions are generally more dilute than those of the precipitation process, which means that the volumes of solution that are treated in the process and in the barrens recovery are higher than the precipitation methods;
(b) the kinetics of the solvent extraction for two of the key separations, palladium and gold, are slow, which means that hold-up, although better than in the precipitation process, is not as high as in the ion-exchange process;
(c) the solvents are highly flammable and introduce significant risk of fire into the refinery;
(d) entrainment of organics in the aqueous phases, particularly where the next step involves oxidants, is undesirable; and,
(e) the primary separation of palladium by a secondary amine, while better than the precipitation techniques, co-extracts small amounts platinum, ruthenium and iridium, which increases the complexity of the secondary purification of platinum.

37.12. IMPALA PLATINUM'S ION-EXCHANGE PROCESS

The solvent-extraction processes were developed in the 1970s and 1980s. A later development is the process developed by Impala Platinum in the late

1990s. This process is primarily based on ion-exchange technology. In this section, the process at Impala Platinum Refineries is described in detail. An overview of the refinery is presented first, after which each of the separations is discussed in more detail.

37.12.1. Overview of the Impala Platinum Process

An overview of the refining process employed at Impala Refineries in Springs, South Africa, is shown in Figure 37.10.

The concentrate, which is the converter matte that has been leached, contains about 65% PGMs. This concentrate is initially dissolved almost to completion in hydrochloric acid and chlorine.

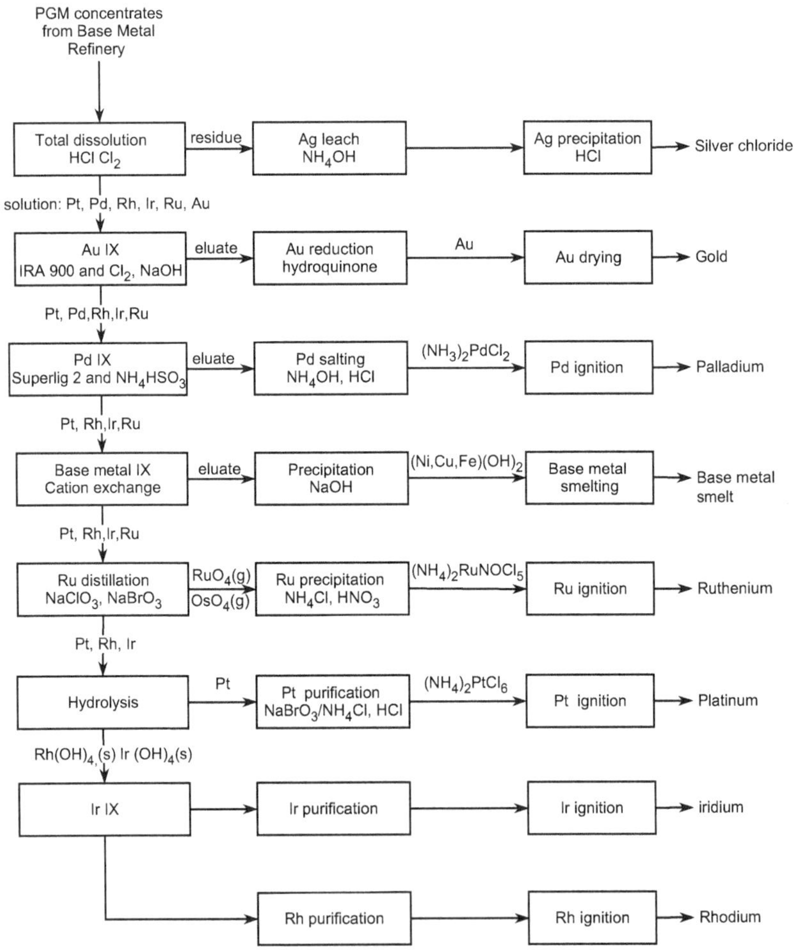

FIGURE 37.10 An overview of Impala's precious metals refinery at Springs, South Africa.

The residue from the dissolution is leached in ammonium hydroxide. Hydrochloric acid is added to precipitate silver chloride.

Gold is extracted from the leach liquor using an ion-exchange process. The loaded gold is stripped from the ion-exchange media and precipitated with a reductant to produce gold with a purity of about 99.95%. This gold is sold to the Rand Refinery, based in Germiston, South Africa.[2]

Palladium is removed by an ion-exchange step that uses molecular recognition technology. The ion-exchange raffinate contains less than 1 mg/L of palladium, and less than 1% of the platinum in the feed is co-extracted with the palladium. This is a key step, because it extracts all of the palladium very selectively.

Copper, nickel, iron and other base metals are removed from the palladium raffinate using a cation-exchange process. These base metals are eluted from the ion-exchange media, precipitated from solution, and, together with other refinery residues, are smelted as a matte. This matte is returned to the base metals refinery.

The raffinate from the cation-exchange process is boiled down, and sodium chlorate and sodium bromate are added to oxidize the ruthenium and osmium to their tetroxides. Air is drawn through the reactor by vacuum, and ruthenium tetroxide is stripped from the aqueous phase into the gas phase. Ruthenium and osmium are recovered by scrubbing the gas with hydrochloric acid. Ruthenium is precipitated from solution as the ruthenium nitrosyl salt, $(NH_4)_2RuNOCl_5$. This salt is ignited to produce pure ruthenium metal.

The remaining metals in solution are platinum, iridium and rhodium. The pH of the solution that remains after ruthenium and osmium distillation is raised to about 6, and the iridium and the rhodium precipitate as hydroxides. This slurry is filtered. The filtrate, containing mainly platinum, is boiled to dryness to eliminate excess acid, redissolved in demineralized water, oxidized and neutralized with alkalis. Platinum is finally precipitated using ammonium chloride to form ammonium hexachloroplatinate, $(NH_4)_2PtCl_6$. This salt is then ignited to form pure platinum metal.

The hydrolysis cake, which contains mostly rhodium and iridium hydroxides, is redissolved. The rhodium and iridium are precipitated using an organic amine, resulting in an organic salt. This reagent selectively precipitates iridium and rhodium, leaving any base metals in solution. The precipitate is redissolved, oxidized and iridium is extracted using an ion-exchange resin. Iridium is stripped from the resin, purified, precipitated with ammonium chloride and ignited to produce pure metal.

Rhodium in the raffinate is precipitated with the same organic amine and ignited to produce pure rhodium.

---

2. Rand Refinery, a private company held by a number of major gold producers, is the largest refiner of gold in the world.

This completes the overview of the ion-exchange process. Each of the separation steps at the Impala Platinum refinery at Springs is described in more detail in the next sections.

37.12.2. Concentrate

The concentrate from the base metals refinery contains approximately 65% PGMs. The total concentration of copper, nickel, iron and bismuth is, on average, less than 10%. The remainder of the concentrate is oxygen and silica. A large proportion of the PGMs in the concentrate is oxidized during caustic leaching in the base metals refinery.

The impurities that create specific problems are polymerized silica, iron spinels and various alloys encountered in the ore, all of which may occlude or include PGMs and result in poor PGM dissolution efficiencies. Polymerized silica may cause poor filtration after the initial dissolution.

Lead in the concentrate can result in problems if it dissolves initially and moves to gold ion exchange, where it fouls the resin.

37.12.3. Dissolution of the Concentrate

The initial step in the process is the total dissolution of the concentrate. This is achieved by pressure leaching of the concentrate using chlorine gas in a hydrochloric acid solution. The operating temperature is between 65°C and 70°C, and the pressure is 0.7 bar. The liquid-to-solid ratio is 4:1. The concentration of hydrochloric acid is 3 M.

A Grade 2 titanium, jacketed vessel with a volume of 2.5 m^3 is used for the dissolution step. A photograph of the vessel is shown in Figure 37.1. A pitched-blade turbine is used to agitate the slurry and to ensure good induction and dispersion of the chlorine gas.

Because there is only a small headspace in the reactor, the chlorine may become depleted as reaction proceeds. As a result, the reactor is flushed with chlorine once for every 2 h to refresh the chlorine in the headspace.

Dissolution occurs over 8 h. The dissolution reaction is exothermic, requiring cooling. However, as the reaction progresses toward completion, the rate of reaction slows, so that the requirement switches from cooling to heating by steam.

The concentration of gold in solution is used to monitor the progress of the reaction, because it dissolves most slowly.

The dissolution operation, shown in Figure 37.11, occurs in two steps. The objective of the first step is to dissolve as much of the PGMs as possible, while leaving the lead and silver as insoluble residue.

The contents of the reactor are de-chlorinated and allowed to cool overnight. This serves to promote the precipitation of lead and silver.

The slurry from the first step is filtered in a filter press. The slurry is filtered and the filtrate is transferred to gold ion exchange.

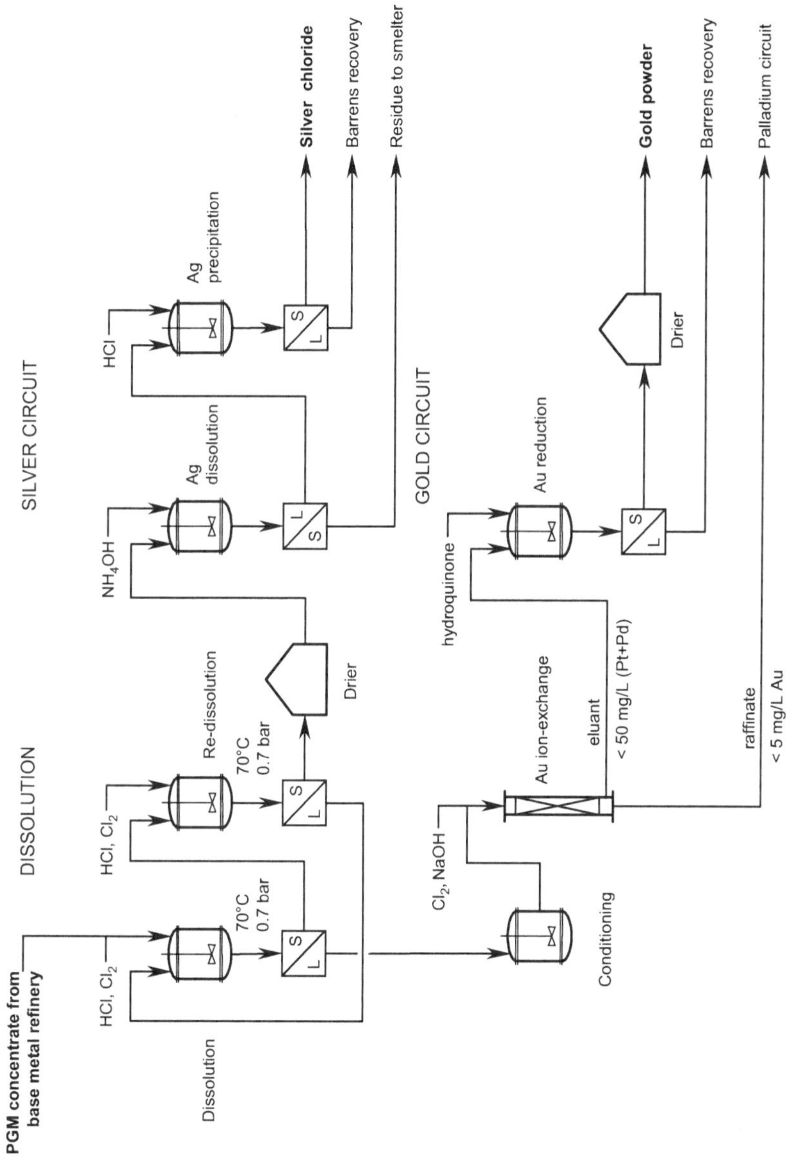

FIGURE 37.11 Total dissolution, and silver and gold circuits at Impala's precious metals refinery at Springs, South Africa.

The filter cake is re-pulped in water and hydrochloric acid for the second dissolution step using chlorine gas as the oxidant.

In the second dissolution step, the remaining OPMs (rhodium, iridium and ruthenium) are leached. The second dissolution step allows for the recovery of entrained PGMs in the first residue and ensures complete dissolution of the more refractory OPMs.

The concentration of hydrochloric acid in the second dissolution step is balanced between two requirements. If the acid is too low, there is the possibility of distilling ruthenium from the dissolution reactor. If the acid is too high, there is the possibility of producing a hexavalent form of osmium, which does not completely distill later in the process, resulting in impure ruthenium.

The slurry from the second step is filtered in a filter press. The filter cake is thoroughly washed to reduce soluble losses and is transferred to silver recovery. The filtrate is recycled to the first dissolution step.

37.12.4. Removal of Silver

The residue from the chlorine leach is mixed with a solution containing ammonium hydroxide. This results in the dissolution of the silver. The slurry is filtered. Hydrochloric acid is added to the filtrate, and the silver precipitates as silver chloride, which is sold. The filtrate is transferred to barrens recovery, which is a central waste recovery process.

The filter cake contains a refractory residue. It consists mainly of lead, iron and chromium spinels and other refractory materials. It is smelted in an electric arc furnace with iron, forming an iron grain. Lead and amphoterics (such as selenium, tellurium and arsenic) mainly report to the slag in the furnace. It is important that lead and the amphoterics do not return to the refinery. The iron grain is atomized and recycled to the base metals refinery or externally toll refined.

37.12.5. Extraction and Production of Gold

The solution from the dissolution step contains all the PGMs and gold. The gold is extracted from this solution using an ion-exchange resin. The objectives of this separation step are to produce a raffinate that contains less than 5 mg/L Au in solution, while at the same time ensuring that the contamination of platinum and palladium in the gold eluate is less than 50 mg/L.

A column packed with the ion-exchange resin is used. A photograph of the columns is shown in Figure 37.3. The resin, which is commercially available, is a polymer bead with a diameter of about 200 μm. The columns are about 2-m high with a diameter of 0.3 m. The columns are arranged in a lead, middle and trail configuration. Once the resin in the column is loaded, it is eluted with a chlorinated alkaline solution.

Precipitated silver and lead particles from the dissolution step may become trapped in the packed column. As a result, the column is eluted in counter-current flow so that any silver and lead particles in the bed are removed with the eluant.

The eluate is collected in glass-lined vessel, where it is boiled down to one tenth of its initial volume. Once boil-down is complete, gold is precipitated from solution by adding a reductant, such as hydroquinone. This results in a gold precipitate with a purity of 99.95% that is easily filtered. The slurry is filtered and the filter cake of gold powder is dried. The gold powder is sold. The filtrate is transferred to barrens treatment.

37.12.6. Extraction and Production of Palladium

Palladium is removed from the gold IX raffinate using molecular-recognition technology (Black, Izatt, Dale, & Bruening, 2006; Izatt, Dale, & Bruening, 2007). The objective of the separation is to qualitatively and quantitatively remove palladium from the solution. In other words, all the palladium must be removed from solution, and nothing else must be removed with the palladium.

The palladium circuit used at the Springs refinery is shown in Figure 37.12.

The specifications are that less than 1 mg/L palladium remains in the raffinate, and that less than 1% of the platinum in the feed stream is co-extracted with the palladium. Palladium tends to deport throughout the process if it is not removed quantitatively, increasing inventories and unit costs. For this reason, the stringent separation specifications are required.

This separation is made in columns that are about 0.5 m in height and 0.3 m in diameter. The columns are arranged in a lead and trail configuration. The columns are shown in Figures. 37.13 and 37.14.

The ion-exchange medium is SuperLig 2, a molecular recognition-type resin. It consists of an active ligand mounted on a high-surface area silica bead. This ion-exchange medium is highly selective for palladium at the chosen conditions and the kinetics of extraction are fast, hence the small column size.

The extraction is sensitive to temperature, redox potential and acidity. As a result, the solution is carefully conditioned prior to extraction. This conditioning ensures that the concentration of the palladium, the acidity and the redox potential are correct. The redox potential range is important because if it is too low, platinum may co-extract. If the redox potential is too high, Ir^{4+} forms in solution. Ir^{4+} reacts with the ligand attached to the silica substrate, causing it to detach, resulting in the degradation of the ion-exchange medium.

The first-pass yield is very high. In the unlikely event of a batch failing, however, recycling is easily accomplished.

Palladium loaded on the medium in the column is eluted with ammonium bisulfite. Ammonium hydroxide is added to the eluate. HCl is then added to the solution, and palladium is precipitated as the diamino palladous dichloride, $(NH_3)_2PdCl_2$. This precipitation removes 98% of the palladium from solution. The remaining palladium in solution is recycled.

The palladium salt is ignited to produce pure palladium sponge. The raffinate is transferred to solution boil down, which is discribed in the next section.

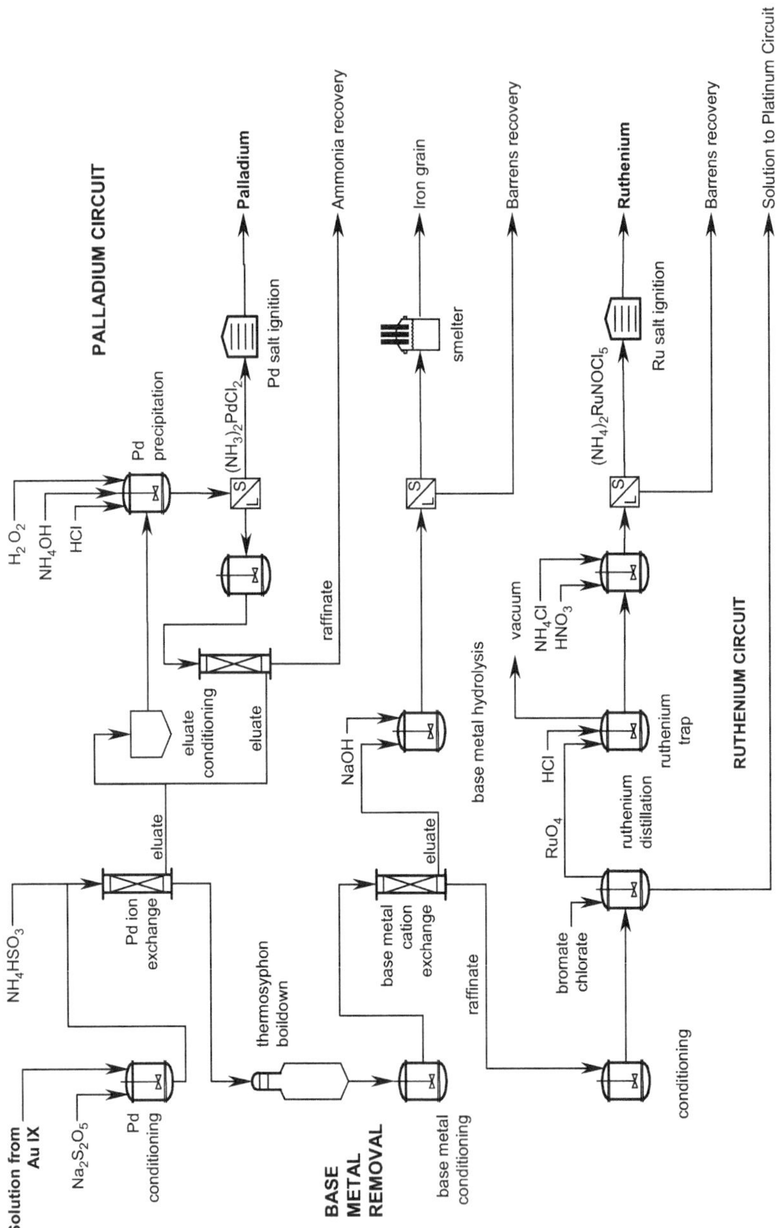

FIGURE 37.12 The palladium, base metal removal and ruthenium circuits at Impala's precious metals refinery at Springs, South Africa.

FIGURE 37.13 The ion-exchange columns used for extracting palladium from solution containing, platinum, palladium, rhodium, iridium, ruthenium, osmium and base metals, such as copper, iron and nickel. The columns are loaded with SuperLig 2, which is highly selective for palladium. The columns are small in size because the kinetics of extraction are fast. *Photograph courtesy of Impala Platinum.*

FIGURE 37.14 A close-up of the ion-exchange columns, showing an extraction front in the column on the right-hand side. This front indicates palladium that has loaded onto the SuperLig 2. The column on left-hand side is fully loaded. *Photograph courtesy of Impala Platinum.*

37.12.7. Solution Boil Down

The loading on the ion-exchange media is, in general, relatively low. As a result, the volumes of eluate are relatively high and the concentrations of PGMs in solution are relatively low. Boiling down the solution is a common method of concentrating these solutions.

The raffinate from the palladium extraction is boiled down in a thermosiphon evaporator. The steam economy is 0.4 parts steam to 1 part solution. Both water and acid are recovered from the evaporator. The remaining solution is transferred to base metal removal.

37.12.8. Removal of Base Metals

Prior to ruthenium distillation, approximately 95% of the base metals are removed by cation exchange.

The base metals are eluted from the column and precipitated from the eluate as hydroxides, which are smelted with iron to form an iron grain. The iron grain is sent to the base metals refinery.

37.12.9. Distillation of Ruthenium

After base metal removal, ruthenium is removed from solution by the distillation of ruthenium tetroxide, RuO_4. The objective of this step is to remove ruthenium so that its concentration in solution is less than 100 mg/L. The ruthenium circuit is shown in Figure 37.12.

The distillation of ruthenium occurs in 1.6-m^3 glass-lined vessels. The size of the batches for ruthenium distillation is approximately 30 kg. The reactor is equipped with a Rushton turbine, made of glass, to ensure good mass transfer. The system is under vacuum (–0.3 bar), and air is drawn through the vessel.

Sodium chlorate and sodium bromate are added to oxidize the ruthenium in solution to ruthenium tetroxide, RuO_4, which is volatile. RuO_4 is stripped from the solution into the gas phase. The gas moves through a glass heat exchanger, where it is cooled slightly. Care is taken to prevent the formation of liquid ruthenium tetroxide, RuO_4, or ruthenium dioxide, RuO_2.

The gas moves from the heat exchangers to two hydrochloric acid traps, which are smaller glass-lined vessels filled with dilute hydrochloric acid. Ruthenium tetroxide is reduced in this first trap. Another trap containing sodium hydroxide captures the chlorine and bromine gases.

Nitric acid and ammonium chloride are added to the solution from the traps containing the ruthenium. This results in the volatilization and capture of the small amounts of osmium tetroxide that may be present and the precipitation of a ruthenium nitrosyl salt, $(NH_4)_2RuNOCl_5$.

The steps involved in the distillation of ruthenium and osmium are shown in Figure 37.15.

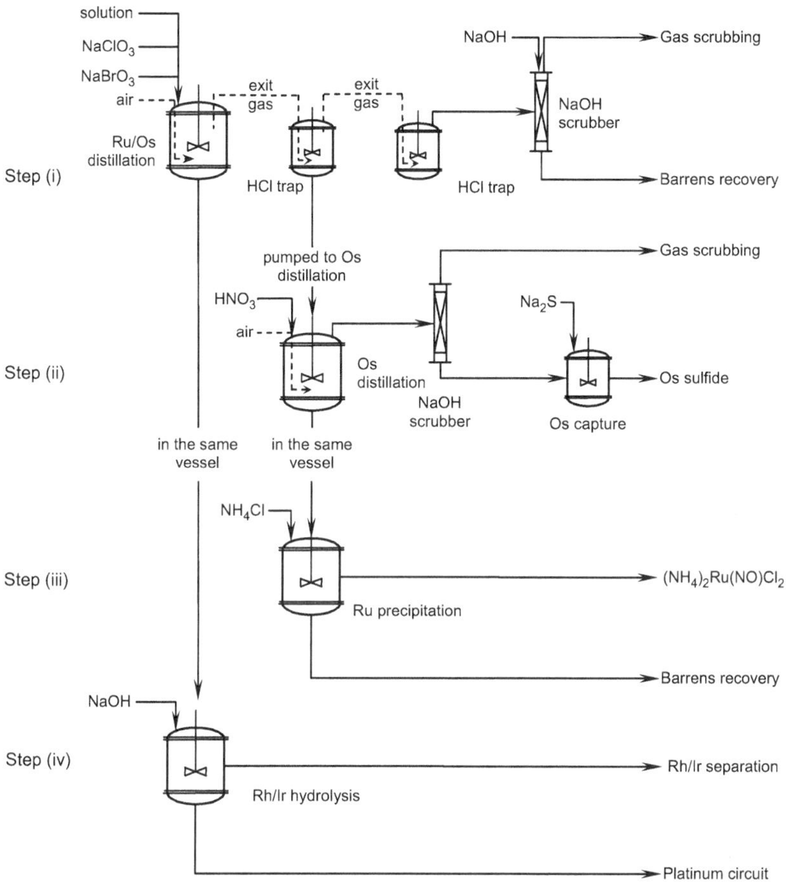

FIGURE 37.15 Distillation of ruthenium and osmium. Step (i): Ruthenium and osmium are oxidized to RuO_4 and OsO_4, and withdrawn from the reactor by a stream of air. They are both reduced in the first trap by hydrochloric acid. Step (ii): Once distillation is complete, the solution in the trap is pumped osmium distillation. Nitric acid is used to oxidize osmium to OsO_4. The osmium is scrubbed from gas using sodium hydroxide and collected in a trap. At the end of osmium distillation, sodium sulfide is added to this trap to precipitate osmium sulfide. Step (iii): After osmium distillation, ammonium chloride is added to the solution to precipitate $(NH_4)_2RuNOCl_5$. Step (iv): Sodium hydroxide is added to the distillation vessel to precipitate rhodium and iridium hydroxides. The slurry is filtered. The cake goes to the rhodium–iridium circuit, and the solution goes to the platinum circuit.

The ruthenium nitrosyl salt is ignited to produce pure ruthenium sponge. Osmium is precipitated from the hydroxide trap as a sulfide.

The solution remaining in the still after ruthenium distillation is complete contains platinum, rhodium, and iridium. The next step is the removal of the rhodium and iridium.

37.12.10. Removal of Rhodium and Iridium

The objective of this step in the process is to remove rhodium and iridium from solution so that the combined concentration of rhodium and iridium remaining in solution is less than 40 mg/L. The circuit for this part of the process is shown in Figure 37.16.

Rhodium and iridium are precipitated from solution as $Rh(OH)_4$ and $Ir(OH)_4$ by hydrolysis using sodium bicarbonate at a pH greater than 5.

The slurry is filtered in a membrane filter press. The filtrate contains mainly platinum, with minor amounts of ruthenium, iridium and rhodium. The filter cake contains less than 1% platinum.

The filter cake is transferred to the rhodium and iridium circuit, while the hydrolysis solution is transferred to the platinum circuit.

37.12.11. Production of Platinum

Platinum is recovered from the hydrolysis solution, which contains about 40 mg/L rhodium and iridium, some nickel, and about 30% of the tellurium, present in the original feed to the refinery. Rhodium and iridium are further removed from the solution by repeating the hydrolysis step using sodium bicarbonate, sodium hydroxide and sodium bromate.

The solution is boiled to dryness in a glass-lined vessel. The dry contents in the vessel are redissolved with demineralized water and the pH of the solution is raised to a value of 8 at a high redox potential. The resultant slurry is filtered. Ammonium chloride is added to the solution, causing ammonium hexachloroplatinate to precipitate. This salt is centrifuged to about 1% moisture in order to remove all the mother liquor containing unprecipitated tellurium and other impurities and to assist in the mechanical conveying of the salt.

The salt is ignited to produce platinum sponge with a purity of greater than 99.95%.

37.12.12. Production of Rhodium and Iridium

Rhodium and iridium are produced from the hydrolysis cake. The cake is dissolved in a glass-lined vessel using hydrochloric acid. Residual platinum that is present in the cake is removed by adding sodium metabisulfite and ammonium chloride. These solids are filtered from the solution and ignited to produce an impure platinum sponge that is retreated.

The filtrate from the platinum removal step is treated with a long-chain organic amine to precipitate an organic salt of rhodium and iridium, while the base metals remain in solution. The organic precipitate of rhodium and iridium is redissolved using aqua regia. The solution is conditioned by adding chlorine gas. This conditioning ensures that the iridium is present as Ir^{4+}. The

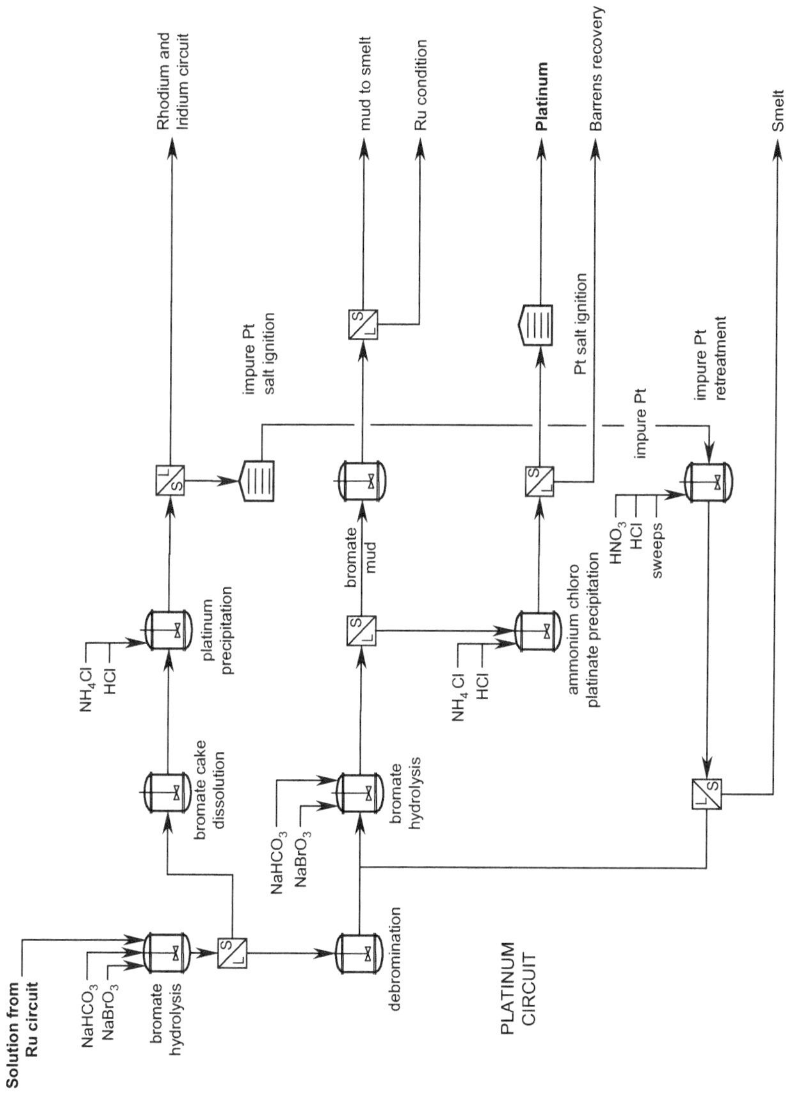

FIGURE 37.16 The rhodium and iridium removal circuit and the platinum circuit at Impala's precious metals refinery at Springs, South Africa.

conditioned solution is passed through an ion-exchange column, where Ir^{4+} loads selectively. The loaded iridium is eluted from the column and precipitated as ammonium chloroiridate, $(NH_4)_2IrCl_6$. This iridium salt is purified to remove residual amounts of impurities, particularly rhodium, and ignited to produce iridium sponge.

The raffinate from the ion-exchange column is precipitated using an organic amine, filtered, and ignited to produce rhodium sponge.

37.13. APPRAISAL OF THE ION-EXCHANGE PROCESSES

The advantages of the ion-exchange process are as follows:

(a) the selectivity of the ion-exchange steps is much higher than the selectivity of the precipitation methods, especially for the most important separation, which is the removal of palladium from the main platinum stream;
(b) the first-pass yields are significantly higher than the precipitation processes;
(c) the kinetics of ion-exchange are fast, leading to a relatively small refinery; and,
(d) the handling of materials, and hence exposure by refinery workers to allergenic platinum and rhodium chemicals, is greatly reduced.

The disadvantages of the process are as follows:

(a) the solutions are generally more dilute than those of the precipitation process, which means that there are a large number of steps devoted to evaporation; and,
(b) the consumption of energy, primarily as a result of evaporation, is high.

37.14. SUMMARY

The feeds to the precious metals refineries are highly concentrated, containing between 15% and 70% in PGMs.

There are three broad categories of process for the refining of PGMs based on the technique used for the separation of platinum and palladium:

(a) precipitation processes, used by Lonmin Platinum and Krastsvetmet;
(b) solvent-extraction processes, used by Johnson Matthey, Anglo American Platinum, Vale, and Heraeus; and,
(c) ion-exchange process, used by Impala Platinum.

This classification is based on the method of separating platinum and palladium, the two most abundant of the PGMs.

While each of these processes is different, they are all based on the same broad principles. The first of these is that the separation of any metal has three areas of activity: primary separation, secondary purification and reduction to metal.

REFERENCES

Asamoah-Bekoe, Y. (1998). *Investigation of the leaching of the platinum-group metal concentrate in hydrochloric acid by chlorine.* (MSc thesis). Johannesburg: University of the Witwatersrand.

Barnes, J. E., & Edwards, J. D. (1982). Solvent extraction at Inco's Acton precious metal refinery. *Chemistry and Industry, 5*, 151–155.

Black, W. H., Izatt, S. R., Dale, J. B., & Bruening, R. L. (2006). The application of molecular recognition technology (MRT) in the palladium refining process at Impala and other selected commercial applications. In *IPMI Conference 2006*. International Precious Metals Institute.

Cramer, L. A. (2008). What is your PGM concentrate worth? In M. Rogers (Ed.), *Proceedings of Platinum in Transformation, Third International Platinum Conference* (pp. 387–394) SAIMM.

Izatt, S. R., Dale, J. B., & Bruening, R. L. (2007). The application of molecular recognition technology (MRT) to refining platinum and ruthenium. In *IPMI conference 2007*. International Precious Metals Institute.

Niezborala, M., & Garnier, R. (1996). Allergy to complex platinum salts: A historical prospective cohort study. *Occupational Environmental Medicine, 53*(4), 252–257.

SUGGESTED READING

Al-Bazi, S. J., & Chow, A. (1984). Platinum metals – Solution chemistry and separation methods. *Talanta, 31*, 815–836.

Anderson, C. G., Newman, L. L., & Roset, G. K. Platinum group metal bullion production and refining. In A. L. Mular, D. N. Halbe & D. J. Barratt (Eds.), *Mineral processing plant design, practice, and control: proceedings, Vol. 2*, (pp. 1760–1777). SME.

Barnes, J. E., & Edwards, J. D. (1982). Solvent extraction at Inco's Acton precious metal refinery. *Chemistry and Industry, 5*, 151–155.

Benner, L. S., Suzuki, T., Meguro, K. & Tanaka, S. (Eds.), (1991). *Precious metals science and technology* (pp. 375–398). International Precious Metals Institute.

Bernadis, F. L., Grant, R. A., & Sherrington, D. C. (2005). A review of methods of separation of the platinum-group metals through their chloro-complexes. *Reactive and Functional Polymers, 65*, 205–217.

Brits, J. H., & Deglon, D. A. (2007). Palladium stripping rates in PGM refining. *Hydrometallurgy, 89*, 253–259.

Charlesworth, P. (1981). Separating the platinum group metals by liquid–liquid extraction. *Platinum Metals Review, 25*(3), 106–112.

Charlesworth, P. (2002). Developments in platinum-group metals refining and the utilization of solvent extraction technologies. In K. C. Sole, P. M. Cole, J. S. Preston & D. J. Robinson (Eds.), *Proceedings of the international solvent extraction conference ISEC 2002* (p. 14). SAIMM.

Cleare, M. J., Charlesworth, P., & Bryson, D. J. (1979). Solvent extraction in platinum group metal processing. *Journal Chemical Technology Biotechnology, 29*, 210–224.

Cleare, M. J., Grant, R. A., & Charlesworth, P. (1981). *Separation of the platinum group metals by use of selective extraction techniques*. In *Extractive metallurgy, 81*. Institute of Mining and Metallurgy. 34–41.

Cramer, L. A. (2001). The extractive metallurgy of South Africa's platinum ores. *JOM, 53*(10), 14–18.

Demopoulos, G. P. (1986). Solvent extraction in precious metal refining. *JOM, 38*, 13–17.

Demopoulos, G. P., Chang, Y., Benguerel, E., & Riddle, M. (2002). Opportunities and challenges in developing a solvent extraction process for rhodium based on the use of stannous chloride. In K. C. Sole, P. M. Cole, J. S. Preston & D. J. Robinson (Eds.), *Proceedings of the international solvent extraction conference ISEC 2002* (pp. 908–915). SAIMM.

Demopoulos, G. P., Pouskouleli, G., & Ritcey, G. M. (1986). *A novel solvent extraction system for the refining of precious metals. International Solvent Extraction Conference, ISEC, 86 (11–16 September 1986)*. Dechema.

Dhara, S. C. (1984). Solvent extraction of precious metals with organic amines. In V. Kudryk, D. A. Corrigan & W. W. Liang (Eds.), *Precious metals: mining, extraction and processing* (p. 199). TMS.

Du Toit, Z. (2006). *Simulation of a palladium extraction circuit*. (MSc Thesis). South Africa: University of Stellenbosch.

Edwards, R. I., & Te Riele, W. A. M. (1983). Commercial processes for precious metals. In T. C. Lo, M. H. I. Baird & C. Hanson (Eds.), *Handbook of solvent extraction* (pp. 725–732). John Wiley & Sons.

Grant, R. A. (1989). The separation chemistry of rhodium and iridium. In *Precious metals recovery and refining*. (p. 7). International Precious Metals Institute.

Grant, R.A., Burnham, R.F., & Collard, S. (1990). The high efficiency separation of iridium from rhodium by solvent extraction using mono n-substituted amide. In *International solvent extraction conference ISEC '90 (18–21 July 1990)*, Japan (pp. 961–966).

Grant, R.A., & Drake, V. A. (2002). The application of solvent extraction to the refining of gold. In *Proceedings of the international solvent extraction conference, ISEC 2002*, SAIMM, Johannesburg (pp. 940–945). 18–21 July 1990.

Grant R.A., & Murrer, B. A. (28 January 1987). Solvent extraction process for separation of precious metals. European Patent Application 0210004 (to Matthey Rustenburg Refiners).

Grant, R. A., & Woollam, S. F. (2008). Development of a computer simulation of the platinum solvent extraction at Precious Metals Refiners (PMR). In *International solvent extraction conference, ISEC 2008* (pp. 415–420). CIM.

Harris, G. B. (1993). *A review of precious metals refining. Precious metals*. (pp. 351–374). International Precious Metals Institute.

Lea, R.K., Edwards, J.D., & Colton, D. F. (June 28, 1983). Process for the extraction of precious metals from solutions thereof. U.S. Patent No. 4,390,366.

Mooiman, M. (1993). The solvent extraction of precious metals - A review. *Precious Metals, 17*, 411–434.

Muir, D. (1982). Solvent extraction of precious metals. *Chemistry in Australia* 348–350.

Muir, D. (1990). Solvent extraction of precious metals. In *AMMTEC/Henkel solvent extraction symposium*. AMMTECH/Henkel.

Reavill, L. R. P., & Charlesworth, P. (1980). The application of solvent extraction to platinum group metals refining. *International Solvent Extraction Conference, ISEC, 80*, 80–93.

Rimmer, B. F. (1974). Refining of gold from precious metal concentrates by liquid–liquid extraction. *Chemistry and Industry, 2*, 63–66.

Rimmer, B. F. (1989). Refining of platinum group metals by solvent extraction. In B. Harris (Ed.), *Precious Metals 1989* (pp. 217–226). International Precious Metals Institute.

Sole, K. C. (2008). Solvent extraction in the hydrometallurgical processing and purification of metals. Process design and selected applications. In M. Aguilar & J. L. Cortina (Eds.), *Solvent extraction and liquid membranes: Fundamentals and applications to new materials* (pp. 141–198). CRC Press.

Sole, K. C., Feather, A. M., & Cole, P. M. (2005). Solvent extraction in southern Africa: An update of recent hydrometallurgical developments. *Hydrometallurgy, 78*, 52–78.

Woollam, S. F., & Grant, R. A. (2008). The degradation of the ketoxime LIX-84-I and its impact on the solvent extraction of palladium. In *International solvent extraction conference, ISEC 2008* (pp. 281–286). CIM.

Part IV

Recycling Nickel, Cobalt and Platinum-Group Metals

Chapter 38

Recycling of Nickel, Cobalt and Platinum-Group Metals

About 40% of the global consumption of nickel, 20% of the global consumption of cobalt and 25% of the global consumption of platinum-group metals were made by recycling end-of-use consumer products (Jollie, 2010; Kuck, 2010; Shedd, 2010). Examples of recycled end-of-use products are kitchen sinks made of nickel-bearing (austenitic) stainless steel, batteries that contain cobalt and catalysts for the reduction of automobile emissions, which contain palladium, platinum and rhodium. Additional production of metal is obtained by recycling manufacturing wastes, such as casting wastes, cutting wastes and flawed fabrications.

Recycling is advantageous in every respect. Some of the advantages of recycling are the following:

(a) recycling slows depletion of the natural resources;
(b) recycling uses up to 90% less energy than metal-from-ore production, that is, it avoids mining, concentration and (possibly) smelting (Vanbellen & Cintinne, 2007);
(c) recycling avoids mine waste products; and,
(d) recycling slows the amount of landfill space required.

This chapter discusses the recycling of nickel, cobalt and platinum-group elements. The objectives of the chapter are to describe technologies used for the recycling nickel-, cobalt- and platinum-group metals.

The following three examples of metal recycling are examined:

(a) recycling of platinum-, palladium- and rhodium-bearing catalysts for the reduction of automobile emissions;
(b) recycling of nickel-bearing stainless steel; and,
(c) recycling of cobalt-bearing batteries.

38.1. RECYCLING OF PLATINUM-GROUP METALS FROM AUTOMOBILE CATALYST

About 30% of platinum, 50% of palladium and 85% of rhodium are used in the catalytic converters of cars and trucks. They catalyze the production of carbon

Extractive Metallurgy of Nickel, Cobalt and Platinum-Group Metals. DOI: 10.1016/B978-0-08-096809-4.10038-3

FIGURE 38.1 Ceramic car/truck catalyst substrate blocks. They are mostly made of cordierite, $Mg_2Al_4Si_5O_{18}$, which has a low coefficient of thermal expansion (Kendall, 2004). Their pores are coated with nanoparticles of platinum, palladium and rhodium. The platinum-, palladium- and rhodium-coated substrates are recycled for the recovery of these metals after their host vehicles have been scrapped (Figures 1.6 and 38.2).

dioxide and nitrogen from noxious hydrocarbons, carbon monoxide and NO_x in engine exhaust gases (refer to Chapter 31 for a description of autocatalysts).

The converters mostly consist of an outer can, made of ferritic stainless steel, and an inner ceramic block, which has a high surface area. The surfaces of the internal pores are coated with a mixture of alumina and fine particles of platinum-group metals (Figures 31.1 and 38.1). The converters usually last the lifetime of a car and are recycled when the car is scrapped.

Car manufacturers produce huge quantities of the same car with the same catalytic converter, for example, each 1995 Toyota Corolla has the same catalytic converter. Each converter contains the same amount of platinum, palladium and rhodium, so each has the same value to the recycler. This allows the recycler to pay cash immediately for every intact 1995 Toyota Corolla converter without slow and costly chemical analyses by the buyer and seller. Of course, periodic checks of catalyst mass and chemical composition are made.

There are two main routes for the recycling of automobile catalysts. The first of these is to return the spent catalyst to the smelter of a primary production of platinum-group metals, such as Impala Platinum. The second route is to build a custom smelter for the recovery of the platinum-group metals.

38.1.1. Recycling to a Primary Producer of Platinum-Group Metals

The most logical option for recycling is to send end-of-use material back to the smelter of the primary producer, such as Impala Platinum. This process entails the following steps:

(a) collecting end-of-use automobiles;
(b) dismantling them into major components, including batteries and catalytic converters;
(c) decanning the converters by shearing the cans crossways into two halves;
(d) collecting and crushing the ceramic block, which contains between 0.1% and 0.3% PGMs; and,
(e) charging the crushed ceramic to a primary PGM concentrate smelting furnace.

After smelting, the platinum-group elements that originate from the automobile catalyst follow the platinum-group metal flowsheet, which consists of smelting, converting, leaching and precious metals refining. A schematic diagram of the flowsheet is shown in Figures 1.6 and 32.1.

An example of this type of business is the venture between A-1 Specialized Services & Supplies and Impala Platinum. A-1 is the largest lot consolidator and dry processor of salvage automotive catalytic converters in the world. A-1 sources feed material from hundreds of suppliers, ranging from cottage-type collectors of recyclable automotive parts to vehicle manufacturers. The material collected and consolidated by A-1 is shipped to Impala Platinum in Rustenburg, South Africa, which then processes the platinum-group metals to final product. This business is probably the single biggest activity in the recycling of automobile catalysts.

Decanned ceramic substrate catalyst is a perfect feed for the smelting furnaces of the primary producers of platinum-group metals. The ceramic substrate joins the gangue minerals from the concentrate feed to form slag while the platinum-group elements coalesce with descending molten sulfide droplets to join the matte phase below the slag (see Chapter 35).

The catalyst is a sulfur-free source of platinum-group elements, which minimizes the production of sulfur dioxide per tonne of production of platinum-group metals, which is usually beneficial.

The disadvantage of smelting catalyst in the smelters of primary producers is that the smelters are mostly in South Africa while the major sources of scrap are on other continents. Another disadvantage might be the stringent requirements for detailed metal accounting, particularly in light of the internal auditing controls required by the accounting authorities.

The disadvantage of location has led to the building of purpose-built recycle smelters around the world.

38.1.2. Recycling to Secondary Autocatalyst Smelting

Several smelters for processing spent autocatalysts have been built around the world. Some of these refiners are given in Table 38.1.

TABLE 38.1 PGM Recycle Smelters and Refineries Around the World

Smelters	Description
Hoboken, Belgium (Umicore)	Base and precious metals smelter
Hanau, Germany (Heraeus)	Precious metals smelter
Kosaka, Japan (Dowa Mining)	Catalyst recycle smelter
Niihama, Japan (Sumitomo)	Copper smelter and refinery, nickel-cobalt refinery, precious metals refinery
Rustenburg, South Africa (Impala Platinum)	Primary PGM smelter with catalyst recycle
Anniston, USA (Multimetco)	Catalyst recycle smelter
Stillwater, USA (Norilsk)	Primary PGM smelter with catalyst recycle
Williston, USA (Sabin)	Catalyst recycle smelter
Refineries (produce PGMs and/or chemicals)	
Sao Paulo, Brazil (Umicore)	Recycle refinery
Hoboken, Belgium (Umicore)	Feed is from Hoboken smelter
Royston, England (Johnson Matthey)	Primary and recycle refinery
Acton, England (Vale Limited)	Primary and recycle refinery
Hanau, Germany (Heraeus)	Feed is from Hanau smelter
Hong Kong (Heraeus)	Recycle refinery
Badia Al Pino, Italy (Chimet)	Recycle refinery
Rome, Italy (BASF)	Recycle refinery
Chikusei, Ibaraki, Japan (Furuya)	Recycle refinery
Niihama, Japan (Sumitomo)	Feed is from Niihama base metal plants
Sodegaura, Chiba, Japan (Tanaka KK)	Recycle refinery
Springs, South Africa (Impala Platinum)	Feed is from Impala Rustenburg smelter
Balerna, Switzerland (Valcambi)	Recycle refinery
Castel San Pietro, Switzerland (PAMP)	Recycle refinery
Mendrisio, Switzerland (Argo-Hereaus)	Recycle refinery
Tainan, Taiwan (Solar Applied Technology)	Recycle refinery
Los Angeles, USA (Heraeus)	Recycle refinery
Newark, USA (Heraeus)	Recycle refinery

TABLE 38.1 PGM Recycle Smelters and Refineries Around the World—cont'd

Smelters	Description
Rochester, USA (Sabin)	Feed is from Williston smelter
Seneca, USA (BASF)	Recycle refinery
Wayne, USA (Johnson Matthey)	Recycle refinery
Neuchâtel, Switzerland (Metalor)	Recycle refinery

The process that is used to recover the PGMs entails the following steps:

(a) continuously feeding free-flowing, crushed, decanned catalyst plus particulate collector iron, silicon, phosphorous, nickel alloy, lime and carbon into a plasma arc furnace at about 1700°C;
(b) continuously tapping slag from the furnace;
(c) intermittently tapping ferroplatinum, containing 10–20% PGMs, through a separate taphole; and,
(d) collecting offgas and capturing the contained dust in a baghouse.

Clearly, this is in addition to the steps (a)–(d) presented in the previous section.

A flowsheet for a secondary catalyst smelter is shown in Figure 38.2. The slag is cooled, crushed and remelted in a slag-cleaning furnace to recover entrained ferroplatinum. The ferroplatinum from both furnaces is crushed and sent to a hydrometallurgical precious metals refinery, such as those discussed in Chapter 37.

The captured dust is combined with other dusts from around the smelter and also sent to a hydrometallurgical precious metals refinery.

The recoveries of platinum, palladium and rhodium from catalytic converter ferroplatinum are estimated to be about 96%. The purity of the final product is obviously the same as that of mined platinum-group metals, because the refining of the ferroplatinum is performed in the same hydrometallurgical refinery.

A small amount of automobile catalyst consists of a washcoat of alumina and platinum-group metal slurry on a honeycomb of steel foil. These converters are crushed with heavy chains and smelted with the ceramic catalyst by the same process as that shown in Figure 38.2.

38.1.3. Recycling of Other Types of Platinum-Group Metal Catalyst

Ceramic- and carbon-supported platinum-group metal catalysts are used extensively in the petroleum refining and petrochemical industries. These

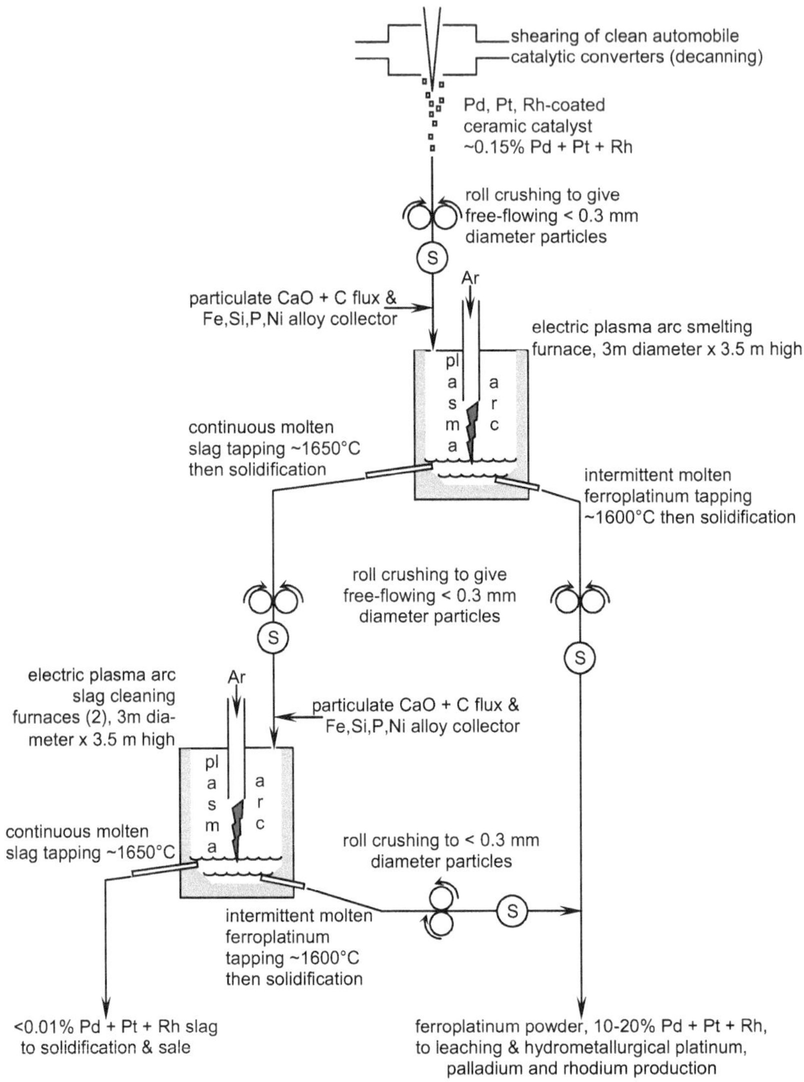

FIGURE 38.2 Secondary catalytic converter recycle smelter. The plasma-arc electric furnaces are notable. Other smelters use carbon electrode furnaces and fuel-fired furnaces. A key feature is the collector alloy that melts and collects platinum-group metals as it descends to the ferroplatinum layer dust is collected in baghouses throughout the smelter (not shown). It is sent to a hydrometallurgical plant for metals recovery. (S) = sampling point for chemical analysis. Accurate sampling and chemical analyses are essential for controlled, profitable recycling.

catalysts are often contaminated with hydrocarbons, coke and water, which must be removed before smelting.

This removal is usually done in rotating kilns adequately equipped with after-burners and dust collection systems. This type of recycling is described by Jacobsen (2002, 2007, 2010).

Platinum-group metals may also be recycled from the catalysts used in the chemical industry by hydrometallurgical techniques (Izatt & Mansur, 2006). One such process entails the following steps:

(a) acid leaching of PGMs and rhenium from the recycle materials fed to the plant;
(b) solid/liquid separations; and,
(c) recovery of the individual metals from the pregnant solution by molecular recognition technology (Izatt, Izatt, Dale, & Bruening, 2010) and precipitation.

Success of the process depends on efficient leaching of metals from the recycle plant inputs. This may restrict its application to materials in which the surface is accessible.

38.2. RECYCLING OF NICKEL IN STAINLESS STEEL

Nearly 90% of all production of nickel ends up in ferrous alloys, as shown in Table 2.1. At the end of their useful lives, objects made from these alloys are recycled. The flowsheet of the recycling process is shown in Figure 38.3. As can be seen from this flowsheet, the scrap is mostly remelted to make new alloy alike in composition to the recycled object. The process is entirely pyrometallurgical.

38.2.1. Recycle Materials in Stainless Steel Production

As shown in Figure 38.3, new stainless steel is made from the following ingredients:

(a) stainless steel scrap;
(b) primary iron, ferronickel and ferrochrome; and,
(c) high-purity recycle steel.

The proportions of these materials are chosen to produce the target stainless steel composition and quantity at the lowest possible cost (that is, with 'the lowest cost mix').

38.2.2. Types of Scrap

The main categories of stainless steel scrap are internal, fabricator and reclaimed end-of-use scrap.

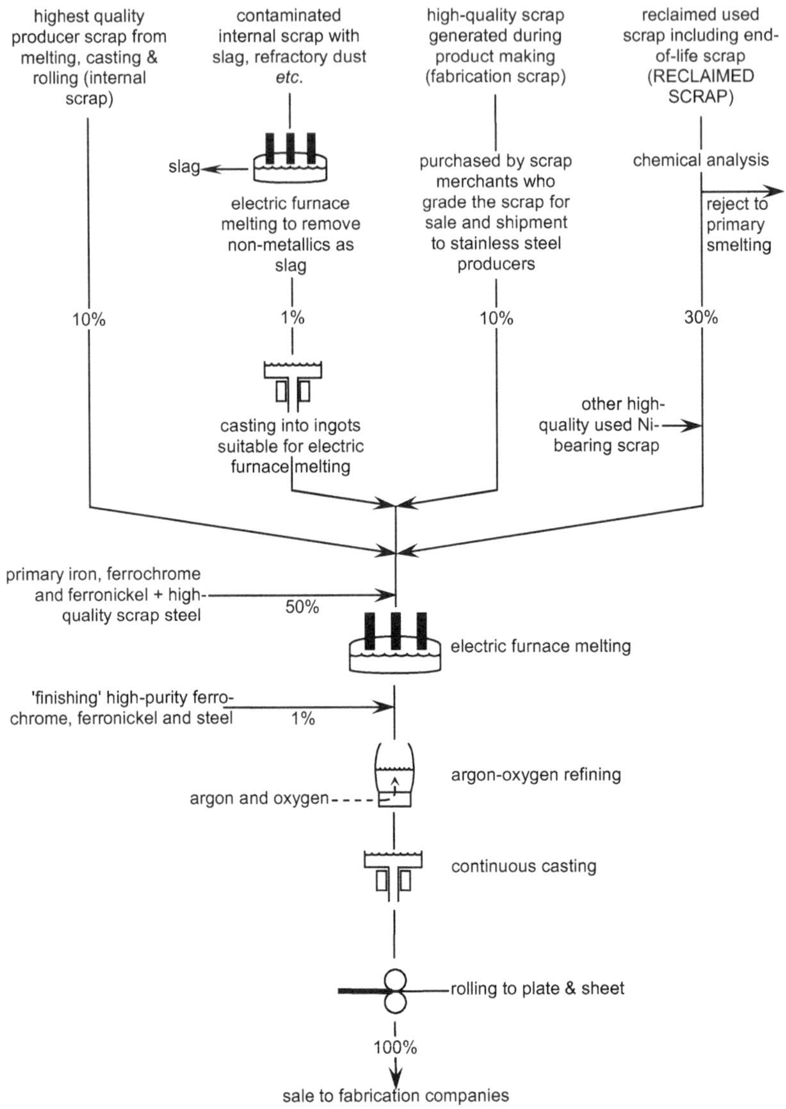

FIGURE 38.3 Stainless steel recycle flowsheet. The numerical values vary considerably. In some cases, most of the feed is end-of-use scrap.

The most easily recycled scrap is internal scrap from the producer. This is because the producer knows the degree to which the scrap has been contaminated and knows the best place to introduce it into the smelting flowsheet. For example, if it is very pure, it might be introduced just before argon-oxygen refining.

38.2.3. Chemical Analysis

Categorization of scrap stainless steel composition has been greatly simplified by the development of hand-held analyzers that utilize energy-dispersive X-ray fluorescence (Niton, 2010). These analyzers allow individual pieces of scrap to be quickly analyzed and sorted.

38.2.4. The Stainless Steel Production Process

The main process in the flowsheet shown in Figure 38.3 is the melting of the stainless steel scrap and other ingredients. The melting is always done with electric furnaces equipped with carbon electrodes. In this case, there is too little slag to generate the heat by slag resistance. As a result, most of the heat is generated by electric arcs in the gas phase between the melting scrap and the electrodes.

The process sequence consists essentially of the following steps:

(a) charging the furnace with cold scrap;
(b) bringing the electrodes down near the scrap surface (power on) and drawing arcs from electrodes to the scrap; and,
(c) raising the applied power, melting the charge and decanting the slag off.

The molten stainless steel is then transferred to an argon-oxygen decarburizing furnace. The operation in this furnace removes carbon down to the required level of less than 0.03% C, which is often needed for stainless steel. Other impurities are also removed.

The argon and oxygen are usually introduced into the melt through submerged tuyeres. The reaction is given as follows:

$$\underset{\substack{\text{in molten stainless}\\ \text{steel from scrap}\\ \text{and ferroalloys}}}{2C(\ell)} + \underset{\substack{\text{in oxygen/argon}\\ \text{gas mixture}}}{O_2(g)} \xrightarrow{1600^\circ C} \underset{\substack{\text{dilute in}\\ \text{CO/argon}\\ \text{gas mixture}}}{2CO(g)} \quad (38.1)$$

The role of the argon is to dilute the carbon monoxide, CO, in the product gas, thereby favoring (by Le Chatelier's principle) the oxidation of carbon over the oxidation of iron, chromium and nickel. Vacuum is also used to the same effect. Alloying ingredients are added toward the end of decarburization or subsequently in ladles.

The final step in stainless steel refining is continuous casting of the molten product into long rectangular or circular shapes. These are subsequently cut into pieces, rolled and sent to manufacture.

Their quality is equal to that of virgin stainless steel because stainless steel scrap contains few contaminants (such as copper) that cannot be removed during melting and oxidation.

38.3. RECYCLING OF COBALT

The two most important applications of cobalt are in alloys and in rechargeable batteries (see Table 24.1). Objects made from the alloys are recycled much as described above for stainless steel, but on a much smaller scale. Many of the alloys that contain cobalt are used in critical applications, for example, in aircraft engines, so that extremely good-quality control measures must be applied.

About 30% of cobalt consumption is in rechargeable batteries. Cobalt is used in all three of the most common rechargeable batteries, that is, in nickel-cadmium, in nickel-metal hydride and in lithium-ion batteries (Cobalt Development Institute, 2010).

At the end of their lives, the batteries are smelted in (i) primary nickel smelters; or, (ii) purpose-built recycle (secondary) smelters.

38.3.1. Recycling of Cobalt to Primary Smelters

Primary smelters of nickel concentrates can treat a large fraction of end-of-use battery (and other cobalt) scrap, because the overall cobalt recycle is only about 20 000 tonnes/year. However, the two requirements that must be met are as follows:

(a) the hydrocarbon components of the scrap must be burnt in an oxidizing atmosphere at high temperatures to avoid the formation of dioxin, furan and other volatile organic compounds; and,
(b) the operation must have a reducing furnace to minimize loss of cobalt to slag.

The first requirement is accomplished by burning off hydrocarbons before the scrap is charged to the smelting furnace. The second requirement is accomplished by including coke or coal in the smelting furnace charge or by having a separate slag-cleaning furnace. Electric furnaces are particularly good for slag cleaning.

The cobalt that is charged into the smelting furnace reports to the matte phase. The recycled cobalt then follows the same path as the primary cobalt, through refining and on to metal or chemical products. These operations were discussed in Chapters 23–27.

38.3.2. Secondary Smelting

End-of-use rechargeable batteries are also recycled through purpose-built secondary smelters. Typical compositions of rechargeable batteries are given in Table 38.2.

A flowsheet of one such process (Meekers, Hageluken, & Van Damme, 2009; Tygat & Van Damme, 2009) is given in Figure 38.4. This process entails the following steps:

(a) charging as-received rechargeable batteries and alloy scrap to a melting furnace;
(b) melting the charge under reducing conditions;

TABLE 38.2 Compositions of Two Recyclable Rechargeable Batteries*

	Li-ion Battery	Li-polymer Battery
Component	Composition, %	
Steel or Fe	35[a]	1
Plastic	10–20	3
Ni	1–2	2
Cu	7–8	16
$LiCoO_2$	25–27	35
Carbon	10	15
Al	35[a]	15
Others	12–18	23

[a]*Depending on use of steel or aluminum cans.*
**Meekers et al., 2009*

(c) tapping alloy and slag from the furnace;
(d) sending the slag to cement manufacture or iron blast furnace fluxing;
(e) granulating the product alloy with water to a size of ~100 μm;
(f) leaching the resulting alloy particles in sulfuric acid using oxygen;
(g) removing copper, iron, manganese and zinc from the resulting aqueous solution by solvent extraction and precipitation;
(h) separating the purified pregnant solution into aqueous cobalt and nickel streams, usually by solvent extraction;
(i) crystallizing nickel sulfate, $NiSO_4$, from the nickel stream, then producing battery-grade nickel hydroxide, $Ni(OH)_2$, from the precipitated nickel sulfate; and,
(j) producing battery-grade cobalt oxide, Co_3O_4, from the cobalt stream.

The recoveries of cobalt are 60–85%, comparable to the primary smelter recycling. Product quality is equal to primary product quality.

38.4. SUMMARY

Nickel, cobalt and platinum-group metals are recycled extensively. Examples discussed are the recycling of nickel-containing stainless steel, cobalt-containing batteries and platinum-, palladium- and rhodium-containing catalytic converters for automobile.

The first process in the recycle plant is most often smelting and converting, which makes a final product alloy, for example, stainless steel, or concentrates

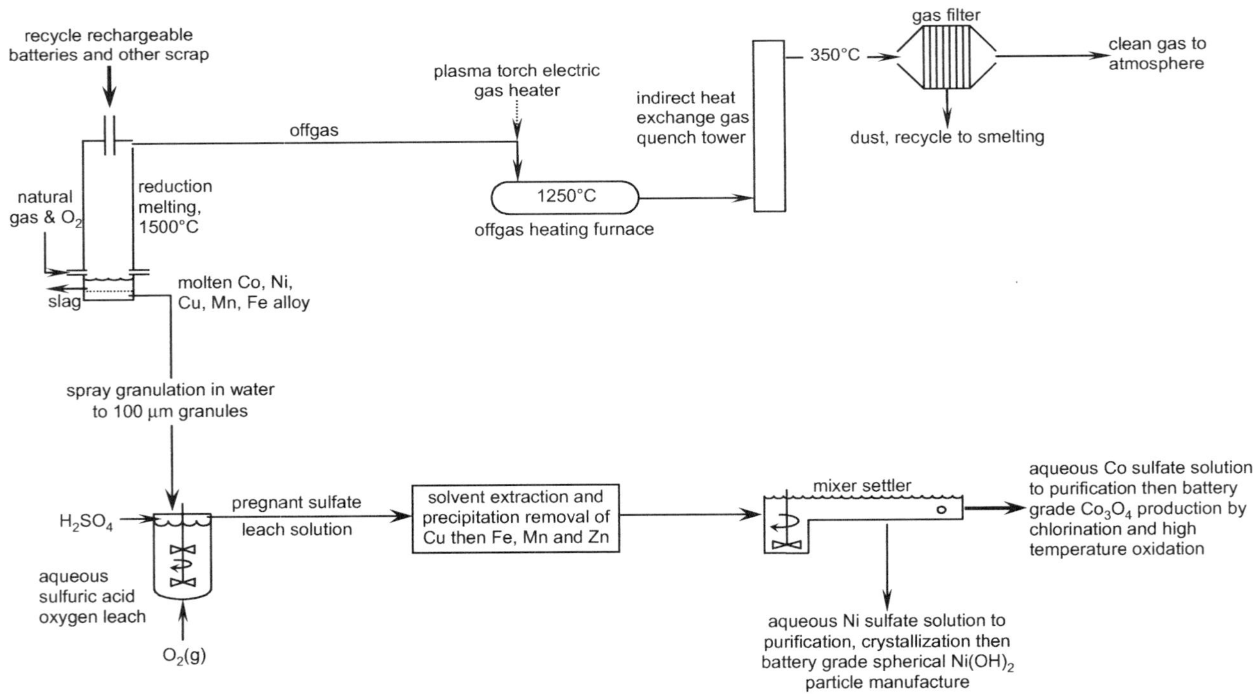

FIGURE 38.4 Secondary cobalt-nickel battery recycle smelting flowsheet. The major steps are alloy production, leaching, solution purification and production of cobalt and nickel chemicals. The melting furnace offgas is simultaneously heated to 1250°C to decompose the hydrocarbons then quenched to 350°C to prevent the formation of dioxins, furans and other volatile organic compounds.

the recycle metal in sulfide matte or metal alloy ready for hydrometallurgical refining.

The purity of the refined recycle metals or chemicals is equal to that of the primary products (that is, products from the ore).

REFERENCES

Cobalt Development Institute. (2010). Rechargeable batteries.

Izatt, S. R., Izatt, N. E., Dale, J. B., & Bruening, R. L. (2010). MRT applications in copper refining: bismuth removal from copper electrolyte. In *International Conference, Copper 2010, Vol. 5, Hydrometallurgy, 1941–1956*. GDMB.

Izatt, S. R., & Mansur, D. M. (2006). Environmentally friendly recovery of precious metals from spent catalysts. In *International Precious Metals Institute Petroleum Seminar* (pp. 1–8). IMPI.

Jacobsen, R. T. (2002). *The Sabin arc smelter at Williston, North Dakota*. Scottsville, New York: Sabin Metal Corporation.

Jacobsen, R. T. (2007). *A brief overview of the secondary precious metals industry*. Scottsville, New York: Sabin Metal Corporation.

Jacobsen, R. T. (2010). *The Sabin metal refinery at Scottsville, New York, U.S.A*. Scottsville, New York: Sabin Metal Corporation.

Jollie, D. (2010). *Platinum 2010*. Johnston Matthey.

Kendall, T. (2004). 30 years in the development of autocatalysts. In *Platinum 2004*. Johnson Matthey.

Kuck, P. H. (2010). *Nickel*. Washington, DC: United States Geological Survey.

Meekers, C. E. M., Hageluken, C., & Van Damme, G. (2009). Green recycling of EEE: special and precious metal recovery from EEE [Electronic and Electric Equipment]. In S. Howard (Ed.), *EPD Congress 2009* (pp. 1131–1136). TMS.

Niton. (2010). *Scrap metal sorting in the recycling industry*. Thermo Fisher Scientific Inc.

Shedd, K. G. (2010). *Cobalt*. Washington, DC: United States Geological Survey.

Tygat, J., & Van Damme, G. (2009). *End of life management and recycling of rechargeable lithium ion, lithium polymer and nickel metal hydride batteries*. Umicore.

Vanbellen, F., & Chintinne, M. (2007). "Extreme makeover": UPMR's Hoboken plant. In *2007 European Metallurgical Conference* (pp. 1–9). GDMB.

SUGGESTED READING

Hageluken, C. (2007). Closing the loop – recycling of automotive catalysts. *Metall, 61*, 24–39.

Hageluken, C. (2009). *Recycling of precious metals – current status, challenges, developments*. In *Proceedings of EMC 2009, Vol. 2*. GDMB. 473–486.

Meekers, C. E. M., Hageluken, C., & Van Damme, G. (2009). Green recycling of EEE: special and precious metal recovery from EEE [Electronic and Electric Equipment]. In S. Howard (Ed.), *EPD Congress 2009* (pp. 1131–1136). TMS.

Nickel Institute. (2010). Nickel Institute and recycling.

Phoenix Automotive Cores. (2010). Spent auto catalyst recycling at Multimetco. Lecture given at short course: Recycling Metals from Industrial Waste (organizer Queneau, P).

Appendix A

Ferronickel Smelting of Non-Tropical Laterite Ores

About 40 000 tonnes of nickel contained in ferronickel are produced from ancient (Golightly, 1979) eastern European non-tropical laterite ores (Bergman, 2003; Warner *et al.*, 2006). The Ni in these ancient ores occurs mainly in chlorite minerals, for example, $(Mg,Fe,Al,Ni)_3(Si,Al)_2O_5(OH)_4$ (Rigopoulos, Kostika, Zevgolis, & Halikia, 2004).

These ores are mined by surface mining methods. As received by smelters, they typically contain 1–2% Ni (Warner *et al.*, 2006). They are much drier than tropical laterites. They typically contain approximately 15% moisture as compared to about 30% moisture in tropical ores. They do not require drying before calcination and smelting (Zevgolis, 2004; Zevgolis & Livanou, 2005).

Data for smelting these ores are provided by Bergman (2003) and Warner *et al.*, 2006).

REFERENCES

Bergman, R. A. (2003). Nickel production from low-iron laterite ores: process descriptions. *CIM Bulletin, 96*, 127–138.

Golightly, J. P. (1979). Nickeliferous laterites: a general description. In D. J. I. Evans, R. S. Shoemaker & H. Veltman (Eds.), *International Laterite Symposium* (pp. 3–23). Littleton, CO: American Institute of Mining, Metallurgical and Petroleum Engineers.

Rigopoulos, E., Kostika, I.-P., Zevgolis, E. N., & Halikia, I. (2004). Contribution to the recycling of rotary kiln dust in nickeliferous laterite reduction, reductive behavior of the dust. In W. P. Imrie, D. M. Lane & S. C. C. Barnett, et al. (Eds.), *International Laterite Nickel Symposium – 2004* (pp. 479–492). TMS.

Warner, A. E. M., Diaz, C. M., Dalvi, A. D., et al. (2006). JOM world Nonferrous Smelter Survey, Part III: laterite. *Journal of Metals, 58*, 11–20.

Zevgolis, E. N. (2004). The evolution of the Greek ferronickel production process. In W. P. Imrie, D. M. Lane & S. C. C. Barnett, et al. (Eds.), *International Laterite Nickel Symposium – 2004* (pp. 619–632). TMS.

Zevgolis, E. N., & Livanou, S. C. (2005). Nickel recovery from metallurgical slags. In J. Donald & R. Schonewille (Eds.), *Nickel and cobalt 2005 challenges in extraction and production* (pp. 273–284). CIM.

SUGGESTED READING

Larco. (2010). Available from. http://www.larco.gr/smelting_plant.php.

Appendix B

Caron Process for Processing Nickel Laterites

The Caron Process treats mostly limonite-type laterite ore (~1.4% Ni, 0.1% Co) to produce nickel oxide sinter (~90% Ni, 10% O) and other products. It produces ~70 000 tonnes of nickel per year (Table B.1). It also produces ~5000 tonnes of cobalt in various intermediate products (Table B.1).

The process is energy intensive (Rodriquez, 2004). It also has low nickel recoveries of the order of 75%. Further adoption of the process is unlikely.

The remainder of this appendix outlines the main steps of the process.

B.1. CARON PROCESS STEPS

See Boldt and Queneau (1967), Chen, Jak, and Hayes (2009), Prado (2004), Reid and Fittock (2004), Rodriguez (2004), Figures B.1 and B.2.

1. Ore drying/grinding

Objectives: Remove 95% of mechanically entrained water from the ore and grind the dried product to ~75 μm, so that it flows steadily and evenly through the following hearth roasters (Figure B.1).

Equipment: Cocurrent rotary kilns using combustion gas at 1000°C to vaporize the water.

2. Ore reduction roast (Figures B.1 and B.2)

Objective: Reduce nickel and cobalt minerals to metallic Ni–Co–[low Fe] alloy in preparation for subsequent dissolution of the Ni and Co by aqueous NH_3, CO_2, air leaching.

Equipment: Counter-current hearth roasters (760°C) using oil/air combustion gas for reduction. Molar CO/CO_2 and H_2/H_2O ratios in reducing gas are ~1 to reduce Ni and Co in laterite to alloy – and its Fe to magnetite. NiOOH reduces more easily than FeOOH (thermodynamically).

Representative Reactions:

$$\underset{\text{in goethite}}{2NiOOH(s)} + \underset{\text{in reducing gas}}{3H_2(g)} \xrightarrow{760^\circ C} \underset{\text{in Ni, Co alloy}}{2Ni(s)} + \underset{\text{in roaster offgas}}{4H_2O(g)} \quad \text{(B.1)}$$

$$\underset{\text{in goethite}}{2CoOOH(s)} + \underset{\text{in reducing gas}}{2CO(g) + H_2(g)} \xrightarrow{760^\circ C} \underset{\text{in Ni, Co alloy}}{2Co(s)} + \underset{\text{in roaster offgas}}{2CO_2(g) + 2H_2O(g)} \quad \text{(B.2)}$$

TABLE B.1. Operating Details of Four Caron Reduction Roast-NH_3, CO_2, O_2 Leach Plants. All Treat Mostly Laterite Ore to Produce Nickel Oxide Sinter and Other Products. Total Productions are ~70 000 Tonnes of Nickel and ~5000 Tonnes of Cobalt Per Year

Location	% Ni in ore	% Co in ore	Products	Ni recovery, %	Co recovery, %	Production, tonnes of Ni in sinter/year	Production, tonnes of Co/year
Yabulu, Australia	1.5	0.15	Ni oxide sinter, ~90% Ni, 10% O and Co oxyhydroxide, ~66% Co intermediate	80	50	~12 000	~1000 in Co oxyhydroxide
Niquelandia, Brazil			Ni,Co carbonate is sent to an electrorefinery for nickel and cobalt metal production	60	60	23 000 in carbonate intermediate	1400 in carbonate intermediate
Nicaro, Cuba	1.3	0.1	Ni oxide sinter, ~90% Ni, 10% O and Co/Ni sulfide precipitate 60% Ni, 40% Co plus sulfur	75	40	20 000	~1100 in sulfide precipitate
Punta Gorda, Cuba	1.3	0.1	Ni oxide sinter, ~90% Ni, 10% O and Co/Ni sulfide precipitate 60% Ni, 40% Co plus sulfur	75	40	20 000	~1100 in sulfide precipitate

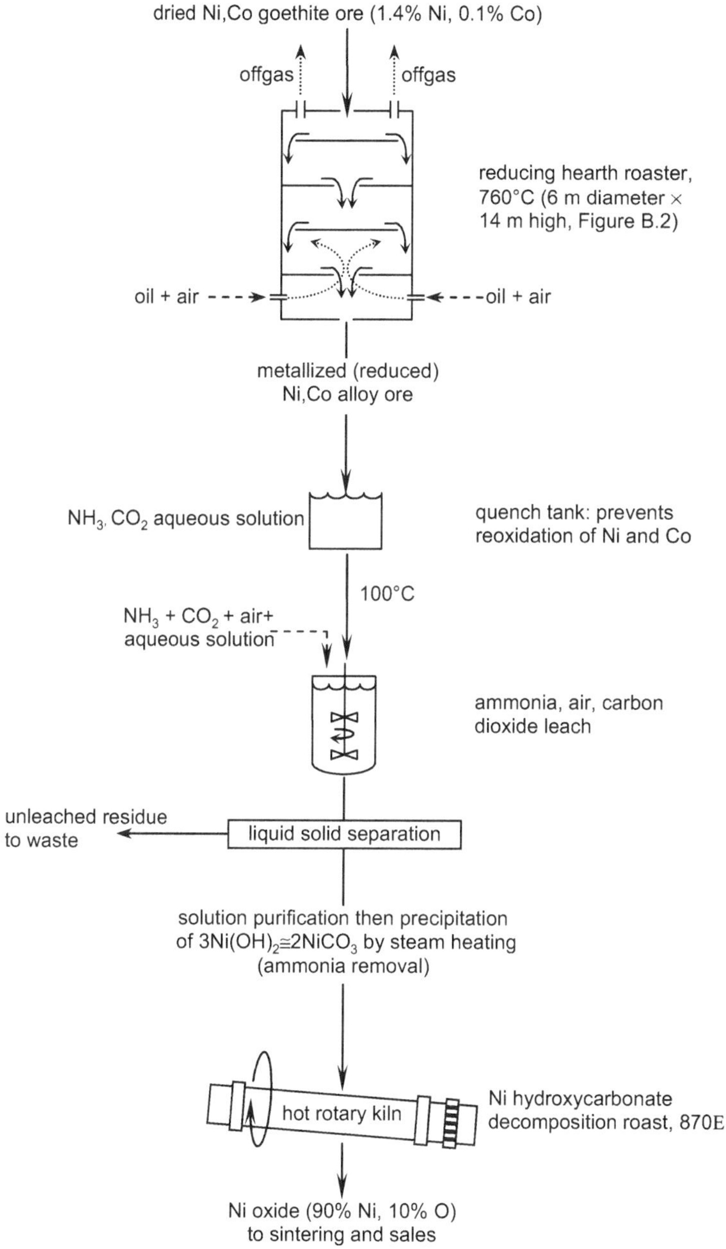

FIGURE B.1 Caron reduction-leach precipitation-process flowsheet for producing Ni oxide sinter (Boldt & Queneau, 1967; Reid & Fittock, 2004). The starting material is laterite ore ~1.4% Ni, 0.1% Co. The product is Ni oxide sinter ~90% Ni, 10% O. Co is also produced as described in Step 11 (above). Figure B.2 shows a cutaway view of a multiple hearth roaster.

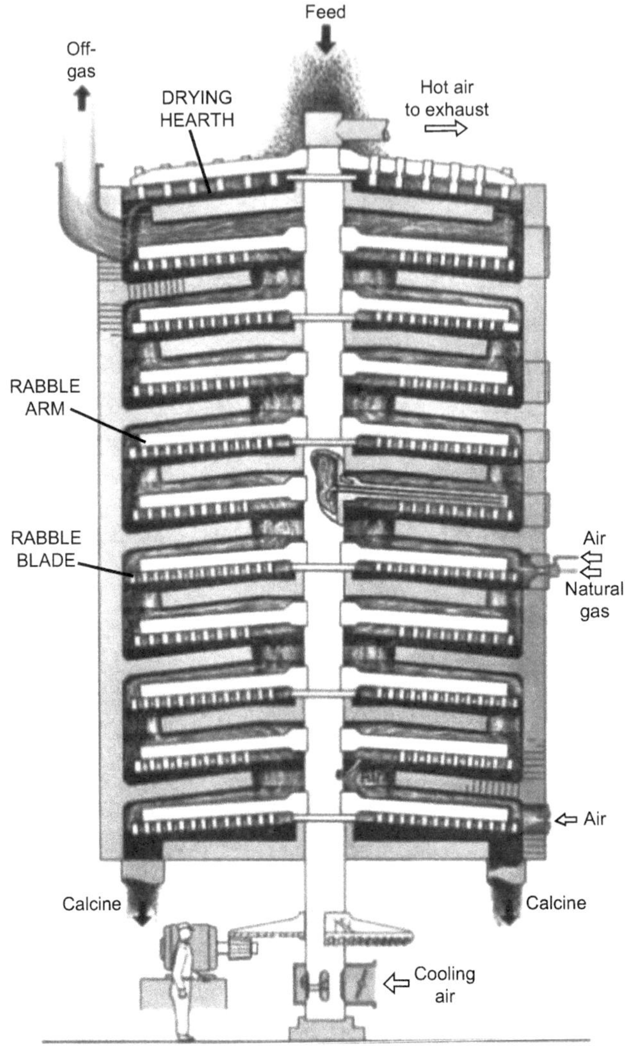

FIGURE B.2 Cutaway view of multiple hearth roaster. Dry goethite ore is fed to the top hearth. It is raked to the outside by an air-cooled rotating rake. There it falls to the second hearth where it is raked to the inside where it falls to the third hearth *etc*. Reducing gas rises counter-currently up the roaster, all the while reducing nickel and cobalt in the laterite to a nickel-copper alloy and FeOOH to magnetite. The reduced product goes to quenching, then leaching. (The inward and outward raking is achieved by different rake angles.) *Drawing from Boldt and Queneau, 1967, courtesy Vale.*

$$\underset{\text{in laterite}}{6FeOOH(s)} + \underset{\text{in reducing gas}}{H_2(g)} \xrightarrow{760^\circ C} \underset{\text{magnetite}}{2Fe_3O_4(s)} + \underset{\text{in roaster offgas}}{4H_2O(g)} \quad (B.3)$$

$$\underset{\text{in laterite}}{6FeOOH(s)} + \underset{\text{in reducing gas}}{CO(g)} \xrightarrow{760^\circ C} \underset{\text{magnetite}}{2Fe_3O_4(s)} + \underset{\text{in roaster offgas}}{CO_2(g) + 3H_2O(g)} \quad (B.4)$$

3. Calcine Quenching in NH_3/CO_2 aqueous solution

Objective: Avoid re-oxidation of nickel and cobalt, begin leaching of Ni–Co alloy.

Equipment: Ore cooling (to 150°C) tubes followed by quench tanks. Product quench slurry is ~100°C.

4. Calcine Leaching

Objective: Dissolve Ni and Co from alloy into aqueous $NH_3 + CO_2 + O_2$ (from air) solution in preparation for Ni carbonate precipitation.

Equipment: Leach and solid/liquid separation tanks.

Representative Reaction:

$$\underset{\text{in Ni-Co alloy}}{Ni(s)} + \underset{\text{bubbled into aqueous solution}}{6NH_3(g) + CO_2(g) + 0.5O_2(g)} \xrightarrow{\sim 100^\circ C} \underset{\text{dissolved in aqueous solution}}{Ni(NH_3)_6^{2+} + CO_3^{2-}} \quad (B.5)$$

5. Solid liquid separation

Objective: Separate unleached solids from Ni, Co rich solution.

Equipment: Thickeners and filters.

6. Solution purification

Objective: Remove impurities (e.g. Co, Cu, Fe, Zn).

Methods: Solvent extraction, sulfide precipitation (Reid & Fittock, 2004).

7. Precipitation of basic Ni carbonate ($3Ni(OH)_2 \cdot 2NiCO_3$)

Objective: Produce high-purity solid ($3Ni(OH)_2 \cdot 2NiCO_3$) product.

Method: Remove ammonia from solution by counter-current steam heating in cylindrical multi-tray stripping towers (Boldt & Queneau, 1967).

Representative Reaction:

$$\underset{\text{aqueous solution}}{5Ni(NH_3)_6^{2+}(aq) + 2CO_3^{2-}(aq) + 6OH^-(aq)} \xrightarrow{\text{steam heating}} \underset{\text{solid basic Ni carbonate}}{(3Ni(OH)_2 \cdot 2NiCO_3)(s)} + \underset{\text{ammonia (recycle)}}{30NH_3(g)} \quad (B.6)$$

8. Decomposition of $3Ni(OH)_2 \cdot 2NiCO_3$

Objective: Produce nickel oxide powder.

Method: Rotating kiln heating in neutral atmosphere at up to 1300°C.

Representative Reaction:

$$\underset{\text{basic Ni carbonate precipitate}}{(3Ni(OH)_2 \cdot 2NiCO_3)(s)} \xrightarrow{900^\circ C} \underset{\text{powder}}{5NiO(s)} + \underset{\text{offgas}}{3H_2O(g) + 2CO_2(g)} \quad (B.7)$$

9. Sintering of NiO powder

Objective: Sinter a mixture of NiO + coal to form Ni-enriched sinter pieces, 1–2 cm diameter ready for steelmaking use.

Method: High-temperature traveling grate sintering machine (Boldt & Queneau, 1967).

10. Product

1–2 cm diameter sinter, 85–90% Ni, remainder mostly oxygen. Nickel recovery is ~75%.

11. Cobalt production

Objective: Precipitate cobalt oxyhydroxide (Reid & Fittock, 2004) and/or cobalt-nickel mixed sulfide (Rodriguez, 2004) from aqueous cobalt impurity stream. Cobalt recovery, 40–50%.

REFERENCES

Boldt, J. R., & Queneau, P. (1967). *The winning of nickel.* Longmans. pp. 425–437.

Chen, J., Jak, E., & Hayes, P. C. (2009). Factors affecting nickel extraction from the reduction roasting of saprolite ore in the Caron process. In J. Liu, J. Peacey & M. Barati, et al. (Eds.), *Pyrometallurgy of nickel and cobalt 2009, Proceedings of the International Symposium* (pp. 449–461). The Metallurgical Society of the Canadian Institute of Mining, Metallurgy and Petroleum.

Prado, F. L. (2004). Sixty years of Caron: current assessment. In W. P. Imrie, D. M. Lane & S. C. C. Barnett, et al. (Eds.), *International laterite nickel symposium – 2004, process development for prospective projects* (pp. 593–598). TMS.

Reid, J. G., & Fittock, J. E. (2004). Yabulu 25 years on. In W. P. Imrie, D. M. Lane & S. C. C. Barnett, et al. (Eds.), *International laterite nickel symposium – 2004, process development for prospective projects* (pp. 599–618). TMS.

Rodriquez, R. I. (2004). Reduction in energy cost in Cuban Caron process plants. In W. P. Imrie, D. M. Lane & S. C. C. Barnett, et al. (Eds.), *International laterite nickel symposium – 2004, process development for prospective projects* (pp. 657–664). TMS.

SUGGESTED READING

Boldt, J. R., & Queneau, P. (1967). *The winning of nickel.* Longmans.

Appendix C

Flash Cooling of Autoclaves

The slurry from sulfuric acid laterite leaching leaves its autoclave at approximately 245°C. The vapor pressure of this slurry is about 35 bar. It is cooled to 100°C by allowing it to boil in three sequential flash tanks. This allows subsequent slurry processing (e.g. neutralization, solid/liquid separation) to be done in open tanks.

Flash cooling consists of injecting the slurry from the autoclave at 245°C into a large flash tank (Figure C.1). The pressure in the gas space of the flash tank is less than the equilibrium vapor pressure of the hot slurry, so the slurry boils and cools itself by the reaction:

$$H_2O(\ell) \rightarrow H_2O(g)$$

which is highly endothermic.

C.1. INDUSTRIAL PROCESS

The industrial process proceeds as follows:

1. Hot slurry is injected through the top pipe, propelled by the difference in pressure between the autoclave and the flash tank;
2. The slurry impinges on the bottom impingement block in a turbulent manner;
3. The slurry boils, causing steam to depart through top steam exit;
4. Unboiled slurry departs through side exit, which is always submerged in slurry – this is ensured by keeping the tank about one third full by adjusting the slurry choke valve on the exit pipe; and,
5. The steam pressure in the tank is controlled by adjusting the rate at which steam departs the tank. This in turn controls:
   (a) the rate at which steam vaporizes;
   (b) the rate at which heat is extracted from the slurry; and,
   (c) flash tank temperature and pressure.

Increasing the rate at which steam leaves the flash tank (by opening the control valve on the steam exit pipe) increases steam departure rate, hence the vaporization rate of steam and hence cooling rate and product temperature – and *vice versa*.

The main control measurements are:

(a) slurry level; and,
(b) flash tank pressure.

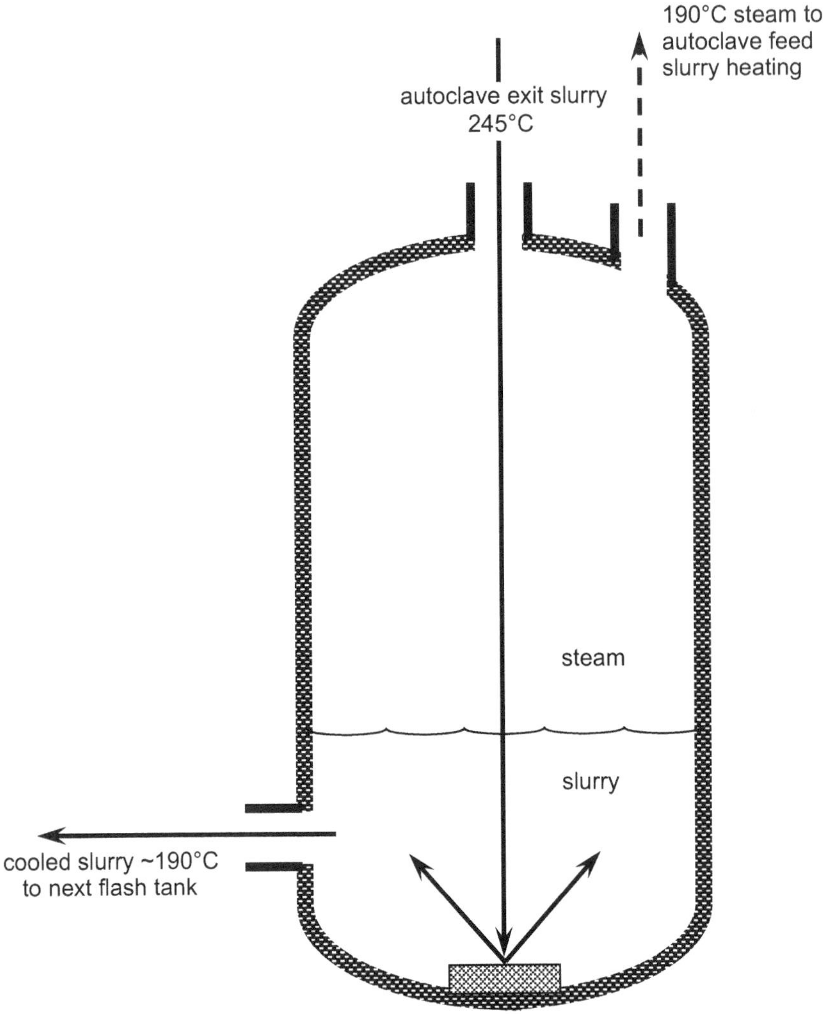

FIGURE C.1 Sketch of flash tank for cooling autoclave exit slurry from 245°C to ~190°C. The tank is steel, lined with acid resistant refractory. It is about 4-m diameter and 8-m tall (inside) and treats ~300 m^3 of slurry per hour. Note the impingement block at the bottom. It stops the incoming slurry from wearing through the tank wall.

Typical pressures and equivalent temperatures in two industrial goethite leach plants are:

	Ravensthorpe, Australia		Coral Bay, Philippines	
	Pressure, bar	Temperature, °C	Pressure, bar	Temperature, °C
Autoclave	41	245	40	245
High-pressure flash tank	13	190	17	200
Medium-pressure flash tank	3.5	140	6	160
Low-pressure flash tank	1	100	1	100

C.2. USE OF THE STEAM PRODUCT

The steam from the flash tanks is used to heat the feed slurry to the autoclave. There are three slurry heaters, each using steam from one of the flash tanks. The first heating is done with the lowest temperature steam. The final highest temperature ($\sim$190°C) is obtained by heating with the highest temperature flash tank steam. This arrangement produces a high-temperature slurry and efficient energy utilization.

The mass of autoclave exit slurry is larger than the mass of autoclave feed slurry – because steam and sulfuric acid are added during leaching (see Chapter 11). This ensures that there is enough steam to heat the feed slurry to 190 or 200°C.

Appendix D

Counter-Current Decantation of Leaching Slurries

All laterite leach plants neutralize the autoclave exit slurry, then wash nickel-rich pregnant leach solution from their unleached and precipitated solids. The objectives of the washing are to:

(a) separate nickel-rich pregnant leach solution from unleached and precipitated solids; and,
(b) maximize nickel recovery to the clarified pregnant solution.

Figure D.1 shows a simplified flowsheet of the process, which:

(a) produces clarified pregnant leach solution containing <0.2 g/L of solution from 45% solids slurry; and,
(b) recovers 99% of leached nickel to the clarified solution.

D.1. PROCESS DESCRIPTION

The process entails counter-current flow of wash water and leach slurry through seven (sometimes six) settlers (see Figure D.1). The cleanest water washes nickel from the reject solids in the last settler (number 6 in Figure D.1). The wash water then proceeds down through the other settlers in Figure D.1, increasing in nickel and remaining low in entrained solids.

The keys to successful operation are:

(a) thorough mixing of wash water and solids before each settler (as shown);
(b) quiescent settling of washed solids in each settler; and,
(c) adequate flocculent addition.

The process is usually called counter-current decantation.

D.2. DESIGN AND CONTROL

Nickel recovery to pregnant solution increases with increasing (Page, 1976):

(a) wash water flowrate/solids flowrate ratio (both kg/h); and,
(b) number of settling stages.

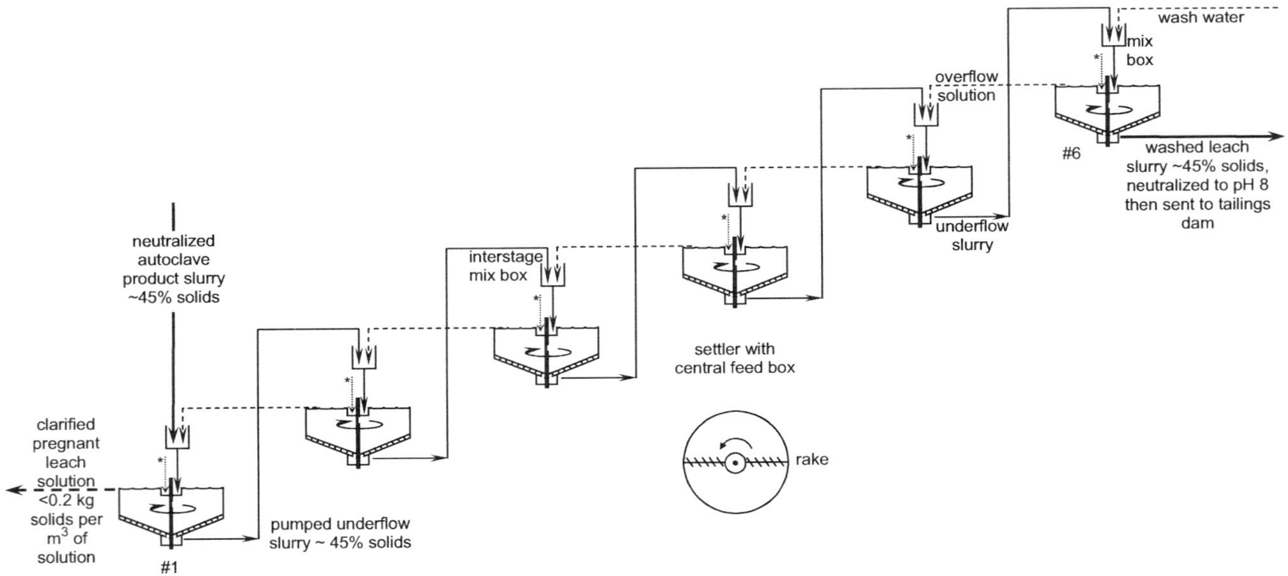

FIGURE D.1 Simplified flowsheet for washing dissolved nickel from the autoclave exit solids. Six settlers are shown. The more normal number for goethite leaching is seven. *Flocculent is added to each autoclave, as shown. A residue wash plant is typically designed to recover 99% of dissolved Ni to clarified pregnant leach solution, left. Settlers are typically 70 m diameter and 10 m high. Wash water mass flowrate/slurry mass flowrate ratio is ~1. Flocculent input rate is up to 150 g of solid flocculent per tonne of residue. Dissolved nickel recovery to pregnant solution increases with number of settlers and wash water/feed slurry ratio.

Of course capital, operating and maintenance costs increase with increasing number of settlers and dilution of nickel in pregnant solution increases with increasing ratio of the flowrate wash water to the flowrate solids.

Seven settlers at a mass ratio of wash water to feed slurry of about 1.0 are optimum for laterite leaching operations.

D.3. FLOCCULENTS

Polymer flocculent (usually polyacrylamide) is always used to accelerate residue settling. The flocculent is dissolved in water, allowed to soak, then diluted once again to about 0.025% in solution before adding it to the slurry. Total solid flocculent addition is up to 150 g/tonne of solid residue.

These procedures cause very fine residue particles (~ 10 μm) to be incorporated in large flocs (up to 300 μm), which rapidly sink.

Point of addition can considerably affect the effectiveness of the flocculent.

REFERENCE

Page, P. M. (1976). A simple equation for CCD calculations. In R. Thomas, (Ed.), *E/MJ operating handbook of mineral processing* (pp. 165–166). McGraw-Hill.

SUGGESTED READING

Klepper, R. P. (2009). High density thickeners in CCD circuits: case study. In J. J. Budac, R. Fraser & I. Mihaylov, et al. (Eds.), *Hydrometallurgy of nickel and cobalt 2009* (pp. 195–207). The Metallurgical Society of the Canadian Institute of Mining, Metallurgy and Petroleum.

Appendix E

Recovering Nickel-, Copper-, Cobalt- and Platinum-Group Elements from Slags

The slags produced during the smelting and converting of nickel-, cobalt- and platinum-group metal (PGM)-bearing concentrates always contain significant quantities of these elements (see Table E.1). These slags are often treated for metal recovery, especially when the sum of the nickel, copper and cobalt contents (%Ni %Cu + %Co) is greater than 0.5% or the PGM content % PGMs > 10 g/tonne.

The dominant recovery techniques are (Table E.2):

(a) reduction and settling of molten slag in purpose-built electric slag-cleaning furnaces (Table E.3); and,
(b) settling of molten converter slag by recycle to the smelter's primary electric smelting furnace.

Settling is also occasionally done in fuel-fired furnaces.

E.1. SETTLING MECHANISMS

Nickel, cobalt and copper occur in molten slags (i) in tiny entrapped matte droplets; and, (ii) as oxides dissolved in the silicate slags themselves. The droplets need to be settled under quiescent conditions. The dissolved elements need to be reduced from the slag, reacted with sulfide additions and settled.

PGMs are dissolved in entrapped matte droplets (rather than as oxides in slag), so they only need to be settled.

E.1.1. Promoting Settling

Efficient settling is promoted by:

(a) reducing conditions, provided by coke or coal;
(b) fully molten low viscosity slag;
(b) adding collector sulfide solids (i.e. concentrates) to the slag surface;
(c) keeping the slag as quiescent as possible; and,
(d) long residence times and avoidance of short circuiting of slag in and out.

TABLE E.1 Treatment Methods for Recovering Nickel and PGMs from Smelting and Converting Slags

Slag description	Nickel concentrations in smelting and converting slags, %	Ni-from-slag removal method	Nickel-in-slag concentrations after nickel-from-slag removal
Laterite smelting			
Ferronickel-from-laterite smelting slag	0.1–0.2	Discard	
Matte-from-laterite smelting slag	0.15	Discard	
Matte-from-laterite converting slag	~0.6	Slag containing more than 0.6% Ni is solidified then recycled through calcination kilns to electric furnace smelting	0.15
Sulfide smelting			
Electric furnace smelting slag	0.1–0.4	Discard	
Flash furnace smelting slag	0.6–1.5	Discard or send to electric settling furnace	0.1–0.6

With electric furnace appendage	0.2	Discard	
Direct- to low-Fe matte smelting slag (Harjavalta, Finland)	4	Send to electric settling furnace	0.1
Converting slag	1–6 (increases as %Ni in matte increases)	Send to electric settling furnace or recycle to smelting furnace	0.1–0.6
PGM smelting	**PGM concentration, g/tonne**		**PGM concentrations after treatment, g/tonne**
Electric furnace smelting slag	~8	Discard or solidification/flotation	~4
Converting slag	~40	(a) Recycle to smelting furnace; (b) send to electric slag settling furnace; or, (c) send to solidification/flotation	4–8

Element-in-slag concentrations before and after treatment are shown.

TABLE E.2 Summary of Slag-cleaning Methods Used in Sulfide Smelting*

Smelter	Type of smelting furnace	Converter slag treatment method	Smelting furnace slag treatment method
BHP Billiton: Kalgoorlie, Australia	Outokumpu flash with electrodes	Recycled molten to smelting furnace	Discard
BCL: Selebi Phikwe, Botswana	Outokumpu flash	Sent molten to 2 electric slag-cleaning furnaces	Sent molten to 2 electric slag-cleaning furnaces
Votorantim: Fortaleza, Brazil	Outokumpu flash DON process	No converting	Sent molten to 1 slag-cleaning furnace
Vale: Sudbury, Canada	Inco flash	Recycled molten to smelting furnaces	Discard
Jinchuan: Gansu, China	Outokumpu flash with electrodes	Sent molten to 2 electric slag-cleaning furnaces	Sent molten to 2 electric slag-cleaning furnaces
Boliden: Harjavalta, Finland	Outokumpu flash DON process	No converting	Sent molten to 1 electric slag-cleaning furnace
Norilsk: Nadezda, Russia	Outokumpu flash	Sent molten to 4 slag-cleaning furnaces	Sent molten to 4 electric slag-cleaning furnaces
Xstrata: Falconbridge, Canada	Electric furnace	Sent molten to fuel-fired slag-cleaning furnace	Discard
Vale: Thompson, Canada	Electric furnace	Recycled molten to smelting furnaces	Discard
Norilsk: Norilsk, Russia	Electric furnace	Sent molten to 1 electric slag-cleaning furnace	Discard
Norilsk, Pechenga Russia	Electric furnace	Recycled molten to smelting furnace	Discard
Anglo Pt: Mortimer, South Africa	Electric furnace	No converting	Discard

TABLE E.2 Summary of Slag-cleaning Methods Used in Sulfide Smelting*—cont'd

Smelter	Type of smelting furnace	Converter slag treatment method	Smelting furnace slag treatment method
Anglo Pt: Polokwane, South Africa	Electric furnace	No converting	Discard
Anglo Pt: Waterval, South Africa	Electric furnace	Granulated then sent to electric slag-cleaning furnace (Figure 30.1)	Granulated then sent to milling/flotation
Impala Pt: Rustenburg, South Africa	Electric furnace	Granulated then sent to milling/flotation	Granulated then sent to milling/flotation
Lonmin Pt: Marikana South Africa	Electric furnace	Granulated then sent to milling/flotation	Discard
Northam Pt: Northam South Africa	Electric furnace	Recycled to smelting furnace	Discard
Norilsk: Stillwater, USA	Electric furnace	Granulated then recycled to smelting furnace	To primary concentrator
Zimplats: Selous Zimbabwe	Electric furnace	Recycled molten to smelting furnace	Discard

**Warner et al., 2007*

The collector solids melt and 'collect' the entrapped matte droplets as they descend to a matte layer beneath the slag. The reductants reduce Ni, Co and Cu from the oxide slag and allow the reduced metals to dissolve into the descending sulfide drops.

E.2. SLAG SOLIDIFICATION, GRINDING AND FLOTATION

Slags bearing platinum-group elements are treated by molten slag settling as described above. They are also treated for PGM recovery by:

(a) water granulation of the slag to coarse beach sand size;
(b) grinding the granules to ~75 μm; and,
(c) isolation of liberated PGM-bearing matte droplets in a sulfide concentrate by froth flotation (see Table E.4).

TABLE E.3 Details of Three Electric Slag-cleaning Furnace Operations*

Smelter	BCL, Selbi Phikwe, Botswana	Harjavalta, Finland	Waterval, Rustenburg, South Africa
Slag feed	Molten flash furnace slag, 2600 tonne/d molten converter slag, 250–400 tonne/d	Molten flash furnace (DON) slag 530 tonnes/ day	Solid granulated continuous converter slag, ~450 tonnes/day
Composition, %Ni	Flash furnace slag, 1.5% Ni converter slag, 5% Ni (estimated)	4% Ni	40 g/tonne PGM (estimated)
Products			
Molten matte	43% Ni to market	50% Ni, to granulation and leach	granulated and recycled to converter
Molten slag	0.36% Ni, to discard	0.11% Ni, to discard	~8 × 10^{-4}% Pt group elements, to granulation then grinding/ flotation (Table E.4)
Electric furnace details			
Number	2	1	1
Shape	Round	Round	Round
Size, m	8 m diameter × 6 m high	9 m diameter × 5 m high	12 m diameter
Electrodes	3 Self-baking	3 Self-baking	3 Self-baking
Diameter, m	0.9	1.3	1.4
Consumption		0.8 kg/tonne of slag	
Electrical			
Maximum power, megawatts	2	7	23
Average operating voltage, V	75	210	200–800

TABLE E.3 Details of Three Electric Slag-cleaning Furnace Operations*—cont'd

Smelter	BCL, Selbi Phikwe, Botswana	Harjavalta, Finland	Waterval, Rustenburg, South Africa
Average electrode current, A	13 000	5000	45 000–60 000
Power consumption, kWh/tonne of slag	41	172	~600
Operating details			
Reductant	Coal, 52% C	Metallurgical coke, 30 kg/tonne of slag	Metallurgical coke (with silica)
Collector solids	Solid flash furnace matte, 2 kg/tonne slag	Concentrate, 2 kg/tonne of slag	Concentrate
Slag residence time, h	2.5	2	
Offgas	To atmosphere	De-dusted in a baghouse then sent to stack	Quenched and scrubbed to remove dust and SO_2

Jacobs, 2006; Warner et al., 2007

TABLE E.4 Details of Waterval Smelter Slag Flotation PGM Recovery Plant* (see Figure 35.1)

Company and location	Anglo American Platinum, Rustenburg, South Africa
Feed	(1) Granulated electric smelting furnace slag (2) Granulated electric slag-cleaning furnace slag
Granule size	Like coarse beach sand
Design feed rate	100 tonnes/h of granulated slag
Grinding	
Equipment	Closed circuit ball mill and cyclone

(*Continued*)

TABLE E.4 Details of Waterval Smelter Slag Flotation PGM Recovery Plant* (see Figure 35.1)—cont'd

Company and location	Anglo American Platinum, Rustenburg, South Africa
Product size	60–75% passing 75 μm diameter
Flotation	
Equipment	Two 30 m^3 cylindrical Outotec mechanical flotation cells and seven 8 m^3 cubic mechanical cells (in series)
Feed	Slag-water slurry, 33% solids
Total slurry retention time in flotation cells	~0.8 h
Reagents	Sodium isobutyl xanthate collector, polyglycol ether frother
Products	
Concentrate (froth)	Combined froth from all flotation cells is thickened and filtered to form concentrate which is re-smelted with primary concentrate
Tailing (underflow)	From last cell, thickened and sent to tailings dam
PGM recovery from slag to concentrate, %	~60% for smelting furnace slag, 80+% for electric slag-cleaning furnace slag

Jacobs, 2006; Warner et al., 2007

Recovery of PGMs by this technique is 60–90%. The process works because the PGMs occur almost entirely dissolved in matte rather than as oxides in slag.

REFERENCES

Jacobs, M. (2006). Process description and abbreviated History of Anglo Platinum's Waterval Smelter. In R. Jones (Ed.), *Southern African Pyrometallurgy 2006 International Conference* (pp. 17–34). South African Institute of Mining and Metallurgy.

Warner, A. E. M., Diaz, C. M., Dalvi, et al. (2007). JOM world nonferrous smelter survey part IV: nickel: sulfide. *JOM, 59*, 58–72.

Appendix F

Electrorefining of High-Purity Nickel from Cast Impure Ni Alloy and Ni Matte Anodes

Electrorefining of cast impure nickel anodes to pure nickel cathodes was once a common way of producing high-purity nickel. It is still used by Norilsk at Norilsk and Pechenga. It is well described by Boldt and Queneau (1967a). It has largely been replaced by electrowinning (Chapter 25).

However, electrorefining remains in the form of Ni matte anode refining in several places around the world, for example:

Thompson, Canada (60 000 tonnes of nickel per year [Oliver & Wrana, 2009])
Jinchang, China (100 000 tonnes of nickel per year).

The chemistry of the process is described by Boldt and Queneau (1967b). This appendix outlines the industrial process, using Thompson, Canada, as the example. The main steps of the process are as follows.

1. Molten matte: 76% Ni, 19% S, 3% Cu is produced by roasting, electric furnace smelting and converting of Thompson and other Ni sulfide concentrates.
2. This molten matte is carefully poured into open horizontal molds to give flat matte anodes 1.1 m long, 0.7 m wide and 0.045 m thick with cast support lugs.
3. The anodes are loaded (long dimension vertically) into electrolytic cells (electrolyte in place) inside lug-supported woven polypropylene bags, 27 anodes per cell.
4. Simultaneously, thin high-purity nickel 'starting sheet' cathodes are loaded into the electrolytic cell, each inside a woven canvas cloth compartment, 26 cathodes per cell. Anodes and cathodes are all hung on copper conductor bars.
5. A ~3 V DC potential is applied between the anodes and cathodes, causing direct current to flow; Ni^{2+} to dissolve from the anodes; and nickel metal to plate on the 'starting sheet' cathodes, that is:

$$\underset{\text{in impure anodes}}{Ni(s)} \xrightarrow{\text{at anode, 50°C}} \underset{\text{in impure chloride anolyte which overflows the cell}}{Ni^{2+}(aq)} + \underset{\text{into external electric circuit}}{2e^-} \quad (F.1)$$

and

$$\underset{\text{in pure chloride catholyte from anolyte purification plant}}{Ni^{2+}(aq)} + \underset{\text{from external electric circuit}}{2e^-} \xrightarrow{\text{at cathode, } 50^\circ C} \underset{\text{plated on high-purity nickel starting sheets}}{Ni(s)} \quad (F.2)$$

6. Anolyte is purified in an electrolyte purification plant before being returned to the cathode bags as pure catholyte. The catholyte is returned to the cathode compartments in such a way that flow is always away from the cathodes into the anolyte, preventing catholyte contamination. Physically, this is indicated by a 'head' of catholyte in the cathode compartments above the anolyte.
7. After 10 days, the enlarged cathodes are pulled and replaced with new starting sheets. After 16 or 17 days, the anodes and their bags are pulled and replaced with new anodes and bags. The cathodes are washed and sent to market. The anode scrap and solid corrosion products (mainly sulfur) inside the anode bags are treated for Ni-, Co-, Cu-, Ag-, Au- and Pt-group metal recovery (Oliver & Wrana, 2009).

More details are given in Tables F.1–F.3; in Boldt and Queneau (1967b); and in Oliver and Wrana (2009).

TABLE F.1 Composition of Nickel Produced by Electrorefining Matte Anodes*

Element	Mass%
Ni + Co	99.99
Co	0.048
Cu	0.008
C	0.002
Fe	0.0004
S	0.0002
As	0.0009
Pb	0.0004
Zn	0.0001

Vale Inco, 2010

TABLE F.2 **Details of Thompson, Canada, Matte-to-pure Nickel Electrorefining Tankhouse***

Item	Value
Date of construction	1961
Capacity	~60 000 tonnes nickel per year
Number of electrolytic cells	684
Cell dimensions	0.9 m deep, 5.8 m long, 1.6 m wide
Construction	Concrete, lined with fiber reinforce plastic
Anodes per cell	27
Size	1.1 m long, 0.7 m wide and 0.045 m thick
Composition	76% Ni, 19% S, 2.5% Cu
Mass at beginning of refining	250 kg
Anode to anode center line spacing	0.2 m
Typical life in cells	16–17 days
Cathodes per cell	26
Facial dimensions	1 m × 0.7 m
Typical plating time	10 days
Mass at beginning of plating	~5 kg
Mass at end of plating	~90 kg[a]
Composition	99.9 + % Ni
Electrical	
Applied potential	3 V, increasing to 6 V as solids build up on anode faces over time
Cathode current density	~250 A/total m^2 of area, both sides
Electrolyte	Compositions are given in Table H.3
Temperature	50°C
Flowrate	0.7 m^3/h/cell

[a]*0.02 kg 'crowns' are also produced on patterned, non-conductive, oven-set epoxy-covered stainless steel cathodes (Oliver & Wrana, 2009).*
**Oliver & Wrana, 2009*

TABLE F.3 Anolyte and Catholyte Compositions[a] in the Thompson, Canada, Matte-to-nickel Refinery*

Element	kg/m^3 of anolyte	kg/m^3 of catholyte
Ni	75–85	75–85
Cu	0.53	0.0003
Co	0.23	0.022
Fe	0.022	0.0001
As	0.03	0.0005
Pb	0.0008	0.0003
Chlorides	50–60	50–60
Boric acid	7–8	7–8
pH	1.5	3

[a]*A proprietary organic compound is added to improve plated nickel morphology and purity.*
**Oliver & Wrana, 2009*

REFERENCES

Boldt, J. R., & Queneau, P. (1967a). *The winning of nickel*. Longmans Canada Ltd. pp. 339–355.

Boldt, J. R., & Queneau, P. (1967b). *The winning of nickel*. Longmans Canada Ltd. pp. 362–369.

Oliver, B. M., & Wrana, D. (2009). Recent and forthcoming operational improvements in Vale Inco's Thompson nickel refinery. In J. J. Budac, R. Fraser & I. Mihaylov, et al. (Eds.), *Hydrometallurgy of nickel and cobalt 2009* (pp. 601–610). CIM.

Vale Inco. (2010). *Vale Inco electrolytic nickel squares [cut cathodes]*. www.vale.com.

Appendix G

Top Blown Rotary Converter

Top blown rotary converters (TBRCs) (Figure G.1) are used to:

(a) melt and oxidize (convert) Fe and S from granulated Pt-element-rich electric furnace Fe–Ni–Cu–S matte – in preparation for matte leaching/refining;
(b) melt, oxidize and reduce Ni alloy, sulfide and oxide solids to produce impure Ni alloy – ready for carbonyl refining to high-purity nickel; and,

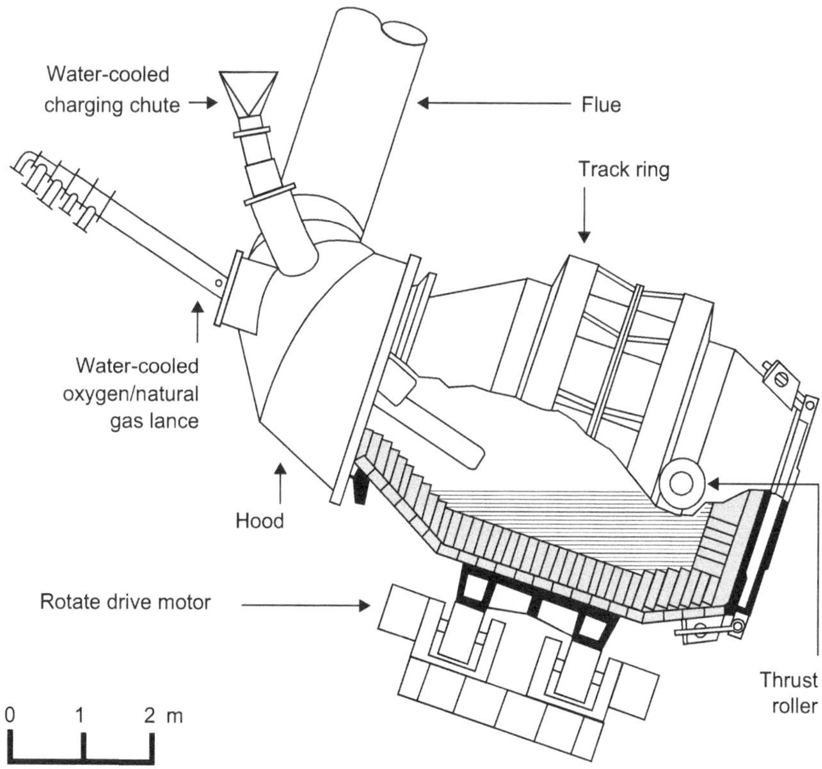

FIGURE G.1 Sketch of top blown rotary converter in Sudbury, Canada (Wiseman, Bale, Chapman, & Martin, 1988). It operates at 1450–1650°C and produces about 50 tonnes of impure nickel metal in 3–4 h. It normally operates 18° from horizontal at ~10 rpm.

(c) extract platinum group metals, gold and silver from refractories into molten metallic lead in preparation for metal production.

G.1. DESCRIPTION (TABLE G.1)

The TBRC:

(a) rotates during melting, oxidation and reduction;
(b) tilts to allow feeding and pouring; and,
(c) blows hydrocarbons, air and oxygen through a stationary (but moveable) lance into the rotating converter.

Hydrocarbon combustion provides heat for the process, air and oxygen oxidize Fe and S; and incompletely combusted hydrocarbon gas removes oxygen from the charge.

G.2. MELTING AND CONVERTING PGM-RICH MATTE AT STILLWATER, MONTANA

The TBRCs at Stillwater:

(a) melt solidified PGM-rich electric furnace matte; and,
(b) oxidize the iron and sulfur in the molten matte to produce molten low-low matte – ready for leaching to product Ni–Cu–Co solution and PGM-rich residue (~60%).

This residue is the principal product of the smelter. It is sent to a precious metals refinery for high-purity metal production.

It does the same job as Peirce-Smith and Ausmelt converters do in South African Pt smelters (see Chapter 35).

Appraisal

Stillwater's electric furnace matte production is too small to support a Peirce-Smith converter. The TBRC allows a batch of granulated matte to be accumulated, then melted and converted.

G.3. PREPARING CARBONYL REFINERY FEED FROM NICKEL METAL, SULFIDE AND OXIDE SOLIDS AT SUDBURY, CANADA

The TBRCs in Sudbury are used to:

(a) melt nickel metal, sulfide and oxide solids;
(b) desulfurize the resulting molten product; and,
(c) reduce the desulfurized product to impure nickel, ready for carbonyl refining.

They inject their requisite hydrocarbon fuel, air and oxygen through a lance.

TABLE G.1 Details of top blown rotary converters used in nickel and platinum group metal production

Company and location	Vale Inco, Sudbury, Canada	Stillwater Mining, Montana	Acton Pt Refinery, London, England
Number of TBRCs	2	2	1
Usage	(a) Melting metal, sulfide and oxide solids; (b) oxidizing then reducing the melted charge	(a) Melting granules of Pt-group-element-rich electric furnace matte; (b) oxidizing Fe and S from the melted matte	(a) Melting custom Pt-group-element-rich feed; (b) adjusting composition of melted feed by oxidation/reduction
Product	Impure Ni alloy	Low Fe, Pt-group element-rich matte	Pt-group-rich alloy
Destination	Carbonyl refining to high-purity nickel	(a) Leached; (b) residue (60% Pt-group elements) is sent Pt-group metal refineries	Aqua regia leach then sequential platinum group metal production
TBRC details			
Dimensions	~3 m diameter × 6 m long inside dimensions (Figure G.1)	8 m diameter × 1.5 m long inside dimensions	0.25 m^3 working volume
Rotational speed	~10 rotations/minute	8–10 rotations/min	2 rotations/minute
Operating temperature	1450–1650°C		
Batch size	40–70 tonnes		
Tap to tap time	3–4 h		

G.4. MELTING PGM SOLIDS IN PLATINUM-GROUP METAL REFINERIES

PGM refinery TBRCs can melt, then adjust the composition of a multiplicity of feed solids much as described in Section G.3. They are used to make leach feeds for subsequent leaching and metal-from-aqueous solution recovery.

G.5. TBRC PROBLEMS

The main problem with top blow rotary converters has been bearing, track rim and drive wheel wear. It is minimized by avoiding:

(a) overly heavy feed batches;
(b) excessive rotational speed; and,
(c) unbalanced converter operation due to localized refractory wear or accretion build-up.

REFERENCE

Wiseman, L. G., Bale, R. A., Chapman, E. T., & Martin, B. (1988). Inco's Copper Cliff nickel refinery. In G. P. Tyroler & C. A. Landolt (Eds.), *Extractive metallurgy of nickel and cobalt* (pp. 373–390). TMS.

SUGGESTED READING

Diaz, C. M., Landolt, C. A., Vahed, A., et al. (1988). A review of nickel pyrometallurgical operations. In G. P. Tyroler & C. A. Landolt (Eds.), *Extractive metallurgy of nickel and cobalt* (pp. 211–239). TMS.

Donald, J. R., & Scholey, K. (2005). An overview of Inco's Copper Cliff operations. In J. Donald & R. Schonewille (Eds.), *Nickel and cobalt 2005, challenges in extraction and production* (pp. 463–464). CIM.

Mroczynski, S. A. (2009). TBRC slag flux control at the Copper Cliff nickel refinery. In J. Liu, J. Peacey & M. Barati, et al. (Eds.), *Pyrometallurgy of nickel and cobalt 2009, Proceedings of the International Symposium* (pp. 293–304). CIM.

Musu, R., & Bell, J. A. E. (1979). P.T. Inco's Indonesian nickel project. In D. J. I. Evans, R. S. Shoemaker & H. Veltman (Eds.), *International laterite symposium* (pp. 300–322). Society of Mining Engineers of the AIME.

Thoburn, W. J., & Tyroler, P. M. (1979). Optimization of TBRC operation and control at Inco's Copper Cliff nickel refinery. In R. E. Johnson (Ed.), *Copper and nickel converters* (pp. 274–290). AIME.

Appendix H

Nickel Carbonylation Free Energies and Equilibrium Constants

Industrial ambient pressure carbonylation, that is:

$$\mathrm{Ni}(s) + 4\mathrm{CO}(g) \rightarrow \mathrm{Ni(CO)_4}(g) \qquad (22.1)$$

approaches equilibrium at: 50°C (323 K)

Carbonyl decomposition (Equation 19.2) approaches equilibrium at: 240°C (513 K)

Equilibrium calculations over this temperature range require thermodynamic data between 300 and 600 K.

This appendix:

(a) provides literature values of the standard free energies of formation for $CO(g)$ and $Ni(CO)_4(g)$;
(b) develops a linear equation that represents the standard free energy of reaction for Reaction (19.1) as a function of temperature; and,
(c) calculates equilibrium constants with (b)'s linear equation.

The free energy data are from Chase (1998). Chase provides standard free energies of formation for the reactions:

$$\mathrm{Ni}(s) + 4\mathrm{C}(s) + 2\mathrm{O_2}(g) \rightarrow \mathrm{Ni(CO)_4}(g) \quad \Delta_\mathrm{f} G^\circ_{\mathrm{Ni(CO)_4}(g)} \qquad (H.1)$$

and

$$\mathrm{C}(s) + {}^1/_2\mathrm{O_2}(g) \rightarrow \mathrm{CO}(g) \quad \Delta_\mathrm{f} G^\circ_{\mathrm{CO}(g)} \qquad (H.2)$$

$\Delta G^\circ_{\mathrm{Ni}(s)\,+\,4\mathrm{CO}(g)\rightarrow\mathrm{Ni(CO)_4}(g)}$ is determined by subtracting 4 × Equation (H.2) from Equation (H.1), which gives:

$$\mathrm{Ni}(s) \rightarrow \mathrm{Ni(CO)_4}(g) - 4\mathrm{CO}(g)$$

or:

$$\mathrm{Ni}(s) + 4\mathrm{CO}(g) \rightarrow \mathrm{Ni(CO)_4}(g)$$

for which:

$$\Delta G^\circ_{\mathrm{Ni}(s)\,+\,4\mathrm{CO}(g)\,\rightarrow\,\mathrm{Ni(CO)_4}(g)} = \Delta_\mathrm{f} G^\circ_{\mathrm{Ni(CO)_4}(g)} - 4 \times \Delta_\mathrm{f} G^\circ_{\mathrm{CO}(g)} \qquad (H.3)$$

Chase's 300–600 K $\Delta_f G^\circ$ values (MJ/kmol) for Equations (H.1)–(H.3) are:

T, K	$\Delta_f G^\circ_{Ni(CO)_4(g)}$	$\Delta_f G^\circ_{CO(g)}$	$\Delta G^\circ_{Ni(s)+4CO(g)\rightarrow Ni(CO)_4(g)} = \Delta_f G^\circ_{Ni(CO)_4(g)} - 4\times\Delta_f G^\circ_{CO(g)}$
300	−587.290	−137.328	−37.978
400	−582.983	−146.338	2.369
500	−579.297	−155.414	42.359
600	−575.934	−164.486	82.010

Application of Microsoft Excel's 'Slope' and 'Intercept' functions to the left and right columns of the above table gives the linear equation:

$$\Delta G^\circ_{Ni(s)+4CO(g)\rightarrow Ni(CO)_4(g)} = 0.4000 \times T(K) - 157.8 \text{ MJ/kmol of } Ni(CO)_4(g) \tag{H.4}$$

This equation has a correlation coefficient (adjusted R^2) of 0.99998 indicating that it is a high degree of prediction over the temperature range of 300–600 K.

This equation is used for all of this book's Ni carbonyl calculations.

H.1. EQUILIBRIUM CONSTANTS

The equilibrium equation for Reaction (19.1) is:

$$K_E = \frac{P^E_{Ni(CO)_4(g)}}{a^E_{Ni} \times \left(P^E_{CO(g)}\right)^4} \tag{H.5}$$

where:

K_E = equilibrium constant, dependent only on temperature.

$P^E_{Ni(CO)_4(g)}$ and $P^E_{CO(g)}$ = equilibrium partial pressures of $Ni(CO)_4(g)$ and $CO(g)$ in bar.

Note that equilibrium constants are unitless. In addition, the right-hand side of Equation (H.5) should be written in terms of activity or fugacity. Both the activity and the fugacity are unitless. The fugacity of a gas is given by the partial pressure of the gas divided by the standard pressure of 1 bar.

$P^E_{Ni(CO)_4(g)}$ and $P^E_{CO(g)}$ in Equation (H.5) are related to gas composition by:

$$P^E_{Ni(CO)_4(g)} = X^E_{Ni(CO)_4(g)} \times P_T \tag{H.6}$$

and:

$$P^{E}_{CO(g)} = X^{E}_{CO(g)} \times P_T \tag{H.7}$$

where X^E is the equilibrium mole fraction of each gas and P_T is the total equilibrium gas pressure, bar. Equations (H.6) and (H.7) assume ideal gas behavior.

Lastly, a^{E}_{Ni} is the equilibrium thermodynamic activity of Ni in the alloy with which the equilibrium gas is in contact, unitless. It is assumed here to be 1, though it could be assumed to be the mole fraction of Ni in the alloy feed.

At 1 bar pressure, Equations (H.6) and (H.7) become:

$$P^{E}_{Ni(CO)_4(g)} = X^{E}_{Ni(CO)_4(g)}$$

$$P^{E}_{CO(g)} = X^{E}_{CO(g)}$$

which alter Equation (H.5) to:

$$K_E = \frac{X^{E}_{Ni(CO)_4}}{1 \times \left(X^{E}_{CO}\right)^4} \tag{H.8}$$

where 1 is the specified value of a^{E}_{Ni}.

H.2. K_E AS A FUNCTION OF TEMPERATURE

Equation 19.1's equilibrium constant is related to its free energy of reaction by the equation:

$$\ln(K_E) = \frac{-\Delta G^{\circ}_{T_E}}{(R \times T_E)} \tag{H.9}$$

where $\Delta G^{\circ}_{T_E}$ is the standard free energy change [MJ/kg-mole of $Ni(CO)_4(g)$] for producing $Ni(CO)_4(g)$ from $Ni(s)$ and $CO(g)$ at equilibrium temperature T_E and where R (gas constant) = 0.008314 MJ/kg-mole of $Ni(CO)_4(g)$.

Equations (H.4) and (H.8) are combined to give:

$$\ln(K_E) = \frac{-(0.4000 \times T_E(K) - 157.8)}{(R \times T_E)} \tag{H.10}$$

H.3. EQUILIBRIUM CONSTANT FOR CARBONYLATION REACTION (19.1) AT 50°C

The equilibrium constant for Reaction (19.1) at 50°C is calculated by the equation:

$$\ln\left(K_E^{323K}\right) = \frac{-(0.4000 \times 323 - 157.8)}{(0.008314 \times 323)} = 10.7 \tag{H.11}$$

from which:

$$K_E^{323K} = K_E^{50°C} = e^{10.7} = 4 \times 10^4 = \left\{ \frac{X^E_{Ni(CO)_4(g)}}{\left(X^E_{CO(g)}\right)^4} \right\}_{50°C} \quad (H.12)$$

as noted in Section 19.1.

H.4. EQUILIBRIUM CONSTANT FOR DECOMPOSITION REACTION (19.2) AT 240°C

Equation (19.2)'s free energy of reaction is the negative of Reaction (19.1)'s free energy of reaction, that is:

$$\Delta G^\circ_{Ni(CO)_4(g) \rightarrow Ni(s)+4(CO(g))} = -0.4000 \times T(K) + 157.8 \text{ MJ/kg-mole of Ni(CO)4}(g) \quad (H.13)$$

The equilibrium constant for this reaction at 513 K (240°C) is, therefore, calculated by the equation:

$$\ln\left(K_E^{513K}\right) = \frac{-(-0.4000 \times 513 + 157.8)}{(0.008314 \times 513)} = 11.1 \quad (H.14)$$

from which:

$$K_E^{513K} = K_E^{240°C} = e^{11.1} = 7 \times 10^4 = \left\{ \frac{\left(X^E_{CO(g)}\right)^4}{X^E_{Ni(CO)_4(g)}} \right\}_{240°C} \quad (H.15)$$

as noted in Section 19.1.

H.5. 70 BAR CARBONYLATION CALCULATIONS

The above calculations can be repeated for 70 bar carbonylation by substituting 70 for P_T in Equations (H.6) and (H.7).

Decomposition is always done at atmospheric (~1 bar) pressure.

REFERENCE

Chase, M. W. (1998). *NIST-JANAF thermochemical tables* (4th ed.). American Chemical Society and American Institute of Physics.

Index

A

B

C

D

E

F

G

H

K

L

M

Q

R

T

Kolibri 360 GmbH
Ferdinand-Jühlke-Straße 7
99095 Erfurt, Deutschland
produktsicherheit@kolibri360.de